AF342269

Scanning Microscopy for Nanotechnology

Scanning Microscopy for Nanotechnology
Techniques and Applications

edited by

Weilie Zhou
University of New Orleans
New Orleans, Louisiana

and

Zhong Lin Wang
Georgia Institute of Technology
Atlanta, Georgia

Weilie Zhou
College of Sciences
University of New Orleans
New Orleans, Louisiana 70148

Zhong Lin Wang
Center of Nanotechnology and
 Nanoscience
Georgia Institute of Technology
Atlanta, Georgia 30332

Library of Congress Control Number: 2006925865

ISBN-10: 0-387-33325-8   e-ISBN-10: 0-387-39620-9
ISBN-13: 978-0-387-33325-0   e-ISBN-13: 978-0387-39620-0

Printed on acid-free paper.

9 8 7 6 5 4 3 2 1

springer.com

Contributors

Robert Anderhalt
Ametek EDAX Inc.
91 McKee Drive,
Mahwah, NJ 07430

Anzalone, Paul
FEI
5350 NE Dawson Creek Drive
Hillsboro, OR
97124-5793

P. Robert Apkarian
Integrated Microscopy and
  Microanalytical Facility
Department of Chemistry
Emory University
1521 Dickey Drive
Atlanta GA 30322

A. Borisevich
Oak Ridge National Laboratory
P.O. Box 2008
Oak Ridge, TN 37831

Daniela Caruntu
Advanced Materials Research Institute
University of New Orleans
New Orleans, LA 70148

Gabriel Caruntu
Advanced Materials Research Institute
University of New Orleans
New Orleans, LA 70148

M.F. Chisholm
Oak Ridge National Laboratory
P.O. Box 2008
Oak Ridge, TN 37831

Lesley Anglin Compbell
Advanced Materials Research
  Institute
University of New Orleans
New Orleans, LA 70148

M. David Frey
Carl Zeiss SMT Inc.
1 Zeiss Drive
Thornwood, NY 10594

Pu Xian Ga
School of Materials Science and
Engineering, Georgia Institute of
  Technology
Atlanta, GA 30332-0245

A. Lucille Giannuzzi
FEI
5350 NE Dawson Creek Drive
Hillsboro, OR
97124-5793

Rishi Gupta
Zyvex
1321 North Plano Road
Richardson, Texas 75081

David Joy
University of Tennessee
Knoxville, TN 37996

Jianye Li
Department of Chemistry
Duke University
Durham, NC 27708-0354

Feng Li
Advanced Materials Research Institute
University of New Orleans
New Orleans, LA 70148

Jie Liu
Department of Chemistry
Duke University
Durham, NC 27708-0354

Xiaohua Liu
Department of Biologic and Materials
    Sciences
Division of Prosthodontics
University of Michigan
1011 N. University
Ann Arbor, MI 48109-1078

A.R. Lupini
Oak Ridge National Laboratory
P.O. Box 2008
Oak Ridge, TN 37831

Peter X. Ma
Department of Biologic and Materials
    Sciences
Division of Prosthodontics
University of Michigan
1011 N. University
Ann Arbor, MI 48109-1078

Tim Maitland
HKL Technology Inc
52A Federal Road, Unit 2D
Danbury, CT 06810

Joe Nabity
JC Nabity Lithography Systems
Bozeman, MT 59717

Charles J. O'Connor
Advanced Materials Research Institute
University of New Orleans
New Orleans, LA 70148

M.P. Oxley
Oak Ridge National Laboratory
P.O. Box 2008
Oak Ridge, TN 37831

Y. Peng
Oak Ridge National Laboratory
P.O. Box 2008
Oak Ridge, TN 37831

Steve Pennycook
Oak Ridge National Laboratory
P.O. Box 2008
Oak Ridge, TN 37831

Richard E. Stallcup II
Zyvex
1321 North Plano Road
Richardson, Texas 75081

Scott Sitzman
HKL Technology Inc
52A Federal Road, Unit 2D
Danbury, CT 06810

K. Van Benthem
Oak Ridge National Laboratory
P.O. Box 2008
Oak Ridge, TN 37831

Brandon Van Leer
FEI
5350 NE Dawson Creek Drive
Hillsboro, OR
97124-5793

M. Varela
Oak Ridge National Laboratory
P.O. Box 2008
Oak Ridge, TN 37831

Peng Wang
Department of Biologic and Materials
    Sciences
Division of Prosthodontics
University of Michigan
1011 N. University
Ann Arbor, MI 48109-1078

Xudong Wang
Center for Nanoscience and
    Nanotechnology (CNN)
Georgia Institute of Technology
Materials Science and Engineering
    Department
771 Ferst Drive, N.W.
Atlanta, GA 30332-0245

Zhong Lin Wang
Center for Nanoscience and
    Nanotechnology
Georgia Institute of Technology
Materials Science and Engineering
771 Ferst Drive, N.W.
Atlanta, GA 30332-0245

Guobao Wei
Department of Biologic and Materials
    Sciences
Division of Prosthodontics
University of Michigan
1011 N. University
Ann Arbor, MI 48109-1078

John B. Wiley
Department of Chemistry and
    Advanced Materials Research
    Institute
University of New Orleans
New Orleans, LA 70148

Weilie Zhou
Advanced Materials Research Institute
University of New Orleans
New Orleans, LA 70148

Mo Zhu
Advanced Materials Research Institute
University of New Orleans
New Orleans, LA 70148

Preface

Advances in nanotechnology over the past decade have made scanning electron microscopy (SEM) an indispensable and powerful tool for analyzing and constructing new nanomaterials. Development of nanomaterials requires advanced techniques and skills to attain higher quality images, understand nanostructures, and improve synthesis strategies. A number of advancements in SEM such as field emission guns, electron back scatter detection (EBSD), and X-ray element mapping have improved nanomaterials analysis. In addition to materials characterization, SEM can be integrated with the latest technology to perform *in-situ* nanomaterial engineering and fabrication. Some examples of this integrated technology include nanomanipulation, electron beam nanolithography, and focused ion beam (FIB) techniques. Although these techniques are still being developed, they are widely applied in every aspect of nanomaterial research. *Scanning Microscopy for Nanotechnology* introduces some of the new advancements in SEM techniques and demonstrate their possible applications.

The first section covers basic theory, newly developed EBSD techniques, advanced X-ray analysis, low voltage imaging, environmental microscopy for biomaterials observation, e-beam nanolithography patterning, FIB nanostructure fabrication, and scanning transmission electron microscopy (STEM). These chapters contain practical examples of how these techniques are used to characterize and fabricate nanomaterials and nanostructures.

The second section discusses the applications of these SEM-based techniques, including nanowires and carbon nanotubes, photonic crystals and devices, nanoparticles and colloidal self-assembly, nano-building blocks fabricated through templates, one-dimensional wurtzite semiconducting nanostructures, bio-inspired nanomaterials, *in-situ* nanomanipulation, and cry-SEM stage in nanostructure research. These applications are widely used in fabricating and engineering new nanomaterials and nanostructures.

A unique feature of this book is that it is written by experts from leading research groups who specialize in the development of nanomaterials using these SEM-based techniques. Additional contributions are made by application specialists from several popular instrument vendors concerning their techniques to characterize, engineer, and manipulate nanomaterials *in-situ* SEM. *Scanning Microscopy for Nanotechnology* should be a useful and practical guide for nano-material researchers as well as a valuable reference book for students and SEM specialists.

WEILIE ZHOU
ZHONG LIN WANG

Contents

1
Fundamentals of Scanning Electron Microscopy

Weilie Zhou, Robert P. Apkarian, Zhong Lin Wang, and David Joy

1. Introduction

The scanning electron microscope (SEM) is one of the most versatile instruments available for the examination and analysis of the microstructure morphology and chemical composition characterizations. It is necessary to know the basic principles of light optics in order to understand the fundamentals of electron microscopy. The unaided eye can discriminate objects subtending about $1/60°$ visual angle, corresponding to a resolution of ~0.1 mm (at the optimum viewing distance of 25 cm). Optical microscopy has the limit of resolution of ~2,000 Å by enlarging the visual angle through optical lens. Light microscopy has been, and continues to be, of great importance to scientific research. Since the discovery that electrons can be deflected by the magnetic field in numerous experiments in the 1890s [1], electron microscopy has been developed by replacing the light source with high-energy electron beam. In this section, we will, for a split second, go over the theoretical basics of scanning electron microscopy including the resolution limitation, electron beam interactions with specimens, and signal generation.

1.1. Resolution and Abbe's Equation

The limit of resolution is defined as the minimum distances by which two structures can be separated and still appear as two distinct objects. Ernst Abbe [1] proved that the limit of resolution depends on the wavelength of the illumination source. At certain wavelength, when resolution exceeds the limit, the magnified image blurs.

Because of diffraction and interference, a point of light cannot be focused as a perfect dot. Instead, the image will have the appearance of a larger diameter than the source, consisting of a disk composed of concentric circles with diminishing intensity. This is known as an Airy disk and is represented in Fig. 1.1a. The primary wave front contains approximately 84% of the light energy, and the intensity of secondary and tertiary wave fronts decay rapidly at higher orders. Generally, the radius of Airy disk is defined as the distance between the first-order peak and

(a)

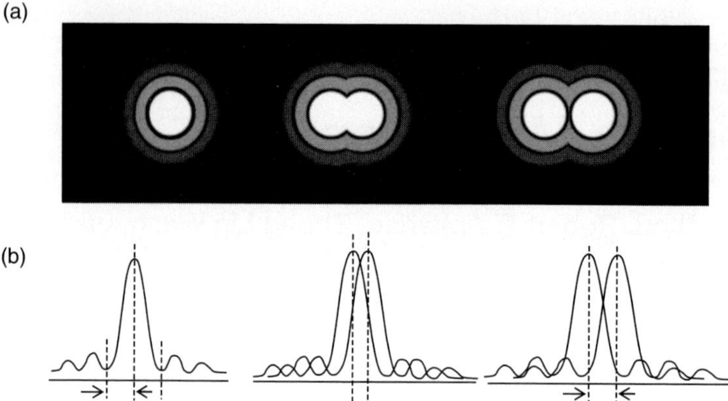

(b)

FIGURE 1.1. Illustration of resolution in (a) Airy disk and (b) wave front.

the first-order trough, as shown in Fig. 1.1a. When the center of two primary peaks are separated by a distance equal to the radius of Airy disk, the two objects can be distinguished from each other, as shown in Fig. 1.1b. Resolution in a perfect optical system can be described mathematically by Abbe's equation. In this equation:

$$d = 0.612\ \lambda / n \sin \alpha$$

where
d = resolution
λ = wavelength of imaging radiation
n = index of refraction of medium between point source and lens, relative to free space
α = half the angle of the cone of light from specimen plane accepted by the objective (half aperture angle in radians)
$n \sin \alpha$ is often called numerical aperture (NA).

Substituting the illumination source and condenser lens with electron beam and electromagnetic coils in light microscopes, respectively, the first transmission electron microscope (TEM) was constructed in the 1930s [2], in which electron beam was focused by an electromagnetic condenser lens onto the specimen plane. The SEM utilizes a focused electron beam to scan across the surface of the specimen systematically, producing large numbers of signals, which will be discussed in detail later. These electron signals are eventually converted to a visual signal displayed on a cathode ray tube (CRT).

1.1.1. Interaction of Electron with Samples

Image formation in the SEM is dependent on the acquisition of signals produced from the electron beam and specimen interactions. These interactions can be divided into two major categories: elastic interactions and inelastic interactions.

Elastic scattering results from the deflection of the incident electron by the specimen atomic nucleus or by outer shell electrons of similar energy. This kind of interaction is characterized by negligible energy loss during the collision and by a wide-angle directional change of the scattered electron. Incident electrons that are elastically scattered through an angle of more than 90° are called backscattered electrons (BSE), and yield a useful signal for imaging the sample. Inelastic scattering occurs through a variety of interactions between the incident electrons and the electrons and atoms of the sample, and results in the primary beam electron transferring substantial energy to that atom. The amount of energy loss depends on whether the specimen electrons are excited singly or collectively and on the binding energy of the electron to the atom. As a result, the excitation of the specimen electrons during the ionization of specimen atoms leads to the generation of secondary electrons (SE), which are conventionally defined as possessing energies of less than 50 eV and can be used to image or analyze the sample. In addition to those signals that are utilized to form an image, a number of other signals are produced when an electron beam strikes a sample, including the emission of characteristic x-rays, Auger electrons, and cathodoluminescence. We will discuss these signals in the later sections. Figure 1.2 shows the regions from which different signals are detected.

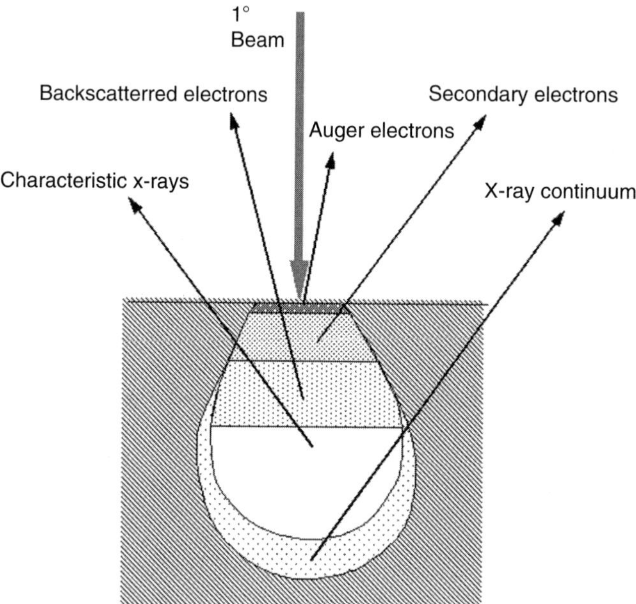

FIGURE 1.2. Illustration of several signals generated by the electron beam–specimen interaction in the scanning electron microscope and the regions from which the signals can be detected.

In most cases when incident electron strikes the specimen surface, instead of being bounced off immediately, the energetic electrons penetrate into the sample for some distance before they encounter and collide with a specimen atom. In doing so, the primary electron beam produces what is known as a region of primary excitation, from which a variety of signals are produced. The size and shape of this zone is largely dependent upon the beam electron energy and the atomic number, and hence the density, of the specimen. Figure 1.3 illustrates the variation of interaction volume with respect to different accelerating voltage and atomic number. At certain accelerating voltage, the shape of interaction volume is "tear drop" for low atomic number specimen and hemisphere for specimens of high atomic number. The volume and depth of penetration increase with an increase of the beam energy and fall with the increasing specimen atomic number because specimens with higher atomic number have more particles to stop electron penetration. One influence of the interaction volume on signal acquisition is that use of a high accelerating voltage will result in deep penetration length and a large primary excitation region, and ultimately cause the loss of detailed surface information of the samples. A close-packed opal structure observed by a field emission scanning electron microscope (FESEM) at different accelerating voltages is shown in Fig. 1.4. Images taken under 1 kV gave more surface details than that of 20 kV. The surface resolution is lost at high accelerating voltages and the surface of spheres looks smooth.

1.1.2. Secondary Electrons

The most widely used signal produced by the interaction of the primary electron beam with the specimen is the secondary electron emission signal. When the primary beam strikes the sample surface causing the ionization of specimen atoms,

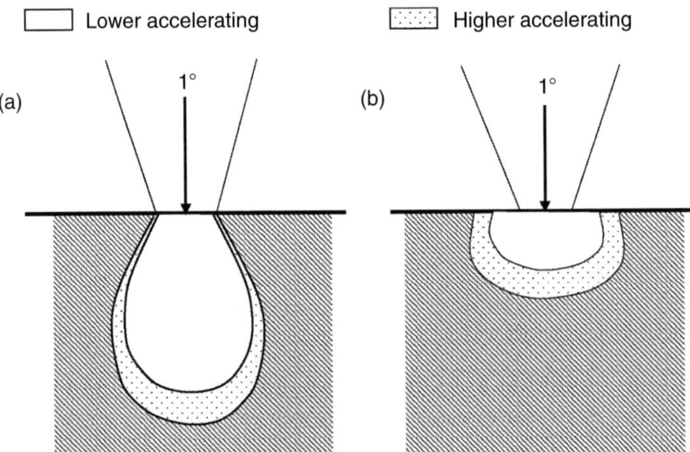

FIGURE 1.3. Influence of accelerating voltage and specimen atomic number on the primary excitation volume: (a) low atomic number and (b) high atomic number.

(a)   (b)

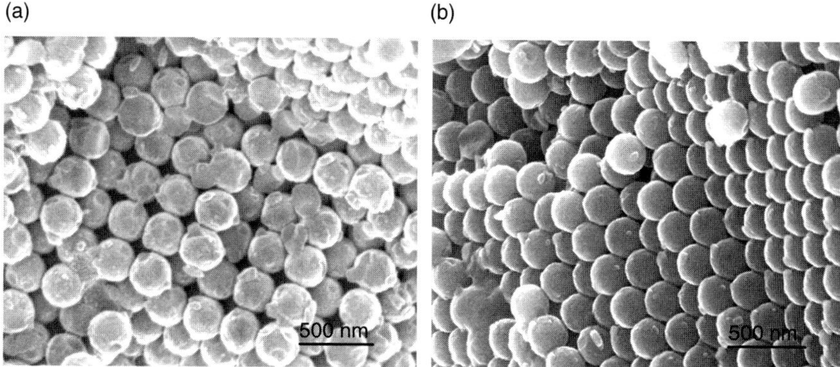

FIGURE 1.4. Scanning electron micrographs of a CaF_2 close-packed opal structure, which are taken under different accelerating voltages: (a) 1 kV and (b) 20 kV.

loosely bound electrons may be emitted and these are referred to as secondary electrons. As they have low energy, typically an average of around 3–5 eV, they can only escape from a region within a few nanometers of the material surface. So secondary electrons accurately mark the position of the beam and give topographic information with good resolution. Because of their low energy, secondary electrons are readily attracted to a detector carrying some applied bias. The Everhart–Thornley (ET) detector, which is the standard collector for secondary electrons in most SEMs therefore applies both a bias (+10 kV) to the scintillator and a lower bias (+300 V) to the Faraday cage, which screens the detector. In order to detect the secondary electrons a scintillator converts the energy of the electrons into photons (visible light). The photons then produced travel down a Plexiglas or polished quartz light pipe and move out through the specimen chamber wall, and into a photomultiplier tube (PMT) which converts the quantum energy of the photons back into electrons. The output voltage from the PMT is further amplified before being output as brightness modulation on the display screen of the SEM.

Secondary electrons are used principally for topographic contrast in the SEM, i.e., for the visualization of surface texture and roughness. The topographical image is dependent on how many of the secondary electrons actually reach the detector. A secondary electron signal can resolve surface structures down to the order of 10 nm or better. Although an equivalent number of secondary electrons might be produced as a result of the specimen primary beam interaction, only those that can reach the detector will contribute to the ultimate image. Secondary electrons that are prevented from reaching the detector will generate shadows or be darker in contrast than those regions that have an unobstructed electron path to the detector. It is apparent in the diagram that topography also affects the zone of secondary electron emission. When the specimen surface is perpendicular to the beam, the zone from which secondary electrons are emitted is smaller than found when the surface is tilted. Figure 1.5 illustrates the effect of specimen topography and the position of detector on the secondary electron signals.

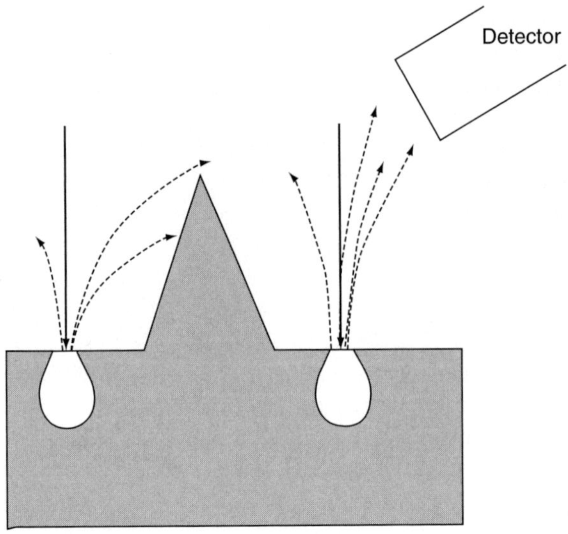

FIGURE 1.5. Illustration of effect of surface topography and position of detector on the secondary electron detection.

Low voltage incident electrons will generate secondary electrons from the very surface region, which will reveal more detailed structure information on the sample surface. More about this will be discussed in Chapter 4.

1.1.3. Backscattered Electrons

Another valuable method of producing an image in SEM is by the detection of BSEs, which provide both compositional and topographic information in the SEM. A BSE is defined as one which has undergone a single or multiple scattering events and which escapes from the surface with an energy greater than 50 eV. The elastic collision between an electron and the specimen atomic nucleus causes the electron to bounce back with wide-angle directional change. Roughly 10–50% of the beam electrons are backscattered toward their source, and on an average these electrons retain 60–80% of their initial energy. Elements with higher atomic numbers have more positive charges on the nucleus, and as a result, more electrons are backscattered, causing the resulting backscattered signal to be higher. Thus, the backscattered yield, defined as the percentage of incident electrons that are reemitted by the sample, is dependent upon the atomic number of the sample, providing atomic number contrast in the SEM images. For example, the BSE yield is ~6% for a light element such as carbon, whereas it is ~50% for a heavier element such as tungsten or gold. Due to the fact that BSEs have a large energy, which prevents them from being absorbed by the sample, the region of the specimen from which BSEs are produced is considerably larger than it is for secondary electrons. For this reason the lateral resolution of a BSE image is considerably worse (1.0 μm) than it is for a secondary electron image (10 nm). But with a fairly large width of escape depth, BSEs carry information about features that are deep beneath the surface. In examining relatively flat samples, BSEs can be used to produce a topographical image that differs from that produced by secondary electrons, because some BSEs are blocked by regions of the specimen that secondary electrons might be drawn around.

The detector for BSEs differs from that used for secondary electrons in that a biased Faraday cage is not employed to attract the electrons. In fact the Faraday cage is often biased negatively to repel any secondary electrons from reaching the detector. Only those electrons that travel in a straight path from the specimen to the detector go toward forming the backscattered image. Figure 1.6 shows images of Ni/Au heterostructure nanorods. The contrast differences in the image produced by using secondary electron signal are difficult to interpret (Fig. 1.6a), but contrast difference constructed by the BSE signal are easily discriminated (Fig. 1.6b).

The newly developed electron backscattered diffraction (EBSD) technique is able to determine crystal structure of various samples, including nanosized crystals. The details will be discussed in Chapter 2.

1.1.4. Characteristic X-rays

Another class of signals produced by the interaction of the primary electron beam with the specimen is characteristic x-rays. The analysis of characteristic x-rays to provide chemical information is the most widely used microanalytical technique in the SEM. When an inner shell electron is displaced by collision with a primary electron, an outer shell electron may fall into the inner shell to reestablish the proper charge balance in its orbitals following an ionization event. Thus, by the emission of an x-ray photon, the ionized atom returns to ground state. In addition to the characteristic x-ray peaks, a continuous background is generated through the deceleration of high-energy electrons as they interact with the electron cloud

(a)   (b)

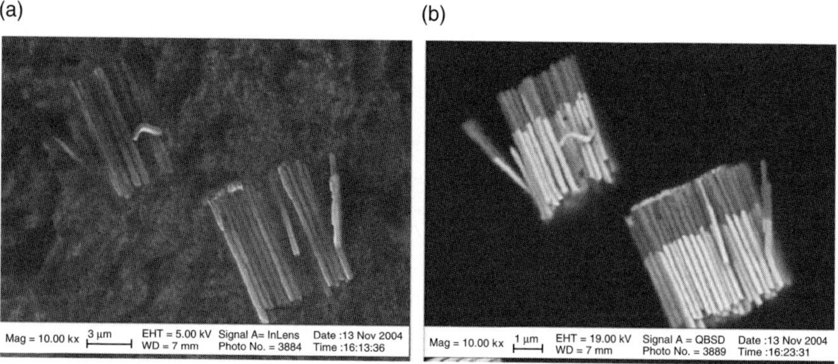

FIGURE 1.6. Ni/Au nanorods images formed by (a) secondary electron signal and (b) backscattering electron signal.

and with the nuclei of atoms in the sample. This component is referred to as the *Bremsstrahlung or Continuum* x-ray signal. This constitutes a background noise, and is usually stripped from the spectrum before analysis although it contains information that is essential to the proper understanding and quantification of the emitted spectrum. More about characteristic x-rays for nanostructure analysis will be discussed in Chapter 3.

1.1.5. Other Electrons

In addition to the most commonly used signals including BSEs, secondary electrons, and characteristic x-rays, there are several other kinds of signals generated during the specimen electron beam interaction, which could be used for microstructure analysis. They are Auger electrons, cathodoluminescence-transmitted electrons and specimen (or absorbed) current.

1.1.5.1. Auger Electrons

Auger electrons are produced following the ionization of an atom by the incident electron beam and the falling back of an outer shell electron to fill an inner shell vacancy. The excess energy released by this process may be carried away by an Auger electron. This electron has a characteristic energy and can therefore be used to provide chemical information. Because of their low energies, Auger electrons are emitted only from near the surface. They have escape depths of only a few nanometers and are principally used in surface analysis.

1.1.5.2. Cathodoluminescence

Cathodoluminescence is another mechanism for energy stabilization following beam specimen interaction. Certain materials will release excess energy in the form of photons with infrared, visible, or ultraviolet wavelengths when electrons recombine to fill holes made by the collision of the primary beam with the specimen. These photons can be detected and counted by using a light pipe and photomultiplier similar to the ones utilized by the secondary electron detector. The best possible image resolution using this approach is estimated at about 50 nm.

1.1.5.3. Transmitted Electrons

Transmitted electrons is another method that can be used in the SEM to create an image if the specimen is thin enough for primary beam electrons to pass through (usually less than 1 μ). As with the secondary and BSE detectors, the transmitted electron detector is comprised of scintillator, light pipe (or guide), and a photomultiplier, but it is positioned facing the underside of the specimen (perpendicular to the optical axis of the microscope). This technique allows SEM to examine the internal ultrastructure of thin specimens. Coupled with x-ray microanalysis, transmitted electrons can be used to acquisition of elemental information and distribution. The integration of scanning electron beam with a transmission electron microscopy detector generates scanning transmission electron microscopy, which will be discussed in Chapter 6.

1.1.5.4. Specimen Current

Specimen current is defined as the difference between the primary beam current and the total emissive current (backscattered, secondary, and Auger electrons). Specimens that have stronger emission currents thus will have weaker specimen currents and vice versa. One advantage of specimen current imaging is that the sample is its own detector. There is thus no problem in imaging in this mode with the specimen as close as is desired to the lens.

2. Configuration of Scanning Electron Microscopes

In this section, we will present a detailed discussion of the major components in an SEM. Figure 1.7 shows a column structure of a conventional SEM. The electron gun, which is on the top of the column, produces the electrons and accelerates them to an energy level of 0.1–30 keV. The diameter of electron beam produced by hairpin tungsten gun is too large to form a high-resolution image. So, electromagnetic lenses and apertures are used to focus and define the electron beam and to form a small focused electron spot on the specimen. This process demagnifies the size of the electron source (~50 µm for a tungsten filament) down to the final required spot size (1–100 nm). A high-vacuum environment, which allows electron travel without scattering by the air, is needed. The specimen stage, electron beam scanning coils, signal detection, and processing system provide real-time observation and image recording of the specimen surface.

2.1. Electron Guns

Modern SEM systems require that the electron gun produces a stable electron beam with high current, small spot size, adjustable energy, and small energy dispersion. Several types of electron guns are used in SEM system and the qualities of electrons beam they produced vary considerably. The first SEM systems generally used tungsten "hairpin" or lanthanum hexaboride (LaB_6) cathodes, but for the modern SEMs, the trend is to use field emission sources, which provide enhanced current and lower energy dispersion. Emitter lifetime is another important consideration for selection of electron sources.

2.1.1. Tungsten Electron Guns

Tungsten electron guns have been used for more than 70 years, and their reliability and low cost encourage their use in many applications, especially for low magnification imaging and x-ray microanalysis [3]. The most widely used electron gun is composed of three parts: a V-shaped hairpin tungsten filament (the cathode), a Wehnelt cylinder, and an anode, as shown in Fig. 1.8. The tungsten filament is about 100 µm in diameter. The V-shaped filament is heated to a temperature of

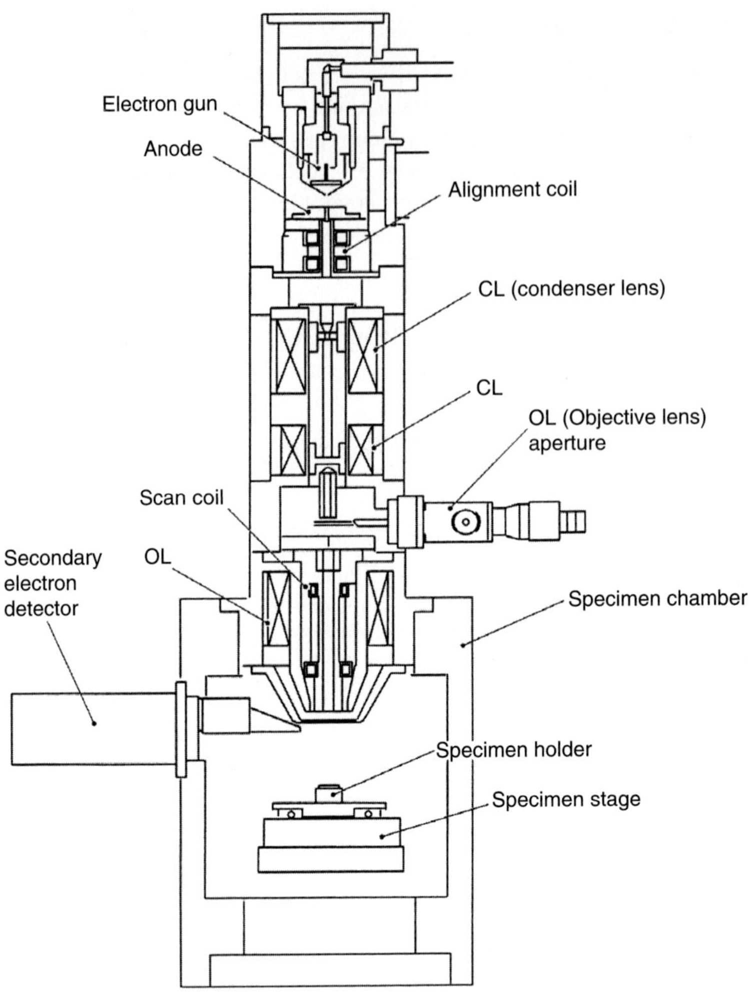

FIGURE 1.7. Schematic diagram of a scanning electron microscope (JSM—5410, courtesy of JEOL, USA).

more than 2,800 K by applying a filament current i_f so that the electrons can escape from the surface of the filament tip. A negative potential, which is varied in the range of 0.1–30 kV, is applied on the tungsten and Wehnelt cylinder by a high voltage supply. As the anode is grounded, the electric field between the filament and the anode plate extracts and accelerates the electrons toward the anode. In thermionic emission, the electrons have widely spread trajectories from the filament tip. A slightly negative potential between the Wehnelt cylinder and the filament, referred to "bias," provides steeply curved equipotentials near the aperture of the

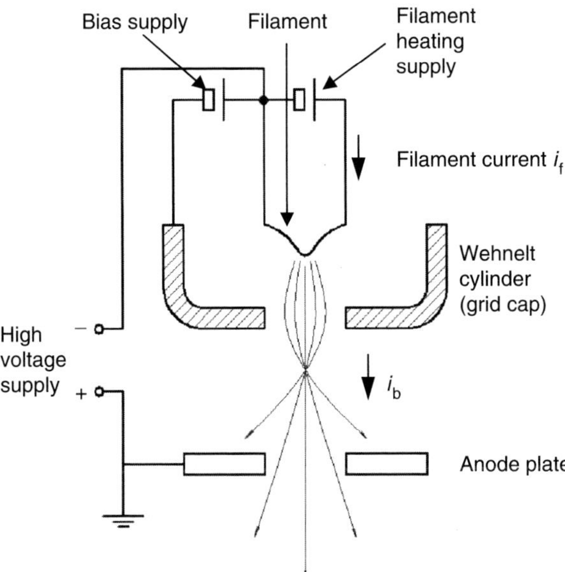

FIGURE 1.8. Schematic of the self-biased thermionic tungsten electron gun. (The effect of the negative bias of the Wehnelt cylinder on the electron trajectory is shown.)

Wehnelt cylinder, which produces a crude focusing of electron beam. The focusing effect of Wehnelt cylinder on the electron beam is depicted in Fig. 1.8.

The electron emission increases with the filament current. There is some "saturation point" of filament current, at which we have most effective electron emission (i.e., the highest electron emission is obtained by least amount of current). At saturation electrons are only emitted from the tip of the filament and focused into a tight bundle by the negative accelerating voltage. If the filament current increases further, the electron emission only increases slightly (Fig. 1.9). It is worth mentioning that there is a peak (known as "false peak") in beam current not associated with saturation, and this character is different from instrument to instrument, even from filament to another. This false peak is sometimes even greater than the saturation point. Its cause remains unexplained because it is of little practical use, but its presence could be the result of gun geometries during filament heating and the electrostatic creation of the gun's crossover. Setting the filament to work at the false peak will result in extremely long filament life, but it also deteriorates the stability of the beam. Overheating the filament with current higher than saturation current will reduce the filament life significantly. The burnt-out filament is shown in Fig. 1.10. The spherical melted end of the broken filament due to the overheating is obvious. The filament life is also influenced by the vacuum status and cleanliness of the gun.

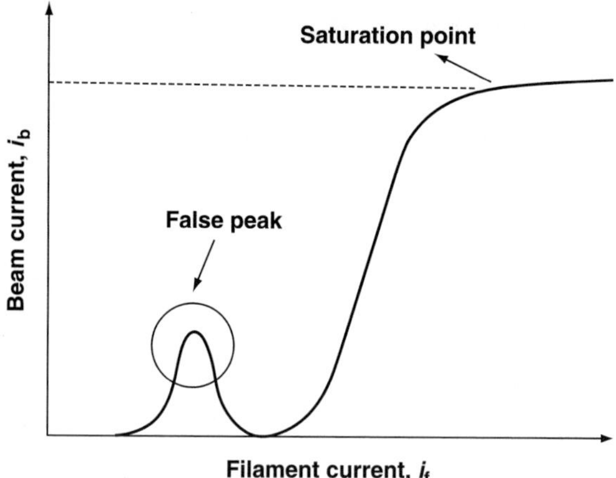

FIGURE 1.9. Saturation of a tungsten hairpin electron gun. At saturation point, majority of the electrons are emitted from the tip of the filament and form a tight bundle by accelerating voltage.

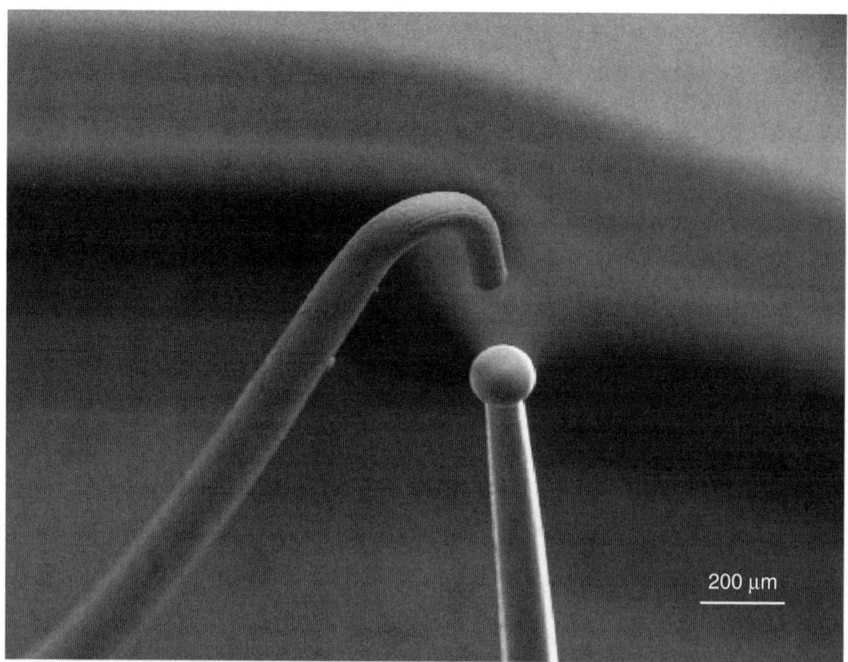

FIGURE 1.10. An SEM image of a "blown-out" tungsten filament due to overheating. A spherical melted end is obvious at the broken filament.

2.1.2. Lanthanum Hexaboride Guns

An alternative for tungsten filament is the LaB_6 filament. This material has a lower work function (2.4 eV) than tungsten (4.5 eV). This means LaB_6 can provide stronger emission of electrons at the same heating temperature. Therefore, LaB_6 electron guns provide 5 to 10× greater brightness and a longer lifetime compared with conventional tungsten guns [4]. Figure 1.11a shows the emitter of a LaB_6 single crystal 100–200 μm in diameter and 0.5 mm long. The crystal is mounted on a graphite or rhenium support, which does not chemically react with the LaB_6 and also serves as the resistive heater to elevate the temperature of crystal so that it can emit electrons. There are several advantages for the use of LaB_6 electron guns. The effective emission area is much smaller than conventional tungsten electron guns, which reduces the spot size of the electron beam. In addition, the electron beam produced by LaB_6 electron guns have smaller energy spread, which means a smaller chromatic aberration and higher resolution of SEM images.

LaB_6 electron source can replace the tungsten electron guns directly in conventional SEMs. However, LaB_6 is readily oxidized at elevated temperatures and the vacuum in gun chamber of conventional electron microscopes is not high enough to avoid contamination on LaB_6 cathode. This reduces the lifetime of the guns significantly. Figure 1.11b shows the details of a used LaB_6 crystal, several contamination spots are easily recognized on its surface. To avoid this situation the chamber electron gun must have a vacuum better than 10^{-8} Torr. Generally, differential pumping of the gun region is needed.

2.1.3. Field Emission Guns

Thermionic sources depend on a high temperature to overcome the work function of the metal so that the electrons can escape from the cathode. Though they are

(a)   (b)

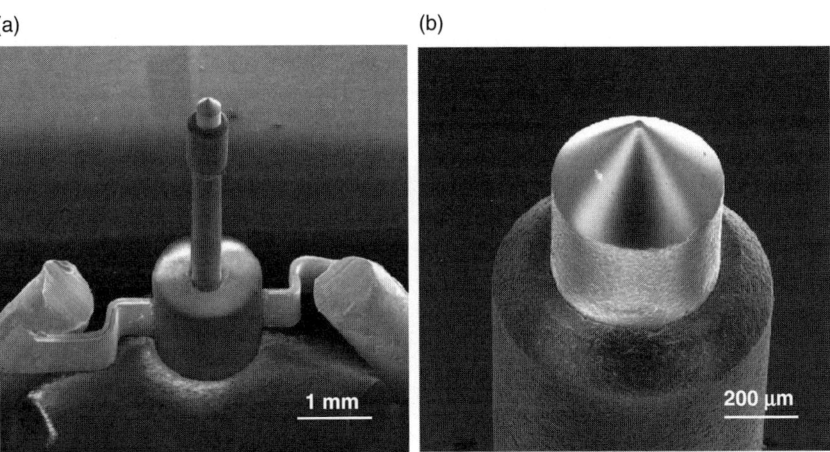

FIGURE 1.11. (a) SEM image of LaB_6 electron gun and (b) a higher magnification image, small contamination spots are easily recognized.

inexpensive and the requirement of vacuum is relatively low, the disadvantages, such as short lifetime, low brightness, and large energy spread, restrict their applications. For modern electron microscopes, field emission electron guns (FEG) are a good alternative for thermionic electron guns.

In the FEG, a single crystal tungsten wire with very sharp tip, generally prepared by electrolytic etching, is used as the electron source. Figure 1.12a and b shows a micrograph of a typical field emission tip and the schematic structure of the FEG. In this system, a strong electric field forms on the finely oriented tip, and the electrons are drawn toward the anodes instead of being boiled up by the filament heating. Two anodes are used in field emission system, depicted in Fig. 1.12c. The voltage V_1 with a few kilovolts between the tip and the first anode is used to extract the electrons from the tip, and the V_0 is the accelerating voltage.

There are three types of FEGs that are used in the SEM systems [5]. One is the cold field emission (CFE) sources. The "cold field" means the electron sources operate at room temperature. The emission of electrons from the CFE purely depends on the electric field applied between the anodes and the cathode. Although the current of emitted electron beams is very small, a high brightness can still be achieved because of the small electron beam diameter and emission area. An operation known as "flashing" in which the field emission tip is heated to a temperature of more than 2,000 K for a few seconds is needed to clean absorbed gas on the tip. The second class is thermal field emission (TFE) sources, which is operated in elevated temperature. The elevated temperature reduces the absorption of gas molecules and stabilizes the emission of electron beam even when a degraded vacuum occurs. Beside CFE and TFE sources, Schottky emitters (SE) sources are also used in modern SEM system. The performances of SE and CFE sources are superior to thermionic sources in the case of brightness, source size, and lifetime. However, SE source is preferred over CFE source because of its higher stability and easier operation. Because the emitting area of SE source is about 100× larger than that of the CFE source, it is capable to deliver more than 50× higher emission current than CFE at a similar energy spread. Further, a larger size of emission source reduces the susceptibility to vibration.

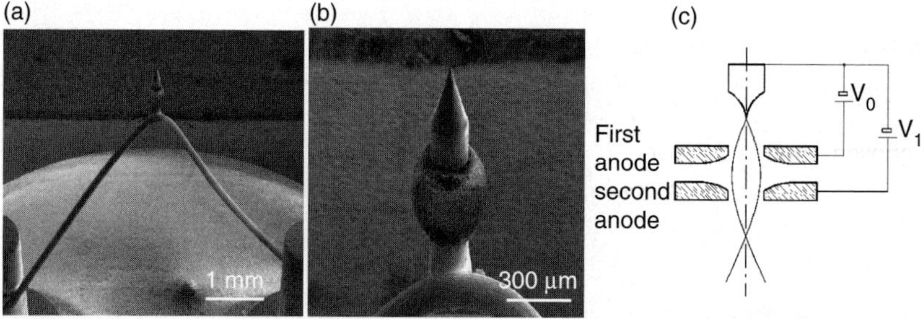

FIGURE 1.12. (a) Field emission source with extreme sharp tip; (b) a higher magnification image; and (c) schematic diagram of a typical field emission electron source. The two anodes work as an electrostatic lens to form electron beams.

Also, electron beam nanolithography needs high emission current to perform a pattern writing, which will be discussed in Chapter 5.

Compared with thermionic sources, CFE provides enhanced electron brightness, typically 100× greater than that for a typical tungsten source. It also possesses very low electron energy spread of 0.3 eV, which reduces the chromatic aberration significantly, and can form a probe smaller than 2 nm, which provides much higher resolution for SEM image. However, field emitters must operate under ultrahigh vacuum (better than 10^{-9} Torr) to stabilize the electron emission and to prevent contamination.

2.2. Electron Lenses

Electron beams can be focused by electrostatic or magnetic field. But electron beam controlled by magnetic field has smaller aberration, so only magnetic field is employed in SEM system. Coils of wire, known as "electromagnets," are used to produce magnetic field, and the trajectories of the electrons can be adjusted by the current applied on these coils. Even using the magnetic field to focus the electron beam, electromagnetic lenses still work poorly compared with the glass lenses in terms of aberrations. The electron lenses can be used to magnify or demagnify the electron beam diameter, because their strength is variable, which results in a variable focal length. SEM always uses the electron lenses to demagnify the "image" of the emission source so that a narrow probe can be formed on the surface of the specimen.

2.2.1. Condenser Lenses

The electron beam will diverge after passing through the anode plate from the emission source. By using the condenser lens, the electron beam is converged and collimated into a relatively parallel stream. A magnetic lens generally consists of two rotationally symmetric iron pole pieces in which there is a copper winding providing magnetic field. There is a hole in the center of pole pieces that allows the electron beam to pass through. A lens-gap separates the two pole pieces, at which the magnetic field affects (focuses) the electron beam. The position of the focal point can be controlled by adjusting the condenser lens current. A condenser aperture, generally, is associated with the condenser lens, and the focal point of the electron beam is above the aperture (Fig. 1.13). As appropriate aperture size is chosen, many of the inhomogeneous and scattered electrons are excluded. For modern electron microscopes, a second condenser lens is often used to provide additional control on the electron beam.

2.2.2. Objective Lenses

The electron beam will diverge below the condenser aperture. Objective lenses are used to focus the electron beam into a probe point at the specimen surface and to supply further demagnification. An appropriate choice of lens demagnification

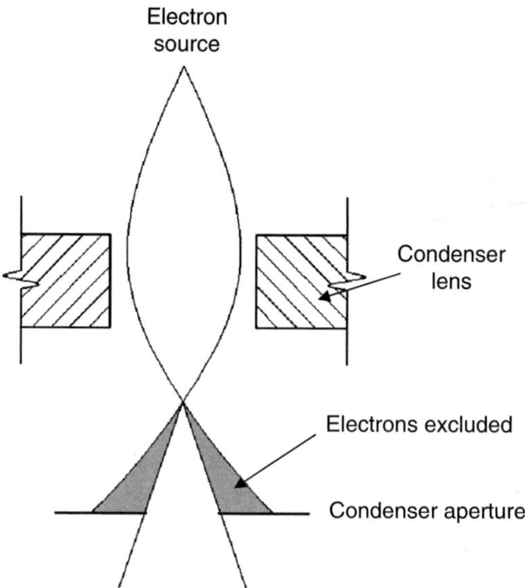

FIGURE 1.13. A diagram showing how the electrons travel through the condenser lens and condenser aperture. Many of the nonhomogeneous or scattered electrons are excluded by the condenser aperture.

and aperture size results in a reduction of the diameter of electron beam on the specimen surface (spot size), and enhances the image resolution.

Three designs of objective lenses are shown in Fig. 1.14 [5]. The asymmetric pinhole lens (Fig. 1.14a) is the most common objective lens. There is only a small bore on the pole piece, and this keeps the magnetic field within the lens and provides a field-free region above the specimen for detecting the secondary electrons. However, this configuration has a large lens aberration. For the symmetric immersion lens (Fig. 1.14b), the specimen is placed inside the lens, which can reduce the focal length significantly. This configuration provides a lowest lens aberration because lens aberration directly scale with the focal length. But the specimen size cannot exceed 5 mm. The Snorkel lens (Fig. 1.14c) produces a strong magnetic field that extends to the specimen. This kind of lens possesses the advantages of the pinhole lens and the immersion lens, combining low lens aberration with permission of large specimen. Furthermore, this configuration can accommodate two secondary electron detectors (the conventional and in-lens detector). The detail detector configurations will be discussed later.

2.3. Column Parameters

It is easy to imagine that spot size and the beam convergence angle α directly relate to the resolution and the depth of focus of the SEM images, but they are influenced by many other parameters such as electron beam energy, lenses

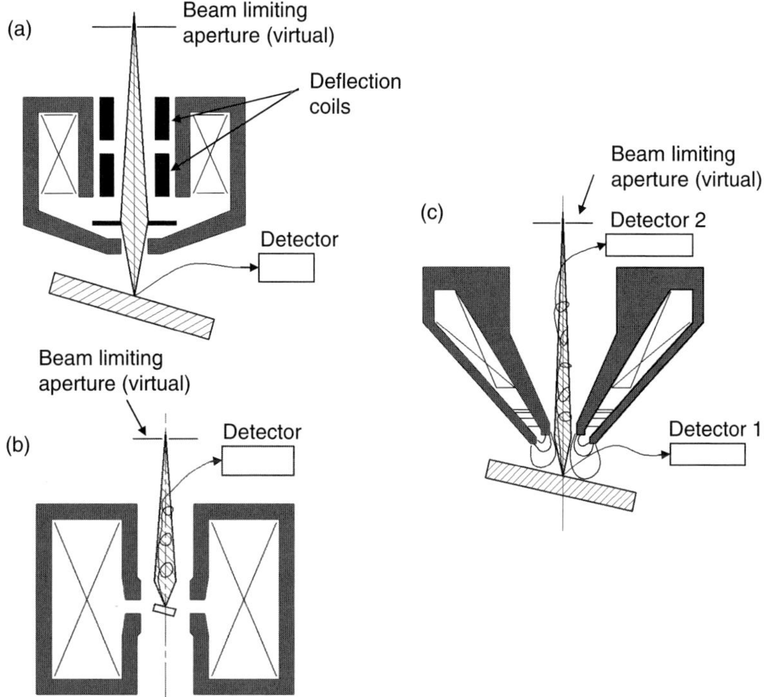

FIGURE 1.14. Objective lens configurations: (a) asymmetric pinhole lens, which has large lens aberration; (b) symmetric immersion lens, in which small specimen can be observed with small lens aberration; and (c) snorkel lens, where the magnetic field extends to the specimen providing small lens aberration on large specimen (Adapted from [5]).

current, aperture size, working distance (WD), and chromatic and achromatic aberration of electron lenses. In this section, several primary parameters that are significant for the image quality will be discussed and a good understanding of all these parameters is needed because these parameters are interdependent.

2.3.1. Aperture

One or more apertures are employed in the column according to different designs of SEM. Apertures are used to exclude scattered electrons and are used to control the spherical aberrations in the final lens. There are two types of aperture: one is at the base of final lens and is known as real aperture; the other type is known as virtual aperture and it is placed in the electron beam at a point above the final lens. The beam shape and the beam edge sharpness are affected by the real aperture. The virtual aperture, which limits the electron beam, is found to have the same affect. The real aperture is the conventional type of aperture system and the virtual aperture is found on most modern SEM system. Because the virtual aperture is far away from the specimen chamber it can be kept clean for a long time, but the aperture alignment becomes a regular operation, as its size is very small. Decreasing the aperture size will reduce the beam angle α for the same WD, resulting in an enhancement of the depth of field (shown in Fig. 1.15) and a decrease of the current in the final probe. Figure 1.16 shows the electron micrograph of branched grown ZnO nanorods taken with different aperture size. Increase of depth of field due to change of aperture size is easily observed, which is emphasized by circles. An optimum choice of aperture size can also minimize the detrimental effects of aberrations on the probe size [6].

2.3.2. Stigmation

The lens defects (machining errors and asymmetry in lens winding), and contamination on aperture or column can cause the cross section of the electron beam profile to vary in shape. Generally, an elliptical cross section is formed instead of a circular one. As a result during operation, the image will stretch along different direction at underfocus and overfocus condition. This imperfection on the electromagnetic lens is called astigmatism. A series of coils surrounding the electron beam, referred to as "stigmator," can be used to correct astigmatism and achieve an image with higher resolution.

Figure 1.17a shows an SEM image with extreme astigmatism. When moving through focus the image stretches first in one direction (Fig. 1.17b) and when the image is in the in-focus position the stretch is minimized (Fig. 1.17a) before it is stretched to another direction (Fig. 1.17c). The astigmatism correction cycle

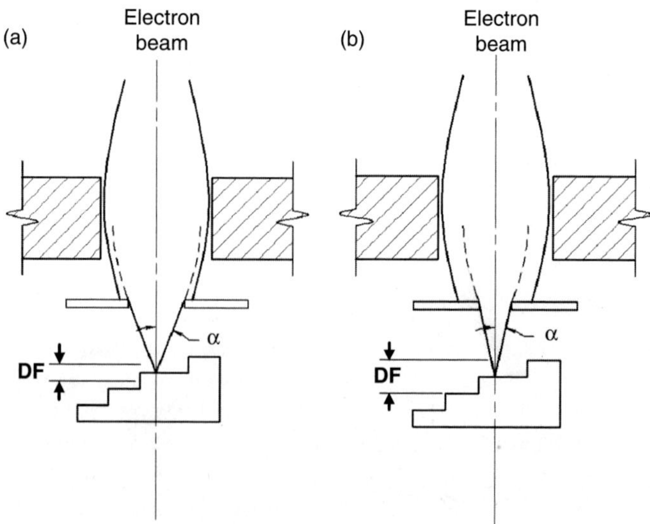

FIGURE 1.15. Small aperture (b) provides enhanced depth of field compared with large aperture (a).

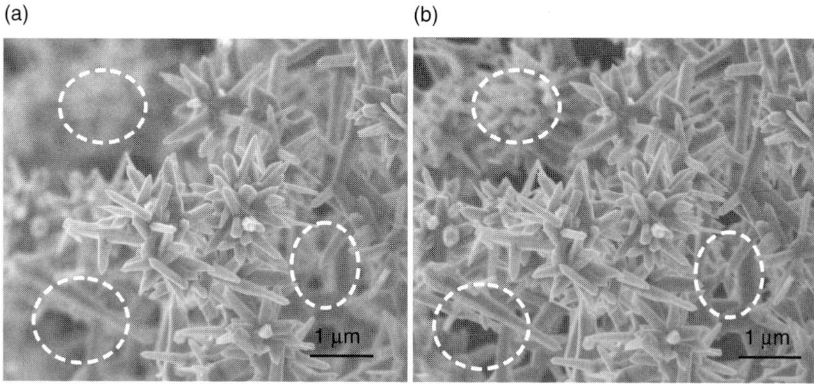

FIGURE 1.16. Electron micrograph of branch grown ZnO nanorods taken with different aperture sizes: (a) 30 μm and (b) 7.5 μm. The enhancement of depth of field is emphasized by circles.

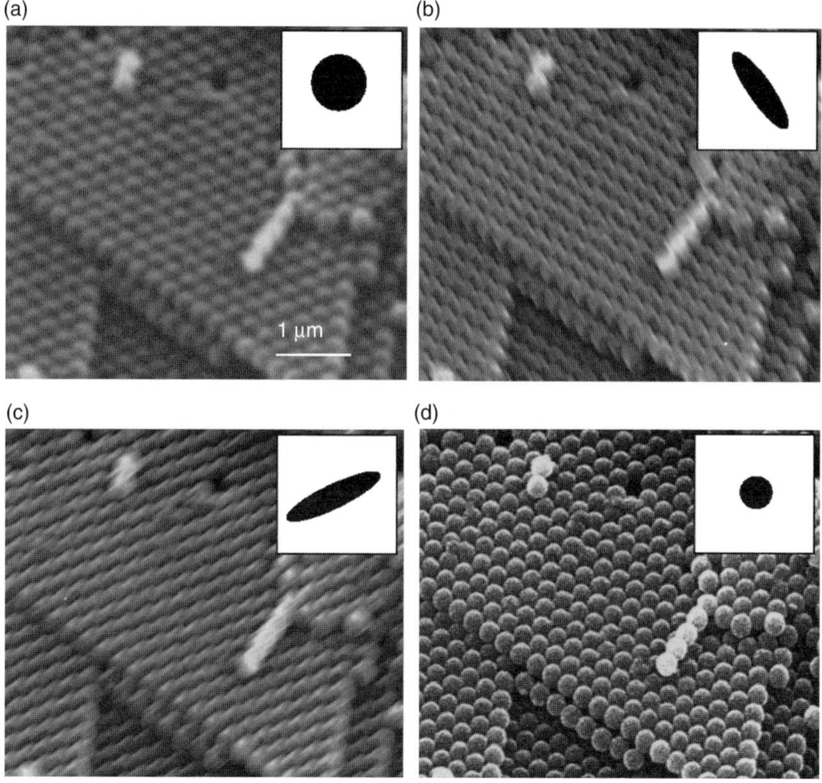

FIGURE 1.17. Comparison of SEM images of opal structure with astigmatism and after astigmatism correction. (a) SEM image with astigmatism in in-focus condition; (b) SEM image with astigmatism in underfocus condition; (c) SEM image with astigmatism in over-focus condition; and (d) SEM image with astigmatism correction. The inset figures are the schematic diagrams of the shapes of probe spots.

(*x*-stigmator, focus, *y*-stigamator, focus) should be repeated, until ultimately the sharpest image is obtained (Fig. 1.17d). At that point the beam cross section will be focused to the smallest point. Generally the compensation for astigmatism is performed while operating at the increased magnification, which ensures the image quality of lower magnification even when perfect compensation is not obtained. However, the astigmatism is not obvious for low magnification observation.

2.3.3. Depth of Field

The portion of the image that appears acceptably in focus is called the "depth of field" [7]. Figure 1.18 shows the effect of a limited depth of field in an SEM image of a SnO_2 nanojunction sample [8]. Only middle part of the image is in the focus, and the upper side and underside of the image show underfocus and over-focus, respectively. An electron beam with a smaller convergence angle α provides a larger depth of field, because the change of spot size is less significant along the beam direction for a sharper electron beam. Besides the aperture size, the WD will also influence the depth of field, which is demonstrated in Fig. 1.19. At a short WD the sample will be scanned with a wide cone of electrons resulting in an image with little depth of field. By contrast, at a longer WD, corresponding to a narrow cone of electron beam results in an enhanced depth of field. However, a long WD does not mean a high resolution. Depth of field is important when we observe a specimen with large topographical variation. In this case, we prefer to use a long WD so that we can bring as much of the image into focus as possible.

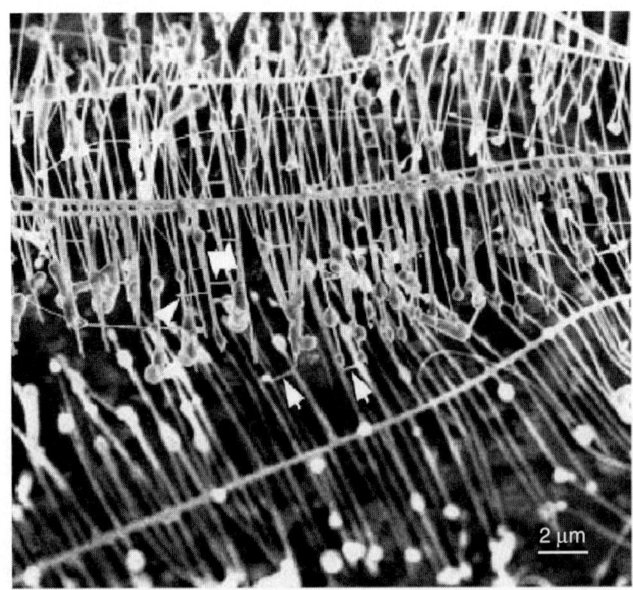

FIGURE 1.18. SEM image of SnO_2 nanojunctions showing depth of field.

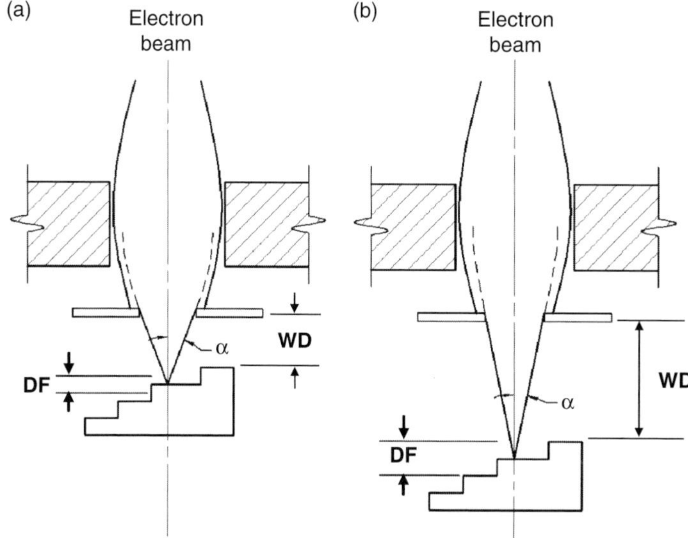

FIGURE 1.19. Beam diagram showing enhancement of depth of field (DF) by increasing working distance (WD). (a) Short working distance and (b) long working distance.

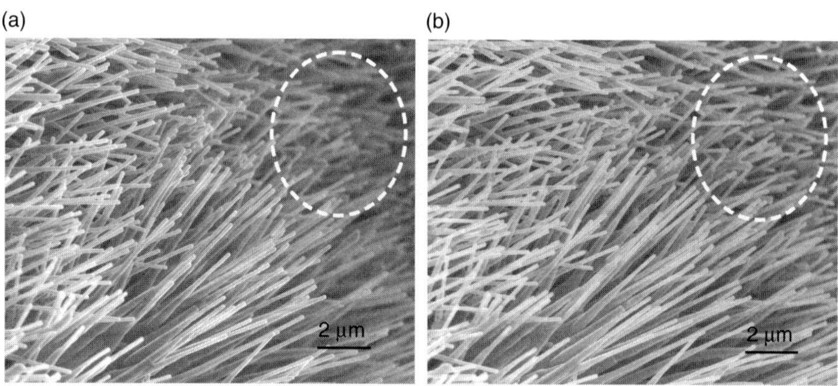

FIGURE 1.20. Well aligned Co-doped ZnO nanowires array fabricated by chemical vapor deposition, showing the enhancement of depth of field by increasing the working distance from (a) 3 mm to (b) 12 mm, which is emphasized by circles.

But if the topography of the specimen is relatively flat, a shorter WD is preferred as depth of field is less important and higher resolution can be achieved by using a shorter WD. Figure 1.20 is the SEM image of well-aligned Co-doped ZnO nanowire arrays fabricated by a chemical vapor deposition method [9], showing the influence of WD on depth of field. The figures are focused on the middle part of the images. By comparing circled part of the two images, the enhancement of depth of field is obvious by increasing the WD from 3 to 12 mm.

2.4. Image Formation

Complex interactions occur when the electron beam in an SEM impinges on the specimen surface and excites various signals for SEM observation. The secondary electrons, BSEs, transmitted electrons, or the specimen current might all be collected and displayed. For gathering the information about the composition of the specimen, the excited x-ray or Auger electrons are analyzed. In this section, we will give a brief introduction about the interactions of the electron beam with the specimen surface and the principle of image formation by different signals.

2.4.1. Signal Generation

The interaction of the electron beam with a specimen occurs within an excitation volume under the specimen surface. The depth of the interaction volume depends on the composition of the solid specimen, the energy of the incident electron beam, and the incident angle. Two kinds of scattering process, the elastic and the inelastic process, are considered. The electrons retain all of their energy after an elastic interaction, and elastic scattering results in the production of BSEs when they travel back to the specimen surface and escape into the vacuum. On the other hand, electrons lose energy in the inelastic scattering process and they excite electrons in the specimen lattice. When these low energy electrons, generally with energy less than 50 eV, escape to the vacuum, they are termed "secondary electrons." Secondary electrons can be excited throughout the interaction volume; however, only those near the specimen surface can escape into the vacuum for their low energy and most of them are absorbed by the specimen atoms. In contrast, the BSEs can come from greater depths under the specimen surface. In addition to secondary electrons and BSEs, x-rays are excited during the interaction of the electron beam with the specimen. There are also several signals that can be used to form the images or analyze the properties of specimen, e.g., Auger electrons, cathodoluminescence, transmitted electrons, and specimen current, which have been discussed in Sections 1.1.5 and 1.1.6.

2.4.2. Scanning Coils

As mentioned in the previous sections, the electron beam is focused into a probe spot on the specimen surface and excites different signals for SEM observation. By recording the magnitude of these signals with suitable detectors, we can obtain information about the specimen properties, e.g., topography and composition. However, this information just comes from one single spot that the electron beam excites. In order to form an image, the probe spot must be moved from place to place by a scanning system. A typical image formation system in the SEM is shown in Fig. 1.21. Scanning coils are used to deflect the electron beam so that it can scan on the specimen surface along x- or y-axis. Several detectors are used to detect different signals: solid state BSE detectors for BSEs; the ET detector for secondary and BSEs; energy-dispersive x-ray spectrometer and

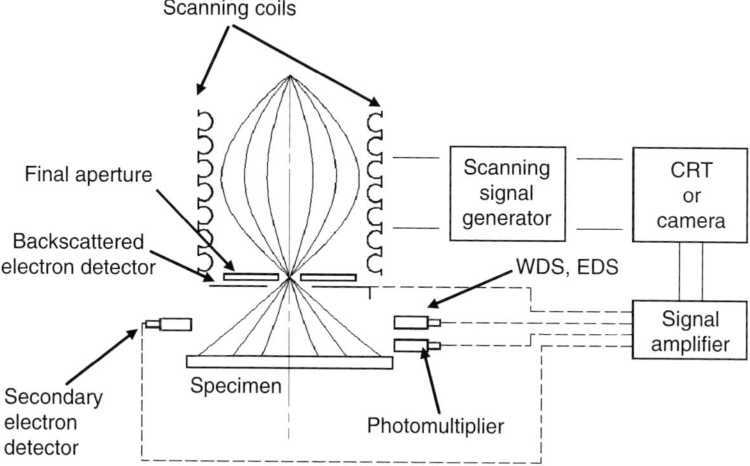

FIGURE 1.21. Image formation system in a typical scanning electron microscope.

wavelength-dispersive x-ray spectrometer for the characteristic x-rays; and photomultipliers for cathodoluminescence. The details of secondary electron detectors will be discussed in Section 2.4.3. The detected signal is also processed and projected on the CRT screen or camera. The scanning process of CRT or camera is synchronized with the electron beam by the scanning signal generator and hence a point-to-point image for the scanning area is produced.

2.4.3. Secondary Electron Detectors

The original goal in building an SEM was to collect secondary electron images. Because secondary electrons are of low energy they could come only from the surface of the sample under the electron beam and so were expected to provide a rich variety of information about the topography and chemistry of the specimen. However, it did not prove to be an easy task to develop a collection system which could detect a small current ($\sim 10^{-12}$ A) of low energy electrons at high speed enough to allow the incident beam to be scanned, and which worked without adding significant noise of its own. The only practical device was the electron multiplier. In this, the secondary electrons from the sample were accelerated onto a cathode where they produced additional secondary electrons that were then in turn accelerated to a second cathode where further signal multiplication occurred. By repeating this process 10 or 20 times, the incident signal was amplified to a large enough level to be used to form the image for display. Although the electron multiplier was in principle sensitive enough, it suffered from the fact that the cathode assemblies were exposed to the pump oil, water vapor, and other contaminants that were present in the specimen chambers of these early instruments with the result that the sensitivity rapidly degraded unless the multiplier was cleaned after every new sample was inserted.

The solution to this problem, and the development that made the SEM a commercial reality, was provided by Everhart and Thornley [10]. Their device consisted of three components: a scintillator that converted the electron signal into light; a light pipe to transfer the light; and a PMT that converts the light signal back into an electron signal. Because the amount of light generated by the scintillator depends both on the scintillator material and on the energy of the electrons striking it, a bias of typically 10 kV is applied to the scintillator so that every electron strikes it with sufficient energy to generate a significant flash of light. This is then conducted along the light pipe, usually made from quartz or Perspex, toward the PMT. The use of a light pipe makes it possible to position the scintillator at the place where it can be most effective in collecting the SE signal while still being able to have the PMT safely away from the sample and stage. Usually the light pipe conducts the light to a window in the vacuum wall of the specimen chamber permitting the PMT to be placed outside the column and vacuum. The conversion of SE first to light and then back to an electron signal makes it possible to use the special properties of the PMT which is a form of an electron multiplier, but is completely sealed and so is not affected by external contaminants. The PMT has a high amplification factor, a logarithmic response which allows it to process signal covering a very large intensity range, is of low noise, responds rapidly to changes in signal level, and is low in price.

This overall arrangement not only offers high efficiency and speed, but is flexible in its implementation, is cheap to construct, and needs little routine attention to maintain peak performance. Consequently, it has been present, in one form or another, in every SEM built since that time. The classic form of the ET detector is that shown in Fig. 1.22a where the sample is placed beneath the objective lens

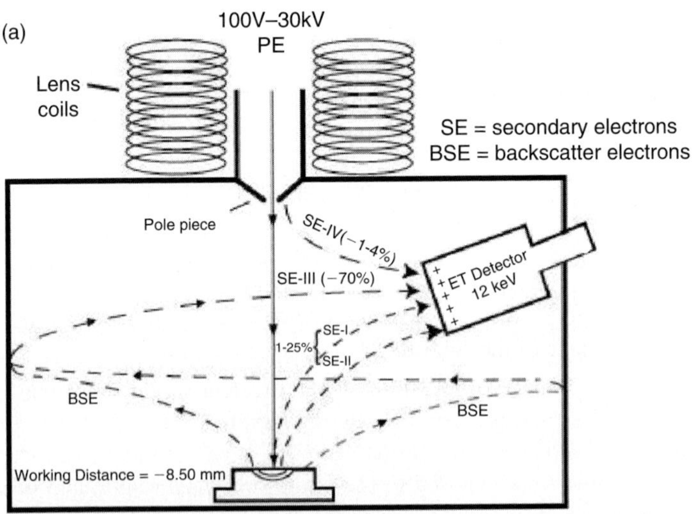

FIGURE 1.22. Three types of different detectors: (a) conventional (below-lens) scanning electron microscope with below-lens ET detector;

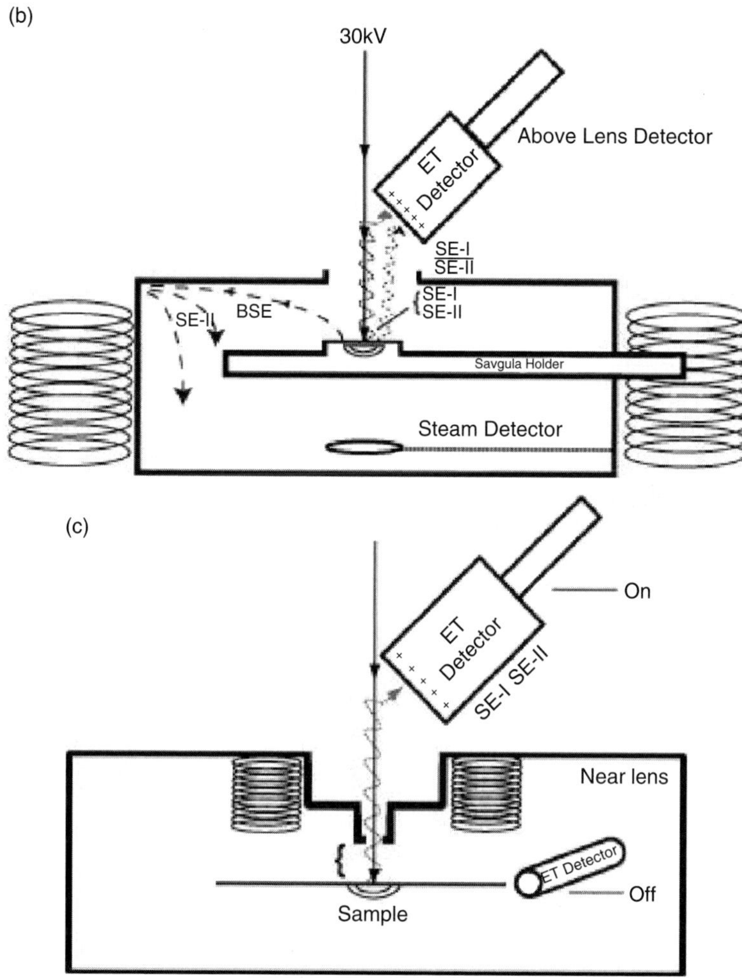

FIGURE 1.22. (*Continued*) (b) condenser/objective (in-lens) scanning electron microscope with above-lens ET detector; and (c) high-resolution (near-lens) scanning electron microscope with above-lens ET detector.

of the SEM and the ET detector is positioned to one side. Because of the +10 kV bias on the front face of the scintillator there is an electrostatic field, of the order of a few hundred volts per millimeter, which attracts the SE from the specimen and guides them toward the detector. At low beam energies, however, a field of this magnitude is sufficient to deflect the incident beam off axis, so a Faraday screen in the form of a widely spaced metal grid is often placed over the scintillator itself to shield the beam. The Faraday screen itself is biased to just 250 or 300 V positive, which is enough to attract many of the emitted SE, but is too low to deviate the beam.

Secondary emission from a horizontal specimen is isotropic about the surface normal, and with the maximum intensity being emitted normal to the surface. Experimental measurements [11] show that this form of the ET detector typically collects about 15–30% of the available SE signal. This relatively poor performance is the result of the fact that many of the SE escape through the bore of the lens and travel back up the column, and also because the asymmetric positioning of the detector only favors collection from half of the emitted SE distribution with a velocity component toward detector. In general this performance is quite satisfactory because of the way in which secondary electron images are interpreted [5]. The viewpoint of the operator is effectively looking down along the beam direction onto the specimen, which is being illuminated by light emitted from the detector assembly. An asymmetric detector geometry therefore results in an image in which topography (e.g., edges, corners, steps, and surface roughness) is shadowed or highlighted depending on the relative position of the feature and the detector. This type of image contrast is intuitively easy and reliable to interpret and produces aesthetically pleasing micrographs.

The main drawback with this arrangement, also evident from Fig. 1.22a, is that the detector will be bombarded not only by the SE1 and SE2 secondary electrons from the specimen carrying the desired specimen information, but also by BSEs from the specimen, and by tertiary electrons (SE3) created by BSE impact on the lens and the chamber walls. Typically at least half of the signal into the detector is from direct backscatters or in the form of SE3 generated by scattering in the sample area. As a result the fraction of SE content from the sample is diluted, the signal-to-noise ratio is degraded, and image detail is reduced in contrast. Although for many purposes it is simply sufficient that the detector produces an adequately large signal, for many advanced techniques it is essential that only specific classes of electrons contribute to the image and in those cases this first type of SE is far from optimum.

In basic SEMs the WD is typically of the order of 12–20 mm so the ET detector can readily be positioned close to the specimen and with a good viewpoint above it. In more advanced microscopes, the WD is often much smaller in order to enhance image resolution and locating the detector is therefore more difficult. Such SEMs often employ a unipole or "snorkel" lens configuration, which produces a large magnetic field at the specimen surface. This field captures a large fraction of the SE emission and channels it back through the bore of the lens and up the column. In order to provide efficient SE imaging the arrangement of Fig. 1.22b is therefore often employed. A standard ET detector is provided as before, for occasions when the sample is imaged at a high WD, or for imaging tilted samples. A second detector is provided above the objective lens to exploit the SE signal trapped by the lens field as first described by Koike [12]. This "upper" or "through the lens (TTL)" detector is a standard ET device and is positioned at 10–15 mm off the incident beam axis. In early versions of TTL detectors the usual 10 kV bias on the scintillator was used to extract the SE signal from the beam path, but at low incident energies the field from the detector was often sufficiently high to misalign the beam. Several SEM manufacturers have now overcome this problem, and introduced an important degree of flexibility and control into the detector system, by positioning a Wien filter just above the lens. The Wien filter consists of a magnetic field (B) at 90° to the direction of the electric field (E) from the detector. This combination of electric and magnetic fields can be adjusted so that the incident beam remains exactly on axis through the lens. However, the returning SE is deflected by the "E×B" fields in a direction, and by an amount, that depends on their energy. By providing electrodes between the E×B field region and the detector, different energies of electrons can be directed by the operator on to the scintillator so that one detector can efficiently collect secondary electrons, BSEs, and all electrons in between those limits.

The upper detector typically collects 70–80% of the available SE signal [10] from the specimen. Because SE3 electrons, generated by the impact of BSE on the lens and chamber walls, are produced well away from the axis of the lens they are not collected by the lens field and do not reach the TTL detector. The upper detector signal is therefore higher in contrast and information content than the signal from the lower detector because the unwanted background of nonspecific SE3 has been eliminated. If the signal is in the upper detector, then the desired contrast effect can in many cases be greatly enhanced by comparison with that available using the lower detector. In most SEMs that use this lens arrangement, both the upper and lower detector can be used simultaneously, because the total budget of SE is fixed by the operating conditions and by the sample, if 80% of the SE signal is going to the TTL detector then only 20% at most is available for the in-chamber ET detector. However, at longer WDs the ability to utilize both detectors can often be of great value. For example, the TTL detector is very sensitive to sample charging effects, but the in-chamber ET detector is relatively insensitive because of the large contribution to its signal from SE3 and BSEs, so mixing the two detector outputs can suppress charging artifacts while maintaining image detail. Similarly, the TTL detector has a vertical and symmetric view of the sample, while the in-chamber detector is asymmetrically placed at the level of the sample. The upper detector therefore is more sensitive to yield effects (e.g., chemistry, electronic properties, and charge) and less sensitive to topography, while the in-chamber detector has the opposite traits.

In the highest resolution SEMs (including TEMs equipped with a scanning system) the specimen is physically inside the lens and is completely immersed within the magnetic field of the lens, so the only access to the SE signal is to collect it using the lens field [11] as shown in Fig. 1.22c. The properties of this detector will be the same as those of the TTL detector described above, but in this configuration there is no opportunity to insert an ET detector at the level of the specimen. The fact that the signal from the TTL detector is almost exclusively comprised of SE1 and SE2 electrons results in high contrast, and high signal-to-noise images that are optimum for high-resolution imaging. The E×B Wien filter discussed above also is usually employed for this type of instrument so that BSE images can also be acquired by appropriate adjustment of the controls.

2.4.4. Specimen Composition

The number of secondary electrons increases as the atomic number of the specimen increases, because the emission of secondary electrons depends on the electron density of the specimen atoms. The production of BSEs also increases with the atomic number of the specimen. Therefore, the contrast of secondary electron signal and BSE signal can give information about the specimen composition. However, BSE signal produces better contrast concerning composition variation of the specimen. Figure 1.23 is a secondary electron micrograph of an electrodeposited nickel mesh on a CaF_2 close-packed opal membrane. The bright spots in the image are nickel [13]. The strong secondary electron emission of nickel is due to its relatively high atomic number.

2.4.5. Specimen Topography

In the secondary electron detection mode, the number of detected electrons is affected by the topography of the specimen surface. The influence of topography on the image contrast is the result of the relative position of the detector, the specimen, and the incident electron beam. A simple situation is depicted in Fig. 1.5, in which the interaction volume by the electron beam and the detector is on the same or on the different sides of a surface island. For the situation illustrated in left side of Fig. 1.5, many of the emitted electrons are blocked by the surface island of the specimen and this results in a dark contrast in the image. In contrast, a bright contrast will occur if the emitted electrons are not blocked by the island (right side of Fig. 1.5). A bias can be applied on the detector so that the secondary electrons from the shadow side can reach the detector.

Surface topography can also influence the emission efficiency of secondary electrons. Especially, the emission of secondary electrons will enhance significantly on the tip of a surface peak. Figure 1.24a illustrates the effective emission region of the secondary electrons with respect to different surface topography. Figure 1.24b

FIGURE 1.23. Secondary electron micrograph of an electrodeposited nickel mesh on a CaF_2 close-packed opal membrane.

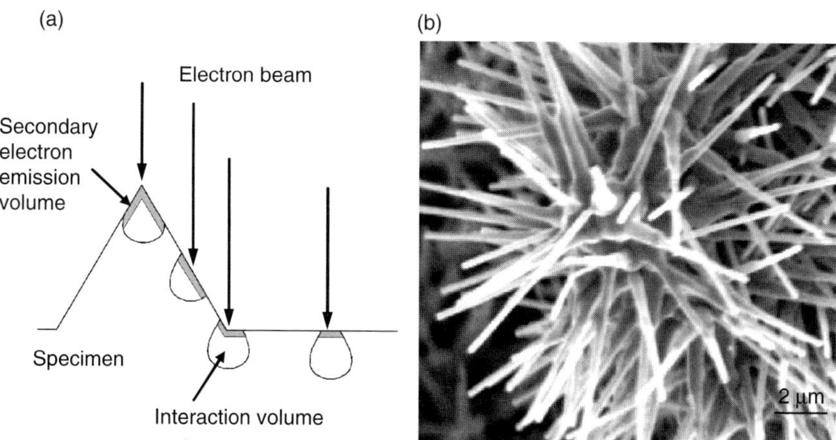

FIGURE 1.24. Enhanced emission on the sharp tips. (a) Schematic of emission enhancement on the tip of a peak; and (b) SEM image of ZnO nanoneedles (emission is enhanced on the tips of the needles).

is a secondary electron micrograph of bundles of ZnO nanoneedles. The enhanced emission is obvious on the tips of the nanoneedles.

Tilting of the specimen, which will change the incident angle of electron beam on the specimen surface, will change the excited region and also alter the effective emission region of secondary electrons. Generally a large tilting angle will contribute to enhanced emission of secondary electrons. The emission enhancement of tilted specimen can be easily understood by Fig. 1.24a at which a slope surface causes the electron beam strike the specimen surface obliquely and enlarges the effective secondary electron emission area.

2.4.6. Specimen Magnification

Magnification is given by the ratio of the scanning length of the CRT image to the corresponding scanning line on the specimen. A change of the size of the scanning area, which is controlled by the scanning coils, will result in the change of the magnification. The WD and the accelerating voltage on the electron beam will also affect the scanning area. However, for many electron microscopes, only the scanning signal is related to the magnification gauge. Therefore, additional calibration is required if accurate magnification is needed.

2.5. Vacuum System

An ultra high vacuum system is indispensable for SEMs in order to avoid the scattering on the electron beam and the contamination of the electron guns and other components. More than one type of vacuum pump is employed to attain the required vacuum for SEM. Generally, a mechanical pump and a diffusion pump are utilized to pump down the chamber from atmospheric pressure.

2.5.1. Mechanical Pumps

Mechanical pumps consist of a motor-driven rotor. The rotating rotor compresses large volume of gas into small volume and thereby increases the gas pressure. If pressure of the compressed gas is large enough, it can be expelled to the atmosphere by a unidirectional valve. A simplified schematic diagram of this type of pump is shown in Fig. 1.25. As the pressure of the chamber increases, the efficiency of the vacuum pump decreases rapidly, though it can achieve a vacuum better than 5×10^{-5} Torr. Avoiding the long pump down times, diffusion pumps are used to increase the pumping rate for pressures lower than 1×10^{-2} Torr.

2.5.2. Diffusion Pumps

The typical structure of a diffusion pump is illustrated in Fig. 1.26. The vaporized oil circulates from top of the pump to bottom. The gas at the top of the pump is transported along the vaporized oil to the bottom and discharged by the

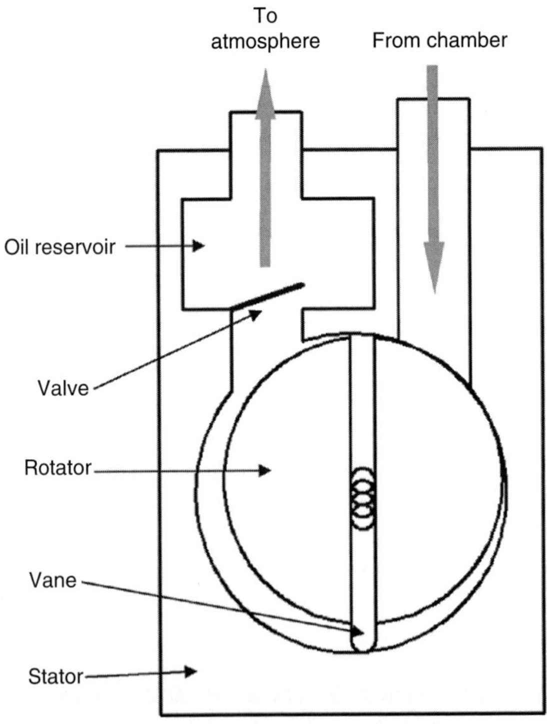

FIGURE 1.25. Schematic of a typical mechanical pump.

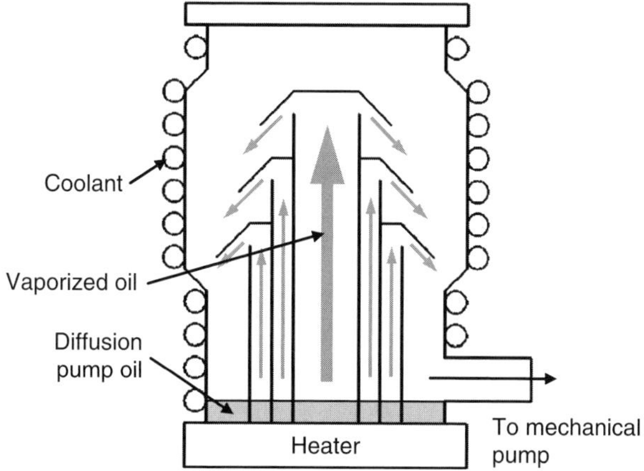

FIGURE 1.26. Schematic of a diffusion pump.

mechanical pump. The heater and the coolants are utilized to vaporize and cool down the oil so that it can be used circularly. The diffusion pump can achieve 5×10^{-5} Torr rapidly; however, it cannot work at the pressure greater than 1×10^{-2} Torr. The mechanical or "backing" pump is used to pump down the chamber to the pressure where the diffusion pump can begin to operate.

2.5.3. Ion Pumps

Ion pumps are used to attain the vacuum level at which the electron guns can work, especially for the LaB_6 guns (10^{-6}–10^{-7} Torr) and field emission guns (10^{-9}–10^{-10} Torr). A fresh surface of very reactive metal is generated by sputtering in the electron gun chamber. The air molecules are absorbed by the metal surface and react with the metal to form a stable solid. A vacuum better than 10^{-11} Torr is attainable with an ion pump.

2.5.4. Turbo Pumps

The basic mechanism of turbo pump is that push gas molecules in a particular direction by the action of rotating vanes. The schematic diagram in Fig. 1.27 shows a small section of one stage in a turbo pump. As the vanes rotate, they push the molecules from the chamber side to the backing pump side and finally go to front pump system. On the one hand, if the molecules were incident from the backing pump side to the chamber side, it will be pushed to the backing pump side. In this way, a preferred gas flow direction is created and a pressure difference is maintained across vane disk. A typical turbo pump contains several rotating disks with the vane angle decreasing at each stage. The pumping speed of the turbo pump decreases rapidly at pressure above a certain level (near 10^{-3} Torr).

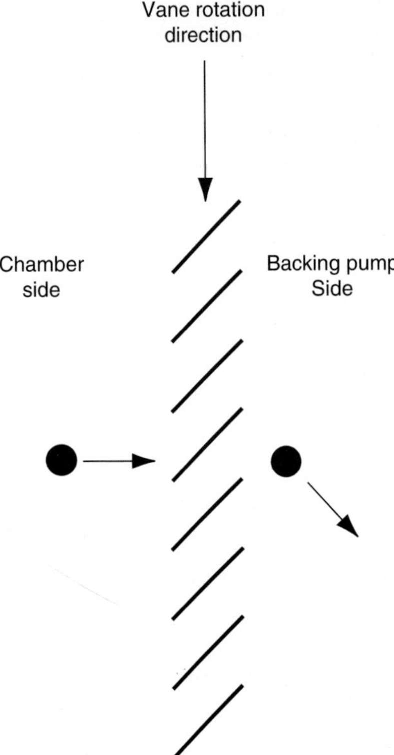

FIGURE 1.27. Schematic of a pumping stage in turbo pump.

A front pump (generally a mechanical pump) is needed for pumping the vacuum before the turbo pump begins to operate.

3. Specimen Preparation

Most nanomaterials, e.g., carbon nanotubes, nanowires, nanoparticles, and nanostructured materials, can be observed by SEM directly through loading them on carbon tape. The sample preparation is quite simple. Some nonconductive nanomaterials, especially bioorganic nanomaterials need metal coating and complicated sample preparation process. In this section, a detailed procedure on bioorganic specimen will be discussed.

3.1. Procedures for High-Resolution SEM of Bioorganic Specimens

The SEM column, gun, and specimen chambers all maintain a vacuum of 10^{-6} Torr or better for performance of the electron optics. The exception to this condition is the specimen chamber of an environmental SEM. High vacuum environments are alien to most forms of life due to the nearly 80% water content of cells and tissues. Even small biomolecules need a hydration shell to remain in a natural state. This incompatibility of specimen fluids such as water with the electron microscope vacuum system necessitates rendering the sample into the solid state, devoid of fluids that would degas in high vacuum and contaminate the microscope. Therefore, all samples placed in an SEM must be dried of fluids in order to be stable for secondary electron imaging. When nanostructural studies of biological or solvated organic systems are planned for the SEM, the science of fluid removal greatly influences the observed structures. The exception to the condition of dried sample preparation is cryofixation and low temperature imaging (cryo-HRSEM) which maintains the water content in the solid state. This topic is covered in Chapter 15.

In this section we will consider necessary steps to render a biological sample preserved in the solid state (dried) with minimal alteration. The preparation needs to be gentile enough so that "significant" structural detail, in the nanometer range, of the solid components can be studied by SEM. During solvent evaporation from a fluid sample into a gaseous atmosphere as in air drying, surface tension forces on the specimen surface are severe. It leads to shrinkage and collapse of structures in the 10^{-3}–10^{-6} m size range making any SEM assessment on a 1–10 nm range impossible. During normal air drying of blood cell suspensions from water, the surface tension forces are great enough to flatten a white blood cell such that no surface features are visible. An interesting exception, the red blood cell has an extremely smooth surface that does not exhibit major structural changes in the submicrometer range after air drying. However, molecular features in the 1–10 nm range are obliterated by surface tension forces.

3.2. *Specimen Fixation and Drying Methods*

Particularly for life sciences, SEM serves to record topographic features of the processed sample surface. Inorganic solids are easily deposited or planed onto a sample support and directly imaged at high magnification exhibiting nanostructural details. The more complex strategy of immobilizing the molecular components and structural features of organisms, organs, tissues, cells, and biomolecules is to chemically fix them into a rigid state with cross-linkers and then treatment with heavy metal salts to enhance the mass density of the components. Biological samples are first "fixed" with glutaraldehyde, a dialdehyde that contains five–carbon atoms. When this molecule is buffered and the biological sample is either perfused or immersed in it, it reacts with the N terminus of amino acids on adjacent proteins thereby releasing $2H_2O$ molecules and cross-linking the peptide chains. Thus, movement of all protein components of the cells and tissues are halted. Biological samples are then "postfixed" with osmium tetroxide (OsO_4) that is believed to interact with the unsaturated fatty acids if lipids serve to halt molecular rotation around these bonds and liberating $2H_2O$. Since OsO_4 contains the heaviest of all elements, it serves to add electron density and scattering properties to otherwise low contrast biological membranes. OsO_4 also serves as a mordant that interacts with itself and with other stains. The use of mordants for high-resolution SEM is not recommended because it binds additional elements to membranes such that at structural resolution of 1–10 nm the observed structures are artificially thickened and the features dimensions not accurately recorded. Mordant methods have been very elegantly employed for intermediate magnification of cellular organelles [14].

3.3. Dehydration and Air drying

Subsequent to fixation, the aqueous content of the sample is replaced with an intermediate fluid, usually an organic solvent such as ethanol or acetone before drying. A series of solvents, e.g., hexamethyldisilazane (HMDS), Freon 113, tetramethylsilane (TMS), and PELDRI II, are sometimes employed for air drying because they reduce high surface tension forces that cause collapse and shrinking of cells and their surface features. Solvent air drying methods are employed in the hopes of avoiding more time-consuming methods. These solvents have very low vapor pressure and some solvents provided reasonable results for white blood cell preservation observed in the SEM at intermediate magnification. However, this method is not appropriate for high-resolution SEM of nanometer-sized structures. Since the drying procedure removes the hydration shell from bioorganic molecules, some collapse at the 1–10 nm range is enviable. The topic of biomolecular maintenance of a hydration shell is covered in Chapter 15.

The most common dehydration schedule is to use ethanol or acetone in a graded series such as 30%, 50%, 70%, 80%, 90%, and 100%, and several washes with fresh 100% solvent. Caution should be exercised not to remove too much of the bulk fluid and expose the sample surface to air. Although biological cells are known to swell at concentrations below 70% and shrink between 70% and 100%, these shape change can be controlled by divalent cations used in the fixative buffers or wash. A superior dehydration method is linear gradient dehydration that requires an exchange apparatus to slowly increase the intermediate fluid concentration [15,16] and serves to reduce osmotic shock and shape change to the specimen.

3.4. Freeze Drying

An outline of processing procedures will now be considered for HRSEM recordings of biologically significant features in the 1–10 nm range. A tutorial on the pros and cons of freeze drying (FD) vs. critical point drying (CPD) was presented when these processes were first being scrutinized for biological accuracy in SEM structural studies [17]. In summary, FD of fixed specimens usually involves addition of a cryoprotectant chemical (sucrose and DMSO) that reduces ice crystal formation but itself may interact with the sample. The fixed and cryoprotected sample is plunge frozen in a cryogen liquid (Freon-22, propane, or ethane) and then placed in an evacuated chamber of 10^{-3} Torr or better and maintained at temperatures of $-35°C$ to $85°C$. If FD method is employed it is best to use ethanol as the cryoprotectant because it sublimes away in a vacuum along with the ice. The FD and subsequent warming procedure that takes place under vacuum requires several hours to a couple of days and produces greater volume loss than with CPD.

3.5. Critical Point Drying

The most reliable and common drying procedure for biological samples is CPD by Anderson 1951 [18]. In this process, samples that have been chemically fixed and exchanged with an intermediate fluid (ethanol or acetone) are then exchanged with a transitional fluid such as liquid carbon dioxide (CO_2), which undergoes a phase transition to gas in a pressurized chamber. The CPD process is not without specimen volume and linear shrinkage; however, there is no surface tension force at the temperature and pressure ($T = 31.1°C$, $P = 1,073$ ψ for CO_2) dependent critical point. Studies in the 1980s showed that linear gradient dehydration followed by "delicate handling procedures" for CPD can greatly reduce the shrinkage measured in earlier studies. Flow monitoring of gas exhausted from the CPD chamber during intermediate transitional fluid exchange and thermal regulation of the transitional fluid from the exchange temperature (4–$20°C$) to the transitional temperature ($31.1°C$) greatly reduces linear shrinkage and collapse. Subcellular structures such as 100 nm diameter surface microvilli or isolated calathrin-coated vesicles retain their near native shape and size. The diversity of biological samples imaged in the SEM is great and therefore the necessary steps for processing isolated molecules with 1–10 nm features may be different than for imaging 1–10 nm structures in the context of complex biological organization such as organelles, cells, tissues, and organs. It is prudent to employ CPD for all HRSEM studies involving bulk specimens; however, FD may serve best for molecular isolates. Following the discussion of appropriate metal coating techniques for nanometer accuracy in HRSEM, a CPD protocol is presented for imaging ~50–60 nm organelles containing 1–10 nm fine structures within the context of a bulk sample (>1 mm^3) by HRSEM and correlate these images with fixed, embedded sectioned STEM materials.

3.6. Metal Coating

Bioorganic specimens are naturally composed of low atomic number elements that naturally emit a low secondary and BSE yield when excited by an electron source. These hydrocarbon specimens also act as an insulator and often lead to the charging phenomena. Since the beginning of biological SEM, precious metals were evaporated onto the specimen in order to render the specimen conductive. Such metals produced conductive specimens, but the heat of vaporization leads to hot metals impinging into the sample surface. Sputter coating these metals (gold, silver, gold/palladium, and platinum) in an argon atmosphere reduced bioorganic specimen surface damage, but still lead to structural decoration with large grain uneven film thicknesses. Precious metals are not suitable for high-resolution SEM because of large grain size and high mobility. That is, as grains of gold or other precious metals are sputtered they tend to migrate toward other grains and enucleate building up the metal film around the tallest features of the sample surface leading to "decorations." Some structures would be over-coated whereas other regions of the sample surface may have no film coverage whatsoever thereby creating discontinuous films. In addition to decorative effects of large grain (2–6 nm) metal films, the grain size of large grain precious metals increase the scattering effect of the primary electrons resulting in a higher yield of SE2 and SE3. See Section 2.4.3 for discussion.

In contrast to decorative metal films, ultra thin fine grain metals (Cr, Ti, Ta, Ir, and W) have very low mobility and monatomic film granularity [19]. These met-als form even "coatings" rather than "decorations" because the atomic grains remain in the vicinity of deposition forming an ultra thin continuous film with a small "critical thickness" often ≤1 nm [20]. The small granularity of these metals greatly increases the high resolution SE1 signal yield because the scattering of the primary electrons within the metal coating is restricted. The resultant images reveal remarkable structural resolution down to a few nanometers with great accuracy because the film provides a continuous 1–2 nm thick coating over all the sample contours [20,21].

3.7. Structural HRSEM Studies of Chemically Fixed CPD-Processed Bulk Biological Tissue

High-resolution SE1 imaging of diaphragmmed fenestrae from blood capillaries have been performed using perfusion fixation and delicate CPD procedures in order to correlate with low-voltage (LV) (25 kV) STEM data [22]. Fenestrae are 50–60 nm wide transcapillary "windows" spanned by a thin filamentous diaphragm and clustered together (sieve plate) within the thin attenuated cyto-plasm of capillary endothelial cells. These dynamic structures have been the sub-ject of structural TEM studies using thin section and platinum replica methods and have been implicated in gating and sorting molecular metabolites into and out of certain tissues [23].

Since the human eye can resolve a 0.2–0.3 mm pixel, a minimum magnifica-tion of 50,000× is necessary for a 0.5 mm image pixel to be easily recognized. A 0.5 mm feature would be equal to 10 nm in an HRSEM image and would enter the range of SE1 signal (Fig. 1.28). Bulk adrenal specimens staged in-lens of an FESEM would produce images with various SE1/SE2 ratios based on the nature of the applied metal film [15,21,24]. At higher magnifications, images containing 1–10 nm surface features, evenly coated with Cr, will contain SE1-dominated contrasts. A high magnification SEM image (Fig. 1.29) reveals the luminal cell surface of a fenestrated adrenocortical endothelial cell after deposition of a 3 nm thick metal film of 60/40 Au/Pd. Such an image lacks significant high resolution SE1 contrast information and structural features in the 1–10 nm range are unavailable because 3 nm grain sizes produce significant electron beam scatter-ing resulting in a higher SE2 signal. Additionally image accuracy is diminished

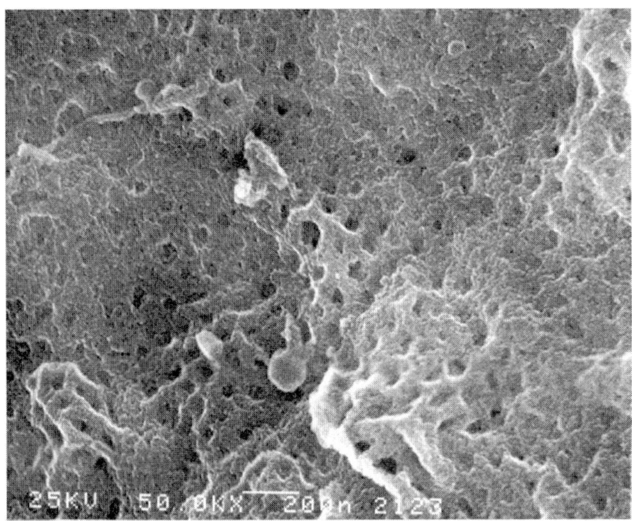

FIGURE 1.28. Intermediate magnification HRSEM (25 kV) of a luminal capillary surface coated with 2 nm Cr film. Note accurate display of numerous 50–60 nm diaphragmmed fenestrae.

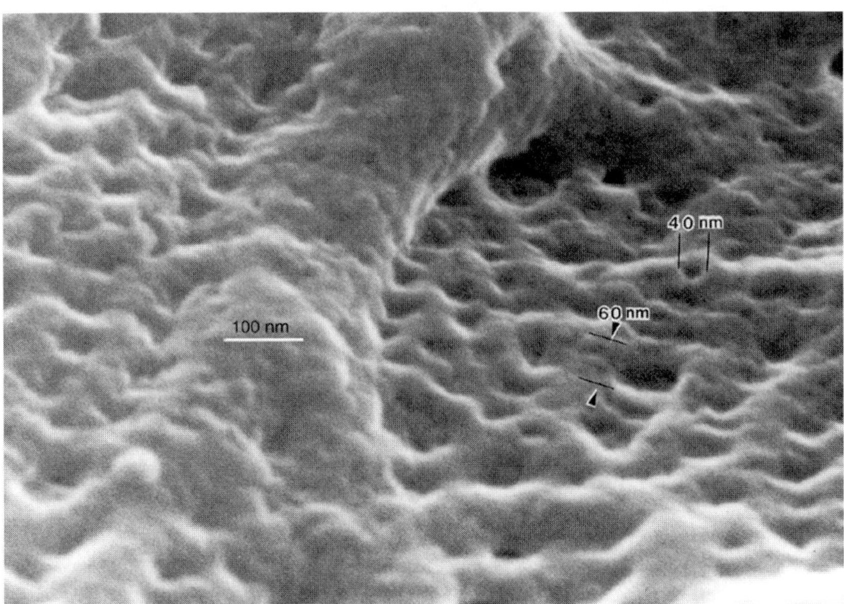

FIGURE 1.29. High magnification SEM (25 kV) of luminal capillary surface decorated with a 3 nm Au/Pd film. Note the uneven film results in a range of fenestral open dimensions and a total lack of 1–10 nm particle contrast.

because the mobility of the metal film during sputtering decorated the fenestrae resulting in a large variation of diaphragm widths.

Thin section specimens of adrenocortical capillaries display thin filamentous diaphragmmed fenestral openings in the 50–60 nm range when imaged by LV-STEM (Fig. 1.30) similarly as those displayed in conventional TEM images. However, little topographic contrast is available to interpret the 3D contours of

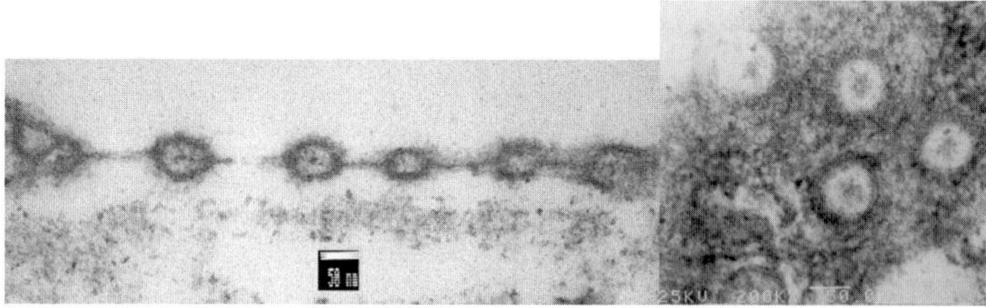

FIGURE 1.30. Low voltage (25 kV) STEM of fenestrae in cross and grazing sections. A thin filamentous structure spans the fenestral opening. Note the lack of topographic contrast or particulate surface features.

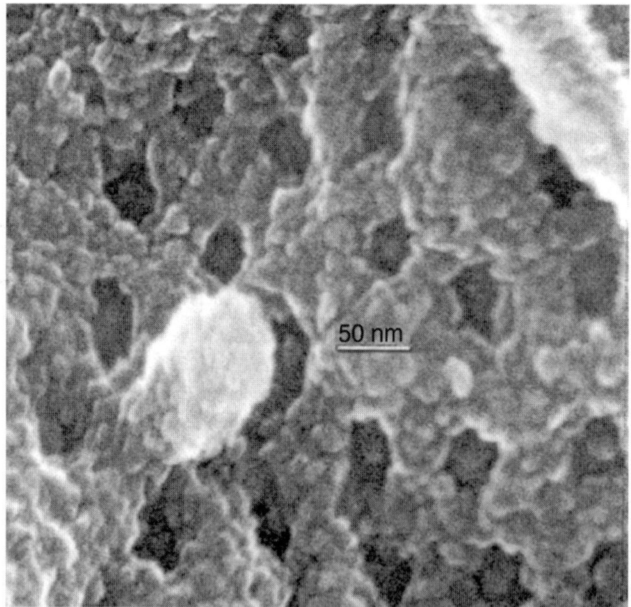

FIGURE 1.31. High magnification HRSEM (25 kV) of the Cr coated (2 nm) fenestrated endothelial cell surface. Openings display various flower-like shapes spanned by a diaphragm and central density. SE1 contrasts provide direct observation of 1–10 nm particulate macromolecular features. These particulate structures probably represent membrane ectodomains that collapse on an atomic level due to the removal of the hydration shell during CPD.

the fenestral walls. Although TEM images of freeze fracture platinum replicas subjected to photographic augmentation suggest fenestral contours have eightfold symmetry, direct recordings of fenestral wall shape were lacking and limited the usefulness of transmitted electron images [23,24].

High magnification (200,000×) HRSEM images collected from 2 nm Cr-coated specimens directly displayed sieve plates containing flower-like ~50 nm fenestral openings were shown in Fig. 1.31. Spherical surface structures in the 5 nm range are easily recognizable. From this example of a bulk CP-dried and Cr-coated specimen, capillary fenestrae were assessed by HRSEM and structural data supported and extended the analysis of fenestral shape by TEM. Caution should be exercised not to overinterpret biological significance of the fine (1–10 nm) surface features in the context of a dried specimen. The hydration shell is removed from the molecular ectodomains of the capillary surface such that there is most certainly atomic collapse within the fine structure. Is the cell surface really covered by fine spherical structures as observed in Fig. 1.31? Most likely not, and structural accuracy in the single nanometer range is therefore limited by the chemical fixation and drying technique.

4. Summary

This chapter gives a quick review of fundamentals of scanning electron microscopy using the nanomaterials as examples. It can help SEM users and nano-materials researchers to master the basic techniques to study nanomaterials in a short time. With the understanding of the basics and knowing the configurations of the microscope, users can easily learn other advanced techniques in this book, such as BSE, advanced x-ray analysis, low voltage imaging, e-beam lithography, focused ion beam, and scanning transmission electron microscopy. Since nanoma-terials do not need complicated sample preparation procedure, nonconductive nanomaterials, especially bioorganic nanomaterials, are explicated in this chapter.

References

1. O. C. Wells, *Scanning Electron Microscopy*, McGraw-Hill, New York (1974).
2. S. Wischnitzer, *Introduction to Electron Microscopy*, Pergamon Press, New York (1962).
3. M. E. Haine and V. E. Cosslett, *The Electron Microscope*, Spon, London (1961).
4. A. N. Broers, in: SEM/1975, IIT Research Institute, Chicago (1975).
5. J. Goldstein, D. Newbury, D. Joy, C. Lyman, P. Echlin, E. Lifshin, L. Sawyer, and J. Michael, *Scanning Electron Microscopy and X-Ray Microanalysis*, 3rd edn, Kluwer Academic/Plenum Publishers, New York (2003).
6. C. W. Oatley, *The Scanning Electron Microscope*, Cambridge University Press, Cambridge (1972).
7. J. I. Goldstein and H. Yakowitz, *Practical Scanning Electron Microscopy*, Plenum Press, New York (1975).

8. Y. X. Chen, L. J. Campbell, and W. L. Zhou, *J. Cryst. Growth*, 270 (2004) 505.

9. J. J. Liu, A. West, J. J. Chen, M. H. Yu, and W. L. Zhou, *Appl. Phys. Lett.*, 87 (2005) 172505.

10. T. E. Everhart and R. F. Thornley, *J. Sci. Instrum.*, 37 (1960) 246.

11. D. C. Joy, C. S. Joy, and R. D. Bunn, *Scanning*, 18 (1996) 533.

12. H. Koike, K. Ueno, and M. Suzuki, *Proceedings of the 29th Annual Meeting EMSA*, G. W. Bailey (Ed.), Claitor's Publishing, Baton Rouge (1971), pp. 225–226.

13. L. Xu, W. L. Zhou, C. Frommen, R. H. Baughman, A. A. Zakhidov, L. Malkinski, J. Q. Wang, and J. B. Wiley, *Chem. Commun.*, 2000 (2000) 997.

14. K. Tanaka, A. Mitsushima, Y. Kashima, T. Nakadera, and H. Osatake, *J. Electron Microsc. Tech.*, 12 (1989) 146.

15. K.-R. Peters, *J. Microsc.*, 118 (1980) 429.

16. R. P. Apkarian and J. C. Curtis, *Scan. Electron Microsc.*, 4 (1981) 165.

17. A. Boyde, *Scan. Electron Microsc.*, 11 (1978) 303.

18. T. F. Anderson, *NY Acad. Sci.*, 13 (1951) 130–134.

19. R. P. Apkarian, *Scanning Microsc.*, 8(2) (1994) 289.

20. K.-R. Peters, *Scan. Electron Microsc.*, 4 (1985) 1519.

21. R. P. Apkarian, *45th Annual Proceedings of the Microscopy Society of America* (1987) 564.

22. D. C. Joy, *52nd Annual Proceedings of the Microscopy Society of America* (1994).

23. E. L. Bearer, L. Orci, P. Sors, *J. Cell Biol.*, 100 (1985) 418.

24. R. P. Apkarian, *Scanning*, 19 (1997) 361.

2
Electron Backscatter Diffraction (EBSD) Technique and Materials Characterization Examples

Tim Maitland and Scott Sitzman

1. Introduction

The term "electron backscatter diffraction" (EBSD) is now synonymous with both the scanning electron microscope (SEM) technique and the accessory system that can be attached to an SEM. EBSD provides quantitative microstructural information about the crystallographic nature of metals, minerals, semiconductors, and ceramics—in fact most inorganic crystalline materials. It reveals grain size, grain boundary character, grain orientation, texture, and phase identity of the sample under the beam. Centimeter-sized samples with millimeter-sized grains, to metal thin films with nanograins may be analyzed. The nominal angular resolution limit is $\approx 0.5°$ and the spatial resolution is related to the resolution of the SEM, but for modern field emission SEMs (FE-SEMs), 20 nm grains can be measured with reasonable accuracy [1]. The macroscopic sample size is dependent on the ability of the SEM's stage and chamber to orient a sample at $70°$ tilt at an appropriate working distance, usually in the range 5 to 30 mm.

1.1. History

The discovery of the fundamental diffraction on which EBSD is based can be traced back to 1928, when Shoji Nishikawa and Seishi Kikuchi (Fig. 2.1) directed a beam of 50 keV electrons from a gas discharge on a cleavage face of calcite at a grazing incidence of $6°$. Diffraction patterns were recorded on photographic plates placed 6.4 cm behind and in front of the crystal, respectively, normal to the primary beam. The patterns were described as "...black and white lines in pairs due to multiple scattering and selective reflection" [2] (Figs. 2.2 and 2.3).

Shinohara [3] and Meibohm and Rupp independently saw the same phenomenon shortly after Kikuchi. Boersch [4], in 1937, produced some excellent patterns (Fig. 2.4) on film. Boersch studied both transmission and backscatter Kikuchi patterns (at 20 kV, ~5° of incidence and up to 162° of acceptance angle) obtained from cleaved, polished, respectively, etched NaCl, KCl, PbS, $CaCO_3$, CaF_2, quartz, mica, diamond, Cu, and Fe surfaces.

FIGURE 2.1. Seishi Kikuchi (*standing*). Thanks to Professor Shun Karato, Yale University, Geology Department. Originally published in Scientific American. Photo credit Nishina Memorial Foundation, courtesy of Hiroshi Ezawa.

FIGURE 2.2. Kikuchi P-pattern from calcite cleavage.

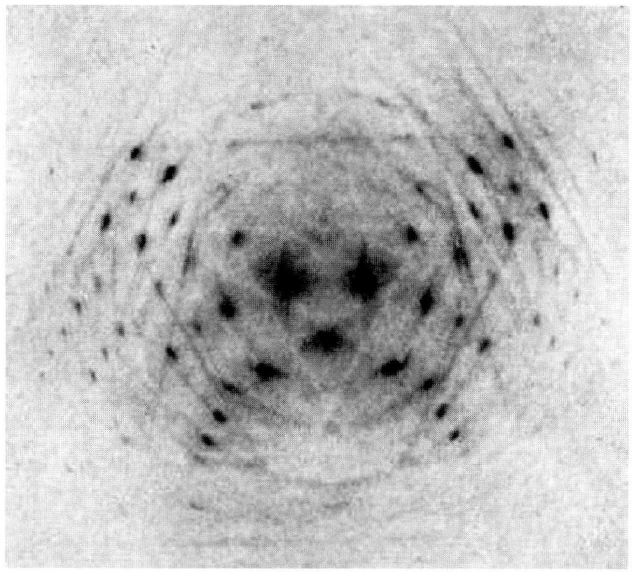

FIGURE 2.3. Kikuchi P-pattern from mica.

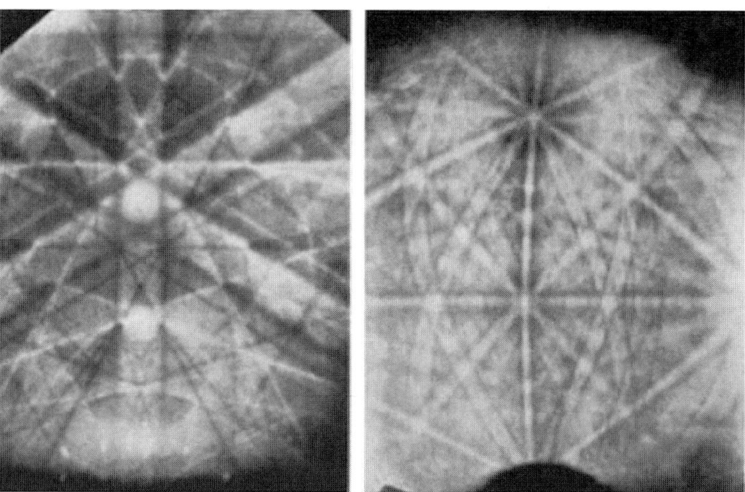

FIGURE 2.4. Boersch 1937 Iron Kikuchi patterns.

Alam et al. [5] in 1954 published their paper entitled "High-Angle Kikuchi Patterns." Both Boersch and Alam et al. used purpose-built vacuum chambers for their work.

The introduction of the commercial SEM, in 1965, allowed greater progress during the decade 1969–1979 with three notable discoveries: selected area channeling patterns (SACP) by Joy et al. at Oxford [6], Kossel diffraction by Biggin and Dingley at Bristol [7], and electron backscatter patterns (EBSP) by Venables and Harland at Sussex [8], the latter where a phosphor screen and TV camera were first used to record the patterns. The term used by Venables, EBSP, has been universally adopted today to refer to the Kikuchi pattern used in an EBSD system. Figure 2.5 shows an early EBSP from Venables [9], showing the ingenious method used to determine the pattern center, critical to accurate analysis of the pattern. The pattern center is defined as the shortest distance from point where the beam strikes the sample to the phosphor screen of the camera. Here Venables arranged three spheres inside the chamber between the sample and the camera. The shadows of the spheres projected onto the camera were elliptical and their major axes could be extrapolated. The intersection of these lines defined the pattern center.

The cross shows the position of the pattern center as predicted by current computer software. Figure 2.6 shows the solution to this EBSP, where the orientation of the BCC cubic unit cell is shown.

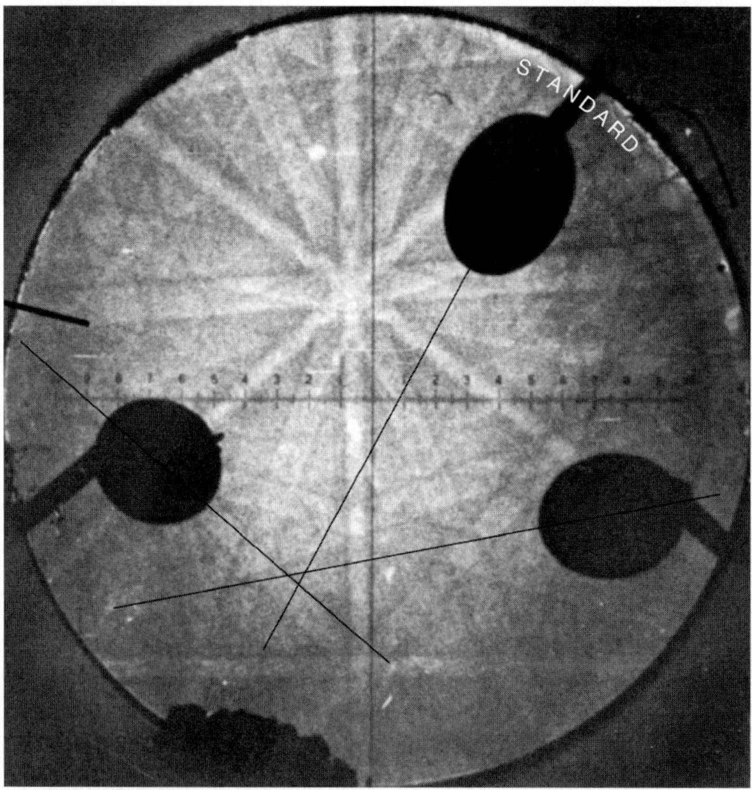

FIGURE 2.5. Venables' early EBSP showing ellipitcal shadows of spheres placed between the sample and the recording TV camera. The intersection of the black lines shows the empirically measured location of the pattern center.

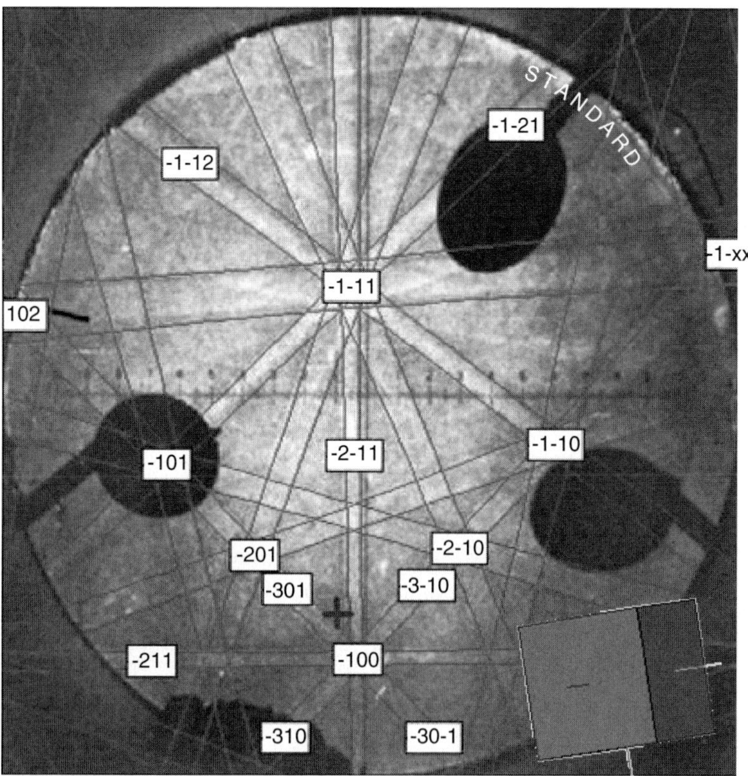

FIGURE 2.6. Venables' early EBSP indexed by modern software. The cross shows the computed location of the pattern center. Graphic inset shows approximate orientation of the crystal with respect to the sample surface.

In 1984, Dingley continued Venables earlier work with a phosphor screen and TV camera but added computer-aided indexing. This was a major step forward and became the model for modern-day EBSD systems.

In 1986, Niels Schmidt (Denmark) wrote software to index electron-channeling patterns (ECP) for all seven-crystal systems when he was a student at Aarhus University, before he went to Risoe National Laboratories. The year 1986 also saw the launch of the first commercial product by Link Analytical (now Oxford Instruments) using Dingley's software and hardware.

ECPs were limited to a spatial resolution of about 10 µm; were difficult to implement (the SEM needed the ability to rock the beam about a point without the need for post lens deflection coils); and were slow. Schmidt adapted his general indexing solution for ECP along with the Dingley/Venables CCD technique to create software to index all seven-crystal systems using EBSPs instead. With this, he founded HKL Technology in 1990.

In 1992, Schmidt's colleagues Krieger-Lassen, Conradsen, and Juul-Jensen made one of the most important breakthroughs in the utility of this technique, the use of the Hough transform. Without this, the ability to automatically detect bands reliably would not be available, relegating the technique to a slow, manual process. The Hough transform was developed in 1962 by Hough [10] as part of a patent to track high-energy particles [11]. Krieger-Lassen et al. used it to automatically identify Kikuchi bands in EBSPs.

In 1993, Brent Adams of Yale coined a new term, orientation imaging microscopy (OIM), to describe the technique of creating an orientation map (OM) of a sample. This is analogous to an EDX map, where instead of using color to represent the spatial distribution of each element present in the sample, it is used to show points of similar orientation.

Shortly thereafter, Dingley and Adams founded TexSEM Laboratory (TSL) in Utah. Thermo (then Noran) were retained to distribute the TSL system, and were responsible for significant numbers in the USA and Japan.

When Dingley and Adams formed TSL, Link/Oxford continued by offering their own EBSD system, as is the case today.

In 1999, EDAX acquired TSL as a part of their strategy to expand their line of analytical instruments. At this time, Thermo Noran had to look elsewhere for an EBSD solution. They turned to two independent players: Robert Schwarzer, of TU Clausthal, who had developed an independent solution for orientation and texture measurements, and Joe Michael, a pioneer of Phase Identification using EBSD at Sandia National Laboratory. Dingley had proposed using EBSD for phase identification in 1989, but Michael and Goehner coupled a slow-scan, high resolution CCD camera to their SEM to obtain sufficient resolution to aid in the indexing of patterns from unknown phases. In this technique, a database of possible phases is used to index the patterns, with the unknown phase being considered identified as the phase from the database that best fits the experimental pattern. Filtering of the database using chemistry and d-spacing is often used to reduce the number of candidate phases to a reasonable number on which to run the indexing algorithm.

Jarle Hjelen, Trondheim, Norway, created many camera designs over these years increasing in sensitivity and speed. Combining his cameras and Schmidt's software, HKL became a key supplier of EBSD systems especially to the geological sciences community who required the low-symmetry indexing algorithms of Schmidt for their mineral studies.

John Sutliff's pioneering work in the industrial application of the EBSD technique at GE Schenectady should not go unmentioned. Using HKL software, Sutliff was one of the first to see the practical utility of the technique in a modern industrial R&D environment.

1.2. How It Works?

EBSD operates by arranging a flat, highly polished (or as-deposited thin film) sample at a shallow angle, usually 20°, to the incident electron beam (Fig. 2.7) (since the SEM stage is often used to tilt the plane of the sample to this shallow angle, the value of stage tilt is often referred to and is typically 70°). With an

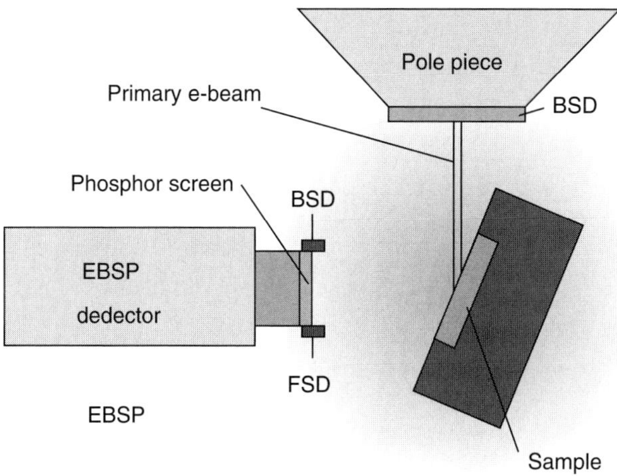

FIGURE 2.7. Schematic arrangement of sample orientation in the SEM. (Adapted from [13].)

accelerating voltage of 10–30 kV, and incident beam currents of 1–50 nA, electron diffraction occurs from the incident beam point on the sample surface. With the beam stationary, an EBSD pattern (EBSDP) emanates spherically from this point.

When the primary beam interacts with the crystal lattice (Fig. 2.8) low energy loss backscattered electrons are channeled and are subject to path differences that lead to constructive and destructive interference. If a phosphor screen is placed a short distance from the tilted sample, in the path of the diffracted electrons, a diffraction pattern can be seen. There are several discussions of the electron interactions involved; in particular Wells, "Comparison of Different Models for the Generation of Electron Backscattering Patterns in the SEM," *Scanning* 21, 368–371 (1999) gives a good descriptions of the competing theories.

The spatial resolution of the technique is governed by the SEM electron optics as in conventional backscattered electron imaging. For high resolution imaging on nanograins, high-performance FE-SEMs are required, along with small samples and short working distances.

The EBSP detector attaches to a free port on the SEM chamber. Ideally, the port will be orthogonal to the stage tilt axis so that the sample may easily be tilted toward the detector at ≈70°, although other orientations are possible. Typically, the port will allow the detector to have a nominal working distance of ~20 mm, since a highly tilted sample necessitates moderate working distances. For small samples, shorter WDs may be attained if the EBSP detector and SEM port allows close proximity to the objective lens. Special detectors are available for less favorable port positions.

The detector is in fact a digital camera. Its CCD chip is illuminated by the phosphor screen that intersects the spherical diffraction pattern. The phosphor

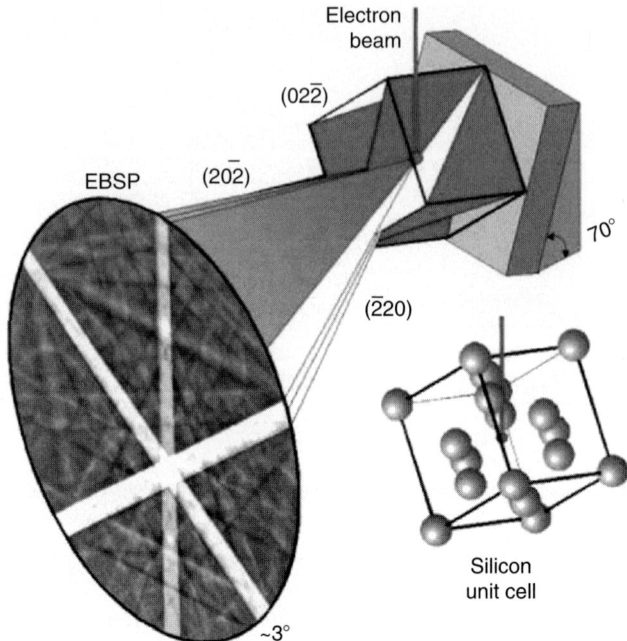

FIGURE 2.8. Electron interaction with crystalline material. (Adapted from [13].)

converts the diffracted electrons into light suitable for the CCD camera to record. With a stationary beam on a point on the sample, an EBSP (Fig. 2.9) is analyzed and in some cases stored. The EBSP is uniquely defined by the lattice parameters of the particular crystal under the beam; by the crystal's orientation in space; the wavelength of the incident of electron bean (which is proportional to the acceleration voltage) and the proximity of the EBSP detector to the sample.

Specialized computer software analyzes the EBSP by detecting a number of Kikuchi bands using an optimized Hough transform. With a *priori* information about the candidate phases under the beam, the software determines all possible orientations with each phase and reports the best fit as the identified phase and orientation. The EBSP is then considered indexed when its orientation and phase are known.

Most SEMs are equipped with EDX spectrometers for chemical analysis by characteristic x-rays produced by the incident electron beam. Today, EDX systems take control of the beam location on the sample using the external scan interface on most SEMs (Fig. 2.10). EBSD requires the same interface to the SEM and thus, for most retrofits to existing systems, a simple electronic method to share this external interface is required. An intelligent switch box is placed between the EDX and the SEM and this arbitrates between the EDX and EBSD systems' access to the SEM. In addition to beam control, for large sample area coverage, integrated stage motion is required. SEM motorized stages are often accessible

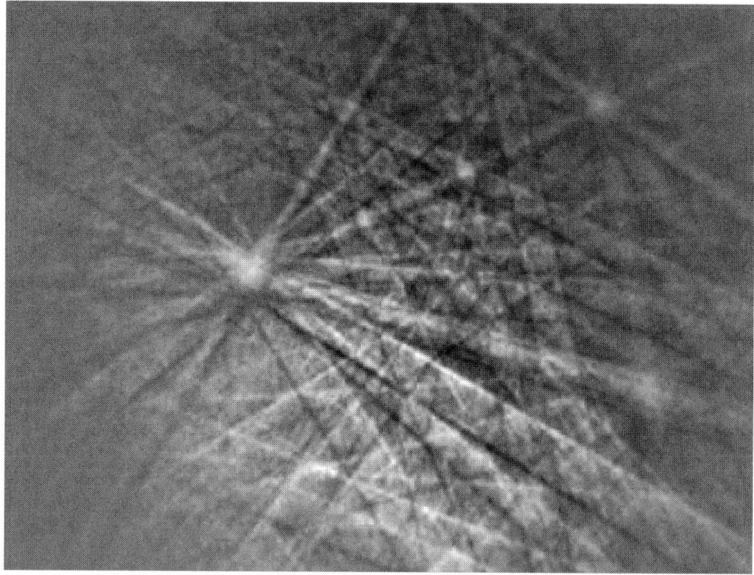

FIGURE 2.9. Example EBSP from Quartz.

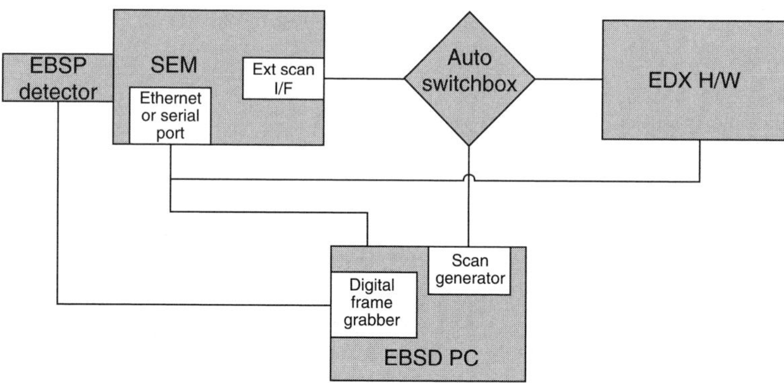

FIGURE 2.10. Schematic of SEM interface.

through an RS232 serial computer interface or Ethernet connection, which can be addressed by the operating software.

As the technique has developed in automated accuracy and overall speed, the ability to scan the beam over multiple points on the sample to create an OM has become practical, and is now the most common method for a microstructural investigation with EBSD. A map is defined by its location, size, and by the sampling step size between points. In this way, the resolution of the map may be adjusted to reveal the grain structure and grain boundary character, depending on the electron beam resolution under the sampling conditions, time available and size of the sample area required. Improvement in speed has been made over the years, from manual indexing rates to 100 automatically indexed patterns/second (Fig. 2.11) [12]. The rate of change approximates an exponential increase since inception (Fig. 2.12).

Detectors used in normal SEM imaging produce a single dimensional output signal. With the beam at a given point on the sample, a signal is recorded and interpreted as the brightness in the output image. However, EBSD gives rise to a 3D pattern emanating from a point that is recorded in 2D on the phosphor screen.

FIGURE 2.11. Brass map acquire at 100 indexed datapoints per second. (Adapted from [12].)

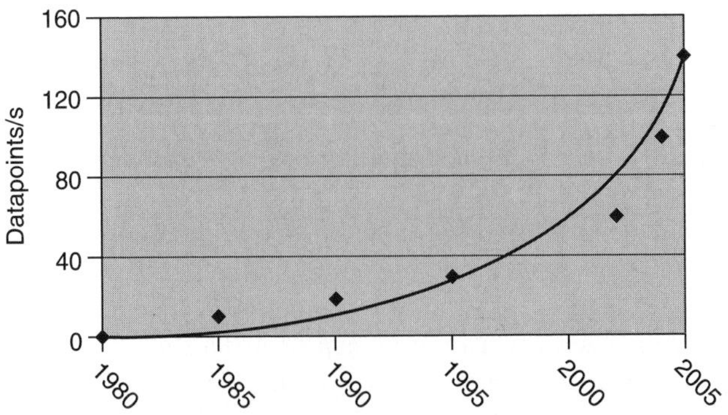

FIGURE 2.12. Mapping speed by year.

Hence, for every point analyzed on a sample, a 2D image of the diffraction pattern is analyzed. Imagine that a 512×512 pixel sample area is analyzed using EBSPs of $512 \times 512 \times 8$ bit pixels in size, $= 512^4 = 64$ GB of uncompressed raw data for one map! In practice, this amount of information is not always stored. Only the position, phase, orientation and some data quality information are stored at each point.

2. Data Measurement

As can be seen in Fig. 2.13, the data set produced by scanning the beam in a regular grid over the sample is comparatively simple. It is a simple database of measurements with each row being a point in the map and each column one of several measured parameters. To create this database, the beam is steered to each pixel point in the map. An EBSP is captured, analyzed, and either the phase and orientation derived or a zero-solution recorded, if the system was unable to measure the orientation. Zero-solutions can come from points where no EBSP is available, where overlapping EBSPs cannot be resolved, or when a new phase is encountered. EBSPs may not be produced when the sample surface deformation is so high that there is no coherent diffraction or when a noncrystalline material is encountered. Overlapping EBSPs occur at grain boundaries when the electron beam diameter is large enough to produce EBSPs from two grains simultaneously. When an EBSP can be solved, its Phase, XY position, orientation, goodness of fit, pattern quality, and other values are recorded.

2.1. Phase

If more than one match unit has been specified by the user, the best fit is listed here. Each match unit contains the information necessary to model the EBSP produced by the expected phase in the sample. The best fit between each match unit

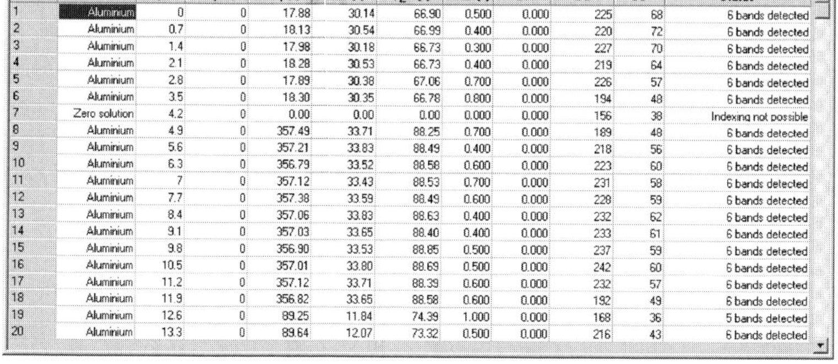

Index	Phase	X (μm)	Y (μm)	φ_1 (°)	Φ (°)	φ_2 (°)	MAD (°)	OI	BC	BS	Status
						Orientations					
1	Aluminium	0	0	17.88	30.14	66.90	0.500	0.000	225	68	6 bands detected
2	Aluminium	0.7	0	18.13	30.54	66.99	0.400	0.000	220	72	6 bands detected
3	Aluminium	1.4	0	17.98	30.18	66.73	0.300	0.000	227	70	6 bands detected
4	Aluminium	2.1	0	18.28	30.53	66.73	0.400	0.000	219	64	6 bands detected
5	Aluminium	2.8	0	17.89	30.38	67.06	0.700	0.000	226	57	6 bands detected
6	Aluminium	3.5	0	18.30	30.35	66.78	0.800	0.000	194	48	6 bands detected
7	Zero solution	4.2	0	0.00	0.00	0.00	0.000	0.000	156	38	Indexing not possible
8	Aluminium	4.9	0	357.49	33.71	88.25	0.700	0.000	189	48	6 bands detected
9	Aluminium	5.6	0	357.21	33.83	88.49	0.400	0.000	218	56	6 bands detected
10	Aluminium	6.3	0	356.79	33.52	88.58	0.600	0.000	223	60	6 bands detected
11	Aluminium	7	0	357.12	33.43	88.53	0.700	0.000	231	58	6 bands detected
12	Aluminium	7.7	0	357.38	33.59	88.49	0.600	0.000	228	59	6 bands detected
13	Aluminium	8.4	0	357.06	33.83	88.63	0.400	0.000	232	62	6 bands detected
14	Aluminium	9.1	0	357.03	33.65	88.40	0.400	0.000	233	61	6 bands detected
15	Aluminium	9.8	0	356.90	33.53	88.85	0.500	0.000	237	59	6 bands detected
16	Aluminium	10.5	0	357.01	33.80	88.69	0.500	0.000	242	60	6 bands detected
17	Aluminium	11.2	0	357.12	33.71	88.39	0.600	0.000	232	57	6 bands detected
18	Aluminium	11.9	0	356.82	33.65	88.58	0.600	0.000	192	49	6 bands detected
19	Aluminium	12.6	0	89.25	11.84	74.39	1.000	0.000	168	36	5 bands detected
20	Aluminium	13.3	0	89.64	12.07	73.32	0.500	0.000	216	43	6 bands detected

Record Browser for Project [Alumin]

FIGURE 2.13. Database of mapped datapoints.

and the experimental EBSP determines the phase and orientation of the point on the sample under the beam.

2.2. Match Unit

A lookup table is used to match the experimental data to a set of crystallographic lattice planes with similar characteristics during crystallographic indexing. The match unit contains the crystallographic indices of Bragg-diffracting lattice planes ("reflectors"), the interplanar angles, the lattice spacing (d) and the intensity of the particular reflectors. The "match unit" is produced utilizing the Kinematical electron diffraction model and contains the following crystallographic parameters:

hkl: The crystallographic indices of Bragg-diffracting lattice planes ("reflectors").

d_{hkl}: The lattice plane spacing of the particular reflectors.

n_{hkl}: The normal vectors to the reflectors.

I_{hkl}: The intensity of the reflectors.

$n_i\, n_j$: The interplanar angles between the reflectors.

From these parameters the interplanar angles ($n_i\, n_j$) are primarily used for indexing. The lattice plane spacing (d_{hkl}) can optionally be applied, whereas the intensity of the reflectors (I_{hkl}) is only used as a threshold value to select the number of reflectors in the match unit.

2.3. Orientation

Orientation is recorded using the Euler angle convention (Euler 1775). Three Euler angles describe a minimum set of rotations that can bring one orientation into coincidence with another. During a measurement, this is the relationship between the EBSP detector and the particular point on the sample being measured under the beam. There is more than one convention but that of Bunge is most common. The three Euler angles: φ_1, Φ, φ_2 represent the following rotations, which are shown schematically in Fig. 2.14.

1. A rotation of φ_1 about the z-axis followed by
2. a rotation of Φ about the rotated x-axis followed by
3. a rotation of φ_2 about the rotated z-axis.

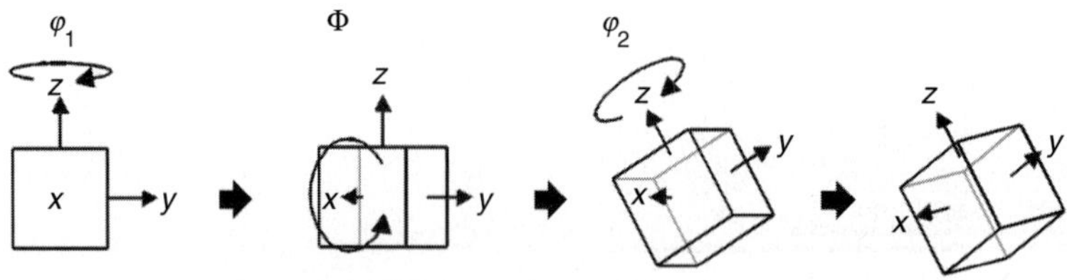

FIGURE 2.14. Euler angle rotations according to Bunge's convention. (Adapted from [13].)

2.4. Mean Angular Deviation

A number that expresses how well the simulated EBSP overlays the actual EBSP. The mean angular deviation (MAD) is given in degrees specifying the averaged angular misfit between detected and simulated Kikuchi bands.

2.5. Band Contrast

Band contrast (BC) is an EBSP quality factor derived from the Hough transform that describes the average intensity of the Kikuchi bands with respect to the overall intensity within the EBSP. The values are scaled to a byte range from 0 to 255 (i.e., low to high contrast). With this scale mapped to a grayscale from black to white, image-like maps can be plotted. These "images" show the microstructure in a qualitative fashion that were used to seeing either in the SEM or light microscope. Because EBSPs along grain boundaries tend to show poor BC they appear dark in a map. Conversely, EBSPs in undeformed regions of a grain appear light (see Fig. 2.15).

2.6. Band Slope

Band slope (BS) is an image quality factor derived from the Hough transform that describes the maximum intensity gradient at the margins of the Kikuchi bands in an EBSP. The values are scaled to a byte range from 0 to 255 (i.e., low to high maximum contrast difference), i.e., the higher the value, the sharper the band.

FIGURE 2.15. Band contrast map on Ni superalloy.

3. Data Analysis

Interrogation and analysis of the acquired data set is often performed away from the SEM with related data processing software that allows a great variety of analyses to be performed: grain size analysis, textural (preferred crystallographic orientation) analysis, and many modes of microstructural visualization and analysis with OMs.

3.1. Grain Size Analysis

EBSD grain size analysis uses changes in crystallographic orientation between neighboring grid points of greater than a defined minimum, typically 10°, to determine the position of grain boundaries (Figs. 2.16 and 2.17). Since the phase and orientation is known at each indexed point on the grid, the exact location and character of each grain boundary is known. Thus, certain grain boundary types such as twins or low-angle grain boundaries may be included or excluded, at the discretion of the analyst.

Classical linear intercept/grain segment detection analyses may be performed by the data processing software on map grids or by using dedicated line scans. Dedicated line scans allow large numbers of grain segments to be included in the analysis using a relatively short EBSD job time (<1 h), while still providing

FIGURE 2.16. Band contrast map with 10° random grain boundaries shown as black 2 pixel-wide lines.

FIGURE 2.17. Band contrast, 10° random grain boundaries and white sigma 3 special boundaries (twins).

robust grain size statistics. Grain area determination from EBSD maps is a more comprehensive method of grain size and shape analysis. Here, the software determines the position of all grain-delimiting boundaries, again using the criteria specified by the operator, and calculates several characteristics of each grain, including area, equivalent circle diameter, aspect ratio (of a fitted ellipse), number of neighbors, and internal deformation by lattice rotation.

3.2. EBSD Maps

As with conventional SEM and optical imaging techniques, EBSD maps can be used to convey visually the basic character of the material's microstructure with 2D information about grain size and shape. However, because the phase and orientation at each pixel in the map is also known, EBSD data processing software can generate an enormous variety of additional visual and analytical information, including overall preferred orientation (texture), prevalence and distribution of grains in specific orientations, phase distribution, state of strain and local variations in residual strain, and character and distribution of grain boundaries.

Maps in EBSD are comprised of "components," or schemes used to color map pixels or boundaries between pixels based on the underlying data recorded. The properties used may be extracted from EBSD-derived properties of the pixel itself, properties of the local group of pixels to which the pixel in question belongs (such as a grain), or information about the pixel by comparison to a neighbor or nonadjacent reference (such as its orientation relative to a reference orientation). In composing an EBSD map, the operator may choose a single component or mixture of components, the latter of which may be combined to form an additive color scheme, such as orientation coloring with the brightness or darkness controlled by the underlying pattern quality. The most commonly used components are pattern quality, phase, orientation, grain boundaries, special grain boundaries, and texture-related components. Other components are used when the material dictates, such as components for displaying state of strain. The basic components are described in more detail below.

3.2.1. Pattern Quality

The most typical is BC, a scalar value measured for each diffraction pattern collected regardless of index result. Essentially, BC is related to the brightness level of diffraction bands above a normalized background, and is affected by the diffraction intensity for a phase, dislocation/crystallographic defect density and orientation. Pattern quality maps are generally grayscale maps that appear similar to coarse SEM images, in part because every point on the map is assigned a brightness based on the pattern quality for that point. Grain boundaries are normally visible as low pattern quality (dark) linear features, and the highly sensitive orientation dependence of BC gives adjacent grains different grayscale values for a clear, SEM micrograph-like microstructural image at the resolution of the EBSD grid. Figure 2.15 is an example of a BC map.

3.2.2. Band Slope

Band slope (BS) is an alternative pattern quality parameter generated when band edge detection is used. This is a measurement of the intensity gradient at the edge of diffraction bands, as determined from peaks in Hough space. BS maps are not as clear as BC in depicting general microstructure because BS is not as sensitive to small orientation changes between grains, but it is more sensitive in general to the state of strain and has been used as a strain differentiator, e.g., to discriminate ferrite from near-cubic but deformed martensite in steels.

3.2.3. All-Euler Orientation Component

The All-Euler (AE) is a basic OM component that uses an Euler angle-based color scale. Euler angles are a set of three angles used to describe the crystallographic orientation of crystals relative to a reference coordinate system (usually defined by the primary SEM stage axes). Here, the value of each Euler angle is individually set to a color scale (normally red, green, and blue for Euler angles φ_1, Φ, and φ_2, respectively), and the three are combined into a single RGB color (see Fig. 2.18). In general, similar colors indicate similar orientations, so the AE-bearing OM is commonly used as a display of general microstructure, since the

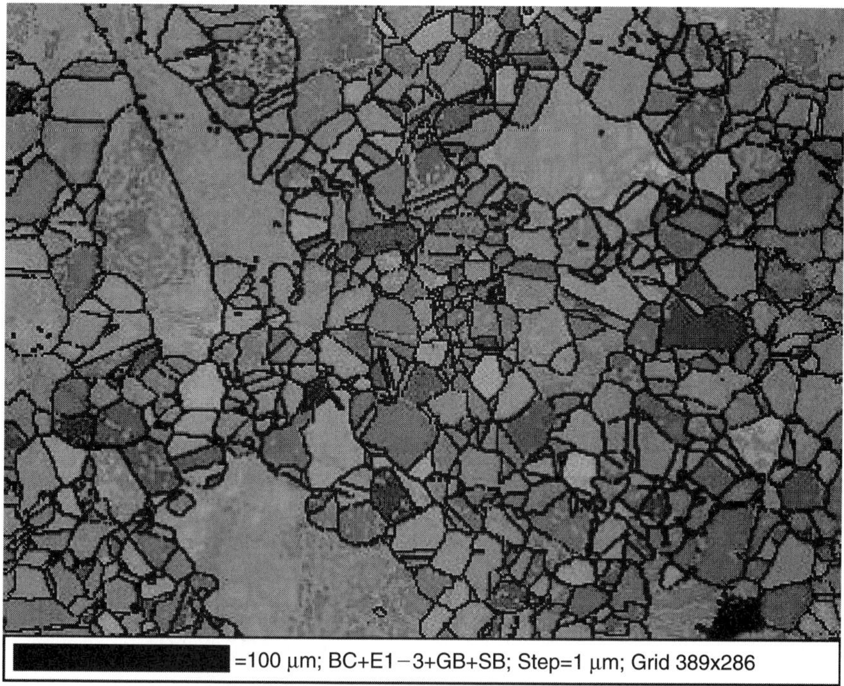

FIGURE 2.18. Ni superalloy showing grain boundaries and all-Euler coloring (see color insert).

basic grain structure and a general idea of strength of texture may be discerned at a glance. However, it is not intuitive to understand the relationship between specific colors and their corresponding orientations, and the AE component is subject to the "wraparound" effect. This happens when one or more of the Euler angles is near a limit, causing the R, G, or B component to vary between maximum and minimum, and showing color speckling where little or no actual orientation change exists. Because of these liabilities, some researchers prefer alternative coloring schemes for the primary EBSD display, most notably the inverse pole figure-based scheme, although all orientation coloring schemes have advantages and disadvantages.

3.2.4. Inverse Pole Figure

Inverse pole figure (IPF) orientation component uses a basic RGB coloring scheme, fit to an inverse pole figure. For cubic phases, full red, green, and blue are assigned to grains whose <100>, <110> or <111> axes, respectively, are parallel to the projection direction of the IPF (typically, the surface-normal direction). Intermediate orientations are colored by an RGB mixture of the primary components, as seen in Figs. 2.19 and 2.20.

Although the IPF orientation map is not susceptible to "wraparound" color speckling as in the total Euler scheme, it has its own limitations. Most notable is

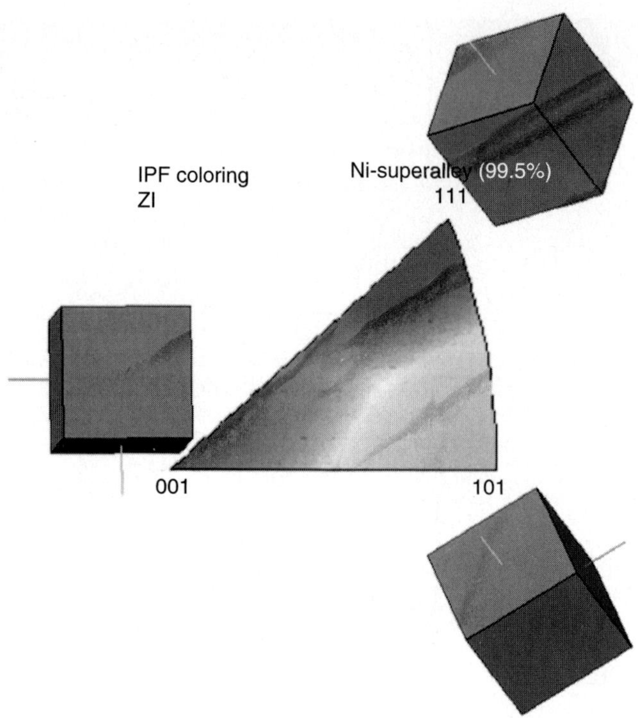

FIGURE 2.19. Inverse pole figure colored map (see color insert).

the coloring of pixels only by the projection-parallel crystallographic axis, independent of rotation about that axis. Thus grains with identical axes parallel to a specified IPF projection direction will have the same color in the IPF-based scheme, but may be in significantly different orientations. For example, two grains with <100> parallel to the surface normal are both colored red, but possess 30° of relative rotation about that axis. IPF-based orientation maps are most useful for displaying materials with strong fiber-textures and for understanding preferred orientations parallel to a sample direction of interest.

3.2.5. Texture Component

Texture component (TC) map is another OM component, but unlike AE and IPF, TC employs a user-determined orientation coloring scheme relative to a specific orientation of interest. Texture components use ideal orientations as references and color each pixel on a map relative to the misorientation between the orientation at that point and the reference. The reference may be set by the analyst to a fiber-texture definition, crystallographic definition (e.g., plane/direction pair), or a set of Euler angles from any point on the map. Commonly, a rainbow scale is

FIGURE 2.20. Surface normal-projected inverse pole figure orientation map (see color insert).

fit to the range of misorientations, the maximum of which may be set by the analyst. These components are used to visually and quantitatively understand the conformity between orientation in the sample and an orientation of interest to the analyst, for example, a deformed primary grain's orientation is the reference and the recrystallized daughter grains within the deformed grain are colored by degree of closeness to that orientation.

3.2.6. Grain Size Coloring Component

Entire grains are colored by their size relative to the range of grain sizes in the map, commonly using a rainbow scale. As with all grain size-related functions in EBSD, the analyst may include or exclude any grain boundary type, such as special boundaries (CSLs, twins) and/or subgrain boundaries, in the grain size analysis and resulting grain size map. An example of the usefulness of this map is in grain reconstructions in Cu thin films for the electronics industry. Here, annealing twins may be disregarded and the nontwinned microstructure revealed and analyzed.

3.2.7. Grain Internal Misorientation Component

Here, grain boundaries are determined and orientation relationships *within* grains are analyzed. Intragranular lattice rotation is a common response of many materials to deformation. It is an expression of plastic strain where dislocation recovery has occurred, forming subcells, and if the subcell size is approximately within the spatial resolution of the SEM and EBSD system, the resultant small changes in orientation may be visualized and analyzed quantitatively. Although the degree of rotation between neighboring pixels depends on the step size used in the EBSD grid as well as the intragranular rotation in the sample, cumulative rotations may be discerned. There are several variations of this map, using different measures of intragranular lattice rotation. Two of the most common are: (a) the kernel-type, whereby each pixel in the map is colored as a function of the degree of orientation change with respect to its neighbors; and (b) the coloring of pixels within grains by the degree of their rotation with respect to a reference orientation within the grain. Quantitative strain analysis by EBSD is not yet common in the materials and geology communities, but studies have shown that this measurement is a strong indicator of degree of plastic strain, and with certain limitations quantitative strain analysis by EBSD may be possible.

3.2.8. Grain Boundary

Grain boundary (GB) component draws grain boundaries between map pixels where there is an interpixel change of orientation greater than an user-defined minimum, usually 2–5° this is because the angular resolution of the technique is limited to ~0.5 at best. When mapping a sample at high speed, this limit may be higher and in some cases reach several degrees. Typically, grain boundaries with misorientations between 2° and 10° are considered subgrain or low-angle grain boundaries and given a specific color, such as silver, whereas boundaries with misorientations >10 are considered random high-angle grain boundaries and are typically colored black. Tools allow percentages of grain boundaries in each category to be compared, and maps possessing this component allow the concentration and distribution of low angle grain boundaries to be determined. If the neighboring pixels are from different phases, phase boundaries may be displayed instead.

3.2.9. Special Boundaries

Special boundaries (SB) are individual user-defined boundaries using an axis-angle definition to identify specific types of boundaries. For example, the $\Sigma3$ CSL/twin typical in copper and nickel-base alloys may be described by a 60° rotation about <111> between neighboring crystallographic domains. Tools are also available to determine the character, prevalence and distribution of special boundaries where the axis/angle definition is not known for a material or processing method.

Most of these map types are demonstrated in the section "Applications." Quantitative analysis of the information depicted in an EBSD map is also usually possible. For a map made of a mixture of pattern quality and grain size coloring components, histograms may be generated describing the overall frequency distribution of the pattern quality parameter, as well as the grain size distribution within the area of analysis. Numbers and percentages of grains and grain areas within different ranges of grain size are easily extracted. These histograms can also display the coloring scheme used for the range of values in the map.

Subsetting is another powerful tool of EBSD, in which selected areas of maps, ranges of measured values or ranges of orientations may be extracted from the complete set of data and independently analyzed and compared. For example, if the grain size analysis indicates a bimodal grain size distribution, the two populations can be analyzed separately with any of the texture or mapping tools available.

4. Applications

Five applications examples are presented, two on samples with grain sizes on the micron-scale, and three with grain sizes on the nanometer scale. The micron-scale group is presented to demonstrate some of the unique analytical capabilities of the technique on strained samples: a strained aluminum alloy and a strained Fe–Al intermetallic alloy, using mapping grid resolutions of 0.7 and 1.0 µm, respectively. The nm-scale group includes samples with grain sizes on the order of tens to hundreds of nanometers: a Pt thin film, a Cu thin film pattern, and an Al thin film using 5, 5 and 10 nm grid resolutions, respectively.

4.1. Friction-Stir Welded Aluminum Alloy

A friction stir welded aluminum alloy (AA2024) was cross-sectioned, polished, and analyzed by EBSD. This analysis used a 772×235 point grid at 0.7 mm steps, placed on the interface between the "nugget" and thermo-mechanically affected zone (TMAZ).

Figure 2.21 is a basic OM with grain boundaries; the "nugget" region is the finer-grained area on the left, the TMAZ is the coarser-grained area on the right. Grain size analysis of these areas gives an average of 4.9 and 13.5 µm, respectively. The orientation colors are related to the surface normal-projected IPF, again where points on the sample with a <111> axis parallel to the surface normal are blue, <110> green, <100> red (see legend) and intermediate orientations have intermediate colors. The variety of colors in the map implies no significant surface-normal parallel texture, verified by the contoured IPF for this projection direction. However, inspection of the orientation distribution function (ODF) indicates a strong brass texture (Fig. 2.22). Grain boundary character and distribution on this map give an indication of the strain state difference between the "nugget" and TMAZ. The TMAZ grains possess considerable subcell structure, indicated by the presence of subgrain boundaries in silver. In contrast, the "nugget" grains are relatively free of subgrain boundaries, with few grains subdivided by silver lines.

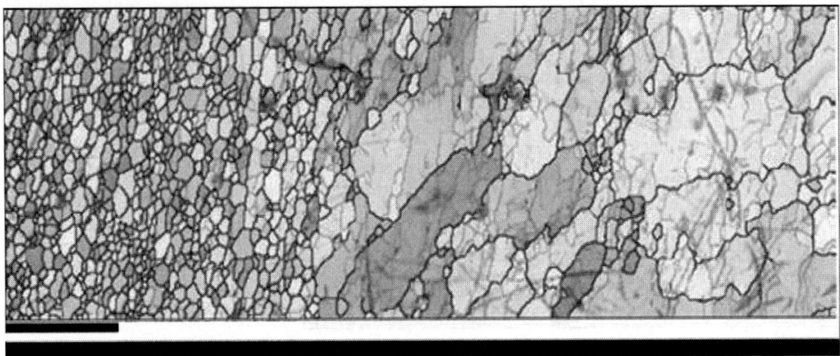

FIGURE 2.21. All euler map of Friction Stir Welded Al AA2024 showing advancing side microstructure. The "nugget" region is the fine grained area left, the TMAZ is the coarser grained area on the right. Scale bar = 100um.

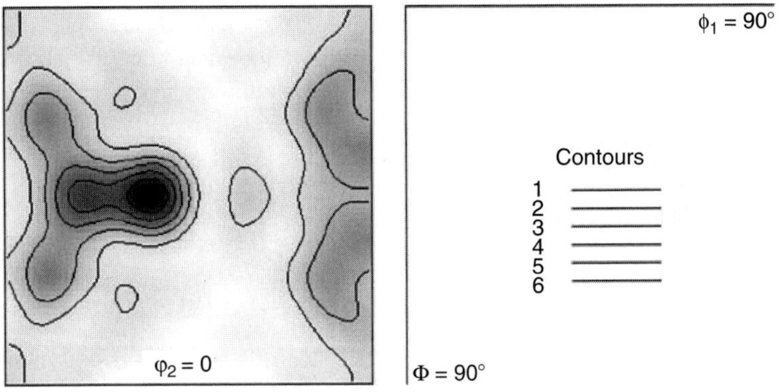

FIGURE 2.22. Slice of ODF at $\varphi_2 = 0°$. Strength and position of contouring indicates strong brass texture.

A map containing an intragranular lattice rotation-based strain analysis component is displayed in Fig. 2.23. Here, the EBSD postprocessing software detected grains, found the most logical reference point within each grain, and colored every pixel by the degree of orientation change (rotation) relative to the orientation of the reference pixel for that grain. The coloring scheme is a blue-to-red rainbow scale, with blue indicating orientations closest to the reference, red furthest, and the range set by the maximum for the map, in this case. The map shows that the TMAZ grains contain a wider variation in color than most "nugget" grains, indicating a higher state of strain on an average, and an interesting subcell structure within the TMAZ grains, bounded by the silver subgrain boundaries. Note that in general it is not known if smaller grain sizes cause seemingly lower degrees of indicated deformation in this type of map, either because of smaller volume per grain or different deformation mechanisms (such as grain boundary

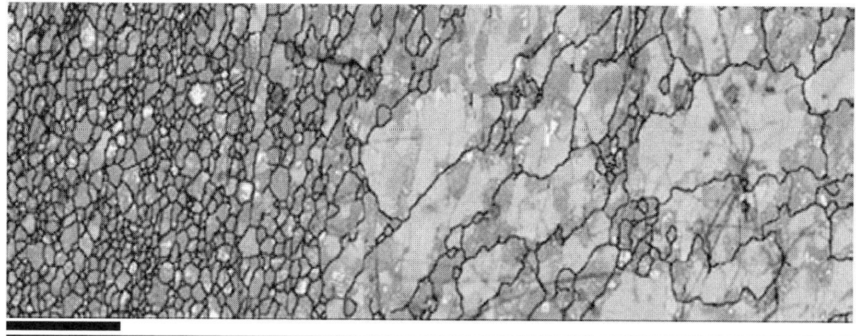

FIGURE 2.23. Grain internal deformation map, based on intragranular lattice rotation relative to a reference point in each grain. Coloring scheme follows a rainbow scale, where blue = smallest rotation from reference orientation, red = 10° of rotation. Scale bar = 100 μm (see color insert).

sliding). However, examinations of strained samples with a range of grain sizes have not yielded a clear correlation between grain size and average intragranular misorientation.

4.2. Deformed Fe–Al Intermetallic Alloy

This example is a single-phase alloy that has been cast, extruded, hot rolled, and cold rolled by 80%. It contains two populations of grains, deformed, and recrystallized, which are differentiated by grain size, orientation, and state of strain. The EBSD mapping used a 563 × 1097 point grid at 1 μm steps.

Figure 2.24 is a basic BC map with grain boundaries where high angle (>10°) boundaries are black, low angle red, and very low angle yellow. The two populations

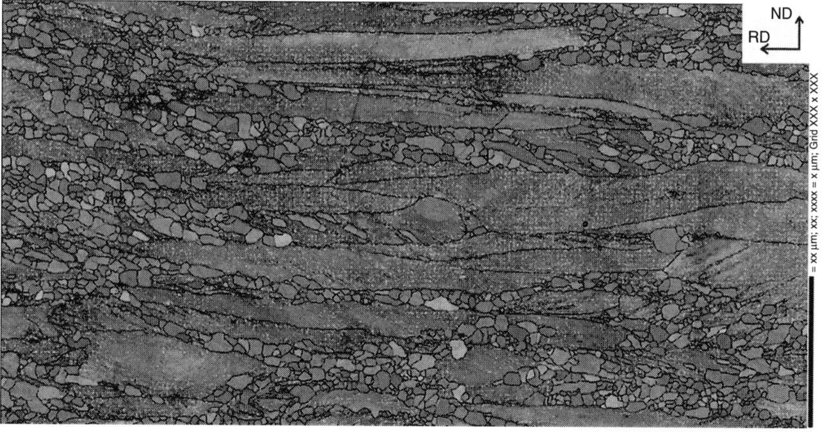

FIGURE 2.24. Pattern quality (band contrast) map, high angle boundaries in black, low angle grain boundaries in red, very low angle in yellow. Vertical black scale bar on right = 200 μm (see color insert).

can be clearly seen in smaller equant grains between large, elongated grains. The abundance of red and yellow boundaries within the larger (deformed) grains and lack of same in the smaller (recrystallized) grains is one indication of a higher degree of strain-related substructure in the deformed grain population.

For Fig. 2.25, all grains have been classified as deformed or recrystallized, in red and blue, respectively. To do this, the EBSD software determines the average intra-granular orientation spread for each grain; the larger this metric for an individual grain, the more substructured the grain is and the more residual strain the grain may be assumed to contain. In the case of this EBSD job, examination of the frequency spread of this value clearly indicated the two populations, and each grain was thus classified. Grain size could also have been used here, but for any given sample, grain size alone may not be a good discriminating factor. Once the grains are classified into a subset, other comparative analyses may be performed, such as area percentage and texture. Here, the recrystallized grains make up 31% of the area of analysis. Comparison of the degree of preferred orientation is shown in the pole figures in Fig. 2.26. Figure 2.27 shows an IPF-based OM, where the IPF projection direction is parallel to the rolling direction (RD). Immediately noticeable is the relative variety of coloring in the smaller grains, implying a weaker or nonexistent preferred orientation, whereas the larger grains have a greenish hue, indicating a stronger <110> ‖ RD fiber texture and corroborating the RD-parallel {110} peak for the deformed grains in the {110} pole figure.

4.3. *Platinum Thin Film*

This experiment characterized a small section of a Pt thin film on a Si substrate (with intermediate SiO_2 and TiO_2 layers), with tens-of-nanometer scale Pt grains. A thermal FE-SEM was operated under high resolution, low probe current

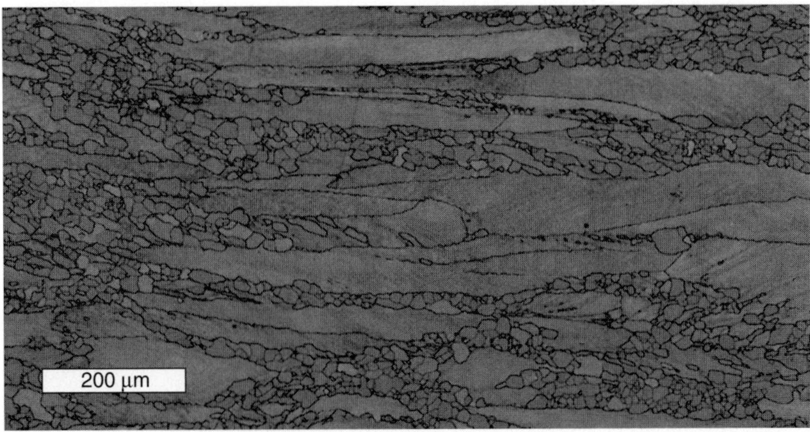

FIGURE 2.25. Map with two classifications of grains: red = deformed, blue = recrystallized, based on degree of internal lattice rotation. High angle grain boundaries are also shown in black. [Adapted from [14].) (see color insert).

(a) Deformed

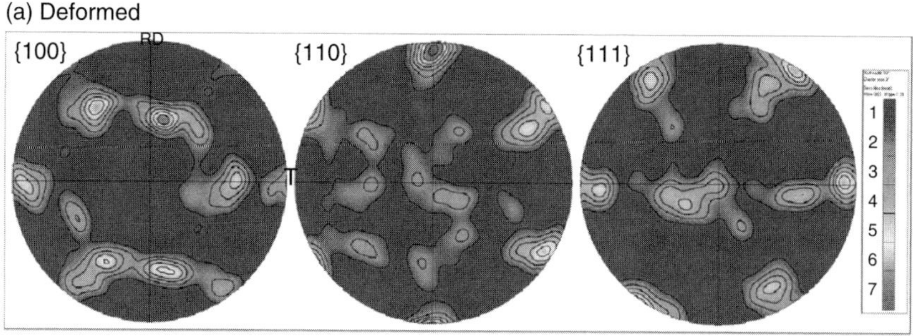

(b) Recrystallized

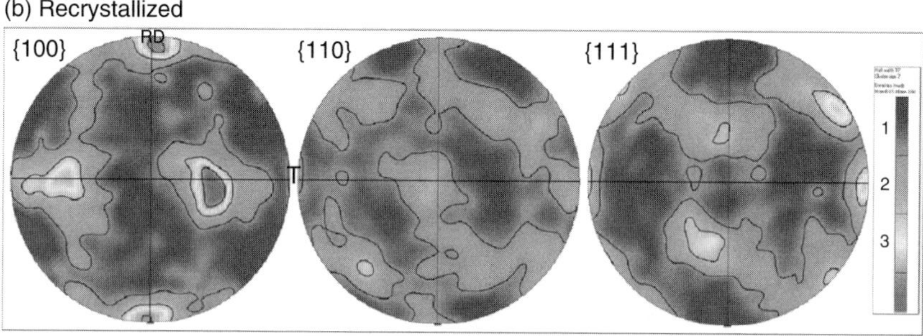

FIGURE 2.26. (a) and (b) Contoured pole figures for the deformed and recrystallized grains, respectively, shown in Fig. 2.25.

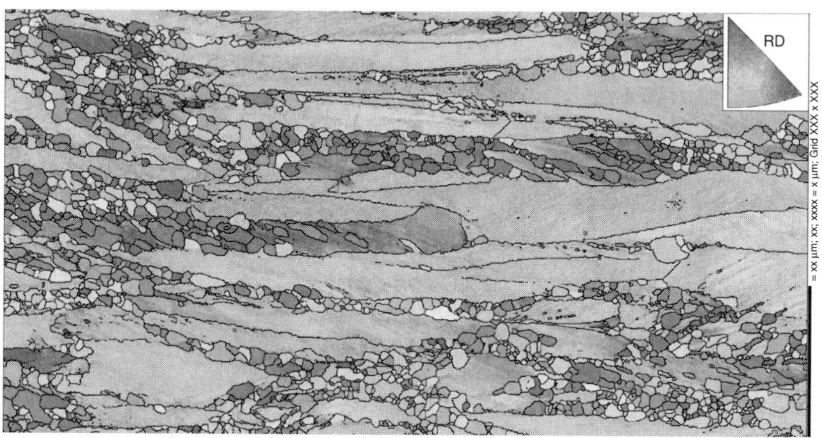

FIGURE 2.27. Rolling direction-projected IPF-based map. Green deformed (large, in this case) grains indicate a strong <110> RD texture not shared by the smaller recrystallized grains. Vertical black scale bar on right = 200 μm.

conditions (25 kV, 0.5 nA) at a fairly small working distance (10 mm) to achieve the required spatial resolution. Further, a high sensitivity digital detector was used. The EBSD grid was 50 × 50 at 5 nm steps.

Figure 2.28a is a pattern quality map of the acquisition area, showing the mostly equiaxed microstructure. Grain boundaries are seen as low pattern quality (darker) regions surrounding higher pattern quality grain cores.

An IPF-based OM is given in Fig. 2.28b projected parallel to the surface normal, with high angle grain boundaries in black and low angle (2–10°) boundaries in silver. The preponderance of blue in the map indicates a strong <111> ‖ surface normal texture, and the lack of consistent coloring in the X-projected (Fig. 2.28c) and Y-projected (Fig. 2.28d). IPF-based maps all indicate that this is a surface normal-parallel *fiber* texture, meaning there is no strongly preferred (rotational) orientation about the primary <111> texture peak. The surface normal-projected IPF and {111} pole figures in Fig. 2.29 corroborates a <111> ‖ surface normal fiber texture.

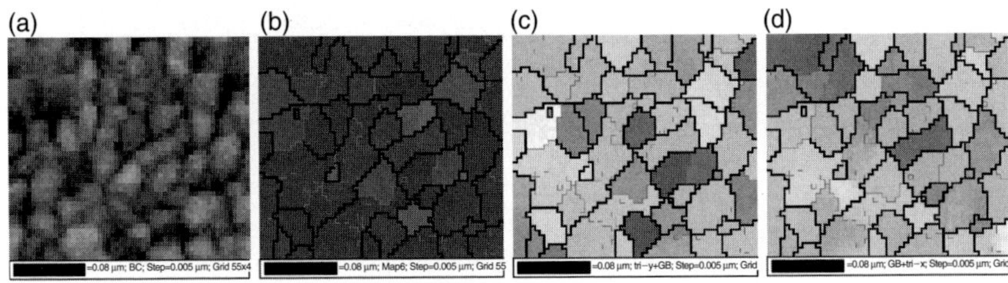

FIGURE 2.28. EBSD maps of Pt thin film. (a) Band contrast (pattern quality) map showing basic grain structure independent of indexing; (b) surface normal-projected IPF-based orientation map, blue color indicating strong <111> ‖ surface normal texture; (c) and (d) are surface plane horizontal and vertical-projected IPF maps, respectively, and show no strong texture in these directions. Scale bar = 80 nm. (Adapted from [15].) (see color insert).

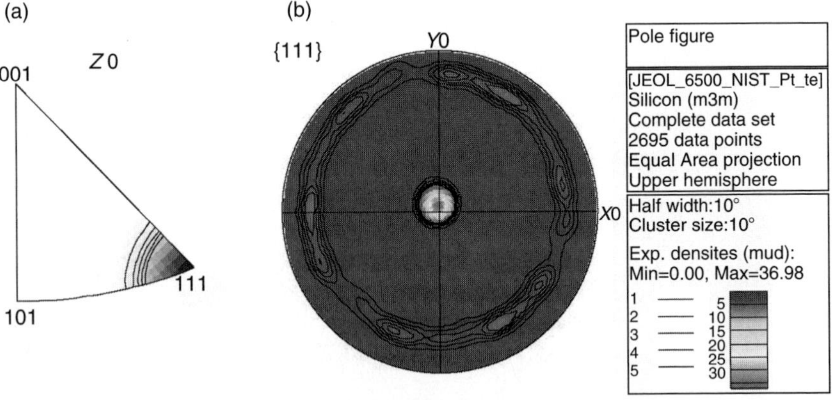

FIGURE 2.29. (a) Surface normal-projected IPF and (b) {111} pole figures, showing strong {111} ‖ surface normal texture. (Adapted from [15].)

The ODF also indicates close compliance of the texture in this area with the $\{111\}$ ∥ surface normal (gamma) fiber texture definition. ODFs are 3D representations of "Euler space," with the three Euler angles (φ_1, Φ, and φ_2) that describe the orientation of a crystal forming the axes. A crystal's orientation is represented as a point within that space. Although not as intuitive to most analysts as the conventional pole figure, the ODF may more clearly reveal subtle and compound textures. Figure 2.30a gives the contoured (coefficient) for this job, showing the strongest texture in a tube-like feature close to centered on the $\Phi - \varphi_2$ plane and elongated parallel to the φ_1 Euler angle, corresponding with the gamma-fiber texture. Offsets between the actual and ideal peak concentration positions, as well as variations in intensity along the length of the tube, can be used to understand the actual texture peak orientation and clustering of orientations within the gamma fiber texture definition, respectively. Figure 2.30b shows a $\Phi - \varphi_1$ cross section, which is a slice along the length of the gamma fiber "tube." The nonuniformity along the length of the "tube" cross section indicates some rotational orientation clustering about the texture peak. These features of the ODF correspond with the peak offset seen in the pole figure in Fig. 2.29b and the nonuniform $\{111\}$ peak distribution in the ring structure in the pole figure. These conclusions, however, are primarily for demonstration, as the grain sampling in this job is too small for a statistically valid textural analysis.

Although grain boundary positions in this map correspond with the positions of anticipated grain boundaries in the BC map, many are low-angle grain boundaries, colored green (Fig. 2.31). See for example the relatively large grain just to the left and below the center of the map. The BC information seemingly clearly shows that this area comprises 6–7 separate grains; however, the low angle grain boundaries in the OM reveal the orientation relationships between neighboring grains in this area to be <10°, forming a typical subgrain-divided primary grain. Although strong fiber textures (strongly preferred orientation of a single crystallographic axis) constrain the possible range of disorientations between grains, increasing the likelihood that random neighboring grains will have low-angle

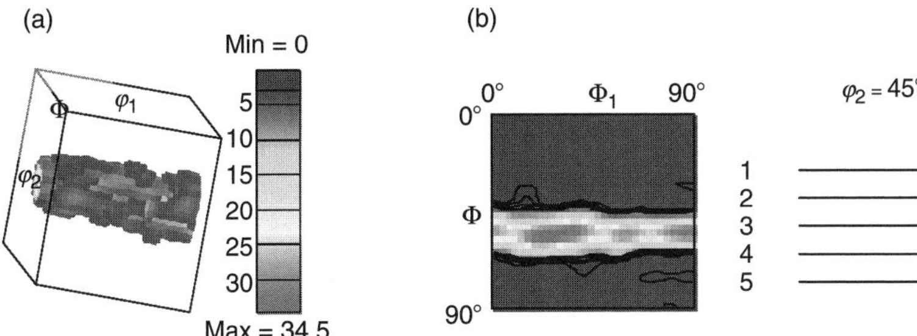

FIGURE 2.30. (a) Orientation distribution function (ODF) coefficient. Tube-shaped feature parallel to φ_1 indicates strong <111> ∥ surface normal texture. (b) $\Phi - \varphi_1$ parallel slice through ODF at $\varphi_2 = 45°$. Discontinuities along "tube" length may be analyzed to detail preferred orientations about <111> within the primary fiber texture. (Adapted from [15].)

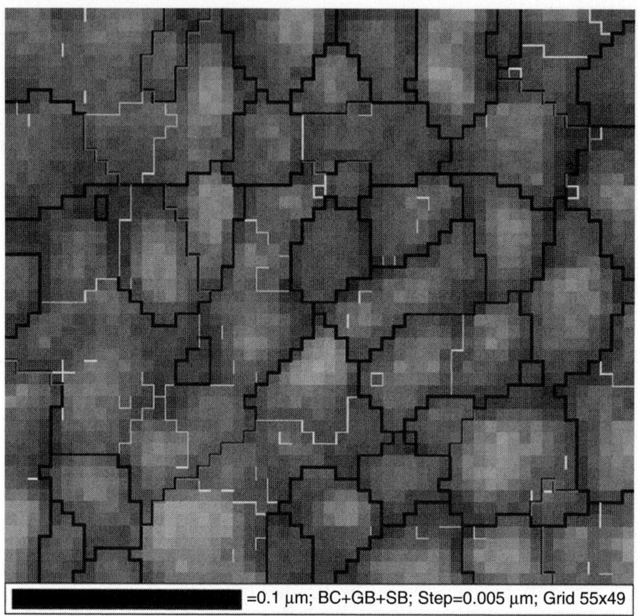

FIGURE 2.31. EBSD map of Pt thin film showing high-angle and low-angle grain boundaries in black and white respectively. Note step size of 5 nm shows that grains of size down to 20 nm in diameter (4 pixels) can be reliably distinguished.

grain boundaries between them, in cubics the <111> fiber texture definition allows the maximum range of 60° for grains to perfectly conform to the textural definition.

Disregarding low angle grain boundaries but including the Σ3 twins (red grain boundaries in Fig. 2.31) as grain-delimiting boundaries, EBSD grain size analysis in this small area revealed 16 grains with a 35 nm grain equivalent-circle diameter average.

4.4. Copper Thin Film

EBSD analysis of thin film copper sheets and interconnects comprises an important and growing segment of modern EBSD use. As the width of interconnects is miniaturized below the single-grain scale, grain boundary character, grain size, and overall texture become increasingly important factors dictating circuit life in service. Each of these characteristics are easily extracted and examined with EBSD. In this example, a set of 500 nm-wide copper lines is examined at two EBSD resolutions, using a thermal FEG-SEM operated at 20 kV, with 1 nA probe current and a 5 mm working distance.

Two jobs are discussed here. The first job used a 537 × 770 point grid at 20 nm steps. Figure 2.32 is an All-Euler OM with random high-angle grain boundaries in black and Σ3 CSLs in red. This shows a typical annealing-twinned copper

FIGURE 2.32. All-Euler orientation map with grain boundaries. Most grains are heavily twinned and span the length of interconnect lines. Drawn box shows location of higher-resolution run. Scale bar = 2 µm (see color insert).

structure, with single grains spanning entire interconnect widths. A strong gamma-fiber (<111> ∥ surface normal) texture is seen in the pole figure set in Fig. 2.33.

Figures 2.34 and 2.35 are grain size maps, in which the smallest detected grains are colored blue, the largest red, and intermediate-sized grains colored with intermediate rainbow colors. Figure 2.34 shows the crystal domain structure using this method where all twins are counted as grain-delimiting boundaries. Figure 2.35 ignores these boundaries in detecting grains, and the resultant "bamboo" structure is revealed, with most grains much larger than interconnect widths, incorporating multiple twin domains.

A second acquisition job was run in this area at a 5 nm step size, using a 395 × 385 point grid. Although some drift/instability caused the lines to appear wavy, the grain and twin domain structures are revealed. Figure 2.36a is a pattern quality map showing the general grain structure and position of grain boundaries. The

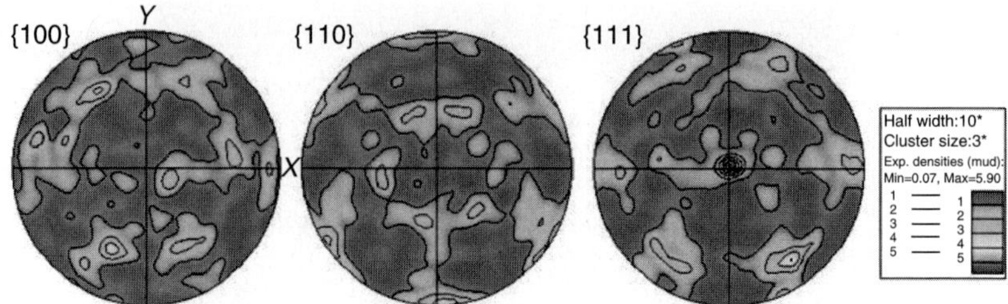

FIGURE 2.33. Contoured pole figure set, displaying strong <111> ‖ surface normal texture (see color insert).

FIGURE 2.34. Grain size (equivalent-circle diameter) map, with all Σ3 twins as grain-delimiting boundaries. Scale bar = 2 µm (see color insert).

All-Euler OM in Fig. 2.36b gives the nature of these boundaries, with all boundaries on the right interconnect in red (twins), and a pair of relatively small grains in the left interconnect bounded by a black (random) high angle grain boundary. The green grain in this pair has grain boundaries in the positions expected from

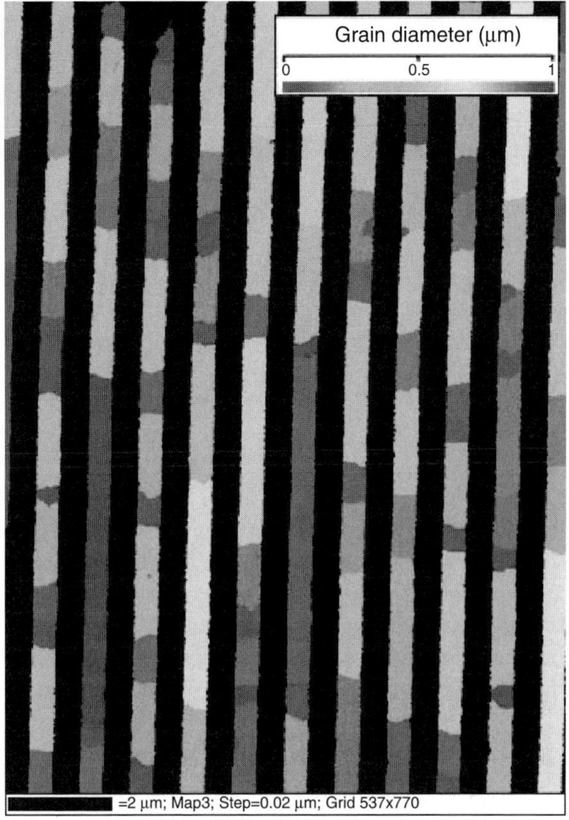

FIGURE 2.35. Grain size (equivalent circle diameter) map, disregarding Σ3 twins as grain-delimiting boundaries. Scale bar = 2 µm (see color insert).

the pattern quality map, but one of them is a finely divided twin domain, the other a low angle grain boundary. Note that twin domains <30 nm wide are resolved in this map.

4.5. Aluminum Thin Film

EBSD data acquisition on this 100 nm-thick film used a thermal FEG SEM operated at 10 kV. Fine-grained aluminum presents a special challenge compared to platinum and copper because of its lower density and lower electron scattering efficiency, impacting both spatial resolution and acquisition speed. Since drift is commonly a factor in high-resolution EBSD, a fast acquisition rate was necessary, so a relatively high beam current (>5 nA) was used at the expense of spatial resolution. These conditions, along with the high sensitivity digital detector, allowed data acquisition in a minimum of time to mitigate the effects of drift. This job used a 500 × 400 point grid at 10 nm steps.

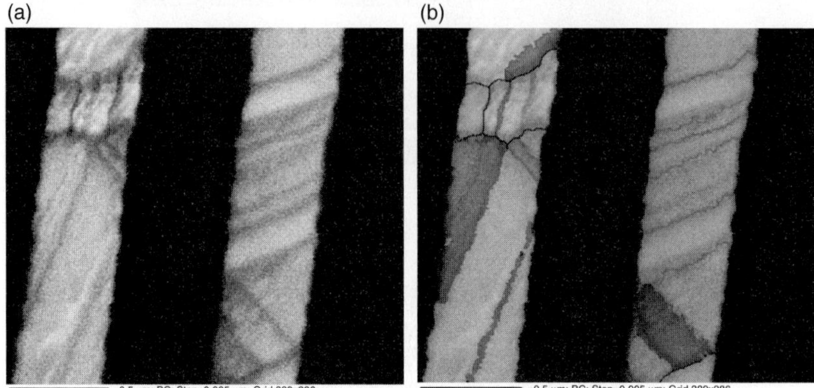

FIGURE 2.36. (a) Pattern quality (band contrast) map; (b) All-Euler orientation map. Wavyness due to instability during acquisition. Note <30 nm twin domain at bottom of left interconnect. Scale bar = 500 nm (see color insert).

The pattern quality map in Fig. 2.37 shows a fine-grained microstructure, with a large proportion of the map dark, indicating poor pattern quality likely associated with a relatively large probe and electron scattering at grain boundaries. The larger grains in the mapping area show as higher pattern quality "islands." Visually, the higher pattern quality grains are ~100 nm in diameter.

FIGURE 2.37. Pattern quality (band contrast) map, showing higher pattern quality grain cores and lower pattern quality (darker) finer grained regions.

Figure 2.38 is a surface normal-projected IPF map. As with the other metal films discussed in this chapter, blue grains are dominant indicating a strong <111> ∥ surface normal texture. Again, the pole figures in Fig. 2.39 corroborate this, and a comparatively uniform <111> ring at 70° from the texture peak indicates good compliance to the ideal gamma fiber texture.

Figure 2.40a further examines this texture. Here, grains with {111} poles within 20° of the surface normal are plotted in color, all others are colored in the

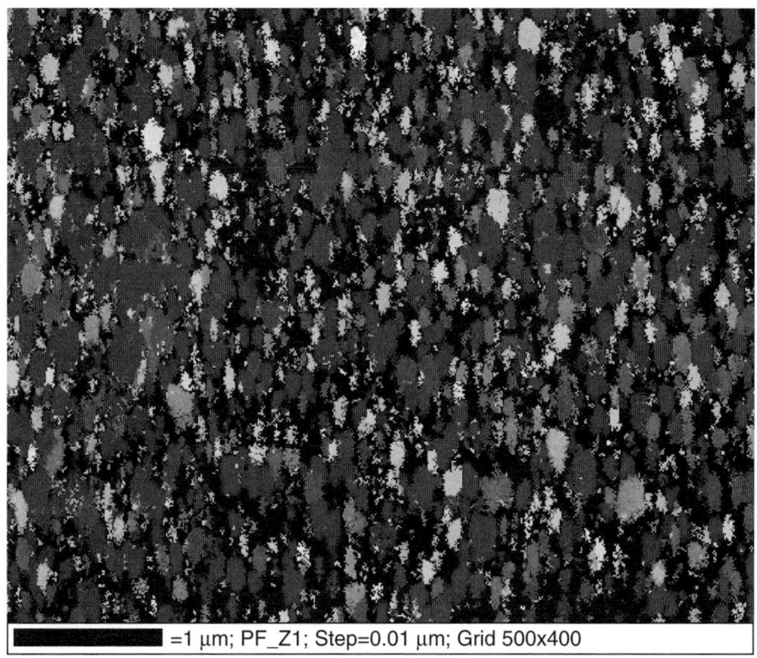

FIGURE 2.38. Surface normal-projected IPF-based orientation map, legend in Fig. 2.17. Predominance of blue grains implies a strong <111> ∥ surface normal texture (see color insert).

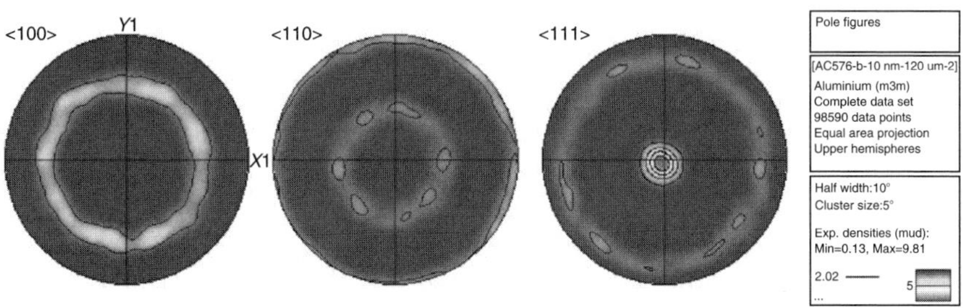

FIGURE 2.39. Standard cubic pole figure set, contoured, showing strong <111> ∥ surface normal fiber texture.

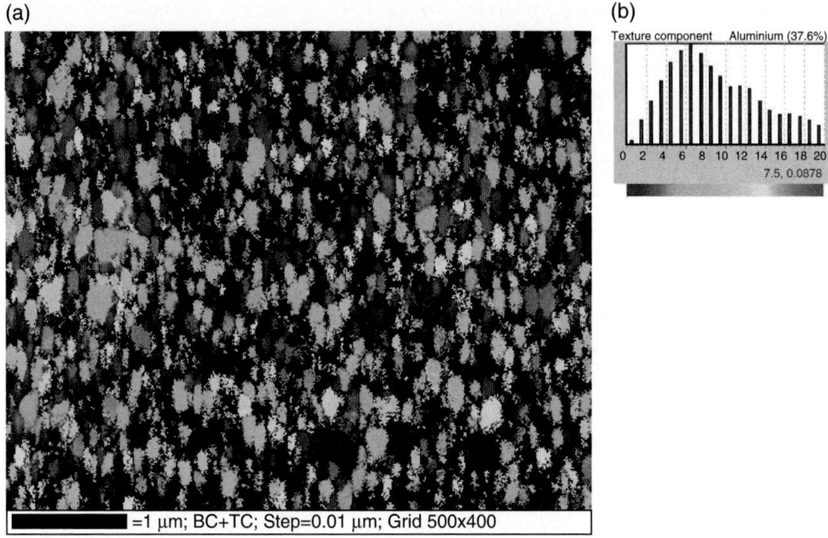

FIGURE 2.40. (a) Orientation map with grain coloring by degrees of misorientation relative to a perfect <111> ‖ surface normal parallelism, with blue = closest to parallel, red = 20° from parallel. (b) Legend showing rainbow scale fitted to 0–20° range (see color insert).

background pattern quality grayscale. Grains oriented most closely to the ideal texture are colored blue, others are colored along the rainbow scale to red with increasing misorientation. The legend, Fig. 2.40b, shows a histogram of the distribution of this misorientation, and indicates that 38% of the map area is within 20° of the ideal orientation.

5. Current Limitations and Future

5.1. Spatial Resolution

At present, spatial resolution of this technique is limited to grain sizes of ~20 nm in diameter. This assumes that a map is created at a sampling step size of 5 nm, and therefore is reasonably conservative in needing 4 pixels to represent a 20 nm grain diameter. The spatial resolution is primarily determined by the SEM and geometry of the sample/lens/EBSD detector relationship. As FE-SEM resolution has improved, so has the minimum grain size visible in EBSD maps. With the advent of aberration corrected FE-SEMs, a concomitant benefit in resolution should be seen.

5.2. Angular Resolution

Angular Resolution is presently limited to ~0.5°. This is dictated by the resolution of the EBSD detector and its position with respect to the sample. Although it is possible to operate at this resolution, current cameras sacrifice speed for additional angular resolution. Since most standard textural analyses need only to resolve a few degrees to determine a grain boundary, the need for improved angular resolution might only be required for strain and small lattice rotation measurements.

5.3. Speed

Speed has improved greatly over the years since automatic mapping was introduced. Speeds of up to 100 indexed points per second are now achievable on well-prepared cubic materials. It is expected that advances in camera design will allow mapping speeds to continue to increase.

6. Conclusion

EBSD is a powerful, quantitative SEM technique that has moved from the province of university materials and geology departments to industrial production control. While its penetration has yet to reach its full potential, the technique is well on the way to becoming yet another detector situated on the chamber of an SEM.

References

1. T. M. Maitland, unpublished work (2005).
2. Figures 2.2–2.4, assistance of Robert Schwarzer is gratefully acknowledged.
3. K. Shinohara, *Sci. Pap. Inst. Phys. Chem. Res.*, 20 (1932/1933) 39.
4. Boersch *Physikalische Zeitschrift*, 38 (1937) 1,000.
5. M. N. Alam, M. Blackman, and D. W. Pashley, *Proc. Roy. Soc.*, 221(1954) 224.
6. D. C. Joy and G. R. Booker, *J. Phys. E: Sci. Instrum.*, 4 (1971) 837.
7. D. J. Dingley, *Proc. Roy. Microsc. Soc.*, 19 (1984) 74.
8. J. A. Venables and C. J Harland, *Phil. Mag.*, 27 (1973) 1193.
9. Used with kind permission of Prof. Val Randle, University of Swansea.
10. P.V.C. Hough, US Patent 3069654 (1962).
11. A. Queisser, *C/C++ Users Journal*, December 2003.
12. Courtesy: Professor Dave Prior, University of Liverpool.
13. A. P. Day et al., *Channel 5 User Manual, HKL Technology A/S*, Hobro, Denmark (2001).
14. P. W. Trimby et al., *Applications Catalogue, HKL Technology A/S*, Hobro, Denmark (2001, 2003).
15. X. D. Han, S. Sitzman, and T. M. Maitland, unpublished work (2005).

3
X-ray Microanalysis in Nanomaterials

Robert Anderhalt

1. Introduction

Traditionally, energy dispersive x-ray spectroscopy (EDS) in the scanning electron microscope (SEM) has been called microanalysis, implying its suitability for materials of micrometer dimensions. EDS was generally performed with a sample matrix and electron beam energy that would produce an interaction volume and resolution which was generally of the order of a few cubic micrometers [1]. Nanoscale structures have one or more features that are an order or magnitude less than a micrometer (<100 nm). Thin-film analysis can provide thickness and chemical data on layers with thicknesses considerably less than 100 nm. The thin film may have one dimension less than 100 nm but the other two dimensions may effectively approach infinity. The analysis of features with two or three dimensions less than 100 nm is considerably more difficult and requires that we modify our analytical procedure or sample preparation. Several case studies are presented in this chapter.

It is possible to reduce our beam interaction volume to less than $100 \times 100 \times 100$ nm by a variety of methods. Reducing the beam energy to 5 keV or less will generally produce the required small interaction volume, but the lower beam energies will produce a spectrum with a lower energy range and more peak overlaps to consider. When an SEM equipped with a tungsten filament source is operated at accelerating voltages that are less than 5 kV, the beam diameter may itself become significantly larger than 100 nm and can become a controlling factor on the relatively poor resolution of the analysis. The SEM with a field emission source would be preferred because beam diameters considerably less than 100 nm are easily attained at low beam energies [2].

The substrate for small or nanoscale particles may contribute much more x-ray information to the spectrum than the particle or feature of interest. If the substrate has no elements in common with a small feature of interest, it may be possible to do a qualitative analysis without ambiguity. Common substrates could include various carbon materials such as adhesives or filters of varying compositions. Carbon thin films on transmission electron microscope (TEM) grids can minimize the contribution of the substrate if the grid is mounted over a hole or is placed in a holder of a detector optimized for scanning transmission electron microscopy (STEM) [2]. Because the substrate is reduced and the sample is extremely thin, many beam electrons will pass through the sample without scattering or being affected in any way and the result may be a very low x-ray count rate. These transmitted beam electrons may produce backscattered electrons (BSEs) or x-rays from surfaces below the sample, which can secondarily excite other areas of the sample, grid, or of the sample holder itself, and it should be possible to separate the stray radiation from the x-rays generated in the feature of interest. Another method to improve the x-ray resolution is to reduce the sample thickness with a focused ion beam (FIB) instrument. Sections may be cut, which have thicknesses of 100 nm or less. Beam spreading is reduced and many, perhaps most, beam electrons will pass through the sample without scattering.

1.1. X-ray Signal Generation

The interaction of the electron beam with the specimen produces a variety of signals. Some signals are used for imaging (secondary electrons, BSEs, transmitted electrons, etc.), but it is the x-ray signal that will be discussed here.

X-ray signals are typically produced when a primary beam electron causes the ejection of an inner shell electron from the sample (Fig. 3.1). An outer shell electron makes the transition to fill this vacancy but gives off an x-ray whose energy can be related to the difference in energies of the two electron orbitals involved. Most commonly we see K, L, or M series x-rays and the series name refer to the orbital from which the original vacancy occurs. It is quite possible that x-rays can occur as a result of a vacancy in the N orbital. N series x-rays are generally very low in intensity and energy (0.15–0.3 keV), and almost never appear as a distinct peak in the spectrum.

The K alpha x-ray results from a K shell electron being ejected and an L shell electron moving into its position. A K beta x-ray occurs when an M shell electron makes the transition to the K shell. The K beta will always have a slightly higher energy than the K alpha and is always much smaller in intensity.

A simplified representation of the electron orbitals is shown in Fig. 3.1 in which a single orbital each is shown for the K, L, M, and N orbitals. The K orbital is relatively simple but the L orbital consists of three suborbitals, the M has five, and the N can have as many as seven suborbitals. When a vacancy occurs in one of the three L suborbitals, an electron can make the transition from any of the five M or seven N suborbitals. Many of these suborbitals have only a few electrons, which means that the x-ray peak will be quite small and will probably be lost in the noise and not detectable by EDS techniques. Some suborbitals that are involved in the creation of the vacancy or which provide the transition electron are only slightly different in their electron energies from an adjacent suborbital. The result is that the x-ray peak is only slightly different from another peak in the spectrum and might not be discernible or resolvable from another larger peak in the spectrum.

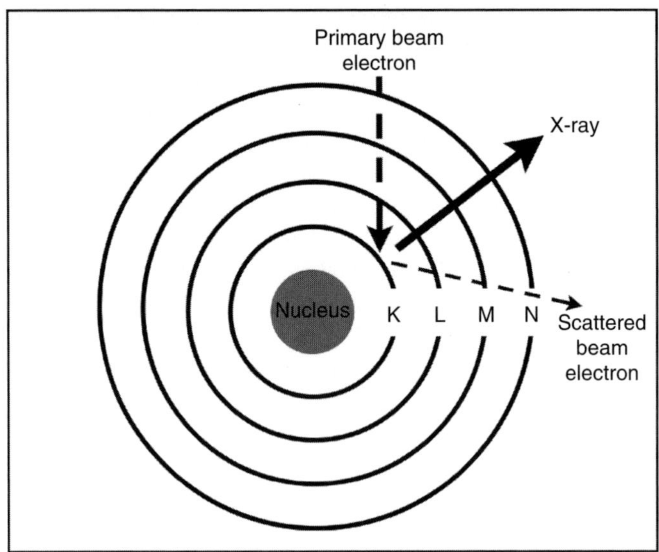

FIGURE 3.1. X-ray generation in a sample from the interaction of high-energy electrons in an electron microscope. The atom has been simplified to show a single orbital for the K through N electron orbitals. Assuming that the primary beam electron has more energy than the K orbital electron, it is possible for an interaction to occur in which the K orbital electron leaves the atom in an ionized and excited state. The vacancy in the K orbital is filled by an electron that makes the transition from an outer orbital. An x-ray with an energy that is characteristic of the element is created and has an energy that is equal to the energy difference of the electrons in the two orbitals. The primary beam electron experiences a loss of energy and a change of direction.

The L series of x-ray peaks result from a vacancy that occurs in the L shell. There can be a variety of beta and gamma peaks that occur at higher energies than the alpha peak. It is no longer so simple as in the K series of x-rays where we can say that the beta peaks originate from electrons making the transition from two orbitals away (e.g., the N suborbitals in this case), but it is always true that the beta peaks will occur at higher energies than the alpha peaks and that the gamma peaks will occur at higher energies than the beta peaks. In the EDS spectrum, the L series will have as many as six resolved peaks. The L alpha x-ray results from a vacancy occurring in the L3 (or the third L suborbital) and the transition electron coming from the M4 or M5 suborbital. It is also possible for the transition electron to come from the M1 suborbital to give what is called the L L peak, which will actually occur at a lower energy than the L alpha. Because the M1 generally has fewer electrons than the M4 or M5, the L L peak will have a lower intensity than the L alpha. There is the special case when the atomic numbers that we might be considering will not have a full compliment of electrons in the M4 or M5 and this means that the L L peak will appear to grow at the expense of the L alpha—in reality, the L alpha becomes smaller. When the M4 and M5 orbitals are almost empty, the L L can be larger than L alpha. There is also the case where the L L peak exists but the L alpha is no longer present because the M4 and M5 orbitals are empty.

The M series of x-ray peaks in the EDS spectrum will originate from a vacancy occurring in one of the M suborbitals. The M series is similar to the L series in many ways except that most observers of EDS spectra will not see M series peaks at energies greater than 3.2 keV. The M alpha for uranium (atomic number = 92) has an energy of 3.165 keV. In theory we could expect greater separation between the M series peaks if there were common elements with atomic numbers greater than 100, but this is not the case in nature. In most instances we may be able to resolve the M alpha, M gamma, and M Z peaks in the EDS spectrum. There are M beta x-rays, but they are generally not resolved from the M alpha peak and make the M alpha appear to be somewhat asymmetric. The M alpha peak occurs as a result of a vacancy in the M5 suborbital with a transition electron coming from the N6 or N7. The M Z results from a vacancy in the M5 with a transition electron coming from the N3, or from a vacancy in the M4 with the transition electron coming from the N2. The M Z peak is similar to the L L in the L series in that it is generally lower in intensity but occurs at a lower energy than the M alpha peak. Also, for elements with atomic numbers lower than 70, the M Z to M alpha intensity ratio increases and the M alpha disappears for atomic numbers less than 57 because the N6 and N7 suborbitals become partially filled and eventually become empty.

In the K, L, and M series of x-ray peaks in the EDS spectrum there are some consistent trends with energy. The alpha and beta peaks become much closer to each other with decreasing energy. Below 2–2.5 keV we might not be able to separate the beta peak from the alpha peak. The beta peak will also become smaller in proportion to the alpha peak at lower energies.

1.2. X-ray Signal Detection

Some portion of the x-rays generated in the sample will escape from the sample. A small fraction of the x-rays leaving the sample will travel toward the EDS detector and be detected as shown in Fig. 3.2. An optimum detector-sample geometry and parameter setup will ensure the best efficiency of spectral data collection with the fewest artifacts.

Once the x-rays are generated in the sample they will have to travel through the sample and whatever coating is on the sample to escape, pass through the detector window, and pass through the thin films on the surface of the detector (the metal layer and the silicon dead layer). Each material or film will effectively filter the x-ray signature by absorbing some part of the signal and allowing other parts to pass through. In most cases the lowest energy part of the signal has the greatest chance of being absorbed. The effects of absorption of the low-energy part of the signal can be minimized by selecting one of the lower beam voltages or a voltage that is not higher than it needs to be. The takeoff angle can also be made higher to increase the sensitivity for the low-energy part of the spectrum.

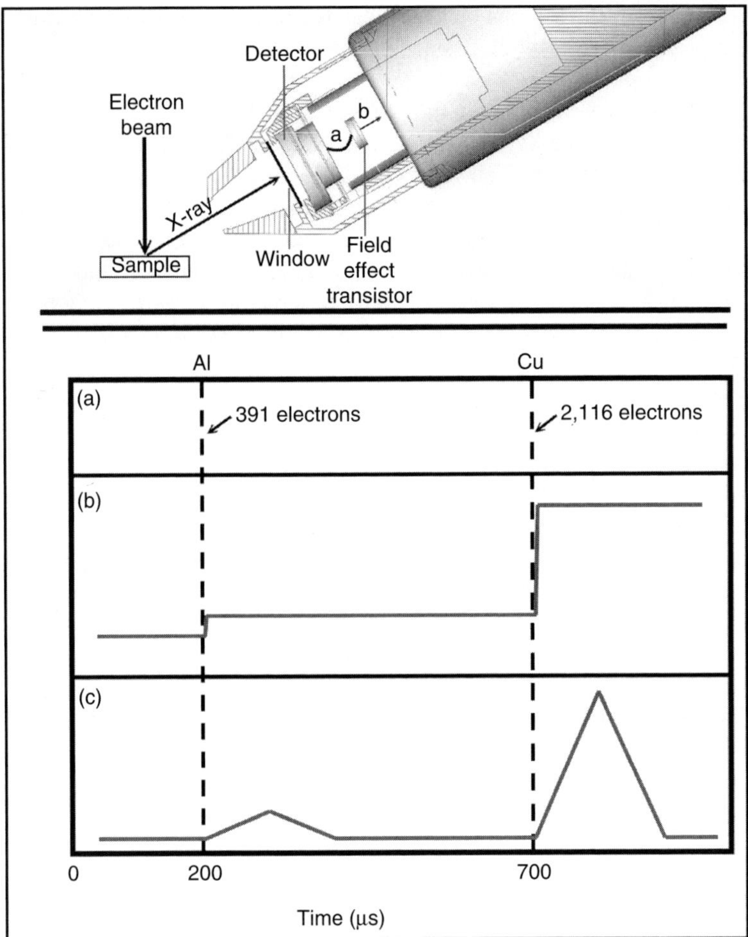

FIGURE 3.2. Signal detection with an EDS detector. The top portion of the figure (Courtesy of EDAX Inc., Mahwah, NJ) is a simplified view of the EDS detector showing the x-ray generated from the sample, which travels through the detector window and then generates electron-hole pairs within the detector. A negative voltage is applied to the front of the detector, which drives the electrons to the back contact and into the FET. The signal leaves the FET and continues to the rest of the amplifier circuitry. (A) The number of generated electrons is shown for two x-rays (Al K alpha and Cu K alpha) that are generated within 500 µs of each other. The location within the detecting unit is shown between the detector and FET as point 'a' in the top figure. (B) The electronic signal without noise showing the detection of each x-ray as a ramp after the FET (the point labeled 'b' in the diagram above). The vertical axis is a voltage (millivolts). (C) The electronic signal without noise showing the two x-rays as amplified triangular pulses whose voltages are measured in mV (not shown with the same scale as in part b). The signal shown is at the moment just before the height of each pulse is converted into energy units (eV). The time axis for each x-ray event has been significantly expanded or amplified as well.

These parameters will be discussed in Section 1.3. Other parts of the detector system that control its sensitivity are effectively constants and these include the window itself as well as the metal layer and dead layer. Their effects on the material phase identification or on the quantification are known and can be corrected.

A simplified view of the detector is shown in Fig. 3.2. Ultimately, there is a preamplifier and an amplifier, which prepare the signal by increasing the signal-to-noise ratio and then calculating its energy from a pulse height based upon the system calibration of peak energies. In the example shown in Fig. 3.2, we are considering the case where an x-ray of aluminum (1.486 keV) is detected and this is followed 500 µs later by an x-ray of copper at an energy of 8.04 keV. If x-rays could be detected consistently at 500-µs intervals, this would be equivalent to a count rate of 2,000 cps (counts per second).

The x-rays that strike a Li-drifted silicon detector will generate one electron-hole pair within the detector for every 3.8 eV of energy of the x-ray. A negative bias voltage is applied to the metal layer to force the electrons to the back contact of the detector and to the input to the field effect transistor (FET). Thus the Al x-ray will create 391 electrons and the Cu x-ray will create 2,116 electrons as shown in Fig. 3.2. This assumes that all electrons will make it to the back contact and that none will recombine with the vacancy or hole, or that none will be lost as a result of detector crystal defects, or be conducted away near the dead layer, or at the sides of the detector active area. The effects that prevent collection of all the generated electrons is collectively referred to as incomplete charge collection, which might be obvious in the spectrum as a low-energy tail of the peaks.

The signal that leaves the detector and enters the FET as a number of electrons (Fig. 3.2 part a) will leave the FET as an electronic signal or ramp (part 'b') and continue to the rest of the amplifier circuitry. The sharp rise or ramp that corresponds to each x-ray event is a very brief event—only of the order of 50 ns. However, the noise has not been shown in Fig. 3.2 (part b), and it must be pointed out that the signal is very small and the noise is very large making it impossible to measure the x-ray energy at this point. A preamplifier/amplifier circuit is used to increase the strength of the signal simultaneously increasing the signal-to-noise ratio. In order to accomplish both purposes, the signal is converted to a triangular pulse whose amplitude is greater but which is now stretched out in time—the horizontal axis in Fig. 3.2. The signal shown is at the moment just before each pulse is converted into energy units and is done using a formula to convert pulse height (mV) to energy (eV or keV), which is learned by the system during the energy calibration procedure.

1.3. EDS Parameters

A number of parameters that are used in collecting EDS spectra are effectively variables that should be controlled in order to efficiently collect data with the fewest artifacts.

1.3.1. Count Rate/Dead Time/Time Constant

The count rate (in counts per second (cps)), dead time, which is generally expressed as a percentage, and the time constant (also known as amplifier time, pulse processing time, shaping time, etc.) are dependent on each other and should be discussed at the same time. The most straightforward of these three parameters is the count rate and is measured in cps. The count rate is controlled by adjusting the SEM to the proper or optimum conditions such as working distance, gun tilt, and shift, perhaps by additionally adjusting and aligning the aperture and finally by adjusting the spot size. The real time or clock time used to collect spectral data is divided into live time and dead time. The live time is that time when the detector/amplifier system is not busy and could collect and process x-ray events. The dead time is the time when the detector/amplifier system cannot collect or process x-rays because it is busy processing an x-ray event or rejecting multiple x-ray events. At 0% dead time (or 100% live time), no spectrum is collected because the system is always available and probably indicates that no x-ray counts are being received. At 100% dead time the system is busy all the time processing, but more likely rejecting x-ray events—this will probably provide no spectrum. At 100% dead time or at high dead times (above 60%), one solution would be to lower the count rate and another would be to change to a faster time constant or amp time. The x-ray counts stored in the spectrum reach a maximum at about 67% dead time. Increasing the count rate beyond this dead time actually yields fewer counts into the spectrum for the same clock time (Fig. 3.3).

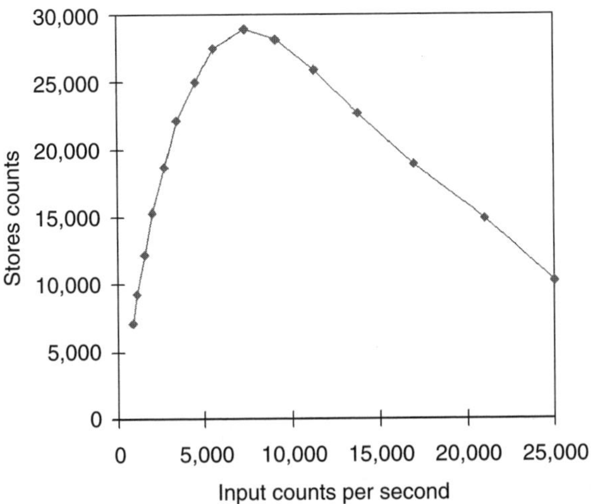

FIGURE 3.3. Plot of input count rate vs. stored counts in a spectrum for one specific time constant. As the count rate increases, the dead time will also increase. The peak in the plot occurs at 67% dead time. The vertical axis represents the total number of counts stored in a spectrum in 10 clock seconds.

For a good-quality spectrum (i.e., good resolution and fewest artifacts), it is best to use one of the longer time constants with a count rate that produces a moderately low dead time—perhaps 20–40%. Extremely high dead times (e.g., >60%) will generate artifacts in the spectrum such as sum peaks, which are also known as pulse pileup events. When some of the older, more analog amplifiers are used at high dead-time conditions, a degradation of resolution, distortions of peak shapes, and a loss of calibration can be expected.

In some instances, higher throughput is more important than good resolution and then it makes sense to use one of the faster time constants with a count rate to provide 20–40% dead time. One of these instances would be when collecting x-ray maps and another might be when doing trace-element analysis without overlapping peaks.

The distribution of pulses shown in Fig. 3.2 (part c) displays two x-ray events where one arrives 500 μs after the other. This is equivalent to 2,000 cps and will probably give a dead time of about 35% using a time constant or amp time of 102 μs. At first, viewing of this situation would appear that the system should be able to process all x-ray events if they arrive at the detector every 500 μs. This would be true if we could somehow schedule the x-ray events to be so regular. There is a substantial component of randomness to when the x-rays will be detected and a certain percentage of them will arrive within 10, 50, or 100 μs of each other, which means that these x-rays will be discarded and will contribute to our dead time.

1.3.2. Accelerating Voltage

The accelerating or beam voltage used should be at least 2× the energy of the highest energy line in our spectrum and no more than 10 to 20× the lowest energy line of interest. The number 10 is used for quantitative applications while the number 20 is relevant for the strictly qualitative applications.

For example, if one is interested in the analysis of a phase containing Fe, Mg, and Si and would like to use the K lines for each, then 15 kV will probably work reasonably well. If, however, the analyst needs to analyze the same three elements plus oxygen as well, then he or she might use 5–10 kV, but it would be best to consider using the L line for the Fe. An alternative might be to use 15 kV but to allow the software to calculate the oxygen by stoichiometry when the sample type warrants this approach.

Why should the beam voltage be at least 2× the highest energy element? Because at lower overvoltages the fraction of the interaction volume where the element can be excited becomes very small and it is not possible to generate many x-rays of that energy—in other words, the result will be a small peak and also a poor sensitivity for that element.

Why should the beam voltage be less than 10 to 20× the lowest energy peak? When the overvoltage number is excessive, the proportion of the interaction volume for which the low-energy x-rays can escape without being absorbed also becomes small. The result is a small peak and in the case of the quantification there will be a strong absorption correction that will magnify the statistical and background-fitting errors in our analysis. Again, the sensitivity of the analysis will not be very good as well if the selected beam voltages are very high.

1.3.3. Sample-Detector Geometry and TakeOff Angle

EDS detectors will typically have a collimator that restricts the view of the detector so that it will not detect x-rays that come from parts of the SEM or sample far removed from where the electron beam strikes the sample. If the detector view or its "line of sight" is considered, a parameter that will be relevant to most detectors can be defined as the "detector normal"—a line drawn through the center of the detector that is perpendicular to the detector surface (Fig. 3.4). The detector normal intersects the electron beam at a specific distance in millimeters below the pole piece. This intersection distance should be the ideal working distance (in millimeters) for collecting x-ray data.

The angle between the intersection of a horizontal plane and the detector normal gives the detector angle or elevation angle. If the sample is at the intersection distance and is not tilted, the detector angle will be equal to the takeoff angle. The takeoff angle can be changed by changing the tilt. When the tilt is changed it becomes essential to know the direction of tilt as compared to the detector normal

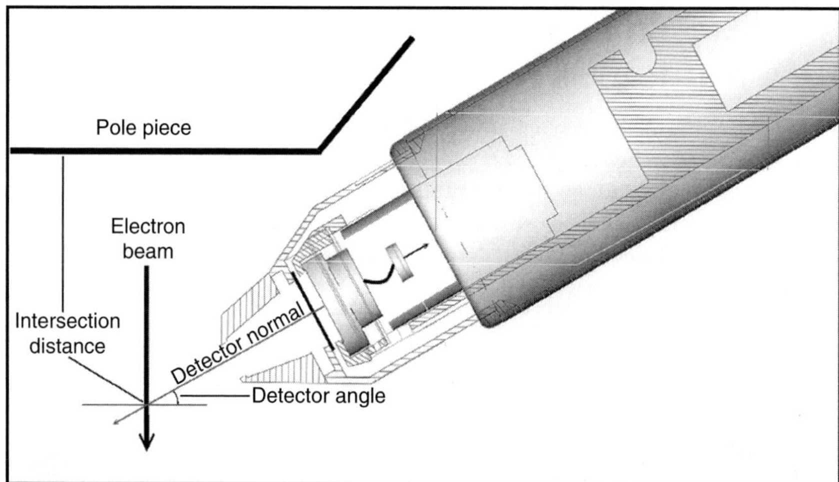

FIGURE 3.4. Detector geometry for an inclined detector and the electron beam. The detector normal intersects the electron beam at a specific distance in millimeters below the pole piece. This intersection distance should be the ideal working distance (millimeters) for collecting x-ray data. The angle between the intersection of a horizontal plane (or line in the figure) and the detector normal gives the detector angle. If the sample is at the intersection distance and is not tilted, the detector angle will be equal to the takeoff angle. The takeoff angle can be changed by changing the tilt. When the tilt is changed, it becomes essential to know the azimuth angle or the direction you are tilting as compared to the detector normal as viewed from above. (Courtesy of EDAX Inc., Mahwah, NJ.)

as viewed from above. This angle may be referred to as the azimuth angle and it must be taken into account to calculate the takeoff angle for tilted samples. Typical takeoff angles will be between 30° and 35°. This angle is a combination of the detector angle, the detector position, sample working distance, and sample tilt. The sensitivity for very low-energy x-rays and/or signals characterized by high absorption can be enhanced by increasing the takeoff angle, which is generally done by increasing the tilt of the microscope. Some inclined detectors (e.g., a detector angle of approximately 35° above the horizontal) do not require sample tilt for most analyses. Horizontal-entry detectors require that the sample be tilted to achieve an optimum takeoff angle.

1.4. X-ray Artifacts

The most significant x-ray artifacts in the EDS spectrum are those that will cause what might be interpreted as a peak. The consequence of interpreting the artifact peak as an element can be severe or embarrassing at a minimum. The major artifacts to be discussed here are the escape peaks, sum peaks, and stay radiation. Because nanomaterials are commonly thinned, or examined as thin films, very fine particles, or wires, the stray radiation can be extremely significant. In nanomaterials, the primary beam electrons may still have high energies as they exit from the sample and can create a variety of signals from other parts of the sample or holder.

An escape peak can form when an x-ray striking the detector will cause the fluorescence of a silicon x-ray. This may result in an x-ray having the energy of silicon (1.74 keV) and another signal having the difference between the original x-ray energy minus the silicon energy (Fig. 3.5). If the silicon x-ray remains in the detector, the two signals are summed and the correct energy is assigned to the peak in the EDS spectrum. If the silicon x-ray escapes from the detector, then the x-ray detected is called an escape peak, which has an energy that is 1.74 keV less than the original x-ray. Only x-rays with energies greater than the critical excitation energy of silicon (1.84 keV) can cause the fluorescence of silicon. The size of the escape peak relative to its parent peak (typically not more than 1% or 2%) actually diminishes at higher atomic numbers as a result of the higher energy x-rays tend to deposit their energy deeper in the detector where the silicon x-rays (if generated) have more difficulty escaping from the detector.

A sum peak can occur if two x-rays are detected at nearly the same time (less than a few tens of nanoseconds) by the EDS detector. It is possible, especially at very high count rates, that two x-rays of the same energy will enter the detector within this time period and be counted as a single x-ray of twice the energy. Compounds can give rise to sum peaks that are the sum of two different energies (a and b of Fig. 3.6). Sum peaks are typically problems at higher count rates and when a spectrum is dominated by a single element. To confirm that the peak in question is really a sum peak, it is possible to overlay our spectrum with suspected sum peaks with a spectrum collected at a lower count. The lower count-rate spectrum might be collected at one fourth to half the count rate of the original

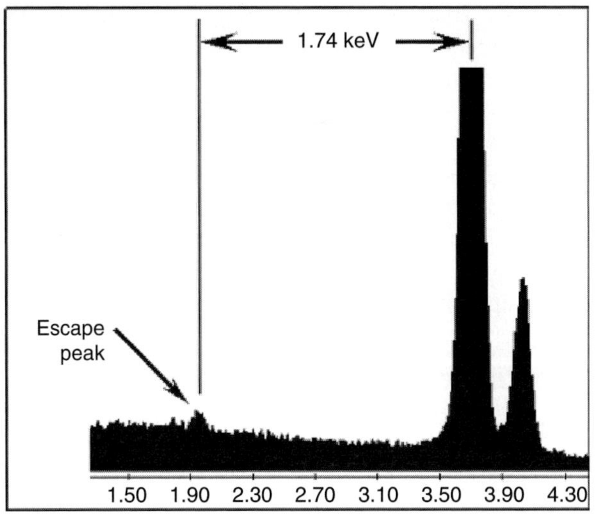

FIGURE 3.5. Escape peaks may form from any peak in the spectrum with energy greater than 1.84 keV. The incoming x-ray into the EDS detector may fluoresce a Si x-ray, which will reduce the total energy of the detected event if the Si x-ray leaves the detector.

spectrum. The peak in the first spectrum is regarded as a sum peak if it is no longer present or diminished in the lower count-rate spectrum. If the peak is undiminished, it represents a real part of the sample. Sum peaks are not very likely in nanomaterials, because we do not expect to have very high count rates from such a thin sample (<<100 nm).

Stray x-rays are those that originate anywhere other than where the primary beam interacts with the specimen and may be produced by a variety of processes. In nanomaterials, stray radiation can be a significant problem because the sample may be relatively thin. This results in a low count rate from the sample itself and

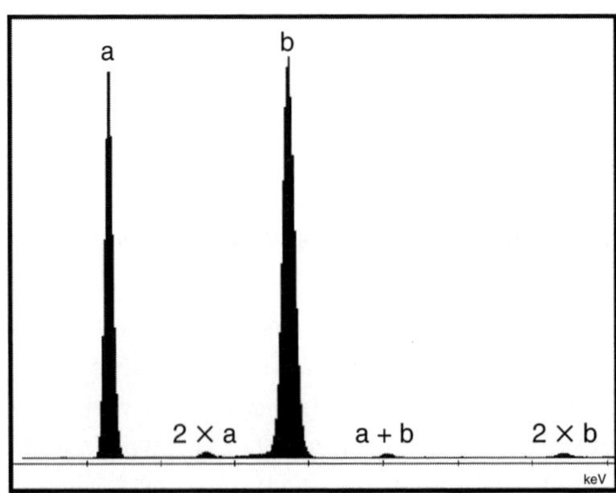

FIGURE 3.6. Sum peaks in an EDS spectrum. A sum peak can be a doubling of an element already present in the spectrum (e.g. "2 × a", or "2 × b") or it can result from the addition of two different elements ("a + b").

FIGURE 3.7. Stray radiation from a nanoparticle sample on a metal grid. The primary beam electron scatters in the sample and strikes a metal surface below creating a BSE or an x-ray. The BSE or x-ray can strike the metal grid or some part of the detector or holder producing stray x-rays. A dedicated STEM detector might minimize the stray radiation by eliminating the surface labeled "metal B" and replacing it with a detector that absorbs the high-energy electron.

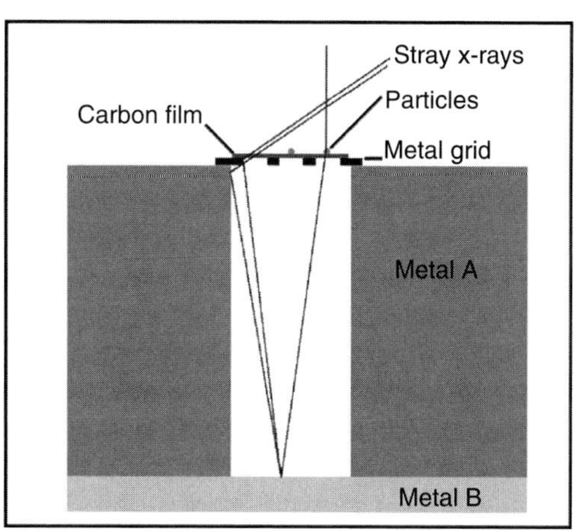

in an energetic primary beam electron that has left the sample or particle. The energetic beam electron can then strike other parts of the sample holder as shown in Fig. 3.7 and create an x-ray or BSE that will strike other parts of the holder that can be viewed by the EDS detector. The peaks in our EDS spectrum that originate from stray radiation when an STEM detector or holder is used will generally be the same even for different types of samples. There are a variety of metals used to create grids and a metal should be chosen that is not expected to be present in the sample, and if possible a metal should be chosen that does not overlap with the elements of interest.

2. Monte Carlo Modeling of Nanomaterials

It is possible to predict the electron beam interaction volume in bulk materials and many other sample types using Monte Carlo simulations [2]. When relatively thick samples ($\gg$1 μm) are analyzed at beam voltages greater than 15 kV, the interaction volumes are typically significantly larger than nanoscale (Fig. 3.8). The dimension of the interaction volume increases with the beam energy and has an inverse relationship with the density and average atomic number of the sample. For very dense materials, nanoanalysis is possible with beam energies in the range of 5–10 keV. For the analysis of less-dense materials, it may be necessary to use beam energies that are 5 keV or less. An inherent difficulty with analysis at such low voltages is that we are forced to use x-ray lines with energies that are typically less than 3 keV where the peak overlaps are significant and some elements will have no alpha lines (Table 3.1), making quantitative and qualitative analysis extremely difficult.

Thin films offer another possibility to image and analyze nanomaterials. Film samples that are approximately 100-nm thick can be created from thicker samples using an FIB instrument. Higher beam energies become practical in low-density

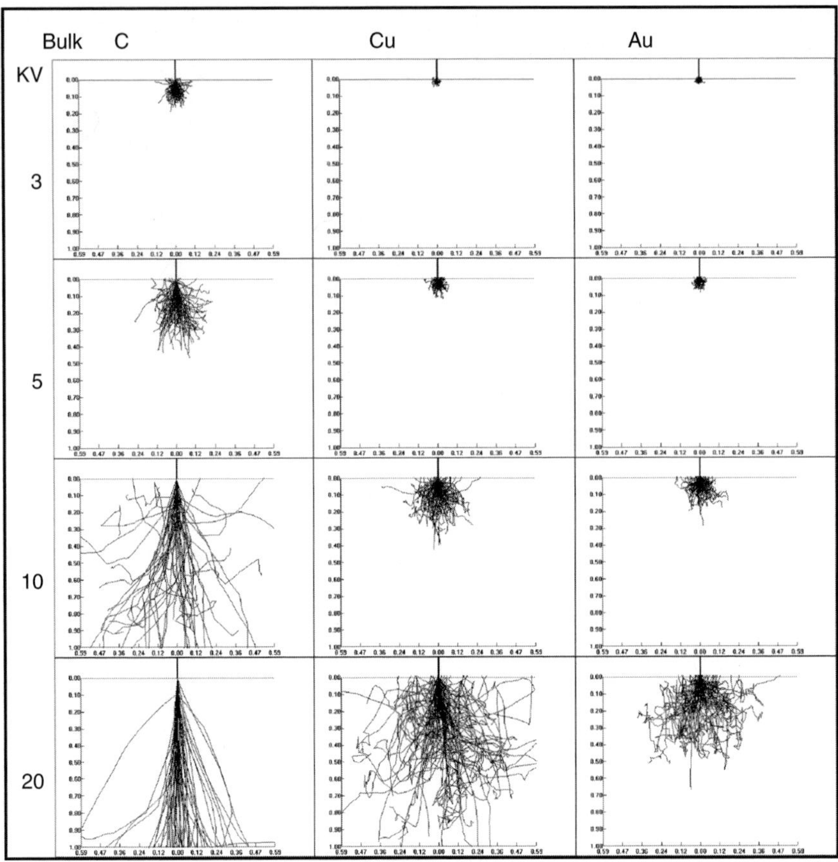

FIGURE 3.8. Monte Carlo simulations of bulk samples of C, Cu, and Au at beam voltages of 3, 5, 10, and 20 kV. The maximum depth shown in all cases is the same (1 μm) and all distances are shown in micrometers.

TABLE 3.1. Atomic numbers (Z) having alpha lines visible by energy range in the EDS spectrum. If a beam voltage of 5 keV is used we cannot reliably expect to excite peaks with energy greater than 3 keV. This means we would see K series peaks from atomic numbers 4 to 18, L series peaks from 21 to 47, and the M series peaks from 57 to 90. Some atomic numbers will not be viewable in the spectrum (19–20 and 48–57)

Energy range (keV)	K alpha (Z)	L alpha (Z)	M alpha (Z)
0–1	4–10	21–29	57–60
1–2	11–14	30–39	61–77
2–3	15–18	40–47	78–90
3–4	19–20	48–53	91–98
4–5	21–23	54–58	
5–6	24–25	59–63	
6–7	26–27	64–68	
7–8	28	69–72	

samples because the amount of horizontal scattering of beam electrons is effectively reduced with a thin sample—we are obtaining our x-ray data from what is really just the top portion of our interaction volume in the bulk sample case. Resolution is greatly enhanced with thin samples (Fig. 3.9) even at higher voltages. Absorption is significantly decreased and many to most primary beam electrons travel through the sample without scattering or scattering only to a low angle. However, because many beam electrons do not scatter at all, our x-ray count rates are relatively low—perhaps only one tenth of what would have been obtained with a bulk sample at the same beam energy.

It is often desirable or necessary to analyze submicrometer particles. The primary beam will generate x-rays from our particle and also from the substrate—quite often the substrate contribution to the spectrum will be dominant. In the analysis of

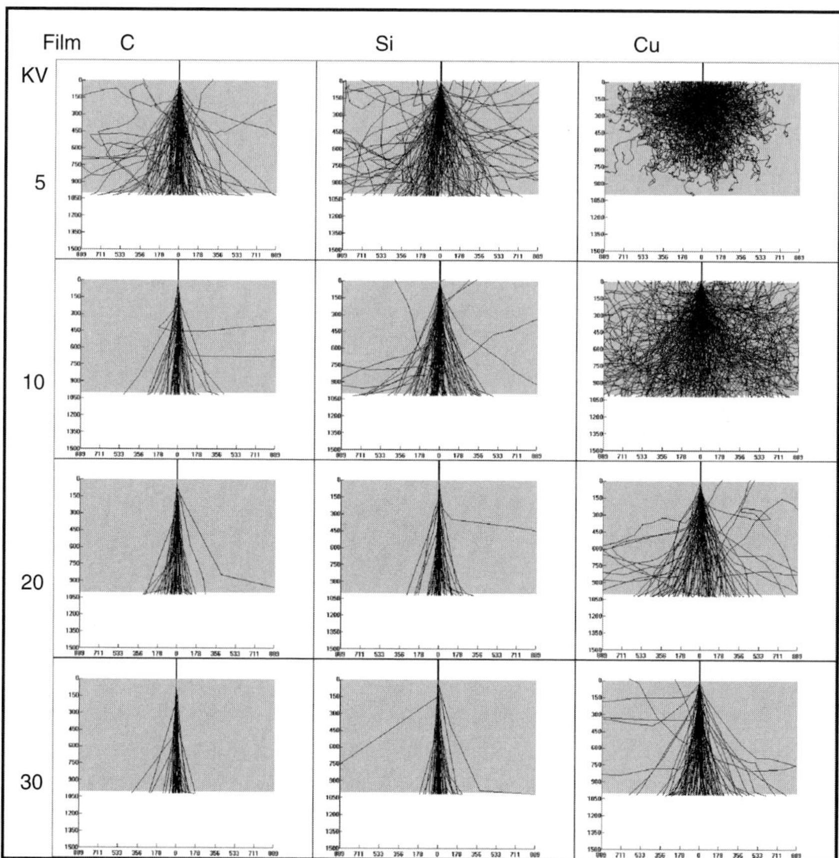

FIGURE 3.9. Monte Carlo simulations of a film (100 nm or 1000 Å thick) of C, Si, and Cu at beam voltages of 5, 10, 20, and 30 kV. All beam electrons are transmitted through the C film at 20 and 30 kV, and through the Si film at 30 kV. All distances are shown in angstroms.

particles we sometimes have the ability to choose our substrate. Ideally, the substrate should not have the same elements that are of interest to us in our particles, should offer contrast to our particles when using the BSE detector, and perhaps have a low x-ray yield. A carbon substrate is often advantageous because carbon is commonly conductive and may also serve as an adhesive. Filters are often used to trap particles and can be used as the substrate; however, they should be smooth and need to be made conductive. Small particles on a thin film have the advantage of offering good contrast, and if the particles and films are thin enough, the conductivity issues may be ignored. Because the substrate is very thin and largely absent, the peak-to-background ratio of the spectrum is quite good. Fig. 3.10 gives several examples of small particles (100 nm) on a 100-nm substrate. TEM grids can be purchased with carbon film of thicknesses about 3–60 nm.

Some carbon films referred to as "lacey" films have some open pores or spaces that will allow some particles to adhere to the carbon fibers but may project over

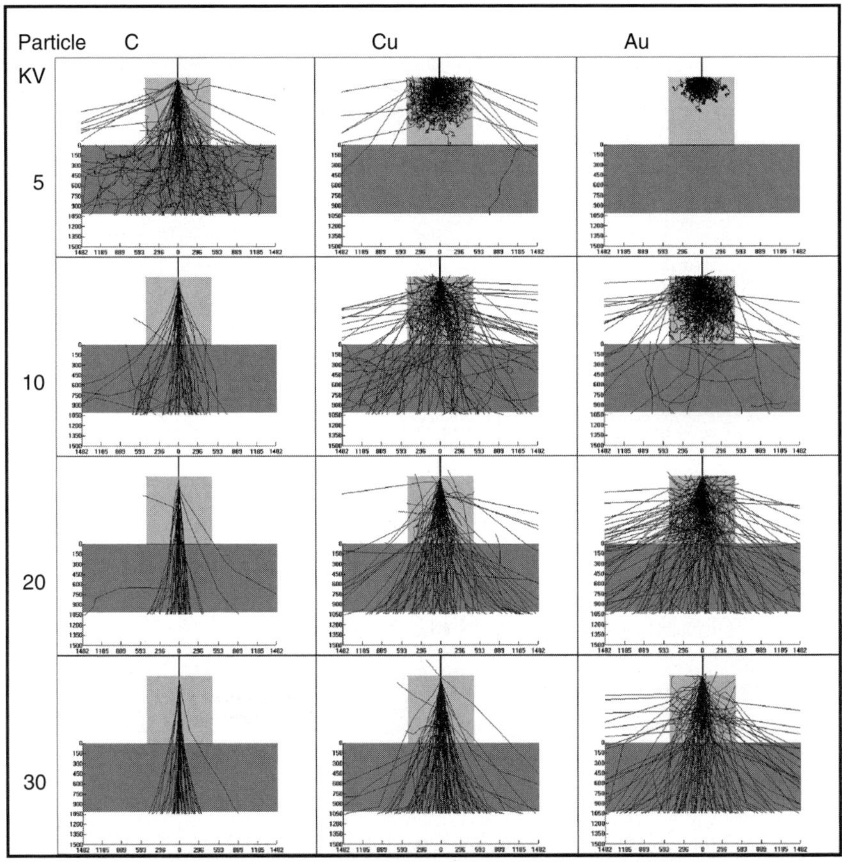

FIGURE 3.10. Monte Carlo simulations of a particle of C, Cu, and Au on a film of Si at beam voltages of 5, 10, 20, and 30 kV. All distances are shown in angstroms.

an open space meaning that the particle has no substrate at all. X-rays generated by the carbon film will be minimal due to the reduced thickness and low density of the carbon film, and x-rays generated under, or sometimes near, the particle will tend to be absorbed by the particle itself.

3. Case Studies

3.1. A Computer Chip

A computer chip has been selected as a sample to illustrate some of the concepts described earlier in the Monte Carlo simulations. A series of x-ray maps at varying beam voltages illustrates very well the decrease in interaction volume and enhanced resolution with lower beam voltages (Fig. 3.11).

Although the resolution of the images and maps is significantly enhanced by using a lower beam voltage in Fig. 3.11, there are still some disadvantages associated with using the lower beam energies. For one, at 5 kV we do not have a sufficient energy to excite Cr K (5.41 keV), which is also present in the sample in the bright areas in the BSE image. It is possible to use the peak for Cr L, but its L alpha (0.57 keV) and L L (0.50 keV) lines are significantly overlapped by the O K line (0.52 keV). Although the higher beam voltages are not as sharp, they do provide a sense of what is in the subsurface. The maps in Fig. 3.11 were created

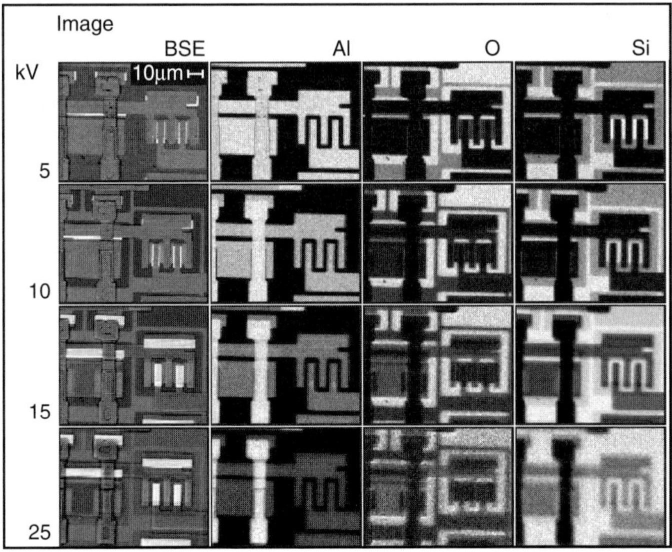

FIGURE 3.11. Series of image (BSE) and x-ray maps at beam voltages of 5, 10, 15, and 25 kV. The maps collected at the lower beam voltages clearly show an enhanced resolution of surface features.

using an SEM with a tungsten source, and at 5 kV the beam diameter can be quite large (>>100 nm). We are probably in the situation where the beam diameter is actually larger than the interaction volume, which means that the resolution in this case is controlled by the beam diameter/spot size and is less controlled by the beam energy and sample density. We can eliminate many of these associated disadvantages by the use of a thin section created by an FIB instrument and by imaging with a field emission gun scanning electron microscope (FEG SEM).

A careful examination of the maps in Fig. 3.11 will allow us to make some interpretations of what features are on the surface of the chip and which are in the subsurface. The Al lines tend to overwrite or obliterate the pattern of the Si and O maps indicating that the Al is on top of the SiO_2 and Si. Some very basic knowledge of computer chip construction allows us to realize that Si underlies these structures and so we might hypothesize that at some point on the sample (Fig. 3.12) we can expect a sequence from the surface downward of Al overlying SiO_2, which overlies Si. When a spectrum is collected from the point shown in Fig. 3.12, we have these elements (O, Al, and Si) as well as a very small peak of Ti. It is likely that the Ti forms a boundary layer to prevent interaction between the Al and the Si, and we can verify its existence by looking at a cross section created in an FIB, which is discussed later. Now, we are concerned with determining the thickness of the Al line, the hypothesized Ti layer, and the SiO_2.

Thin-film analysis [2,3] can be used to construct an estimate of the film thicknesses, assuming that we know the sequence of layers and that we can provide an estimate of the density of each layer. We must also have at least one unique element for each layer. In our case we have Si in the substrate as well as in the SiO_2 layer, but we will assume that we can determine the SiO_2 using the O peak and

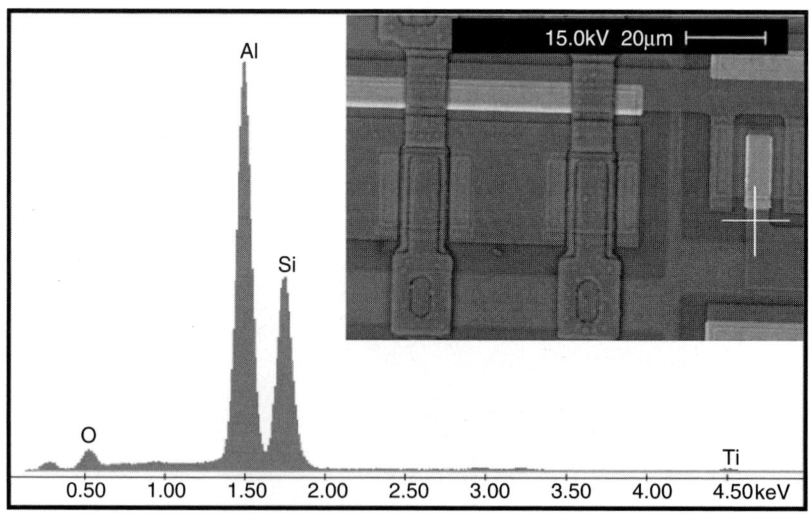

FIGURE 3.12. Spectrum collected from the spot shown in the BSE image at 15 kV. The spectrum contains elemental information from multiple layers.

will calculate the Si in that layer by stoichiometry using the usual valences of Si (+4) and O (−2). Some of the O present in the spectrum will probably be derived from a surface oxidation of the Al line and will detract from our measurement accuracy as well. The accuracy of the thickness estimates is also influenced by our knowledge or lack of knowledge of the densities of the phases present. It is further assumed that the assumptions do not totally invalidate the analysis or that they will at least be a constant in most analyses of this type which will allow for a relative comparison between analyses of different samples of similar film sequences. The K-ratios from the spectrum above (Fig. 3.12) are entered into the thin-film software [3] to provide the results shown in Table 3.2. The Al layer is estimated at 648 nm, the Ti layer at 9 nm, and the SiO_2 layer at 414 nm. The Ti layer is clearly a nanoscale feature but still resolved using thin-film analysis.

An FIB section was created from the chip sample, which is estimated to be about 100-nm thick. The section was mapped at a beam voltage of 20 kV in an FEG SEM (Fig. 3.13). The section was mounted over a hollow cavity to prevent substrate contribution to the map. An area of the image above the uppermost Al layer is tungsten. Before creating the FIB section, the tungsten layer was deposited on the surface to ensure that the surface was preserved during the sectioning process. The tungsten area is shown as the bright area at the top of the SE image. The maps shown were processed from a spectral map data set in order to remove the overlaps as much as possible using deconvolution to produce net intensity maps. The Si map in a conventional intensity or region of interest (ROI) map would have had the overlap with the W M, which is largely removed by this processing. The O map would have an overlap with the Cr L line and this has been significantly diminished but not entirely removed. Because we are able to use a relatively high beam voltage (20 kV), the map for Cr K is shown rather than Cr L. The result is that the Cr map shows no overlap whereas if Cr L had been used, we would expect to have an overlap with the O K.

The Ti map is not shown in Fig. 3.13 but information from the Ti map is shown in the linescan extracted from the spectral map data set in Fig. 3.14. The linescan was selected in a vertical orientation through the map data set such that the line is perpendicular to the interfaces between the various layers. It then becomes possible to sum data from a horizontal array of 9 pixels perpendicular to the linescan to produce the data shown for Al, Ti, Cr, and Pt in the plot in Fig. 3.14. Adding the spectra from the 9 pixels at each point of the linescan has the effect of reducing noise in the data and enhancing the quality of the data displayed.

In the maps (Fig. 3.13), it might look like the Cr layer overlies the Pt layer but this is somewhat ambiguous. It is also difficult to measure the thickness of the Cr and Pt layers, but these ambiguities are largely resolved by plotting the linescan of these data (Fig. 3.14).

The Cr clearly lies on top of the Pt and it is possible to attempt a measurement of these layers by an examination of the width of the half maximum of the Cr and Pt layers. The Cr layer appears to be about 78 nm and the Pt layer about 92 nm. Titanium was hypothesized to underlie the Al in our thin-film example, which was derived from an area of lesser complexity. It would appear that we have Ti at

TABLE 3.2. Analysis of the layer thicknesses of the chip sample using thin-film software [3]

```
Analysis # 1 at 15kU;    Pouchou,Pichoir's Scanning (1990) Model;

Layer 1  :  Al;Ka
Layer 2  :  Ti;Ka
Layer 3  :  Si;Ka , O ;Ka
   "     :  Si is analyzed by stoichiometry.
   "     :  Element:  Si  O
   "     :  Valence:   4  2
Substrate:  Si;Ka


                  Composition        Thickness            K-ratio
                  Weight%  Atom%   ug/cm**2 Angstrom   Cmp.Std Pure El
                  =======  =====   ======== ========   ======= =======
layer 1   Al: 100.00  100.00    174.324   6480.4    -------- 0.49720
sum        : 100.00

layer 2   Ti: 100.00  100.00      4.142     91.4    -------- 0.00960
sum        : 100.00

layer 3   Si:  46.75   33.33    109.805   4143.6    -------- 0.04933
layer 3   O :  53.26   66.67    109.805   4143.6    -------- 0.02310
sum        : 100.00

substrate Si: 313.20  100.00    --------   ------    -------- 0.25110
sum        : 313.20
```

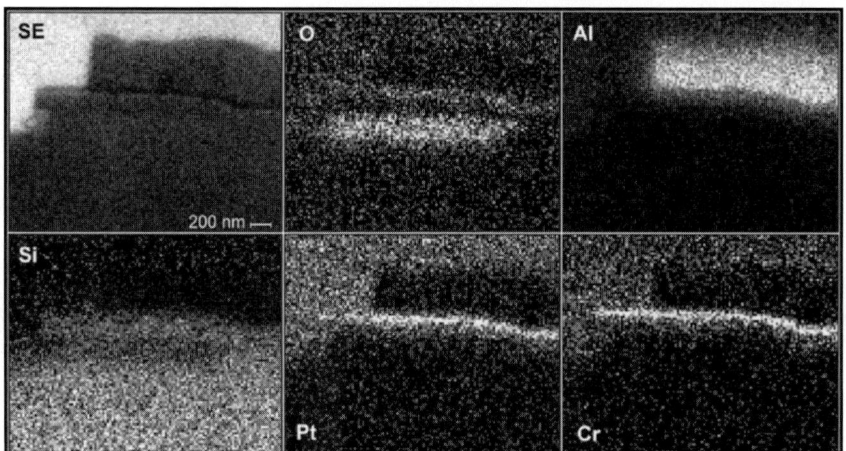

FIGURE 3.13. X-ray maps (O K, Al K, Si K, Pt M, and Cr K) of the chip FIB section collected at a beam voltage of 20 kV. The section was placed over a large cavity or a hole to prevent substrate contribution to the maps.

the base of the Al in the FIB section. It also appears that the Ti forms a boundary in this case separating the Pt and Cr and may also underlie the Pt layer.

3.2. Nanowire

A nanowire sample consisting of ZnO was examined. Spectra were collected at several voltages, and maps were collected at 10, 15, and 25 kV. The ZnO sample consists of wires, which taper from a base that is as much as 2.5 μm in diameter

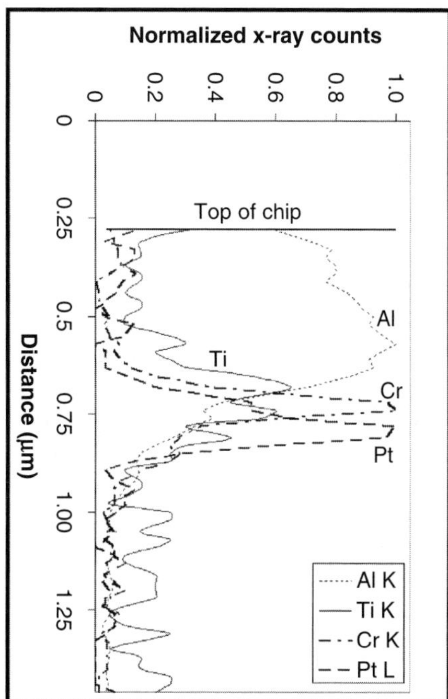

FIGURE 3.14. Linescan extracted from the spectral map data set of the FIB section. The linescan was selected from a vertical orientation in the map from the Al-rich area down to the Si-rich area. The Al, Ti, Cr, and Pt trends are shown normalized to their maximum intensity. Each plotted point in the linescan is a sum of 9 pixels perpendicular to the vertical line.

to 100 nm at the thinnest. Most of the wires have a balloon-shaped tip that is about 0.5-µm wide by 1.0-µm long (Fig. 3.15). Spectra shown in Fig. 3.15 were collected at 10 and 15 kV. At the point the spectra were collected, the wires are about 500 nm in diameter. The 10-kV spectrum shows much less beam penetration in that the carbon peak is significantly smaller.

The maps of the nanowire sample are shown in Fig. 3.16. Although the maps were collected at a variety of beam voltages (10, 15, and 25 kV), the Zn L maps look quite similar and all of the maps show the thinnest part of the wire. At this point the sample is much like a thinned sample or an FIB section, and the spatial resolution is very good. The sample is mounted on a carbon background and at lower voltages appears like a very accurate inverse Zn map with excellent resolution. However, at 25 kV most of the primary beam electrons are transmitted through the wire and strike the carbon background. No pattern (inverse) of the Zn wires is observable in the 25 kV carbon map.

3.3. Nanoparticles

Nanoparticles and submicron particles, in general, are difficult to analyze because we often derive most of the data in their spectra from the substrate or from adjacent areas. As was shown in Fig. 3.10, we may be able to help ourselves by selecting a substrate with a unique composition that is dissimilar from our particles of

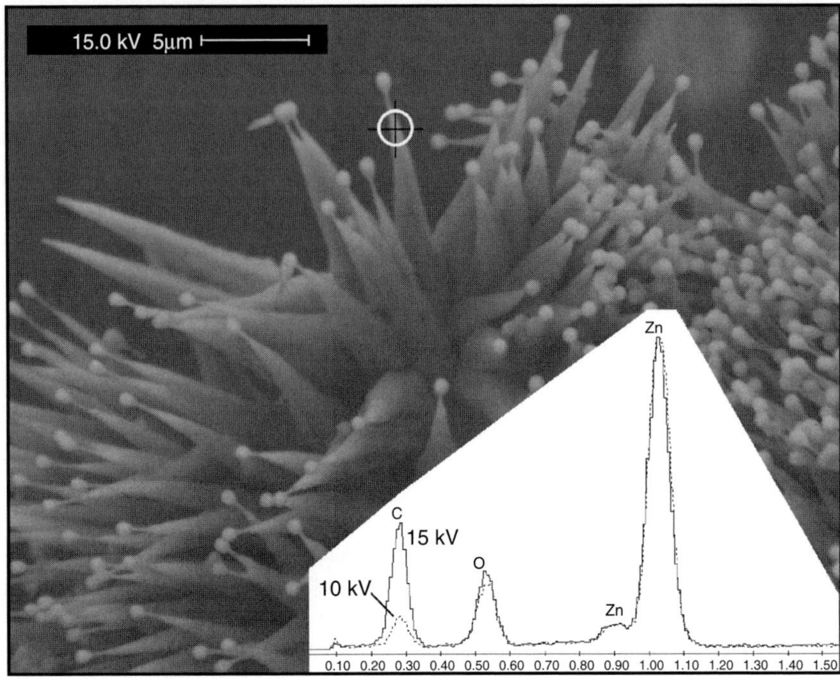

FIGURE 3.15. Secondary electron image of ZnO nanowire sample on carbon tape with spectra collected at 10 and 15 kV. The location of the spectra is shown with crosshairs.

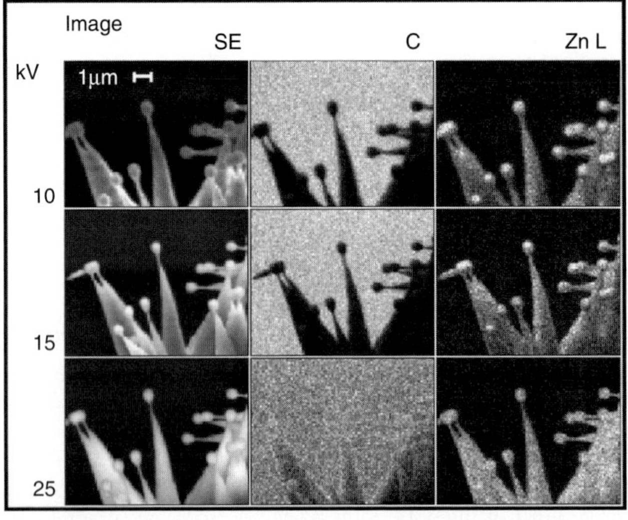

FIGURE 3.16. Maps of ZnO nanowire sample collected at beam voltages of 10, 15, and 25 kV.

interest. When carbon is not significant in our particles, we have many choices including carbon planchets, carbon tape, and filters of many types. For the ultra-fine particles and for the nanoparticles, we can place the particles on a TEM grid with a thin (<<100 nm) polymeric substrate and possibly a discontinuous, lacey carbon film substrate. The grid is suspended over a large cavity, which effectively eliminates the substrate. Because most electrons will be transmitted through the particle and still retain relatively high energies, they will strike whatever surface is below the sample and create x-rays and BSE. Some of these BSE and x-rays will strike the sample, its grid, and parts of the grid-holding mechanism that is viewable by the EDS detector. Stray radiation contributions to our spectrum can be minimized by having a large, deep cavity and some researchers have made the cavity below the grid the entire sample chamber, which provides a spectrum with very little scatter [5].

Nanoparticles of TiO_2 were analyzed manually using a copper grid with a lacey carbon film. The sample grid was placed over a large cavity similar to the setup shown in Fig. 3.17. In this case the metal grid was Cu and "metal A" is aluminum—both contribute to the spectra of the particles on the film. "Metal B" may be a different composition and it typically will not produce a recognizable peak in the spectrum. It is possible to construct such a holder in which metal B will be carbon or some relatively low atomic number material, because this will decrease the yield of x-rays and BSE from this surface. It may also be a good practice to coat all metal surfaces with carbon paint to make it more likely that the high-energy electrons will be absorbed rather than generate x-rays and BSE that will be seen by the EDS detector or that will fluoresce stray x-rays from somewhere else in the sample, grid, or holder. The geometry of the holder shown in Fig. 3.17 does not show a mechanism for holding the grid in place. The grid can be held in place with some carbon paint or carbon tape, which can make the sample semipermanently attached to the holder or it can mean that the sample and grid might be somewhat damaged during the removal of the sample. Some clamping mechanism above the grid can be used to hold the grid in place, but it must

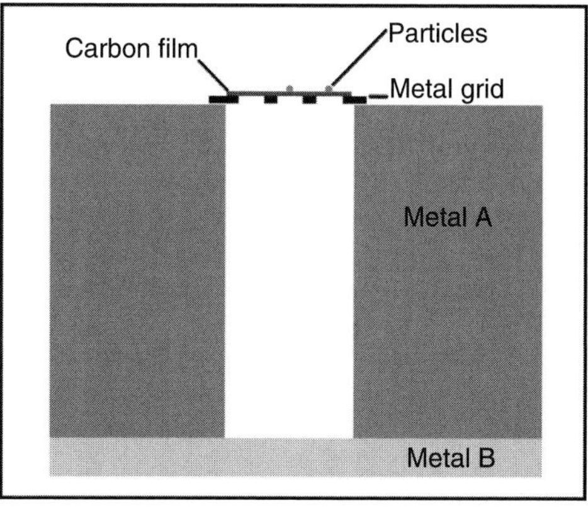

FIGURE 3.17. An STEM holder for the SEM. A cavity is shown below the metal grid. It is possible that there will be some contribution to the spectrum from the grid and perhaps from metal A if the EDS detector can see some part of the sample holder.

allow a clear path to the EDS detector for the x-rays and it should not be a significant source of x-rays that will make interpretation of our particle spectra more difficult.

Many SEMs today have a dedicated STEM detector that will produce a high-resolution bright field image similar to a TEM by capturing the unscattered electrons below the grid, and it is also possible to construct a similar device [2]. The bright field image is inverted or is a negative image of the sample. Some STEM detectors have also been constructed to provide a dark field image, which captures only the scattered electrons passing through our sample and provides an image with a more normal contrast like the SE or BSE image (the particles are bright and the carbon film tends to be dark).

The TiO_2 particles are shown below on the Cu grid substrate and also on a lacey carbon film (on a Cu grid) as well as their EDS spectra (Figs. 3.18 and 3.19). Even though the TiO_2 particles on the Cu substrate are several particles thick, the Cu L peak is the largest peak in the spectrum. By comparison, the TiO_2 particles on a lacey carbon film have a much larger Ti peak though we still see some Cu and Al from stray radiation.

(a)   (b)

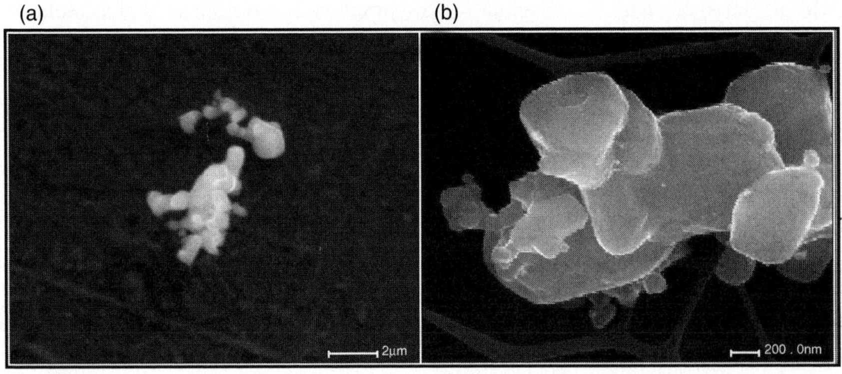

FIGURE 3.18. TiO_2 particles on a Cu substrate (a) and on a lacey carbon film (b).

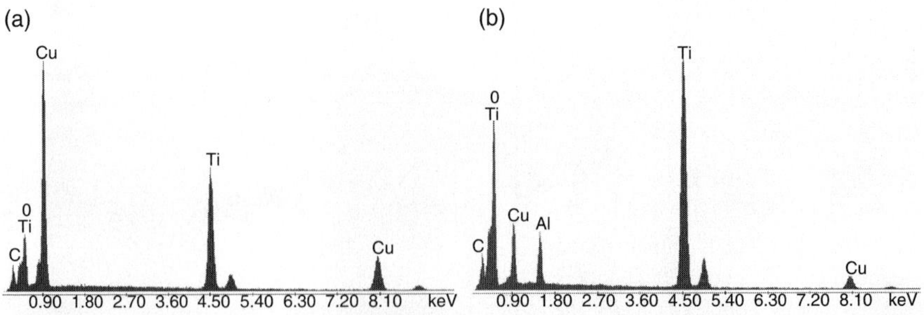

FIGURE 3.19. Spectra collected for TiO_2 particles on a Cu substrate (a) and on a lacey carbon film (b). The lacey carbon film was on a Cu grid underlain by Al (see metal A in Fig. 3.7).

The area shown in Fig. 3.18b was mapped at a beam voltage of 15 kV (Fig. 3.20). The Ti and O maps are derived from the particles directly and show a strong and sharp contrast between the particles and the area surrounding the particles. This map quality is significantly better than what could be expected from maps of the particles on a substrate.

In the C map of Fig. 3.20 the texture of the lacey carbon film is clearly shown, but the carbon displayed at the location of the particles themselves is a result of stray radiation and scattering of the primary beam electrons. The transmitted primary beam electrons create BSE and x-rays from the holder and the area around the particles that ultimately strike the carbon film. The bright areas in the Al map are from the scatter from parts of the holder and detector. The contrast in the Cu map is stray radiation from the Cu grid and also from parts of the sample holder mechanism.

The strong contrast in the Ti and O maps in Fig. 3.20 ultimately originates from a tremendous difference in x-ray count rates between particle (~1350 cps), open space (25 cps), and carbon fiber (70 cps). We can also examine the cps for specific elements in different areas of the sample. The contrast in cps was most extreme in the ROI, which specified the Ti K and was 353 cps on the particle, 1 cps in the open area, and 3 cps on the carbon fiber. The stray radiation was also quite high while the beam was on the particle in that the Cu K ROI was 159 cps. The stray radiation could be minimized further if we increase the depth of the cavity in Fig. 3.17 and also if the diameter of the cavity could be made larger. Another step to improve the quality of the STEM holder would be to use a low atomic number material for metal A and metal B. It might be easiest to use carbon or to coat all metallic surfaces under the sample with carbon paint. This would reduce both the BSE and the x-ray yield from the transmitted electrons and would minimize the presence of x-rays from stray radiation in our spectrum.

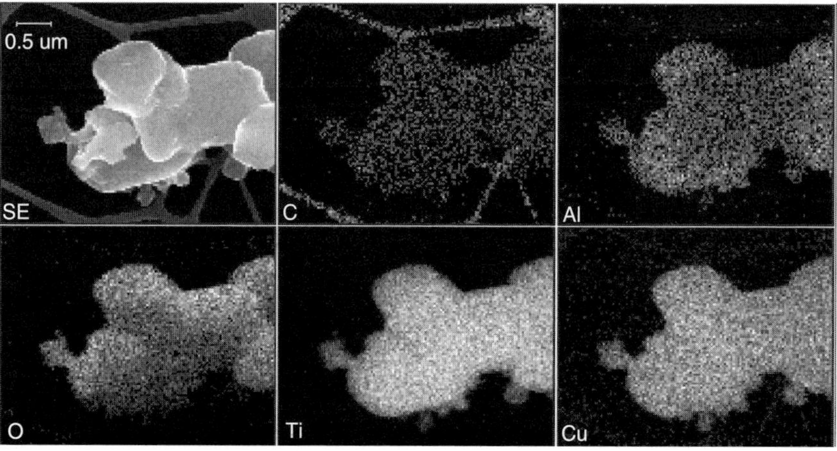

FIGURE 3.20. X-ray maps of TiO_2 particles on a lacey carbon film.

4. Summary

X-ray analysis by EDS in SEM will easily allow us to resolve micrometer-scale features. It is possible for us to reduce our resolution still further by using lower beam energies. However, with the lower beam energies there are also some disadvantages, such as a reduced energy range of the EDS spectrum, more peak overlaps, and incomplete coverage of the periodic table. Another strategy to allow x-ray analysis of nanoscale features would be to reduce the sample thickness (to 100 nm or less), which can be done by preparing FIB sections. The reduced thickness will also reduce the beam spreading and allow us to use higher beam voltages that will also provide a more complete coverage of the periodic table in the EDS spectrum. The FIB section and other nanomaterials such as particles and wires can be mounted on TEM grids, which have a minimal substrate and permits analysis of the nanoscale feature itself with relatively few artifacts.

Acknowledgments. Many people contributed their time and expertise that made this chapter possible or which made it better. Adam Harned showed me what the nanowire sample really looked like. Bill Heady contributed beam time that was necessary for some of the x-ray maps presented. Mike Hernandez helped to prepare the nanoparticle samples and assisted with their mapping. Kevin McIlwrath created the FIB section used in this chapter. Lara Swenson helped to improve the manuscript and did a variety of tasks that allowed me the time to create this chapter. Sara White helped me coordinate with the right people to get everything done. Weilie Zhou at the University of New Orleans provided several of the samples used in this study.

References

1. A. J. Garratt-Reed and D. C. Bell, *Energy Dispersive X-Ray Analysis in the Electron Microscope*, Springer-Verlag, New York (2002).
2. J. Goldstein, D. Newbury, D. Joy, C. Lyman, P. Echlin, E. Lifshin, L. Sawyer, and J. Michael, *Scanning Electron Microscopy and X-ray Microanalysis*, 3rd edition, Kluwer Academic/Plenum Press, New York (2003).
3. R.A. Waldo, in: *Microbeam Analysis*, D.E. Newbury (Eds.), San Francisco Press, San Francisco (1988), p. 310.
4. J. L. Pouchou and F. Pichoir, *Scanning*, 12 (1990) 212.
5. A. Laskin and J. P. Cowin, *Anal. Chem.*, 73 (2001) 1023.

4
Low kV Scanning Electron Microscopy

M. David Frey

1. Introduction

The voltages typically associated with low-voltage scanning electron microscopy are within the range of 5 kV and lower. The low end cutoff is typically the lowest value at which a user's microscope can achieve a usable image. The author of this chapter would like to see a new definition of low kV-accelerating voltage begin within the realm of 2–1 kV. With the advent of new column designs, through the lens (TTL) secondary electron (SE) detectors and the use of electrostatic lens, and similar techniques, the low end of the low kV range is typically reaching well below 500 V, and is often approaching 100 V. These new and tried true techniques are illustrated in Figs. 4.1 and 4.2. A recent development is the use of a retarding field to modulate the landing energy of the primary electrons. By simply applying a negative potential (Vb) to the specimen, a retarding field can be generated between the specimen and a grounded electrode above the specimen (the cold finger or the objective lens in a semi-in-lens type FEG-scanning electron microscope [SEM] instrument). In this simple configuration, the specimen itself is part of a "cathode lens" system. The landing energy (E_L) of the incident electrons can be varied by changing the applied potential:

$$E_L = E_0 - eVb.$$

where E_0 is the energy of the primary electron beam. The nature of the electron–specimen interactions is determined by the electron landing energy E_L. When eVb is equal to or larger than E_0, the incident electrons will not enter the specimen at all; in this particular configuration a mirror image may be formed. Therefore, by varying the potential applied to the sample, we can conveniently extract useful surface information of the sample. The ultimate resolution of ultralow voltage (ULV) SEM images is determined by the combined properties of the probe-forming lens and the cathode lens [1]. The highest resolution secondary images are being obtained using TTL secondary detectors commonly referred to as TTL detectors. These TTL detectors often involve a strong magnetic field projected into the chamber, or a mild electrostatic field. This field is generally used

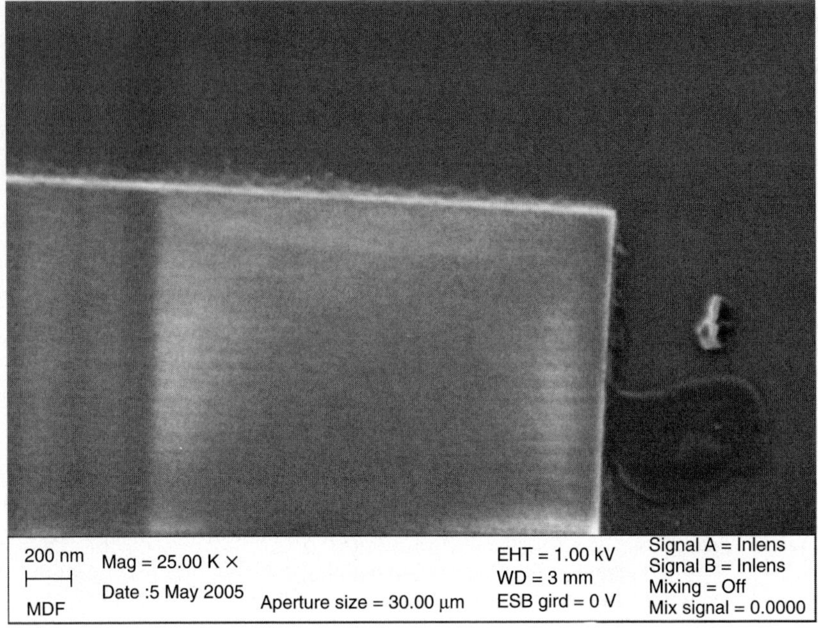

FIGURE 4.1. Secondary electron image of resist on Si at 1 kV uncoated.

FIGURE 4.2. Secondary electron image of resist on Si at 0.1 kV uncoated.

to collect the low-energy SEs, which help to provide the highest resolution images. The type of field and detector positioning above the final lens will often depend upon the manufacturer of the system. Often this field type and detector positioning will have a great effect on image quality, as well as overall system performance. To further this point Joy wrote *Through-the-Lens (TTL) Detectors*, which use the postfield of the lens to collect the secondary signal, have not only a much higher efficiency (DQE ~0.8), but also are more selective in the energy spectrum of the electrons that they accept. In the most advanced design the TTL detector can, in effect, be tuned so as to collect either a pure SE signal or a backscattered signal or some mixture of the two. This provides great flexibility in imaging, and avoids the necessity of a separate BSE detector, and permits backscattered operation at very short working distances. The ultimate goal remains a detector that is both efficient and can be used to select any arbitrary energy window in the emitted spectrum for imaging [2]. Low kV SEM can also have profound effect on the results achieved during elemental analysis. To quote Boyes form of microscopy and microanalysis: the use of lower voltages (Boyes, 1994, 1998) to generate x-ray data for elemental chemical microanalysis improves substantially the relative sensitivity for surface layers, small particles, and light elements. The reduced cross sections for x-ray yield can be partially compensated with improved analysis geometries. There remains a need to achieve a critical excitation energy plus a factor of 20–40% (or overvoltage U, of $1.2 < U < 1.4$), and this limits analysis of complete unknowns to about 6 kV; 5 kV is inadequate for the elements I, Cs, and Ba. The main light elements (B–F) and those 3d series transition metals with strong L lines are still accessible at 1.5 kV and this can be especially important for the analysis of the smallest particles and thinnest overlayers [3].

This chapter will try to address the concepts and practices of low kV SEM, and its use to attain high-resolution micrographs. (What does the term "high resolution" refer to? In the case of current production SEMs, high resolution would be better than 2 nm.)

2. Electron Generation and Accelerating Voltage

The first thing that should be addressed is what exactly is "low kV?" To truly understand the term "low kV" the reader must understand how kV is controlled or set. This will require an understanding of an "electron gun." The basic principles and design of an electron gun and an electron optical column do not vary much from SEM to SEM. The basic design consists of a cathode and anode arrangement. Within this basic design there is an electron source. This source will normally consist of either tungsten or lanthanum hexaboride (LaB6). The electrons are generated by one of a couple of methods. They are either generated thermally or through the use of field emission. In the case of systems where high-resolution imaging is going to be done in the lower kV ranges, the user will more than likely be using a field emission source rather than a thermal source.

In the case of the thermal emitter, the electron source is heated by running a current resistively through a wire. As the current is increased the heating temperature increases. By increasing the temperature of the source, the electrons are liberated due to lowering of the work function of the electron donor material to a point where the electrons are in essence boiled off from the wire. The wire is subjected to temperatures at or close to 2,700 K. This cloud of electrons is gathered up in the Wehnelt cylinder. This Wehnelt cylinder contains the cathode (filament) of the electron gun. The Wehnelt is kept slightly negative in relation to the filament, which provides a focusing effect for the cloud of electrons. The cathode is set to ground potential, 0 V. The voltage of the anode is set to the voltage desired to be the accelerating voltage (e.g., 20 or 1 kV). The anode plate contains an aperture through which the electrons will pass. This difference in voltage provides the potential for acceleration of the electrons down the column [4]. This scenario is called thermal emission (Fig. 4.3). Normally this type of electron emission provides a stable source of electrons, but is frequently associated with lower resolution SEMs (an ultimate resolution of ~3 nm at 20–30 kV).

SEMs that are used at lower accelerating voltages and produce high-resolution images at these "low kV" settings produce their electron beam through the use of one of the two categories of field emission. The two varieties of field emission are thermal and cold. Both have advantages and disadvantages. Generally both are very capable for "low kV," high-resolution imaging. The basic concept behind field emission is an electron source that has been attached to a tungsten wire. This electron source itself is a piece of tungsten. It has a tip radius smaller than 100 nm. As a negative potential is applied to the source, the small tip concentrates this potential. This concentration of potential at the tip subjects it to a high field. This high field allows the electrons to escape from the source material. The two types of field emission guns, cold and thermal (illustrated below), function in more or

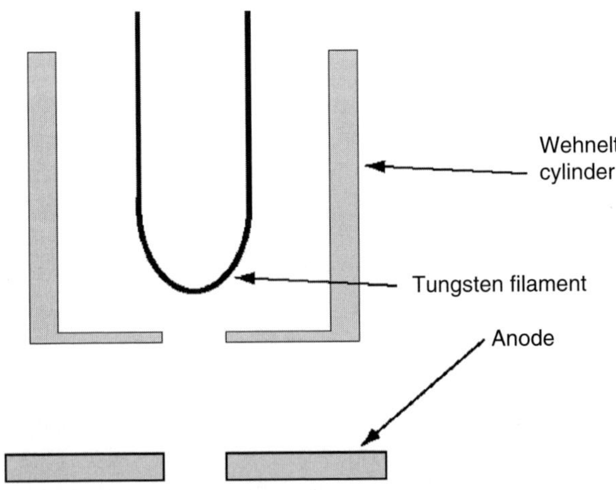

FIGURE 4.3. Thermal emitter.

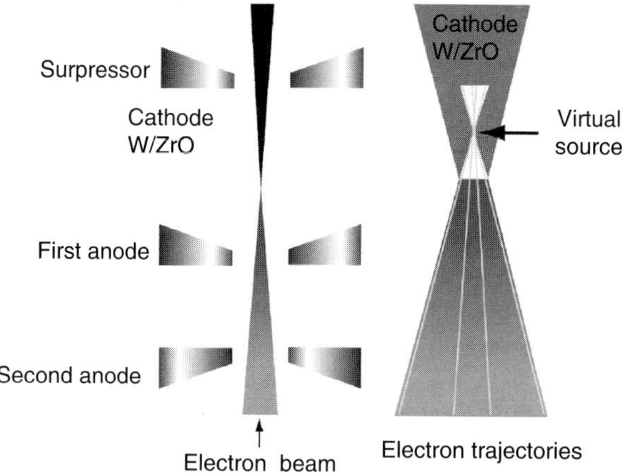

FIGURE 4.4. Schottky field emission gun. (Drawing adapted from Carl Zeiss SMT Literature.)

less the same way in respect to the field generated at the tip. If the tip is held at a negative 3–5 kV relative to the first anode, the effective electric field F at the tip is so strong ($>10^7$ V/cm) that the potential barrier for electrons becomes narrow in width as well as reduced in height by the Schottky effect. This narrow barrier allows electrons to "tunnel" directly through the barrier and leave the cathode without requiring any thermal energy to lift them over the work function barrier. Tungsten is the cathode material of choice since only very strong materials can withstand the high mechanical stress placed on the tip in such a high electrical field. A cathode current density as high as 10^5 A/cm^2 may be obtained from a field emitter compared with about 3 A/cm^2 from a tungsten hairpin filament. In a field emitter, electrons emanate from a very small virtual source (~10 nm) behind the sharp tip into a large semiangle (nearly 20° or about 0.3 rad), which still gives a high current per solid angle and thus a high brightness [5]. A second anode is used to accelerate the electrons to the operating voltage. The only real differences fall into the categories of vacuum requirements are long-term source stability and overall I-probe (probe current) that can be generated by a source. Other considerations in regard to column design will have an effect over low kV performance. Figure 4.4 illustrates Schottky field emission source. Table 4.1 compares the different types of electron sources.

3. "Why Use Low kV"?

There are many interesting reasons why a microscopist or researcher might use low kV. First and foremost would be charge reduction. SEM users have often fought the charging influences of an electron beam. Often users coat their samples with a conductive coating if the samples happen to be of a nonconductive

TABLE 4.1. Electron source comparison

| | Electron source performance comparison | | | |
Emitter type	Thermionic	Thermionic	Cold field emission	Schottky field emission
Cathode material	W	LaB_6	W (310)	ZrO/W (100)
Operating temperature (K)	2,800	1,900	300	1,800
Cathode radius (nm)	60,000	10,000	$\leq$100	$\leq$1,000
Effective source radius (nm)	15,000	5,000	2.5 (a)	15 (a)
Emission current density (A/cm^2)	3	30	17,000	5,300
Total emisson current (μA)	200	80	5	200
Normalized brightness (A/cm^{2*}sr*kV)	1×10^4	1×10^5	2×10^7	1×10^7
Maximum probe current	1,000	1,000	0.2	10
Energy spread at the cathode (eV)	0.59	0.4	0.26	0.31
Energy spread at the gun exit (eV)	1.5–2.5	1.3–2.5	0.3–0.7	0.35–0.7
Beam noise (%)	1	1	10	1
Emisson current drift (%/h)	0.1	0.2	5	<0.2
Minimum operating vacuum (hPa)	$\leq1 \times 10^5$	$\leq1 \times 10^6$	$\leq1 \times 10^{10}$	$\leq1 \times 10^8$
Cathode life (h)	200	1,000	>2,000	>2,000
Cathode regeneration	Not required	Not required	Every 6 to 8	Not required
Sensitivity to external influences	Minimal	Minimal	High	Low

Source: Adapted from Carl Zeiss SMT Literature.

variety. The advantages and disadvantages of coating have long been argued and this author does not wish to wade into this discussion. Many samples that are non-conductive often have a point where they reach equilibrium. This equilibrium is the point where the charge into the sample equals the charge out of the sample. The charge out of the sample will be from SEs, backscattered electrons (BSE), Auger electrons, x-rays, and whatever current is absorbed and then transmitted through the sample to ground. This is referred to as a state of unity. Charging comes in several species, and often exhibits several characteristics. These species are either positive or negative charging. When the sample charges positively the image will appear bright, and when there is a negative charge the image will appear dark. Charging can also be frequently seen as streaks of light or dark on an image. As the charge of a sample changes a user will often also see the image drift across the screen. Users frequently attribute this to stage or sample instability. These two causes can be ruled out as a cause by raising or lowering the kV. If the drift either stops or changes direction, charge balance is the culprit, not stage instability or poor sample mounting. The effect that is being seen as image drift is in fact the buildup of charge in the sample pushing the electron beam away from the area to be imaged; hence the appearance of drift. If the charge on the sample becomes great enough, the result can be a reflection of the primary e-beam by the charged sample causing a scan of the chamber to result. This will happen when the charge of the sample is greater than the charge of the primary beam (e.g., a primary beam of 2.5 kV and a charge on the sample of 20 kV). The SE image that is produced as a result of the excessive charge on the sample (a plastic sphere) can be seen in Fig. 4.5.

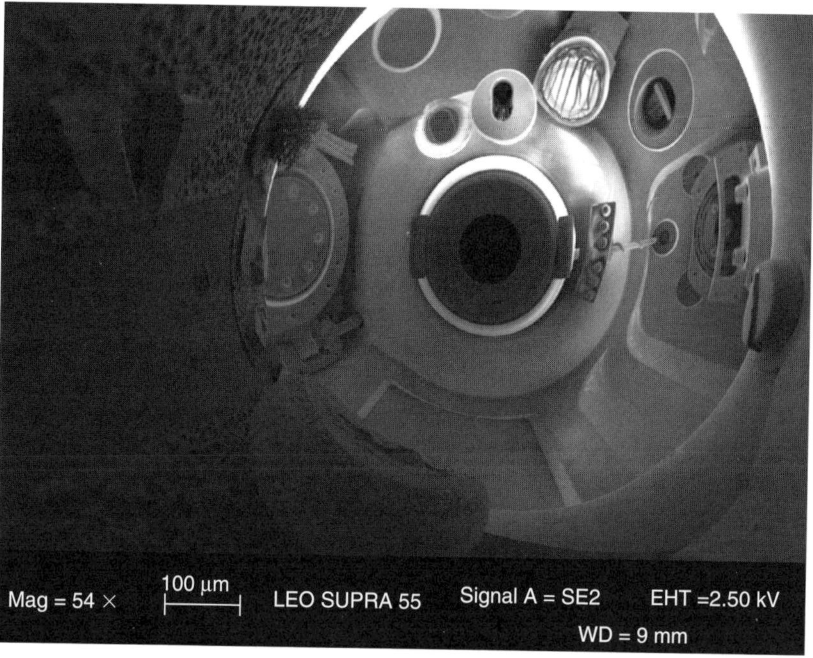

FIGURE 4.5. A secondary electron image formed by charging a plastic sphere at 20 kV, and then imaging the sample at 2.0 kV. The resulting image is formed by the low kV primary beam being reflected about the chamber by the sample that has been charged to a higher kV.

The use of low kV leads to reduced beam interaction volume. It has been noted by Joy and Newbury [6] that as the energy (E_0) of the incident beam is reduced the range (R) of the electrons falls significantly ($R \sim k.E_0^{1.66}$). This reduced interaction volume results in the production of an SE signal from closer to the surface of the sample. Signal development from closer to the surface leads to more surface information. Low kV operation also benefits image quality of low atomic number (low Z) samples. At low kV the penetration depth of the electrons into low Z materials is much less than higher kV electrons on the same samples. Frequently low Z samples will have a ghost-like or semitransparent look at higher accelerating voltages. By lowering the kV of the beam, the resulting image will be less transparent, and display more information about the material. Often this reduction of interaction volume has the result of improving the resolution of x-ray microanalysis. The user must make sure that there is sufficient overvoltage to excite the peaks required for analysis of the samples of interest. It is advisable to do the analysis at higher kV settings to identify the materials present. If the spatial resolution is not satisfactory, a lowering of the accelerating voltage is recommended provided that peaks for elements required for either the mapping or line scans are available. Boyes mentions that the use of low kV and high kV for microanalysis is convergent analysis and will lead to a better understanding of the 3D structure and chemistry. Figures 4.6, 4.7, 4.8, and 4.9 are SiO_2 nanoparticles and Figs. 4.10, 4.11, 4.12, and 4.13 are a Ti fracture sample. These images illustrate how lowering the kV can reveal more surface information. This reduction of interaction volume also has some other effects on the signal being produced. One of these effects is the increase of SE emission. This increase in the emission of SEs is the direct result of the shrinking of the interaction volume. As the interaction volume is brought closer to the surface, the SEs have a better opportunity to escape from the sample since they are being produced in a region where they are more likely to escape than to be reabsorbed by the sample. It is important to remember that SEs are "defined purely on the basis of their kinetic energy; that is, all electrons emitted from the specimen with an energy of less than 50 eV… are considered as SE." SEs are produced as a result of the interaction of the primary electron beam and electrons within the elements of the sample. This interaction is generally a collision between a high-energy electron of the beam, and an electron within the specimen. There is a transfer of energy from the beam electrons to the specimen electrons. This energy transfer results in the final energy of the SE that escapes from the sample. Most SEs have an energy, which is less than 10 eV. SEs are of two types: SE1 and SE2 electrons. The SE1 electrons are generated closer to the surface by the interaction of the primary e-beam and specimen electrons. The SE2 electrons are generated by the interaction of BSEs and

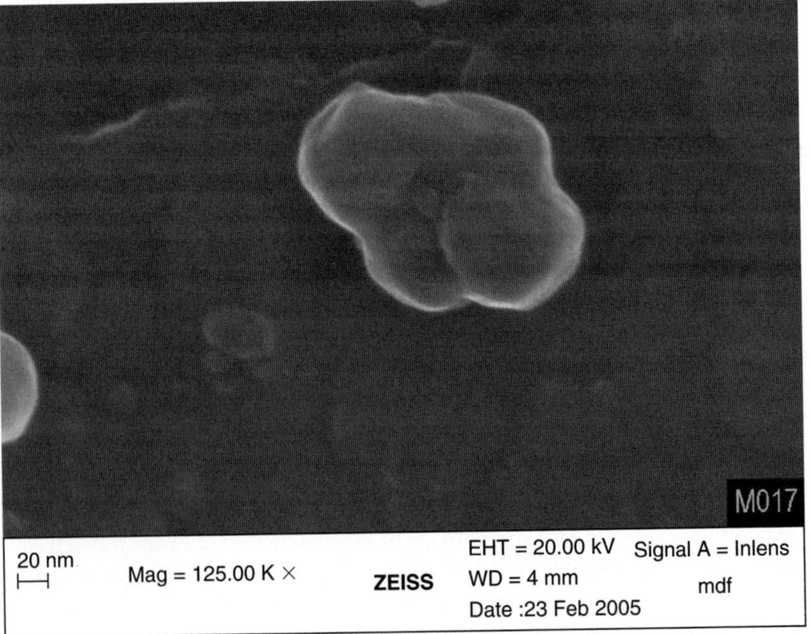

FIGURE 4.6. SiO_2 nanoparticle at 20 kV or an Al stub.

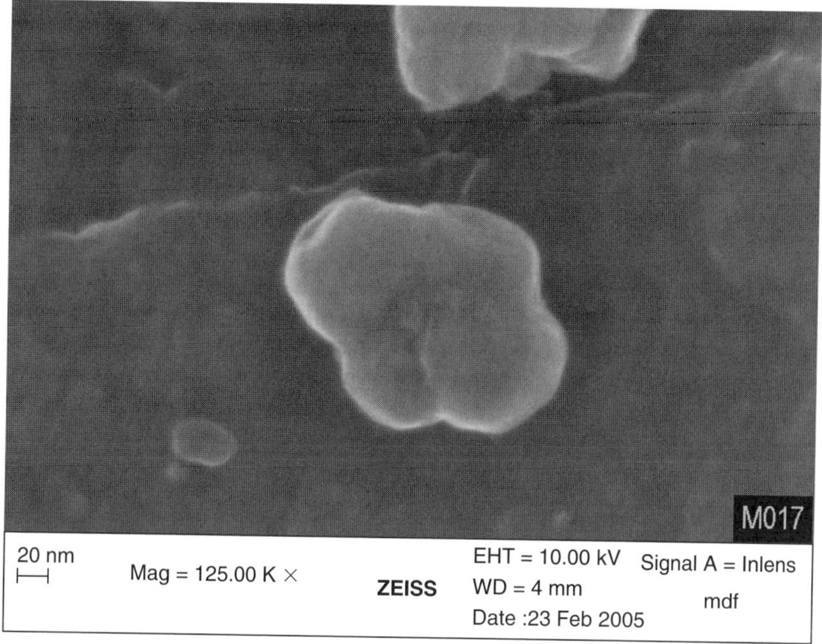

FIGURE 4.7. SiO$_2$ nanoparticle at 10 kV or an Al stub.

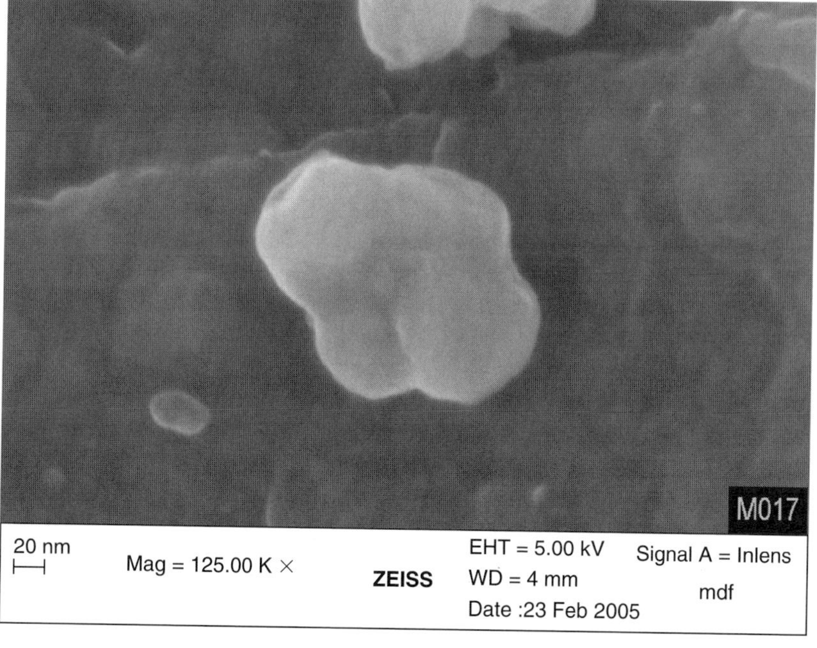

FIGURE 4.8. SiO$_2$ nanoparticle at 5 kV or an Al stub.

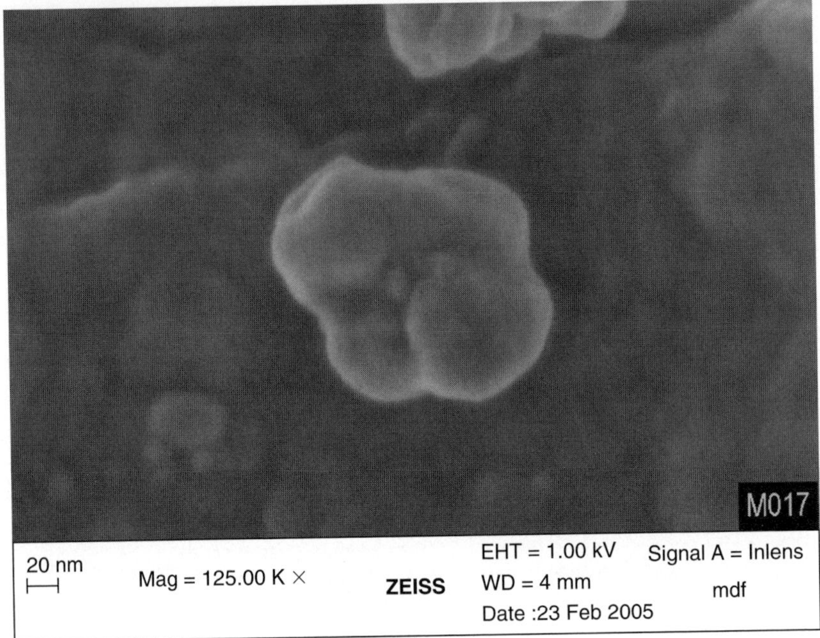

FIGURE 4.9. SiO$_2$ nanoparticle at 1 kV or an Al stub.

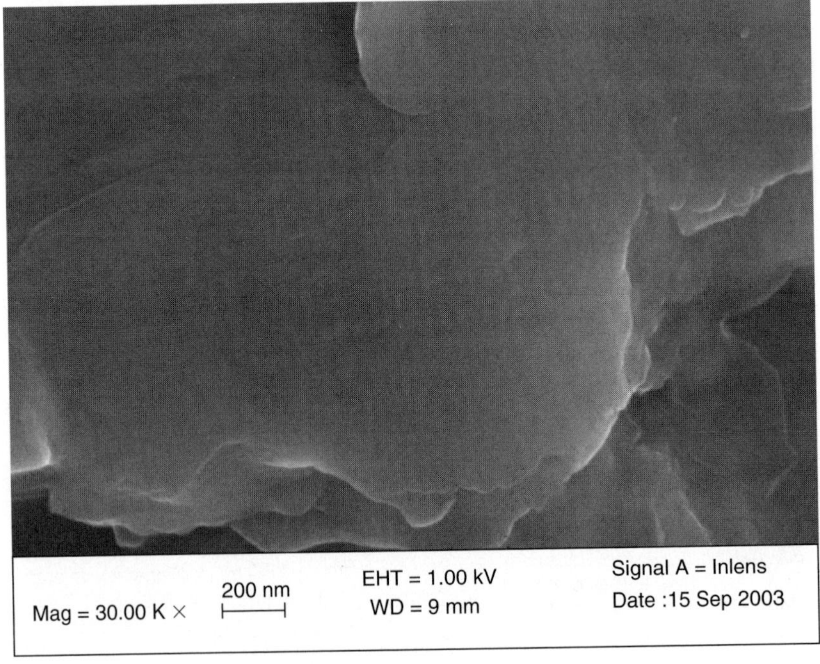

FIGURE 4.10. Ti fracture sample at 15 kV.

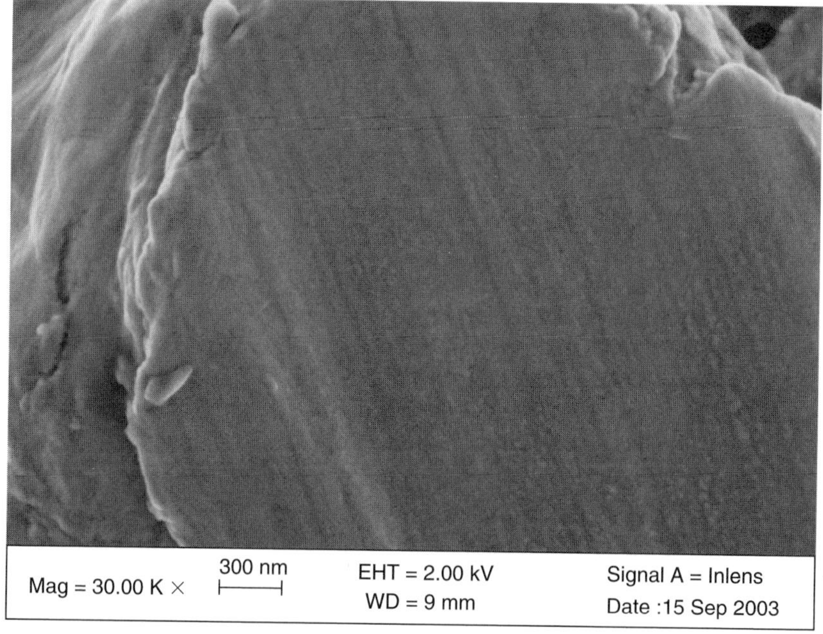

FIGURE 4.11. Ti fracture sample at 2 kV.

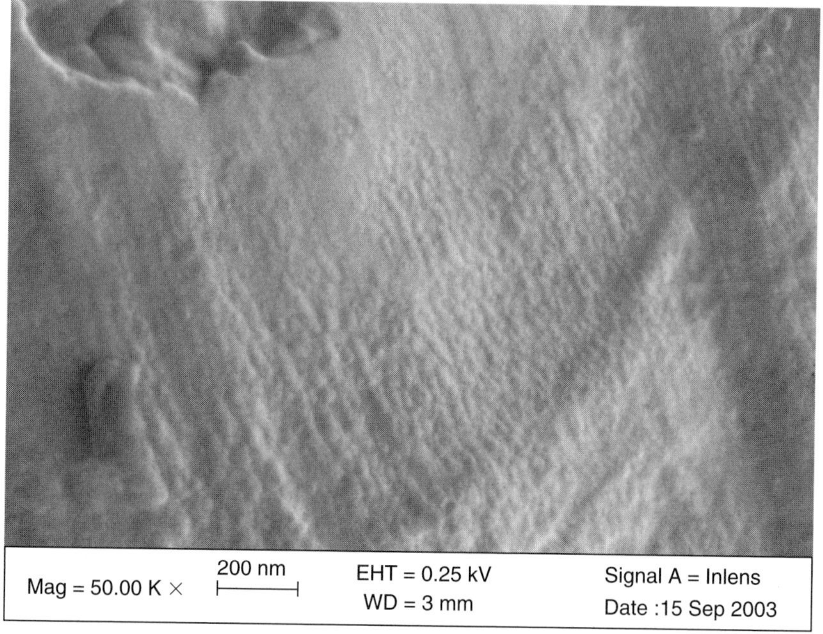

FIGURE 4.12. Ti fracture sample at 0.25 kV.

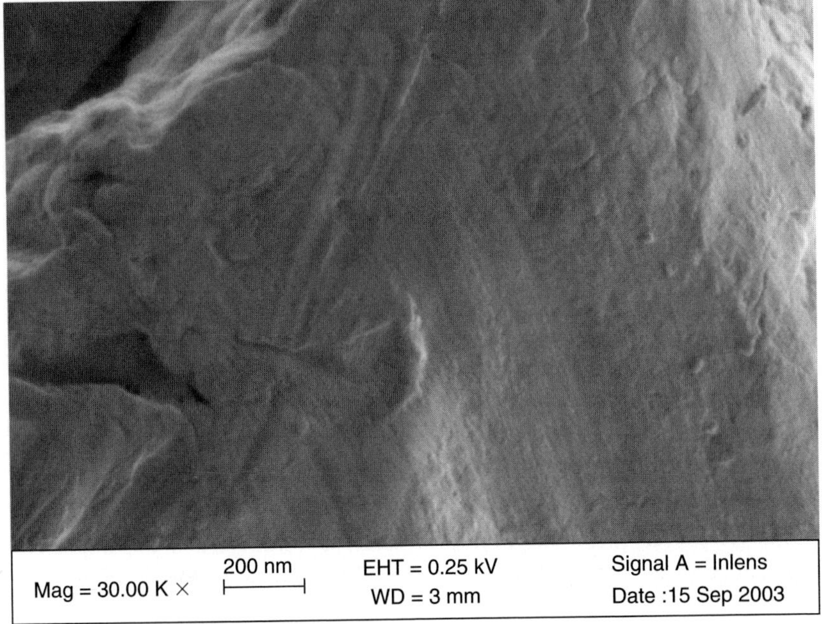

FIGURE 4.13. Ti fracture sample at 0.25 kV.

specimen electrons. The SE2 electrons are of higher kinetic energy, which allows them to escape from deeper within the sample due where they are being generated. These SE2 electrons result in a somewhat lower resolution image when compared with an image with a higher number of SE1 electrons [4]. Another interesting point made by Müllerová and Frank must also be considered:

The active depths of SE and BSE signals become similar and hence both signals are sensitive to the surface cleanliness. Also the widths of the response functions and hence the image resolutions approach each other. The SE signal grows relatively to the BSE one and exceeds this, and at some critical energy no charge is dissipated in the specimen. When the penetration depth of primary electrons approaches, somewhere between 150 and 700 eV— the escape depth of SE—the edge effect disappears together with signal enhancement on inclined facets. Consequently, micrographs appear more "flat," with limited perception of the third dimension. The topographic contrast is more surface-sensitive and from among relief details those just filled by the upper half of the interaction volume are the brightest. Surface films of so defined (or smaller) thickness become opaque, including contamination islands. [7]

4. Using Low kV

The actual practice of using lowered accelerating voltages is not much unlike using high kV. The user of the microscope will have several things they will have to consider. These considerations include no specific order: sample preparation and microscope preparation.

Sample preparation will include the choice of whether or not to coat the sample to make it more conductive, if it is not conductive. A user will also need to consider what type of analysis or imaging they are trying to achieve from the system. Often users will decide on a type of mounting that will accommodate the type of sample they are evaluating. It should be thought that the sample needs to be further processed down the line after evaluation, or is this destructive testing. Are ultra high-resolution images the goal, or are low to moderate resolution images acceptable for this project. Frequently, all of the questions will influence a user's choice of mounting and preparing a sample.

To mount a sample several techniques can be offered. For high-resolution images a user needs to select the most mechanically stable mounting media. The best suggestion is to avoid any type of tape or "sticky tabs." These will often present an image that will display a drift at high magnifications as the tabs or tape out gas in the vacuum. Often the material from which they are made rebounds after a sample has been pressed into the tab or tape. They are not mechanically stable. They do provide a quick mounting solution, and are often excellent for quick and dirty microscopy when an answer is required immediately. An excellent choice for mounting samples well is a carbon paint/dag or silver paint/paste. These types of mounting media do take some time to dry and are not perfect for all samples, but when it comes to a rigid, stable mount they do provide a good solution as long as speed is not of the essence. There are also many different mechanical sample holders such as vices, and spring-loaded holders that are excellent for bulk type samples. When it comes to mounting samples such as particles, carbon nanotubes, and liquid suspensions which can be dried down, the author has found that carbon-coated formvar TEM grids provide an excellent solution. They provide a nice even black background without the topography often encounter with Al or C graphite sample stubs, by mounting the sample on a grid and by placing it into a sample holder that suspends the sample. Artifacts from an SE or BSE signal being returned from the mounting stub will not influence the image being produced of the sample. The image in Fig. 4.14 is on a carbon-coated formvar TEM grid, the image in Fig. 4.15 is on a piece of Si wafer, and the image in Fig. 4.16 is on an Al stub.

The question of coating is one that inspires a great deal of discussion. The first topic is whether or not to coat the sample. If the decision is made to coat the sample, the next topic of discussion will often be what should be used. Often users are restricted to the type of specimen coater they currently own. In most situations this coater will not always be the latest state of the art coater with all of the bells and whistles. This fact frequently influences a user, to make the decision not to coat their sample because of the perceived poor quality of their coater. There is a rational fear of overcoating the sample, and thus obscuring the beautiful details thereafter at the "low kV" settings. The argument can be made that in many cases a user can flash down 1–5 s of a coating with their old out dated coater and still get excellent high resolution results without the artifacts commonly associated

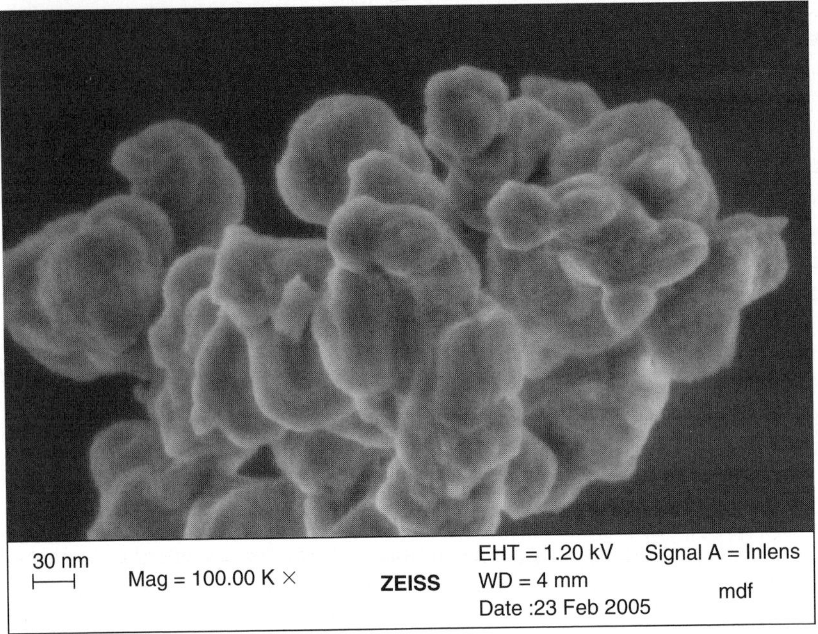

FIGURE 4.14. SiO$_2$ nanoparticles on a carbon-coated formvar TEM grid mounted on an STEM sample holder imaged at 1.2 kV.

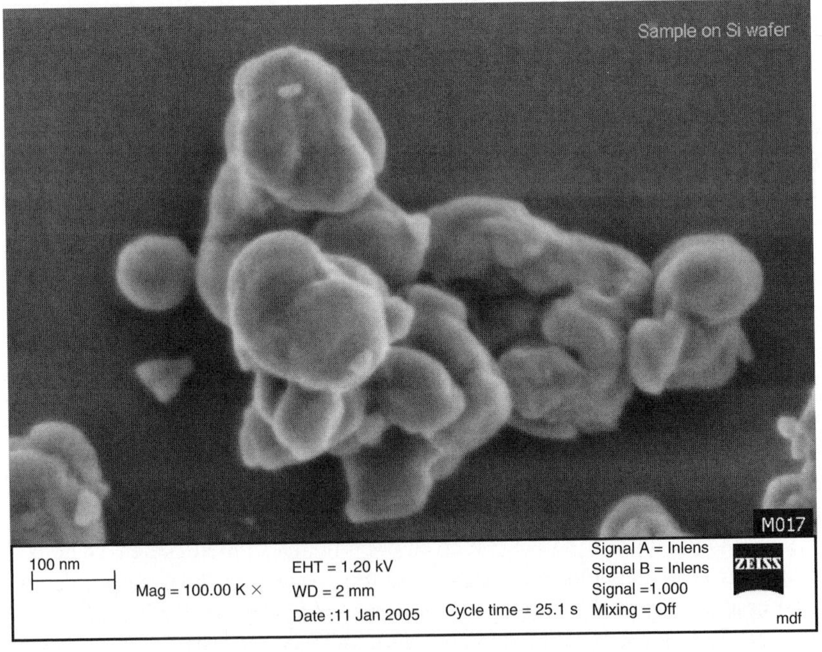

FIGURE 4.15. SiO$_2$ nanoparticles on a piece of Si, imaged at 1.2 kV.

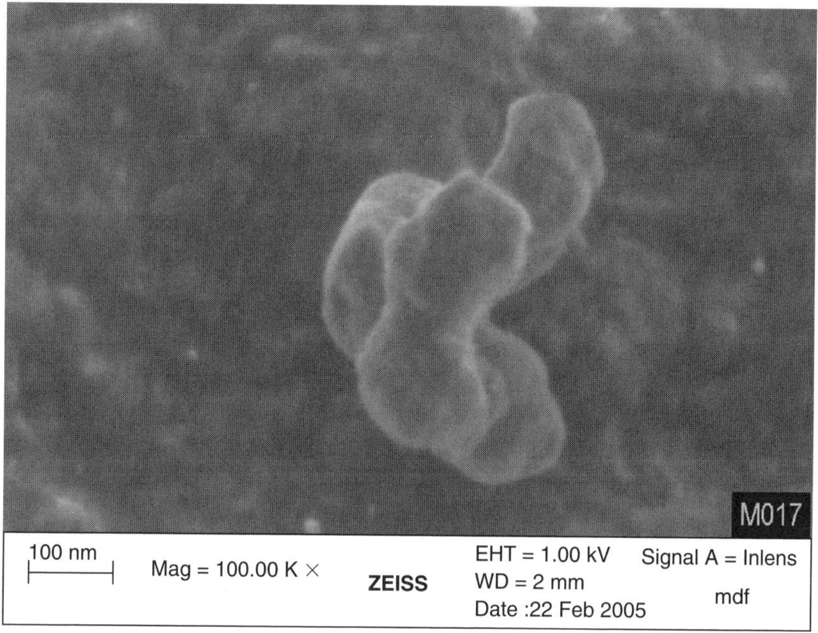

FIGURE 4.16. SiO$_2$ nanoparticles on an Al stub, imaged at 1.0 kV.

with overcoating a sample. This thin or light coating is used to stabilize the sample enough in order to take the sample from impossible to possible. I always suggest a user attempt to image a sample uncoated first, and when all else fails then coating is an apt solution. The user will still have to employ many of the strategies for "low kV" and "low probe current" imaging which have yet to be outlined. Most important will be the proper adjustment of their electron optical column. This of course varies from column to column. In the simplest set of adjustments a user will need to adjust, focus, stigmation and aperture alignment. On more complex columns, user will need to contend with gun alignments, condenser lens adjustments, aperture alignments, biases of the samples, and other adjustments that may influence the landing energy of the beam. To run a microscope at "low kV" and to produce the best quality high-resolution images it is important to be very familiar with the system's controls, and the parameters that need to manipulated to produce the type of image required to tell the specimen's story. Looking to the manufacturer for this information is often the best place to begin.

When approaching an unknown sample, one that the user has never before imaged, it is a good idea to pick an arbitrary starting point. Often this is either a set of conditions where the user has had success with a similar sample, or a configuration where the user feels comfortable with the performance of their microscope. A good choice for the operation of a field emission SEM at low kV might be 1 kV. If the user knows that there are requirements to do x-ray analysis this type of "low kV" setting might not be desired. It is always a good idea to ask if the sample is conductive, or nonconductive. Is it beam or charge sensitive? These types of questions will also have an effect over starting parameters. If the sample is charge or beam sensitive a user will often need to reduce the I-probe (probe current). This will be done in most systems by changing condenser lens settings. On certain microscope designs changing condenser settings can be disregarded, and a smaller aperture will often be all that is needed. Once the baseline for starting conditions has been established, and the sample has been prepared and inserted into the microscope, the user can begin to work out the best conditions for imaging the sample. This is a relatively unglamorous process, one that amounts to trial and error. Beginning with the baseline settings it is good to decide whether or not to increase or decrease the kV setting in response to a less than perfect image. Once this decision has been made the user can experiment with the kV settings until the state of equilibrium has been reached. This experimentation amounts to changing the kV, and observing the response of the sample. If there is more charging of the sample as a result of the choice to increase or decrease the kV setting the user should reverse course, and find a point that is halfway between the starting point, and the last adjustment. If his adjustment does not yield better results then the user may consider going to a point that is greater or less than the starting point by the amount adjusted (e.g., 1, 0.5, 0.75, and 1.5 kV). If the charge appears to be negative, then an increase in kV is what is required, and if the charge appears to be positive, then a reduction in kV is the answer. Often very small difference in kV can result in a charge balance being reached. A good example is shown in Figs. 4.17 and 4.18. These images are of uncoated latex spheres. The difference in kV is 390 V. This point of charge equilibrium was found by stepping through the kV range in 10 V steps until the charge is dissipated. Frequently this trial and error will be a cycle that will eventually result in a charge balance on the sample. Sometime a reduction in I-probe will also be required once the equilibrium has been reached. If it is hard to find this point of equilibrium, coating may be the answer to imaging the sample provided that it is an option. Figures 4.19 and 4.20 illustrate how a small amount of coating can improve the image significantly without influencing the sample. One of the other parameters that often influences the sample charge balance and must be considered is the dwell time of the beam on the sample. Dwell time will often have an effect on the charge balance of the sample. The shorter the dwell time the better for samples which is nonconductive, or suffer from poor conductivity. A drawback to a shorter dwell time is a reduction in the signal produced by the interaction of the beam with the sample. When deceasing the dwell time leads to an image considered unusable to illustrate the characteristics of the sample a user will need to find an appropriate method to improve the image being produced. Normally this will require the use of features specific to the microscope in use. Many systems have the ability to do noise reduction. This is when the image is filtered through the use of averaging to remove the noise that results from the lowered signal production.

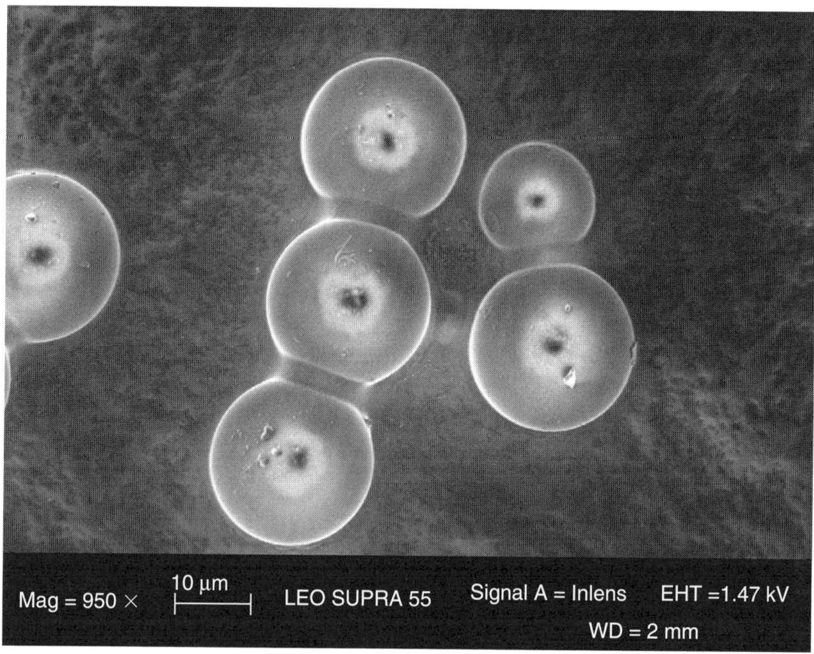

FIGURE 4.17. Uncoated latex spheres at 1.47 kV.

FIGURE 4.18. Uncoated latex spheres at 1.08 kV. A difference of 390 V.

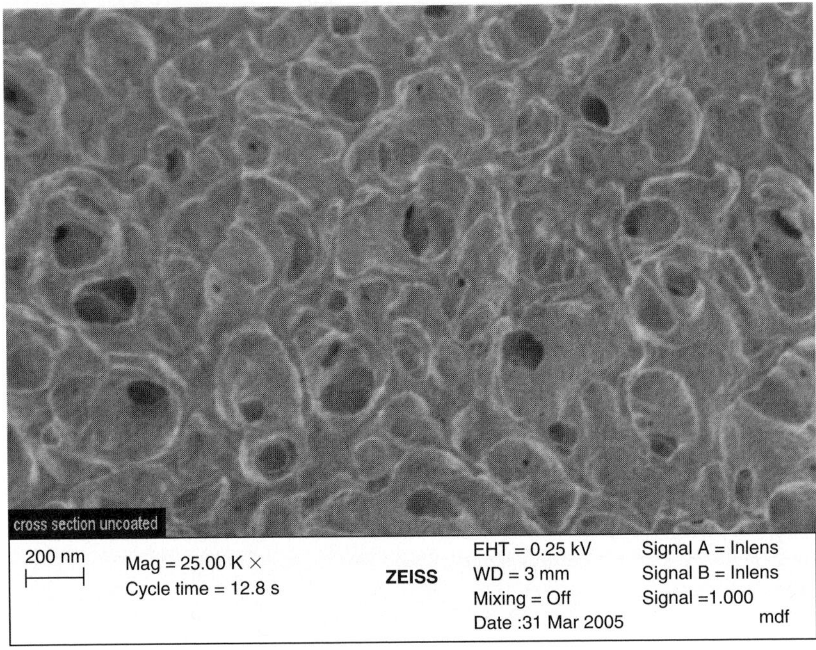

FIGURE 4.19. Uncoated filter sample at 0.25 kV.

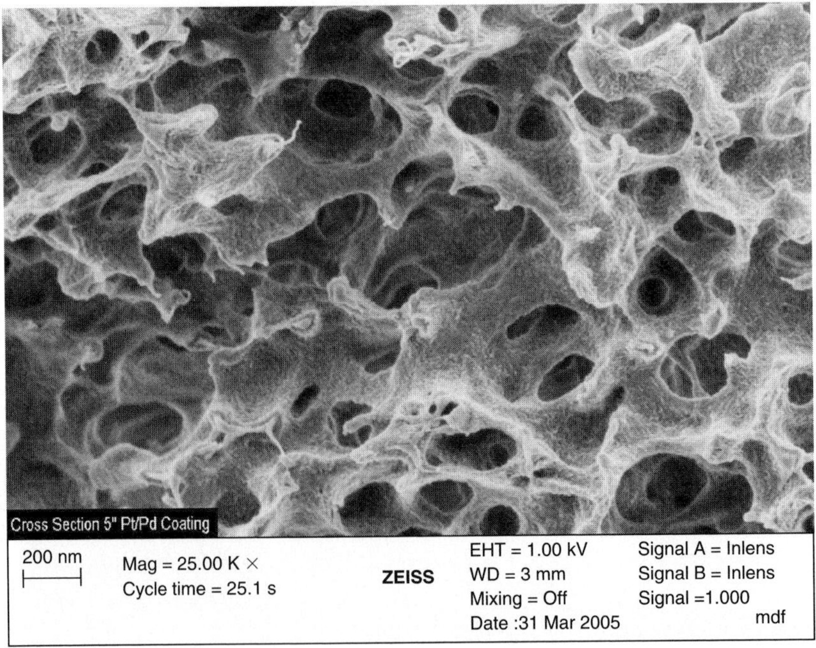

FIGURE 4.20. Filter sample coated for 5 s with Pt/Pd, imaged at 1 kV.

5. Conclusion

Low kV is not a realm where users should fear to tread. It is a gateway to the revelation of new and interesting features. It will help users discover new properties of their nanomaterials, which have been hidden by the overpowering of high kV, high beam penetration, and high-resolution SEM. Dr. Oliver Wells in the May/June 2002 *Microscopy Today* wrote about Jack Ramsey's principle, "There is no best way of doing *anything*" [8]. This is a very fair statement, and one that is very accurate when it comes to the idea of performing scanning electron microscopy at low kV. Low kV is not always the answer to the best possible imaging of every sample. Samples often give the user clues as to where to go for the best imaging and analytical conditions. Users need to keep their eyes open for these clues, take the time to investigate the sample, and be very familiar with the controls and performance of their particular SEM. This familiarity with their system will allow them to exploit relationships of electron microscopy so as to produce the results that will best tell the story of their samples and research. After all the SEM is a tool for making the invisible, visible. Without the skills and understanding of the relationships the results will not help to elucidate the hidden world.

References

1. E. D. Boyes, *Microsc. Microanal.*, 6 (2000) 307.
2. J. Goldstein, D. E. Newbury, D. C. Joy, et al., *Scanning Electron Microscopy and X-Ray Microanalysis*, 3rd edition, Kluwer Academic/Plenum Publishers, New York (2003).
3. J. Y. Liu, *Microsc. Microanal.*, 8(Suppl. 2) (2002).
4. D. C. Joy, *Microsc. Microanal.*, 8(Suppl. 2) (2002).
5. J. Goldstein, D. E. Newbury, D. C. Joy, et al., *Scanning Electron Microscopy and X-ray Microanalysis*, 3rd edition, Kluwer Academic/Plenum Publishers, New York (2003). Supplemental CD.
6. D. C. Joy and D. Newbury, Low voltage scanning electron microscopy, *Microsc. Today*, 10(2) (2002).
7. I. Müllerová and L. Frank, Contrast mechanisms in the scanning low energy electron microscopy, *Microsc. Microanal.*, 9(Suppl. 3), Microscopy Society of America (2003).
8. W. C. Oliver, Low voltage scanning electron microscopy, *Microsc. Today*, 10(3) (2002).

5
E-Beam Nanolithography Integrated with Scanning Electron Microscope

Joe Nabity, Lesely Anglin Campbell, Mo Zhu, and Weilie Zhou

1. Introduction

1.1. Basics of Microscope-Based Lithography

Electron beams have been used for lithography for decades [1,2] and a lithography system can easily be added to nearly all modern electron or ion microscopes, including scanning electron microscope (SEM), scanning transmission electron microscope (STEM), focused ion beam (FIB), and dual SEM/FIB microscope models. Nearly every microscope model will have inputs for external control of the XY beam position and most microscopes will have options for adding a fast beam blanker, which is an optional accessory for lithography. In most cases, the standard microscope stage will be used for lithography applications, and most stages can be controlled through a digital interface. Microscopes with a digital interface will also typically allow external control of column parameters, such as magnification, beam current, and focus as shown in Fig. 5.1.

A very important point is that adding a lithography system does not degrade or limit the functionality or performance of the microscope for imaging applications, because no customization of the microscope is typically required. Consequently, electron microscopes can become very versatile tools for micro and nano fabrication, since the same microscope used for the fabrication can also be used to view the resulting structures.

An SEM is the most common type of microscope used for lithography; however, nearly any system that allows external XY control of a point exposure can be used. Work with scanning tunneling microscope (STM) and atomic force microscope (AFM) lithography has been done [3,4]; however, these microscopes have not been widely used for lithography. In recent years, dual electron and ion beam microscopes have started to become more common for lithography, because a single lithography system can control either the e-beam or the ion beam, thus providing more fabrication capabilities than a single beam system.

120

1.1.1. Electron Source

There are two classifications for the sources in electron microscopes: (1) conventional sources use either tungsten hairpin filaments or lanthanum hexaboride (LaB$_6$) single crystal tips; and (2) field emission (FE) sources use either a cold cathode or a Schottky emitter (see Chapter 1). The latter is also known as a thermal FE source. Because of the lower cost, microscopes with conventional sources have traditionally been the models most widely used for lithography; however, microscopes with thermal FE sources provide both better imaging and better lithography.

When selecting between the two types of conventional sources for lithography, the main considerations are cost, convenience, brightness, and stability. The tungsten filaments have a lifetime of typically 40–200 h, while LaB$_6$ sources typically last significantly longer. Although the replacement cost for a tungsten filament is less than a LaB$_6$ source, the overall cost remains more or less the same. As LaB$_6$ source will have ~3 to 10× more current than a tungsten filament in the same spot size, lithography can be written faster; however, the stability of a LaB$_6$ source is ~3% per hour compared to ~1% per hour or better for tungsten. The ultimate lithography linewidths are basically the same for these sources; but a microscope with a LaB$_6$ source is easier to optimize because the higher brightness allows a higher beam current to be used, while a tungsten source will be more stable.

When selecting a cold cathode FE source vs. a thermal FE source for lithography as the primary application, the choice is simple. A thermal FE source will typically have a stability of ~1% over 3–10 h, while a cold FE source is inherently unstable and can change ±5% in minutes to ±20% or more per hour. Also, a cold FE source requires "flashing" periodically and the beam requires 1–2 h before becoming relatively stable and then typically becomes increasingly unstable as the vacuum in the gun degrades. While the imaging resolution of a cold FE source may be better than a thermal FE source, the instability of a cold FE source is a significant disadvantage for lithography applications. Even so, it can still be worthwhile to use a cold FE SEM for lithography when no other microscope is available. Table 5.1 shows the properties for different SEM electron sources.

1.1.2. Finest Linewidths

The finest linewidths achieved using conventional processing techniques with e-beam lithography typically range from ~10 to ~100 nm, where the microscope is the primary factor that determines the performance. The smaller linewidths are

TABLE 5.1. Comparing properties for common SEM electron sources

Property	Tungsten	LaB$_6$	Cold FE	Thermal FE
Source lifetime	40–300 h	1,000 h to 12 months	4–5 years	1–2 years
Imaging resolution (nm)	3.0–3.5	2.0–2.5	1.0	1.0–1.5
Max. probe current	0.1–10 uA	10 uA	2–10 nA	10–200 nA
Drift (%/h)	~1–2%	2–4%	~5–20%	0.05–1%

commonly produced with a 30 kV FE SEM, a 40 kV tungsten or LaB_6 SEM, or a ≥ 100 STEM, while low performance, low-cost SEMs may have a minimum feature size of 50–100 nm.

Besides the microscope performance, the main factors that determine the ultimate resolution are the choice of resist and substrate, accelerating voltage and beam current, writing field size, and the user's optimization of the microscope. These topics will be discussed in detail in the following sections.

1.1.3. SEM vs. Beam Writer

The typical SEM-based lithography system has many advantages over a dedicated electron beam writing system when research applications are the primary use. The advantages include cost, ease of use, maintenance, and versatility. A typical SEM can produce an accelerating voltage from ~200 eV to 30 kV, and can easily be changed as needed for different applications. Most microscopes can be successfully run by most graduate students after a reasonably short training period, and microscope service contracts generally keep a microscope running well with very little downtime. Commercial beam writers have the advantage when high volume and/or large area applications are required, since they have been specifically designed for such applications.

1.2. SEM Lithography System Considerations

1.2.1. Vector or Raster Writing

During normal image acquisition by an SEM, the beam is rastered from top to bottom of the full image area, where each line in the image is scanned from left to right. It is possible to do lithography with a similar raster if the beam can be blanked as needed as it scans across each raster line. However, the typical microscope-based lithography system uses a vector writing approach, where the beam moves in any direction and scans only the areas to be exposed. In a fully implemented vector writing system, the beam scan direction for sloped lines and circular arcs are along the line or arc and filled areas are not limited to simple XY scanning. In addition, for maximum flexibility, a microscope-based lithography system may provide two independent exposure point spacing parameters, where one will be along the line or arc that is being written, while the second will be in the perpendicular direction.

For a microscope-based lithography system, using a vector writing mode greatly increases the overall writing speed, since the exposed areas only need to be scanned. Using a vector writing mode can significantly reduce the demands on the beam blanker, since only two beam-on/beam-off events are needed for each pattern element. In contrast, a raster writing mode requires very precise blanking, especially when writing narrow lines perpendicular to the raster line scan direction.

A unique capability of the vector writing mode is that pattern writing can be accomplished even when the microscope does not have any blanker at all. This is possible, since the lithography system can jump the beam quickly enough between pattern elements so that only an insignificant dose is applied to the path of the beam during the jump. Writing without a beam blanker is discussed in more detail below.

1.2.2. Writing Speed

Fundamentally, the overall writing speed for any direct write system depends on the beam current, sensitivity of the resist, and the maximum speed at which the beam can be moved across the exposure area. For most SEM systems, the beam current can be varied from ~10 pA to ~10 nA or more, and the current to be used must be selected based on the characteristics of the microscope. For example, some SEMs may write 50 nm lines with 1 nA of current while others may need to be run with less than 50 pA to achieve the same linewidths.

Most commercial microscope-based lithography systems have a maximum step rate for the beam of 3 MHz or higher. However, the typical microscope scan circuitry may be limited to a lower frequency. In general, if the lithography system is faster than the microscope scan coils, the ultimate writing speed will be limited by the materials, exposure conditions, and microscope, not the lithography system. In cases where pattern distortions are caused by the scan coils not "keeping up" with the lithography system, the solution is to reduce the beam current so that a slower writing speed can be used to provide the desired dose.

1.2.3. CAD Interface

A well-designed microscope-based lithography system will provide a powerful CAD program for pattern design. Also, when the GDSII and CIF formats are not the native format of the lithography system, support for importing patterns from these file formats will often be included, since these formats are standard for dedicated e-beam writing systems. In addition, microscope-based lithography systems will often include support for file exchange with DXF and DWG file formats, which are the formats used by AutoCAD and other general purpose CAD programs.

For the greatest flexibility, a microscope-based lithography system will allow ASCII pattern files to be created by any means. Some systems include a programming language that allows the users to write custom programs to automate any complex patterning requirements. For example, if a research application requires a structure to have a shape that is defined by a mathematical function, such as a logarithmic spiral, a programming language built into the CAD program can make it easy to create a custom function that produces the desired pattern in response to the parameters entered by the user.

1.2.4. Alignment

An advanced microscope-based lithography system provides both manual and fully automated alignment. The alignment is performed by imaging selected areas within the writing field and then registering the lithography coordinate system to marks on the sample. Generally, a 2×2 transformation matrix and XY offsets are calculated based on the alignment results. Once calculated, these parameters are used to transform the exposure, so that the exposed pattern elements will be registered to the marks on the sample. Typical alignment accuracies range from 1:1,000 of the writing field to ~1:5,000, with accuracies down to ~20 nm being possible.

When a microscope-based lithography system includes a robust automated alignment feature, the system can use a standard automated microscope stage to get close to the desired location and then automatically align to much higher precision by scanning the registration marks. This allows "step-and-repeat" exposures to be processed with fully automated alignment at each field for tens, hundreds, or even thousands of fields, while using only the standard automated stage found on most modern SEMs.

1.3. SEM Connections

As stated earlier, almost any SEM, STEM, or FIB system can be used for lithography. The basic necessity is that the microscope must have analog inputs for external control of the beam position, where the typical input voltage range will be from ~±5 to ±10 V. Options for microscope-based lithography include an image signal output, a fast beam blanker, automated stage control, and a digital interface to the microscope.

Table 5.2 shows some of the more common electron microscope models that have been used for lithography. Other less common brands that have not been listed include Amray, Camscan, ISI, and Topcon. All models shown have the XY interface as a standard feature or available as an option.

1.3.1. Required: XY Interface and Beam Current Reading

Most modern microscopes have the required XY inputs either as standard or available as an extra cost option. For microscopes that do not have XY inputs, they can usually be added, if the schematics for the microscope are available. In this case, the basic procedure is to add relays to select between the internal scan generator and the external lithography system.

The only other required connection is for reading the beam current. Most microscopes have a single electrical connection to the specimen holder, which can be used to read the current that hits the sample. A Faraday cup can easily be made on most sample holders by drilling a blind hole ~2 mm wide, ~2 mm deep and covering it with a ~3 mm aperture with a 10–100 μm diameter hole. The aperture can be an inexpensive copper aperture or even a used SEM aperture.

A Faraday cup that inserts into the beam path either directly above the sample or higher in the column can be used, but is not required. An advantage of such a mechanism is that the current can be measured without moving the sample; however, care must be taken to ensure that the measured beam current in the cup is the same as the current measured at the sample. For example, a Faraday cup in the column may collect the primary beam and a significant current from stray electrons. In such a case, the excess current collected may be many times higher than the current measured at the sample.

TABLE 5.2. Comparing different models of SEMs

Brand	Models	Source	Blanker	Auto-Stage	Digital Interface
Cambridge	100—300 Series	W, LaB_6	Option or third party	Option or third party	None or serial
FEI/Philips	XL30 LaB_6	LaB_6	Option or third Party	Standard	Serial
FEI/Philips	XL30 FEG, SFEG, ESEM FEG, Sirion	tFE	Option^ or third party	Standard	Serial, Ethernet
FEI	Quanta	W	Option^ or third party	Standard	Ethernet
FEI	Quanta FEG, NanoSEM	tFE	Option^ or third party	Standard	Ethernet
FEI	Nova Nanolab	tFE/Ion	Option^ or third party	Standard	Ethernet
Hitachi	2000 and 3000 Series	W	Option or third party	Option or third party	None
Hitachi	4000 Series (not 4300SE)	cFE	Factory install or third party	Option	Serial
Hitachi	4300SE	tFE	Factory install	Option	Serial
JEOL	840,6300 and 6400	W,LaB_6	Option or third party	Option or third party	None serial
JEOL	5900, 6060, 6360, 6460, 6380, 6480	W,LaB_6	Third party	Option or standard	Ethernet
JEOL	6500F, 7000F	tFE	Third party	Standard	Ethernet
JEOL	6700F, 7400F	cFE	Third party	Standard	Ethernet
Leica/LEO	440, 1400	W,LaB_6	Option	Option	Serial
Tescan	Vega	W,LaB_6	Option	Option	Ethernet
Zeiss	Supra, Ultra,	tFE	Factory install^	Option	Serial
LEO	1500 Series				
Zeiss	EVO	W,LaB_6	Option	Option	Serial
LEO	440, 1400 Series				
Zeiss	1540 Crossbeam	tFE/Ion	Factory install^	Option	Serial

Notes:

1. In general, for older models the more expensive models within a series will give better lithography performance. For newer series, the beam quality is often virtually identical within a series, while other features will determine which models cost more. The microscope companies should always be contacted regarding details on current models, since features and options may change at any time.
2. Bold brand names are currently in business.
3. Usually, a tungsten (W) model may be upgraded to LaB_6. Once upgraded, either W or LaB_6 may be used.
4. "tFE" means thermal field emission. "cFE" means cold cathode field emission.
5. "Option" indicates that the microscope manufacturer has their own blanker or stage that is available as an option on a new microscope and may be available for retrofit of an existing microscope. For most new microscopes, an automated stage will be in the standard configuration, while an electrostatic blanker is rarely included as a standard feature. "^" indicates that the model includes a slow scan coil blanker as a standard feature, however, a fast electrostatic blanker is recommended for lithography use.
6. "Third party" indicates that a company other than the microscope manufacturer offers a compatible blanker or stage. In many cases, a third party blanker or stage will be offered as part of a new microscope purchase. For stages, third party automation packages are almost always available.
7. "Factory install" means that the blanker must be installed at the factory, which makes retrofitting the blanker to an existing microscope very expensive.

1.3.2. Optional: Image Signal

Nearly every microscope will have an image signal output that can be used by the lithography system for image acquisition. While it is not required to have an image signal for basic pattern writing, the image signal is required when the lithography system is used to align to existing marks on the sample. When an old analog microscope is used for lithography, the lithography system can often be used to acquire digital images. This capability will improve the functionality of an older microscope for imaging, in addition to allowing it to be used for lithography.

1.3.3. Optional: Beam Blanking

A microscope will ideally have a beam blanker that has a fast repetition rate, fast rise/fall times, and minimal on/off propagation delays. For most electron microscopes, such blankers are available either from the microscope manufacturer or from third-party vendors. Typical parameters for electrostatic blankers used for e-beam lithography are: repetition rate >1 MHz, rise/fall times <50 ns, and propagation delays <100 ns; however, slower blankers can still be useful. In general, an electrostatic blanker will be completely independent from the microscope user interface and be controlled directly by the lithography system with a TTL compatible on/off voltage.

While a fast beam blanker is certainly desirable for lithography, it is not required to have any blanker at all when using a vector writing system. When no blanker is available, the lithography system can jump the beam between pattern elements fast enough that a negligible dose is received along the path of the beam between the pattern elements. However, when no blanker is used, there are two main issues. One is that between pattern locations the beam will always be hitting the sample. Consequently, care must be taken to consider where the beam is hitting while the stage is being moved. The other is that when the beam is jumped between pattern elements, the scan coils will take some time to settle to the correct position after a long jump. This can result in distortions in the patterns at the starting point of each pattern element, if the beam has jumped a significant distance. In most SEMs, little to no distortion will be seen after a jump of 3–10 μm, but a significant distortion may be observed for longer jumps. A well-designed lithography system can allow the user to minimize the distortions by defining locations where the beam can settle, so that the lengths of the jumps to the desired pattern elements can be minimized. The settle areas will get a dose, however, a flexible lithography system will allow the user to select where the locations are, so that the functionality of the pattern being written will not be degraded.

When a microscope only has a slow beam blanker, the slow blanker can be used during stage moves between exposure fields. Many SEMs have a gun coil blanker that is a standard feature, which can be used this way if the microscope allows external control of the magnetic blanker.

1.3.4. Optional: Automated Stage

In general, having an automated stage for microscope-based lithography is not required. However, for most modern electron microscopes, the standard microscope stage will be automated and can be controlled through a digital interface, such as a serial or Ethernet connection. The advantages of using the standard microscope stage are that there is no additional expense and the versatility and functionality of the microscope are not degraded. Typically, a standard microscope stage will provide an absolute positioning accuracy of a few microns and an "over and back" repeatability of ~0.5–0.1 µm. When coupled with the alignment capability in a well-designed lithography system, a standard SEM stage can get close enough to the desired location for the lithography system to align accurately to marks on the sample.

When patterns must be "stitched" together without any imaging of registration marks in each field, a very accurate stage is required. In that case, most newer microscopes can be retrofitted with stages that use laser interferometer feedback to provide higher accuracy stage positioning. The disadvantages of such stages are that the cost is significant, i.e., around one half the cost of a new FE SEM and more than many W or LaB_6 models, and rotation, tilt, and often height adjustment will not be available.

1.3.5. Optional: Digital Interface

All newer SEM, STEM, and dual beam microscopes will have a digital interface for external control of the microscope parameters. The interface will typically be based on RS232 serial or Ethernet. A well-designed lithography system can take advantage of the digital interface and provide automated control of column parameters, such as magnification, focus, and beam current. Advanced control of microscope accelerating voltage can be possible, however, changing the kV typically requires the user to reoptimize the microscope, thus limiting the usefulness for fully automated control.

A well-designed microscope-based lithography system will also allow any executable to be automatically run at controlled points during the lithography processing, thus giving the user the ability to automate any function of the microscope that can be remotely controlled. For example, automated control of a gas handling system on a FIB microscope could be incorporated into a novel lithography application.

2. Materials and Processing Preparation

2.1. Substrates

Electron beam lithography is widely used to generate submicron or nanoscale structures and the choice of substrate is determined by the application. Normally, any solid substrate can be used with electron beam lithography, including semiconductors (e.g., silicon, Ge, and GaAs), metals (e.g., Au, Al, and Ti), and insulators (e.g., SiO_2, PSG, and Si_3N_4). These materials can either be the substrates themselves or additive thin films on the substrates.

The electron beam lithography technique is most often employed to fabricate electronic or electro-related devices and structures, so silicon is currently the dominant substrate material for fabrication due to its inherent features: (a) well-characterized and readily available; (b) multitude of mature processing techniques available; and (c) intrinsic properties for electrical and electronic applications [5].

When electron beam lithography is performed on an insulating substrate, substrate charging may generate distortion overlay errors [6]. In addition, resist charging may prevent SEM inspection [7]. A simple solution to avoid pattern distortions from charging at high energies (~30 kV) is to deposit a thin layer of metal, such as gold, chrome, or aluminum, on the top of the resist. Electrons travel through the metal layer and expose the resist with reduced scattering [8]. After exposure, the metal layer is removed with the appropriate etchant before the development of the resist. A second method to provide charge dissipation is to coat a layer of conducting polymer under or over the resist [8–10]. Another approach is to apply a plasma process to the resist to increase its electrical conductivity by surface graphitization [11]. The advantage of this approach is its compatibility with industrial processes compared to the first two methods.

2.2. Resists

Electron beam resists are sensitive and are able to be developed by certain developers after exposure. The resists may produce either a positive or negative image compared to the exposed areas. Similar to photoresist, electron beam resists play two primary roles in lithography: (a) precise pattern transfer and (b) formation and protection of the covered substrate from etching or ion implantation [12]. The resists will normally be removed with the completion of these functions. However, in some cases the resists are also employed as a part of device and structure. Important properties of electron beam resists include resolution, sensitivity, etch resistance, and thermal stability [12], which are introduced in this section.

2.2.1. Resolution and Intrinsic Properties

In the process of electron beam lithography, the electrons will travel through the resist and lose energy by atomic collisions which are known as scattering. But, some of the electrons will be scattered back into the resist from the substrate. This phenomenon is known as backscattering. Scattering and backscattering will broaden lines scanned by the electron beam and will contribute to the total dose experienced by the resist.

The effects of scattering will vary with the electron beam energy. At low energies, electrons scatter readily, but travel only small distances after scattering. With higher energies, the scattering rate is lower, while the range after scattering increases. Therefore, two approaches to high resolution can be obtained by high energy with a relatively low applied dosage or low energy with a relatively high applied dosage [13]. In either case, a very tightly focused beam is necessary to obtain small feature sizes.

Another issue for some electron beam lithography applications is the writing speed, which is primarily determined by the sensitivity of the resist and the magnitude of the current used during the writing. The highest resolution resists are usually the least sensitive [8].

The ultimate resolution of the resist is not set by electron scattering [14], but a combination of (a) the delocalization of the exposure process as determined by the range of the Coulomb interaction between the electrons and the resist molecules [15], (b) the straggling of secondary electrons into the resist [16], (c) the molecular structure of the resist, (d) the molecular dynamics of the development process (the tendency of the resists to swell in the developers), and (e) various aberrations in the electron optics. The resolution of resists is also influenced by the proximity effect, which is contributed by the adjacent features during the exposure. In some cases, this effect can significantly degrade exposure patterns which include a large number of closely spaced fine features, or small features placed near larger ones.

Mechanical and chemical properties such as etch resistance, thermal stability, adhesion, solid content, and viscosity are important to pattern transfer. Among these properties, the etch resistance is the most important when using the resist as an etch mask. Etch resistance specifies the ability of a resist to endure the etching procedure during the pattern transfer process. Another important property thermal stability meets the requirement for some specific process like dry etching [13]. As discussed before, e-beam resists are deposited on a variety of substrates including semiconductors (Si, Ge, GaAs), metals (Al, W, Ti), and insulators (SiO_2, Si_3N_4). Good adhesion is necessary to obtain good pattern transfer. Various techniques are used to increase the adhesion between resist and substrate including dehydration bakes before coating [12]: adhesion promoters such as hexamethyl-di-silazane (HMDS) and trimethylsilyldiethylamine (TMSDEA), vapor priming systems, and elevated temperature postbake cycles.

2.2.2. Positive Resists

For positive electron beam resists, the pattern exposed by the electron beam will be removed during development. These resists are usually high molecular-weight polymers in a liquid solvent, and undergo bond breakage or chain scission when exposed to electron bombardment. As a result, patterns of the positive resist exposed become more soluble in the developer solution.

Polymethyl methacrylate (PMMA) is one of the first resists developed for electron beam lithography and still the most commonly used low-cost positive electron beam resist. PMMA has high molecular weight, 50,000–2.2 million molecular weight (MW) (Nano PMMA and Copolymer, PMMA Resist Data Sheet, MicroChem Corp.), and exists in powder form dissolving in chlorobenzene, or the safer solvent anisole. The thickness of the baked resist can be controlled by the coating speed and solid concentration. For example, PMMA (950,000 MW) with 3% concentration in chlorobenzene spun at 4,000 rpm will yield a thickness around 0.3 μm. The specific data for individual resists can be found in their material data sheets regarding different molecular weights, spin speed, concentration, and so on. The exposure dose for a typical electron acceleration voltage of 30 kV is in the range of 50–500 μC/cm^2 depending on the radiation source/equipment, developer, developing time, and pattern density. A common developer is a mixture of methyl isobutyl ketone (MIBK) and isopropyl alcohol (IPA). MIBK/IPA (1:3) is used for the highest resolution and MIBK/IPA (1:1) for the highest sensitivity. The developing time is typically 10–90 s depending on the applications.

The exposure process generates a natural undercut profile in the resists yielding a good geometry for the lift-off technique. More pronounced lift-off geometries may be achieved by using PMMA bilayers composed of two layer of PMMA with different molecular weight. The additional underlying layer of lower molecular weight PMMA requires a lower electron dosage for dissolution and thus gives more undercut. Even more extensive undercut may be achieved by replacing the lower PMMA with a layer of copolymer methyl methacrylate, P(MMA-MAA), or polydimethylglutarimide (PMGI), which are more sensitive to electron dosage. With very high electron dosages, PMMA will be cross-linked and insoluble in acetone, yielding a negative exposure process with acetone as the developer [17].

The ultimate resolution of PMMA has been demonstrated to be less than 10 nm [18]. However, PMMA has relatively poor sensitivity, poor dry etch resistance, and moderate thermal stability [19]. The assistance of the copolymer (PMMA-MAA) results in the better sensitivity image and thermal stability.

Another example of chain-scission resists is poly(1-butene sulfone) (PBS). PBS is a common positive resist for mask making due to its high sensitivity, around 3 μC/cm^2 at 10 kV [20]. However, PBS has poor etch resistance and needs tight control of processing temperature and humidity [21]. Poly(2,2,2-trifluoroethyl-α-chloroacrylate) (EBR-9) [22] is also used for mask making due to its long shelf life, lack of swelling in developer, and large process latitude [8].

ZEP is a new chain-scission positive resist developed on poly (methyl-α-chloroacrylate-co-α-methylstyrene) by Nippon Zeon Co. ZEP provides a high resolution and contrast comparable to PMMA but relatively low dose (8 μC/cm^2 at 10 kV). ZEP also has better etch resistance than PMMA.

The chemically amplified (CA) [23] resists have been developed recently for their high resolution, high sensitivity, high contrast, and good etch resistance [24–28]. Unlike the chain-scission process for PMMA during exposure, positive-tone CA resists are usually functionalized by the acid-catalyzed cleavage of labile blocking groups which protect the acidic functionalities of an inherently base-soluble polymer. Acid is generated in the exposed regions of the resist by radiation-sensitive photoacid generators (PAGs). As an example, a reported CA resist KRS containing a partially ketal-protected poly(p-hydroxystyrene) (PHS) has resolution less than 100 nm, sensitivity 12 μC/cm^2 at 50 kV, and contrast larger than 10 [20].

2.2.3. Negative Resists

Negative electron beam resists will form the reverse pattern as compared to positive resists. Polymer-based negative resists generate bonds or cross-links between polymer chains. The unexposed resists is dissolved during development while exposed resists remain, thus the negative image is formed.

Microposit SAL601 is a commonly used negative-tone chemically amplified electron beam resist with high sensitivity (6–9 μC/cm^2), good resolution (less than 0.1 μm), high contrast, and moderate dry etch selectivity (http://snf.stanford.edu/Process/Lithography/ebeamres.html). The main drawbacks of SAL601 are scumming and bridging between features, particularly in dense patterns, poor adhesion, and a very short shelf life.

An epoxy copolymer of glycidyl methacrylate and ethyl acrylate (COP) is another frequently used negative resist. COP shows very high sensitivities, 0.3 μC/cm^2 at 10 kV [8], because only one cross-link per molecule is sufficient to insolubilize the material. Although COP shows good thermal stability, its resolution is relatively poor, 1 μm, due to strong effect from the solvent swelling of the cross-linked region; and its plasma-etching resistance is also poor [29].

NEB-31 is a relatively new resist for electron beam lithography from Sumitomo Chemical, Inc. NEB-31 exhibits high resolution, 28 nm structures, high contrast, good thermal stability, good dry etch resistance, and long shelf-life [30].

Hydrogen silsesquioxane (HSQ) is a spin-on dielectric material for inter-metal dielectrics and shallow trench isolation in integrated circuit fabrication. Upon electron irradiation, inorganic 3D HSQ undergoes cross-linking via Si–H bond scission. This cross-linking results in an amorphous structure in HSQ similar to SiO$_2$ which is relatively insoluble in alkaline hydroxide developers [31]. HSQ has been employed as a negative resist for electron beam lithography, nanoimprint, and extreme ultraviolet (EUV) lithography [32–35]. As a competitive resist, HSQ has demonstrated high resolution, high contrast, moderate sensitivity, minimum line edge roughness, good etch resistance, high degree of mechanical stability [36,37], etc. A linewidth of about 7 nm with an aspect ratio of 10 has been reported for HSQ with 100 kV electron beam lithography. Dow Corning Corporation provides a series of commercial HSQ named as FOx®-1x and FOx®-2x [38].

Calixarene derivative is another example of high resolution negative resist. Although the dosage required is relatively high, ~20× as that of PMMA, calixarene derivatives are able to generate under 10 nm structures with little side roughness and high durability to halide plasma etching [39]. Calixarene also works with low energy: an optimal resolution (10 nm) at 2 keV is obtained with a reduced electron dose compared to high energy exposure [40].

SU-8 is a chemically amplified epoxy-based negative resist. Due to its good chemical and mechanical properties, SU-8 is normally employed to fabricate high aspect ratio 3D structures and serve as a permanent part of the device with the LIGA (a German acronym that stands for deep-etch x-ray lithography, electroplating, and molding) technique. As an electron beam resist, SU-8 is able to generate sub 50 nm structures with 0.03 nC/cm electron dose [41].

Most photoresists can be exposed by electron beam, although the chemistry is quite different from that of UV exposure [8]. Shipley UV-5 (positive) and UVN-2 (negative) are popular choices for their good resolution and excellent tech resistance.

2.2.4. Other Resists

In addition to the resists mentioned above, a number of metal halide resists are employed to achieve extremely high resolution, including LiF, AlF_2, MgF_2, FeF_2, CoF_2, SrF_2, BaF_2, KCl, and NaCl. For example, LiF_2 has demonstrated a high resolution of under 10 nm with line dose of 200–800 nC/cm [42].

A number of nanoscale structures have been made using the carbonaceous or silicaceous contaminants as the electron beam resist. The contaminations can be deposited on the surface due to (a) oil in the vacuum pumps, (b) organic residue on the sample surface, or (c) delivery to the point of impact of the beam via a capillary needle. With the third method, direct delivery of vapor via a capillary, it is possible to maintain a constant writing rate [43]. Under all deposition conditions, the required electron dose is very high, 0.1–1 C/cm^2. The contamination can be easily cleaned by heating the substrate to about 100°C [8].

Self-assembled monolayers (SAMs) are also used as the resists in electron beam lithography to obtain a high resolution. Since the thickness of the SAM is in the order of 1–2 nm, the scattering effects are negligible in the SAM resists. The molecules composing the SAM can be divided into three different functional parts: a head group that strongly binds to a substrate, a tail group that forms the outer surface of the monolayer, and a spacer linking head and tail [44]. However, the SAM resists demonstrate poor wet or dry etch resistance [45].

2.3. *Spin Coating*

The electron beam resists are normally applied to the substrates by the spin coating technique. A typical spin coating procedure typically involves (a) a dispense step (static or dynamic), and (b) spinning the substrate at high speed. Static dispense is to deposit a small amount of resists on the center of the substrate while stationary, and dynamic dispense is to apply the resists to the rotating substrate at low speed. Static dispense is simple, while dynamic dispense is more effective. After the dispense step, the substrates are accelerated to the final spin speed quickly. A high ramping rate generates better film uniformities than a low one [12]. Given a certain resist and substrate, the resist thickness after spin coating is determined by the spinning parameters including spin speed, spin time, etc. A separate drying step is generally needed after spin coating to further dry the film and improve the adhesion.

3. Pattern Generation

The complete pattern generation process extends from the initial concept for the pattern to the actual writing with the microscope. While the pattern design is relatively straightforward using a CAD interface to define the shapes, sizes, and positions of the pattern elements, there are certain limitations of the writing system that should be considered during the layout. In addition to the design, there are a number of microscope parameters that need to be adjusted in order to properly configure the system such as working distance, accelerating voltage, and beam current or spot size. The final step before writing is to set up the microscope correctly by proper optimization and sample placement.

3.1. Design Guidelines

Using a design program, most commonly a CAD type, any shape can be constructed and interfaced with the writing software. As previously discussed, there are some constraints to the pattern design, such as the size of the writing field and the features themselves. It is also important to consider how the beam is actually exposing the resist and the limits that imposes. Another consideration while designing a pattern is the density of the features. As the beam exposes a region, secondary electrons can partially expose surrounding areas. This phenomenon is known as the proximity effect and is an important factor when writing high density patterns. A high density dot array, for example, will require a lower exposure dose than writing an individual dot or a low density array.

3.1.1. Field Size

Field size is one of the most important restrictions to pattern design, where the size of the writing field is determined by the magnification of the microscope. Fine features can usually be achieved with a writing field size of 50×50 μm^2 to 200×200 μm^2, depending on the type of microscope, where each microscope model will typically have certain magnification values that provide the best signal-to-noise ratio within the electronics of the microscope.

If a large area of fine features is desired, several fields can be positioned so that the field edges align with each other, which is commonly known as stitching. Unless care is taken, the edges of the pattern will mismatch due to irregularities in the stage movement. However, using alignment procedures as outlined in Section 1.2.4, these irregularities can be minimized to as little as 20 nm. Without the alignment, the mismatch can be as great as several microns. Depending on the purpose of the pattern this is often an acceptable amount of error.

3.1.2. Feature Size

Another important restriction to pattern design is the size of the features, which is limited by the resolution of the resist and the optimization of the microscope. Proper microscope optimization and a high resolution resist can routinely yield 50 nm features with most W or LaB$_6$ SEM models, while 20 nm and smaller are routinely produced with thermal FE SEMs.

3.1.3. Point Spacing

To obtain the best results from any lithography system, especially for demanding research applications, it is important to understand how the beam is moved to produce the pattern. To produce the finest lines, a single pass of the beam is typically used, which is composed of adjacent exposure points with a defined center-to-center distance. In this case, the center-to-center distance must be set to produce adequate overlap of the adjacent exposure points, which will typically be one fourth to half of the final linewidth. For wide lines or filled areas, a separately defined line spacing parameter will be used to control the spacing between adjacent passes of the beam.

In a sophisticated lithography system, the user will be able to independently adjust the center-to-center and line spacing parameters, while a more limited system will provide only a single parameter for both. Figure 5.1 shows a schematic of center-to-center distance and line spacing. An application that makes use of independent spacing is when the exposure points are intentionally defined to produce isolated dots in a rectangular array. Using this approach, the dots in the array will be the smallest dots that can be achieved given the system configuration and microscope setup.

3.2. System Configuration

In addition to the design of the pattern, there are system settings that need to be adjusted based on the application. Working distance can have an effect on many aspects of the writing process, such as spot size, interference, and magnification settings. Also, while considering design configurations and material properties, the accelerating voltage and current will be adjusted to define the beam for the intended application.

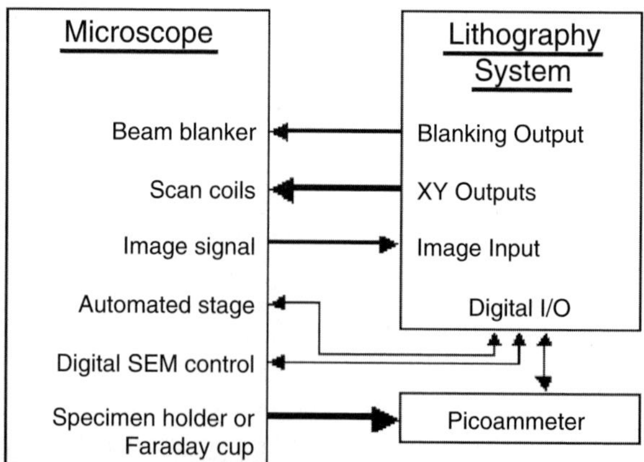

FIGURE 5.1. Bold arrows show required connections, medium arrows show typical connections, and thin arrows show optional connections.

The type of substrate and resist used will have an effect on the optimal exposure dose. For each type and molecular weight of a resist, the typical required dose will vary as shown in Table 5.1. A high Z substrate will have more backscattered and secondary electrons near the surface which will contribute to the overall dose affecting the resist.

3.2.1. Working Distance

The working distance influences the minimum spot size that is achievable; the susceptibility to external interference, and on some microscopes, the optimal magnification setting that should be used. A shorter working distance will improve the resolution of the microscope and will reduce the susceptibility of the beam to external interference. In cases where the microscope environment has magnetic fields, using a shorter working distance may have a dramatic effect on the writing quality. For most microscopes, a working distance between 5 and 10 mm is appropriate for writing fine features.

The working distance can also affect the optimal magnification settings. In all SEMs, the total magnification range will be divided into smaller ranges, where each range will use a different circuit within the microscope scan control electronics. When the scan control electronics change from one subrange to another, it will generally produce a small temporary image distortion and/or an audible click from a mechanical relay. The best signal-to-noise within the scan control electronics will be found at the higher magnification value after a transition. In some microscope models, the magnification values where the transitions occur depend both on the working distance and the accelerating voltage of the beam. For these models, it is often beneficial to use the same working distance for all fine lithography, so that the magnification value of the transition and the optimal magnification will not change.

3.2.2. Accelerating Voltage

Increasing the accelerating voltage will produce a greater electron penetration depth, thus lowering the number of scattered electrons in the resist, which will result in finer linewidths. The accelerating voltage can also affect the magnification settings in conjunction with the working distance previously mentioned. Most SEMs will have a maximum accelerating voltage of 30 kV, which will be the voltage used for most fine writing. When using an STEM, and accelerating voltage of 100 kV to ~300 kV may be used; however, these models typically allow only a very limited sample size, which reduces their versatility for lithography.

3.2.3. Beam Current (Spot Size)

A lower beam current will produce a smaller spot size on the sample than a larger beam current. The smallest spot size can be achieved by using the highest available accelerating voltage and a reasonably low beam current. The typical range of beam current used for fine lithography is 5–50 pA, where the optimal value will vary depending on the model of SEM as well as the filament type. Some of the common ranges are: 5–10 pA for a W filament, 10–20 pA for a LaB_6 filament, and 20–50 pA for a FE microscope. In general, the goal is to use a beam current that is small enough to produce the desired feature sizes, but is also large enough to make the microscope reasonably easy to optimize.

3.3. Microscope Setup

Microscope setup is the final step before writing begins and it consists of two parts: microscope optimization and sample positioning. Fine lithography requires that the microscope be carefully optimized to ensure that the beam has the most favorable settings. The positioning of the sample affects the exposure positioning and axis alignment.

3.3.1. Microscope Optimization

Out of the settings that need to be optimized, focus and stigmatism are the most difficult adjustments to make for novice users. Thus, it is important to be familiar with the microscope and how to properly optimize for normal imaging before attempting to write patterns, where the optimization of the microscope is typically the easiest when using a gold resolution standard. Beginning with a low magnification, the focus and astigmatism should be alternately adjusted until a clear image can be achieved at a field size of approximately 1×1 μm^2 or smaller. If a good image cannot be achieved, the adjustment technique and/or the microscope should be improved before attempting to write the smallest feature sizes. In addition, issues such as filament current, gun alignment, aperture centering (commonly referred to as "wobble"), and lens clearing (also known as hysteresis removal) should be addressed. Since these will depend on the specifics of the microscope model being optimized, it is beyond the scope of this chapter to provide a comprehensive procedure for all microscopes.

After the beam has been optimized on the gold resolution standard, the stage should be so that the surface of the resist is viewed on the edge of the sample away from the writing area. It is recommended to use the stage Z control to focus, since this will physically raise or lower the sample to the proper height without changing the settings of the microscope electronics. This technique removes the requirement that the gold resolution standard be at the same height as the writing surface. A final electronic focus adjustment can be done after the height has been adjusted, since by then only a small change should be necessary, which should not have any adverse effect on the beam optimization.

3.3.2. Sample Positioning

When first learning how to do lithography it is useful to write patterns near some obvious mark, such as a small, thin scratch made by a diamond scribe on the substrate. Placing the pattern near the end the scratch will make the pattern easier to locate after development. Controlling the precise placement of a pattern in reference to other features is described in Section 1.2.4. Also, care must be taken when the patterns are placed far away from the last optimized location because the beam will become defocused if the stage motion significantly changes the height of the sample.

The orientation of the sample relative to the writing axes will affect the positions of the patterns on the sample, thus it is important to align the writing axes with and edge of the sample. This can be achieved by changing the rotation of the stage, adjusting the scan rotation, and/or using positioning features of the lithography software. When moving from one side of the writing sample to another, the vertical position should change less than 1 μm for every millimeter. Without this adjustment, patterns will be written with correct relative position to each other but without proper placement in relation to the sample.

4. Pattern Processing

After a pattern has been exposed in the resist, the subsequent processing will transfer the pattern either to the substrate or to a layer added after the lithography. The processing that is used will depend on the material system and the final structure that is desired. The sections below describe the common processing, as well as common problems that may be encountered. As with any multi-step lithography process, a failure in a later step will always be more significant than a failure in an early step, so avoiding problems becomes increasingly important.

4.1. Developing

During exposure, molecular bonds in the resist are created or broken depending on whether the resist is negative or positive, respectively. In either case, the developing agent dissolves the more soluble areas to produce the desired pattern. If the resist is left in the developer for too long, the less soluble resist areas will also dissolve; however, it is easy to have reasonably consistent results when the development time is on the order of a minute. In addition, the development rate will depend on the developer temperature, so maintaining a controlled developer temperature is recommended. The developer chemicals and development timing will depend on the type of resist used, however, the overall procedure is basically the same. After exposure, the resist-coated substrate is covered with a developing agent for a certain amount of time, rinsed, and dried, typically by blowing with dry N_2.

4.2. Coating and Liftoff

The two common methods of coating are sputtering and evaporation. A significant difference between sputtering and evaporation is the control over the direction of the deposition material. In a typical sputtering system, the deposition is intended to strike the substrate with a wide range of incident angles, while evaporation is usually nearly collimated. This difference is significant because of the shape of the pattern cross section, as shown in Fig. 5.2. The shape is trapezoidal because of forward scattering of the incident beam and because some of the electrons scatter from the substrate and expose the bottom of the resist.

Liftoff is the process of removing the resist and the material that has been coated on top of the resist. The material that has adhered to the substrate will then be left, thus producing the desired pattern. The success of liftoff depends primarily on the adhesion of the coating to the substrate and whether or not the coating covers the sidewalls of the resist. Generally, an evaporated beam will be collimated enough to leave the sidewalls uncoated, while sputtering is much more likely to coat the sidewalls and make the liftoff step more difficult. For liftoff, it is important to know the thickness of the resist and to keep the coating thickness less than approximately two thirds of the resist thickness.

For PMMA, acetone is the solvent typically used during liftoff. The liftoff step with PMMA and acetone can be accomplished in many ways. In general, it is recommended to use the most gentle liftoff process that produces consistently good results. The simplest approach is to let the sample soak in room temperature acetone for about 20 min, or until it can be seen that the coating is floating off. Increasingly aggressive methods include using a squirt bottle, syringe, or ultrasonic cleaner to help remove the metal coating, or even to scrub the sample with a small brush. Heated acetone is sometimes used, however, this should only be done with the proper safety precautions.

4.2.1. Sputtering

In a plasma-magnetron sputtering system, the impact from ions from a plasma cause atoms from a metallic target to be ejected at all angles. The sputtered material coats the specimen which is placed below the target. The wide range of trajectories of the sputtered material is desirable for coating SEM specimens, since the metal will cover most surfaces and prevent charging during imaging. However, for lithography, it is generally undesirable to coat the sides of the resist walls that define the pattern, because this can cause ragged edges after liftoff or sections of the pattern may not lift off at all if the edges are coated with too thick of a layer. For larger patterns, having rough edges may not be a significant issue, however, having clean sidewalls becomes very important when doing liftoff with the very small features. Figure 5.3 shows a schematic representation of sputtering process and a cross section view of a pattern after sputtering has been done.

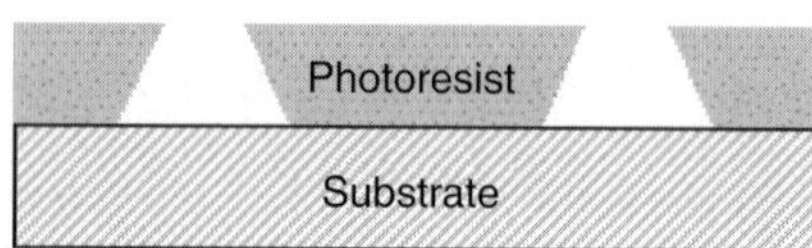

FIGURE 5.2. Schematic showing an undercut cross section of positive resist after exposure.

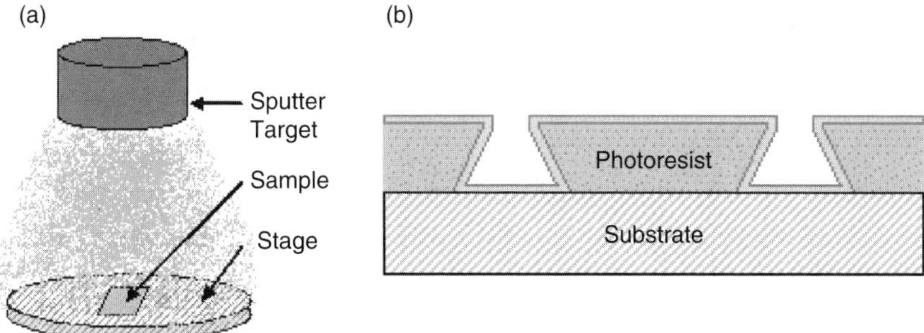

FIGURE 5.3. (a) Is a schematic representation of the sputtering process; and (b) shows a cross section view of a pattern after sputtering coating.

4.2.2. Evaporation

During thermal evaporation, the source material is placed in a boat or a filament coil, which is heated using an electrical current as shown in Fig. 5.4. Electron beam sputtering systems can also be used, which heat a localized spot in the source material using an electron beam. The sample can be placed either above or below the source material depending on the equipment being used. The evaporated material will be nearly collimated, so it is less likely to coat the sides of the pattern as compared to sputtering. This yields cleaner edges after removal of the resist, which is critical in high-resolution patterning. However, even with a collimated deposition, care must be taken to ensure that the deposited material hits at near normal incidence, otherwise the deposited material may coat some of

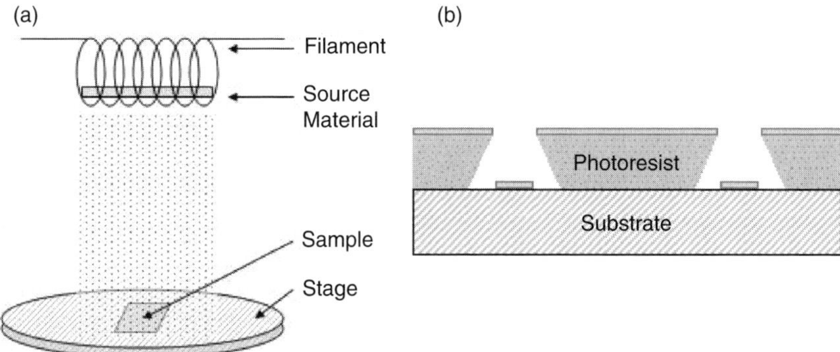

FIGURE 5.4. (a) Shows a schematic of a filament-style evaporation process; a boat is similar except that the sample is above the source material; and (b) shows a cross section view of the pattern after it has been coated using evaporation.

the sidewalls and/or may not reach the bottom of narrow features that have a high aspect ratio between the resist thickness and the feature size.

4.3. Etching

A variety of etching methods can be utilized, including wet chemical etching and reactive ion etching, and the etch may be isotropic or nonisotropic, depending on the method. In general, the resist will be used to protect parts of the substrate, while the exposed areas will be etched away. An important issue when etching is the relative etch rate of the substrate compared to the etch rate of the resist. In some cases when a resist cannot be used as an adequate etch mask, an extra layer will be used as the final etch mask, while the resist will be used to pattern the intermediate layer.

4.4. Pattern Checking and Common Errors

For beginners, it is useful to write a standard pattern on every sample, at least until consistently good results are obtained. The examples to follow are created using the same "wheel" pattern, which is a very effective diagnostic tool, because it allows pattern problems that are caused by poor focus and/or astigmatism in the beam to be easily identified.

4.4.1. Proper Writing

When a pattern is exposed correctly, lines should be straight with crisp edges and have a uniform thickness throughout. The final linewidth from a single line pass of the beam will depend primarily on the resist, beam focus/astigmatism during writing, and the applied line dose. An example of the wheel that has been done correctly is shown in Fig. 5.5. The slightly darker wedges inside of the "wheel" indicate that the area is becoming charged while viewing. This shows that the coating has not covered the sides of the pattern and will most likely lift off cleanly when the resist is removed.

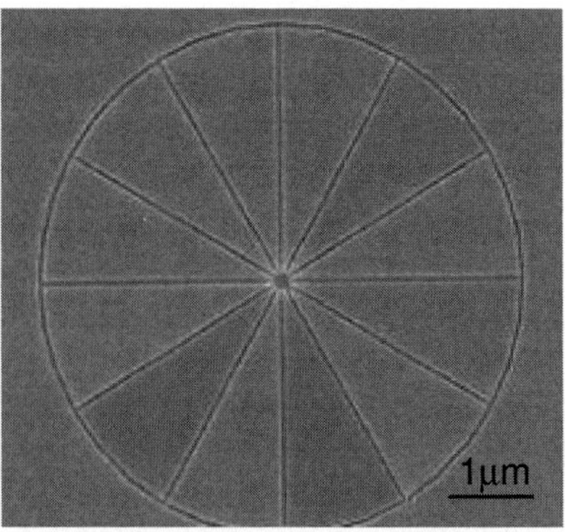

FIGURE 5.5. An example of the wheel pattern that has been written correctly.

4.4.2. Common Problems

Several types of errors can be determined by viewing a diagnostic pattern. Three of the most common errors during pattern generation are astigmatism in the beam, poor focus, and over- and underexposure. Typically, for new users pattern exposure problems are caused by a poor optimization of the beam, but the situation will improve as the user becomes better at running the microscope. Systematic problems, such as line noise, ineffective beam blanking, or general problems with the microscope itself, will usually have a distinctive effect on the outcome of the pattern.

Astigmatism is when the electron beam has an elongated cross section, represented as an oval in Fig. 5.6, as opposed to the ideal circular shape. As the beam moves from point to point forming a line in the direction of the long axis of the oval (vertical in Fig. 5.6a), all of the applied dose hits along the narrow path of the beam. However, when the beam steps in the direction of the short axis of the oval (horizontal in Fig. 5.6a), the dose is applied to a wider area along the line. This effect will produce a 90° asymmetry in the patterns, which is especially easy to identify in the wheel pattern. A schematic showing the dose distribution along the long axis and short axis directions is shown in Fig. 5.6a. An actual exposure of the wheel pattern written in PMMA is shown in Fig. 5.6b. In this case, the elongated beam lines up between the 5–11 o'clock and 6–12 o'clock spokes of the wheel, and the 90° asymmetry is very obvious. This is the classic sign of a pattern written with astigmatism in the beam.

When the beam is not well focused on the surface of the resist, large features will have only a small effect, where the radius on corners will be larger than expected. However, significant changes are caused when narrow lines are written with a beam that is out of focus. In general, poor focus will cause the applied dose for a narrow line to be spread over a larger area than desired. If the line dose is close to the critical dose for writing the smallest line, this broadening will effectively make the line underexposed. When the line dose is sufficiently above the

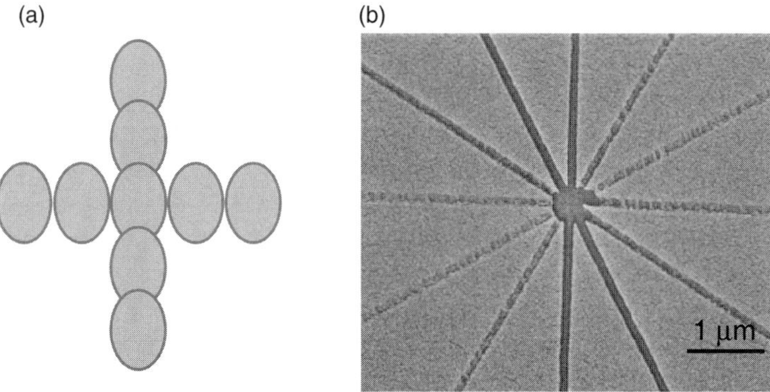

FIGURE 5.6. (a) Schematic showing how the shape of the beam affects the applied dose; and (b) the wheel pattern showing the effect of a beam with astigmatism.

critical dose, the broadening of the beam will make the exposed line larger than desired. In both cases, any junctions of single passes of the beam will effectively receive a double dose, and will generally appear to be "bloomed" out compared to the nearby lines. A picture of the wheel pattern that has been written in PMMA with a slightly degraded focus is shown in Figure 5.7. In this case, the lines of the pattern are slightly underexposed, while the junctions are overexposed. This is the classic sign of a poorly focused beam.

Overexposure causes patterns to become enlarged or in extreme cases a positive resist will receive enough dose to make it cross-link and develop like a negative resist. An image of an enlarged pattern and a positive/negative pattern is shown in Fig. 5.8. In this case, the center white dot is where the PMMA has become cross-linked due to the 12× dose where the 12 lines start at the center, and the entire central area has enlarged from the high dose. When the applied dose is too small, the lines of the resulting exposure will be shallow and/or discontinuous.

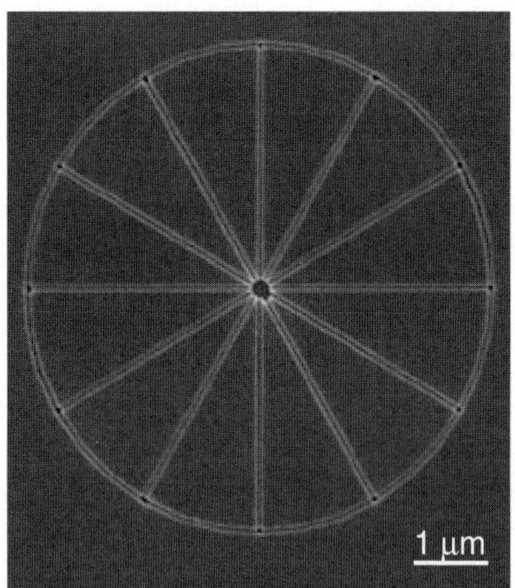

FIGURE 5.7. A wheel pattern that has been written when the beam was out of focus.

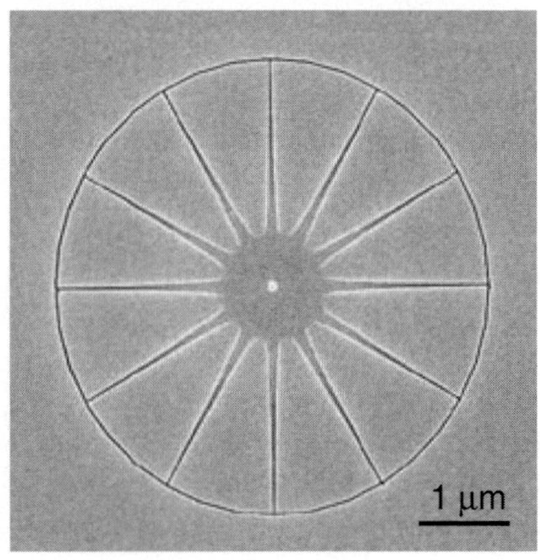

FIGURE 5.8. The wheel pattern written with a high dose.

In general, it is very useful to write an array of a pattern where the applied doses systematically step from a dose that is too low to a dose that is too high. In this way, a new operator can quickly identify the range of resulting structures that are caused by the range of applied doses.

Interference from sources external to the microscope, such as acoustic noise, physical vibrations, or electromagnetic fields, can cause wavy or interrupted lines. In general, the solution is to identify and eliminate the source of the interference or to shield the microscope from the noise. Acoustic noise can be reduced by using acoustic foam on the walls and/or by adding a sound dampening enclosure around the entire column of the microscope. Physical vibration can be minimized by using an air support system for the entire column or simply by adding foam or rubber padding between the column and the floor. Electromagnetic fields that interfere with the beam are often caused by equipment in other rooms or electric lines running through ceilings or floors. These fields can cause the electron beam to deflect at the field frequency, thus causing distortions in the pattern writing. Solutions include moving the microscope to a better location, moving or shielding the source of the interference, installing magnetic shielding (mu-metal) around the column and/or chamber, and installing an active field cancellation system that introduces a magnetic field to cancel the external noise.

5. E-beam Nanolithography Applications in Nanotechnology

Due to its versatility, electron beam lithography is the most common technique used for precise patterning in nanotechnology. Applications include quantum structures, transport mechanisms, solid-state physics, advanced semiconductor and magnetic devices, nanoelectromechanical system (NEMS), and biotechnology. In this section, the applications of electron beam lithography in nanotechnology are demonstrated by four examples in different fields. For each example, the fabrication procedure and device characteristics are discussed.

5.1. Nanotransistors

In the last decade, nanoscale 1D materials and structures have been widely investigated. These structures include semiconductor or metallic nanowires or nanotubes. Electron beam lithography is often involved in either defining the 1D structures directly through top-down nanofabrication [46], forming the catalyst to assist the growth [47], or patterning and wiring the nanowires to nanodevices [48].

As an example, the scanning electron micrograph of a fabricated nanowire Schottky diode is shown in Fig. 5.9. N-type semiconductor ZnO nanowires are synthesized by the vapor phase transport method [49]. A detailed fabrication procedure of this device is shown in Fig. 5.10. The fabrication starts from the Si/SiO$_2$ substrate. Au electrodes for interconnections are first patterned by photolithography. Then ZnO nanowires are deposited on the substrate by IPA dispersion. In the

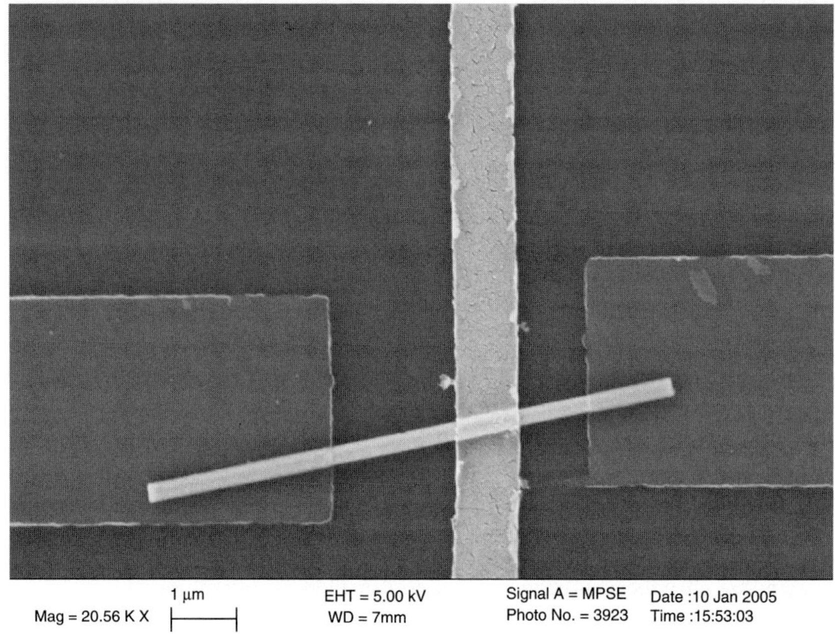

FIGURE 5.9. Scanning electron micrograph of a fabricated ZnO nanowire-based Schottky diode.

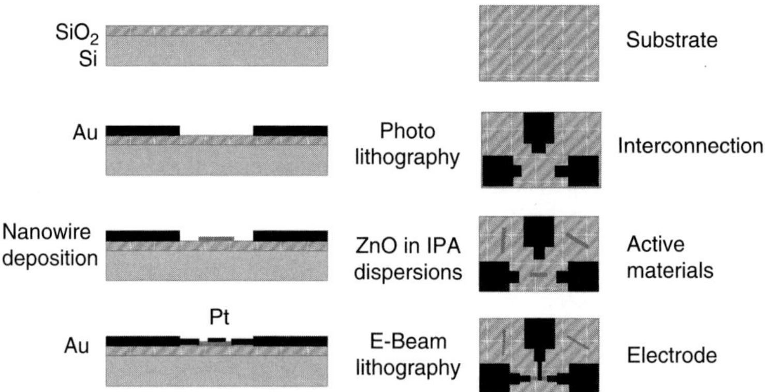

FIGURE 5.10. Step-by-step fabrication procedure of ZnO nanowire Schottky diode.

final step, e-beam lithography is used to connect small leads to the nanowires after aligning to registration marks defined by the photolithography. The Pt electrode deposited over ZnO nanowire forms the Schottky contact with ZnO. The other two electrodes are made by Cr/Au which forms the Ohmic contact with ZnO. Figure 5.11 demonstrates the rectifying characteristics of fabricated Schottky diodes, which shows a rectifying factor around 1.9. Recently, a research

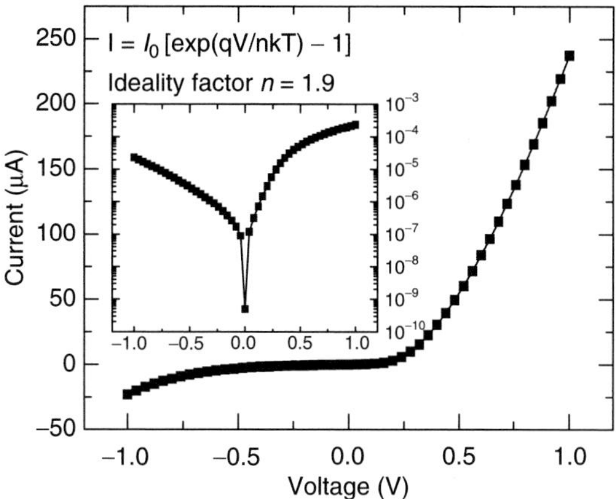

FIGURE 5.11. The I–V characteristics of fabricated ZnO nanowire Schottky diode.

group at Harvard assembled nanowires arrays on patterned electrodes for large size nanoelectronic transistor fabrication [50,51].

5.2. Nanosensors

Semiconducting metal oxides can be used in gas sensing devices and have been extensively studied due to their properties, such as sensitivity to ambient conditions and simplicity in fabrication [52]. Applications in many fields include environmental modeling, automotive applications, air conditioning, and sensor networks. The commercially available gas sensors are made mainly from SnO_2 and In_2O_3 in the form of thick films, porous pellets, or thin films. However, poor long-term stability has prevented wide application of this type sensor. Recent research has been directed toward nanostructured oxides since reactions at grain boundaries and complete depletion of carriers in the grains can strongly modify the materials transport properties [53,54]. Nanowires made of semiconducting metal oxides with a rectangular cross section in a ribbon-like morphology are very promising for sensors because the surface-to-volume ratio is very high [55–57]. Also, the oxide is single crystalline, the faces exposed to the gaseous environmental are always the same, and the small size is likely to produce a complete depletion of carriers inside the nanowires and make the sensor more sensitive.

A ZnO nanowire-based gas sensor is shown in Fig. 5.12a, where the opposite electrodes and comb electrodes are patterned by photolithography and e-beam lithography, respectively. During the second e-beam lithography step, a resist (PMMA) window is opened upon the comb electrodes (Fig. 5.12b). After the deposition of ZnO nanowires from IPA dispersions by the Langmuir-Blodgett technique (Fig. 5.12c), the final liftoff process removes the ZnO nanowires outside the window region along with the PMMA. In this device, the resistance of the nanowires changes with the gas condition and is detected by the I-V characteristics

FIGURE 5.12. The optical micrograph of gas sensor: (a) pattern of electrodes; (b) PMMA window between two electrodes; (c) deposition of nanowires by the Langmuir-Blodgett technique; and (d) liftoff process.

between the two electrodes. Due to the small features of the comb electrode structures that are written by e-beam lithography and the usage of nanowires, the surface-to-volume ratio is largely increased, resulting in an improved sensitivity compared to thin film-based gas sensors. Using the electron beam lithography technique, similar nanoscale sensors have also been fabricated with very high resolution, sensitivity, and density [58–60].

5.3. Magnetic Nanodevices

Magnetic nanostructures have become a particularly interesting class of materials for both scientific and technological explorations. To obtain different magnetic nanostructures, e-beam lithography has been used in combination with other lithography processes. One of the main advantages of this approach is its ability to fabricate well-defined shapes for arbitrary elements and also array configurations. A variety of magnetic elements have been achieved such as dots and lines [61,62], rectangles, triangles [63] and pentagons [64], zig-zag lines [65], rings [66] for the study of the magnetization reversal processes. Moreover, this versatility allows the fabrication of nanodevices, such as non-volatile magnetoresistive magnetic random access memories (MRAM) [67] or "quantum" magnetic disks [68,69]. Magnetoelectronics is considered to be one of the most promising approaches for quantum computation [70,71] and universal memory [72] with ultrahigh density.

As an example, Fig. 5.13 shows an SEM micrograph of 200 nm cobalt zigzag wires patterned using electron beam lithography. The wires are connected for a four-point measurement to four gold electrodes which have been fabricated using photolithography. Nanometric wires were used to determine the magnetoresistance of the domain walls in polycrystalline Co. The magnetic switching processes of an array of Co wires are studied by superconducting quantum interference device (SQUID) at the temperature range from 5 to 300 K. Exchange bias was observed in the Co wires at 5 K, which is responsible for the asymmetric behavior of the magnetoresistance, as shown in Fig. 5.14 [65,73].

5.4. Biological Applications

With the assistance of nanolithography, nanoscale devices have also been fabricated for biological applications [74–76]. One example is the biomolecular motor-powered devices developed by Montemagno et al. [74]. In their study, a system is first established to produce a recombinant biomolecular motor. Then biological molecules are positioned on the fabricated nanostructures and then baseline performance data is acquired. Electron beam lithography is used to pattern a metallic (gold, copper, or nickel) array on a 25 mm cover slip. Subsequently, His-tagged 1 μm microspheres are attached to the metallic surface. These motors show promise as mechanical components in hybrid nano-engineered systems. The results demonstrate the ability to engineer chemical

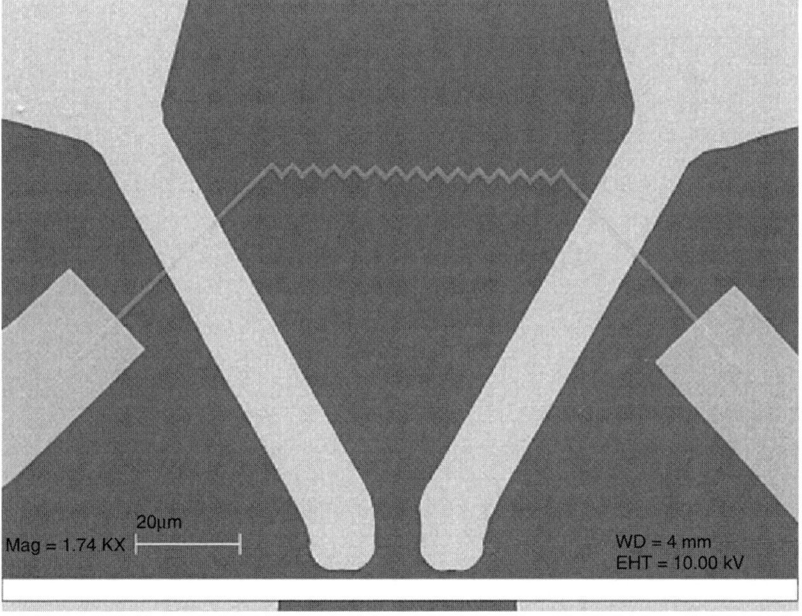

FIGURE 5.13. Co zigzag wires patterned by e-beam lithography.

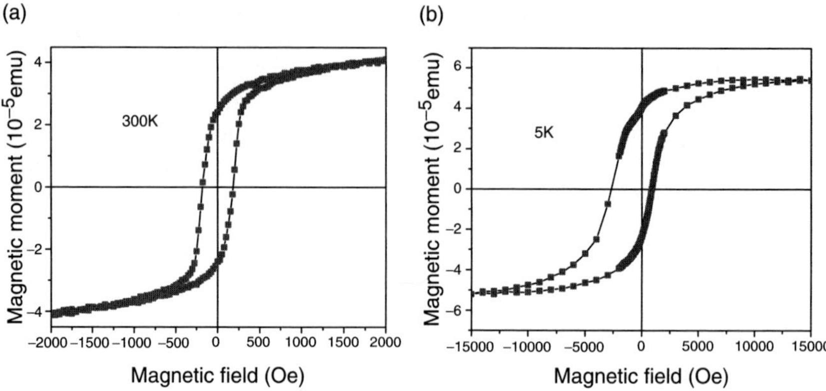

FIGURE 5.14. Field-dependent magnetization of the arrayed Co wires at (a) 5 K and (b) 300 K, respectively.

regulation into a biomolecular motor and represent a critical step toward controlling integrated nanomechanical devices at the single-molecule level.

6. Summary

This chapter introduced technique configurations of electron nanolithography integrated with SEM. Various photo resists and processing preparation were detailed. The step-by-step pattern generation were explicated. Pattern processing and proper writing were also discussed. The applications on nanotransistors, nano-gas sensors, magnetic nanodevices, and biomolecular motor-powered nanodevices have been demonstrated. Electron beam nanolithography shows extreme importance for nanodevice and nanosensor fabrication.

References

1. G. Mollenstedt and R. Speidel, *Phys. Blatter*, 16 (1960) 192.
2. A. N. Broers, in: *Proceedings of the 1st International Conference on Electron and Ion Beam Science and Technology*, R. Bakish (Ed.), Wiley, (1964), pp. 191–204.
3. M. A. McCord and R. F. W. Pease, *J. Vac. Sci. Technol. B*, 4 (1986) 86.
4. A. Majumdar, P. I. Oden, J. P. Carrejo, L. A. Nagahara, J. J. Graham, and J. Alexander, *Appl. Phys. Lett.*, 61 (1992) 2293.
5. G. T. A. Kovacs, *Micromachined Transducers Sourcebook*, WCB McGraw-Hill Stanford Univ. (1998).
6. J. Ingino, G. Owen, C. N. Bergulund, R. Browning, and R. F. W. Pease, *J. Vac. Sci. Technol. B*, 12 (1994) 1367.
7. M. Angelopoulos, J. M. Shaw, K. L. Lee, W. S. Huang, M. A. Lecorre, and M. Tissier, *Pol. Eng. Sci.*, 32 (1992) 1535.
8. P. Rai-Choudhury, *Handbook of Microlithography, Micromachining, and Microfabrication*, SPIE Optical Engineering Press (1997.)

9. M. Angelopoulos, J. M. Shaw, K. Lee, W. Huang, M. Lecorre, and M. Tissier, *J. Vac. Sci. Technol. B*, 9 (1991) 3428.

10. M. Angelopoulos, N. Partel, J. M. Shaw, N. C. Labianca, and S. A. Rishton, *J. Vac. Sci. Technol. B*, 11 (1993) 2794.

11. P. Romand, A. Weill, J. -P. Panabiere, and A. Prola, *J. Vac. Sci. Technol. B*, 12 (1994) 3550.

12. S. Wolf and R. N. Tauber, "Silicon Process for the VLSI Era," Vol. 1: *Process Technology*, Lattice Press, 2nd edition, November (1999).

13. A. N. Cleland, *Foundations of Nanomechanics*, Springer (2002).

14. A. N. Broers, A. C. F. Hoole, and J. M. Ryan, *Microelectron. Eng.*, 32 (1996) 131.

15. M. I. Lutwyche, *Microelectron. Eng.*, 17 (1992) 17.

16. I. Kostic, R. Andok, V. Barak, I. Caplovic, A. Konecnikova, L. Matay, P. Hrkut, A. Ritomsky, *J. Mater. Sci.: Mater. Electron.*, 14 (2003) 645.

17. I. Zailer, J. E. F. Frost, V. Chabasseur-Molyneux, C. J. B. Ford, M. Pepper, *Semicond. Sci. Technol.*, 11 (1996) 1235.

18. M. Khoury and D. K. Ferry, *J. Vac. Sci. Technol. B*, 14(1) (1996) 75–79.

19. A. A. Tseng, K. Chen, C. D. Chen, and K. J. Ma, *IEEE Trans. Electron. Packaging Manuf.*, 26 (2003) 141.

20. D. R. Medeiros, A. Aviram, C. R. Guarnieri, W. -S. Huang, R. Kwong, C. K. Magg, A. P. Mahorowala, W. M. Moreau, K. E. Petrillo, and M. Angelopoulos, *IBM J. R&D*, 45 (2001) 639.

21. K. Nakamura, S. L. Shy, C. C. Tuo, and C. C. Huang, *Jpn. J. Appl. Phys.*, 33 (1994) 6989.

22. T. Tada, *J. Electrochem. Soc.*, 130 (1983) 912.

23. H. Ito, *IBM J. R&D*, 41 (1997) 69.

24. M. Kurihara, T. Segawa, D. Okuno, N. Hayashi, and H. Sano, *Proc. SPIE*, 3412 (1998) 279.

25. T. Segawa, M. Kurihara, S. Sasaki, H. Inomata, N. Hayashi, and H. Sano, *Proc. SPIE*, 3412 (1998) 82.

26. H. Saitoh, T. Soga, S. Kobu, S. Sanki, and M. Hoga, *Proc. SPIE*, 3412 (1998) 269.

27. K. Katoh, K. Kasuya, T. Arai, T. Sakamizu, H. Satoh, H. Saitoh, and M. Hoga, *Proc. SPIE*, 3873 (1999) 577.

28. T. H. P. Chang, D. P. Kern and L. P. Muray, *J. Vac. Sci. Technol. B*, 10 (1992) 2743.

29. M. J. Madou, *Fundamentals of Microfabrication: The Science of Miniaturization*, 2nd edition, CRC Press, 2002.

30. L. E. Ocola, D. Tennant, G. Timp, and A. Novembre, *J. Vac. Sci. Technol. B*, 17 (1999) 3164.

31. M. J. Word, I. Adesida, P. R. Berger, *J. Vac. Sci. Technol. B*, 21 (2003) L12.

32. H. Hamatsu, T. Yamaguchi, M. Nagase, K. Yamasaki, and K. Kurihara, *Microelectron. Eng.*, 41/42 (1998) 331.

33. F. C. M. J. M. van Delft, J. P. Weterings, A. K. van Langen-Suurling, H. Romijn, *J. Vac. Sci. Technol. B*, 18 (2000) 3419.

34. S. Matsui, Y. Igaku, H. Ishigaki, J. Fujita, M. Ishida, Y. Ochiai, H. Namatsu, and M. Komuro, *J. Vac. Sci. Technol. B*, 21(2003) 688.

35. I. Junarsa, M. P. Stoykovich, P. F. Nealey, Y. Ma, F. Cerrina, and H. H. Solak, *J. Vac. Sci. Technol. B*, 23 (2005) 138.

36. W. Henschel, Y. M. Georgiev, and H. Kurz, *J. Vac. Sci. Technol. B*, 21 (2003) 2018.

37. G. M. Schmid, L. E. Carpenter II, and J. A. Liddle, *J. Vac. Sci. Technol. B*, 22 (2004) 3497.

38. H. Namatsu, *J. Vac. Sci. Technol. B*, 19 (2001) 2709.

39. J. Fujita, Y. Ohnishi, Y. Ochiai, and S. Matsui, *Appl. Phys. Lett.*, 68 (1996) 1297.

40. A. Tilke, M. Vogel, F. Simmel, A. Kriele, R. H. Blick, H. Lorenz, D. A. Wharam, and J. P. Kotthaus, *J. Vac. Sci. Technol. B*, 17 (1999) 1594.

41. M. Aktary, M. O. Jensen, K. L. Westra, M. J. Brett, and M. R. Freeman, *J. Vac. Sci. Technol. B*, 21 (2003) L5.

42. W. Langheinrich, A. Vescan, B. Spangenberg, and H. Beneking, *Microelectron. Eng.*, 17 (1992) 287–290.

43. A. N. Broers, A. C. F. Hoole, and J. M. Ryan, *Microelectron. Eng.*, 32 (1996) 131.

44. T. Weimann, W. Geyer, P. Hinze, V. Stadler, W. Eck, and A. Golzhauser, *Microelectron. Eng.*, 57 (2001) 903.

45. A. N. Broers, A. C. F. Hoole, and J. M. Ryan, *Microelectron. Eng.*, 32 (1996) 131.

46. L. Pescini, A. Tilke, R. H. Blick, H. Lorenz, J. P. Kotthaus, W. Eberhardt, and D. Kern, *Nanotechnology*, 10 (1999) 418.

47. N. R. Franklin, Q. Wang, T. W. Tombler, A. Javey, M. Shim, and H. Dai, *Appl. Phys. Lett.*, 81 (2002) 913.

48. Y. Huang, X. Duan, Y. Cui, L. J. Lauhon, K. -H. Kim, and C. M. Lieber, *Science*, 294 (2001) 1313.

49. Y. X. Chen, L. J. Campbell, and W. L. Zhou, *J. Cryst. Growth*, 270 (2004) 505.

50. S. Jin, D. Whang, M. C. McAlpine, R. S. Friedman, Y. Wu, and C. M. Lieber, *Nano Lett.*, 4 (2004) 915.

51. D. Whang, S. Jin, and C. M. Leiber, *Nano Lett.*, 3 (2003) 951.

52. N. Barsan, M. Schweizer-Berberich, and W. Gëpel, *Fresenius J. Anal. Vol.*, 365 (1999) 287.

53. W. Gëpel, *Sens. Actuators A*, 56 (1996) 83.

54. M. Ferroni, V. Guidi, G. Martinelli, E. Comini, G. Sberveglieri, D. Boscarino, and G. Della Mea, *J. Appl. Phys.*, 88 (2000) 1097.

55. E. Comini, G. Faglia, G. Sberveglieri, Z. Pan, and Z. L. Wang, *Appl. Phys. Lett.*, 81 (2003) 1869.

56. D. Zhang, C. Li, S. Han, X. Liu, T. Tang, W. Jin, and C. Zhou, *Appl. Phys. Lett.*, 82 (2003) 112.

57. C. Li, D. Zhang, X. Liu, S. Han, J. Han, and C. Zhou, *Appl. Phys. Lett.*, 82 (2003) 1613.

58. T. Toriyama and S. Sugiyama, *Sens. Actuators A Phys.*, 108 (2003) 244.

59. B. Ilic, H. G. Craighead, S. Krylov, W. Senaratne, C. Ober, and P. Neuzil, *J. Appl. Phys.*, 95 (2004) 3694.

60. F. Patolsky and C. M. Lieber, *Mater. Today*, 4 (2005) 20.

61. J. F. Smyth, S. Schltz, D. Kern, H. Schmid, and D. Yee, *J. Appl. Phys.*, 63 (1988) 4237.

62. J. I. Martin, J. L. Vicent, J. V. Anguita, and F. Briones, *J. Magn. Magn. Mater.*, 203 (1999) 156.

63. B. Khamsehpour, C. D. W. Wilkinson, J. N. Chapman, and A. B. Johnston, *J. Vac. Sci. Technol. B*, 18 (2000) 16.

64. R. P. Cowburn, *J. Phys. D*, 33 (2000) R1.

65. T. Taniyama, I. Nakatani, T. Namikawa, and Y. Yamazaki, *Phys. Rev. Lett.*, 82 (1999) 2780.

66. J. Rothman, M. Klaui, L. Lopez-Diaz, C. A. F. Vaz, A. Bleloch, J. A. C. Bland, Z. Cui, and R. Speaks, *Phys. Rev. Lett.*, 86 (2001) 1098.

67. K. Nordquist, S. Pandharkar, M. Durlam, D. Resnick, S. Teherani, D. Mancini, T. Zhu, and J. Shi, *J. Vac. Sci. Technol. B*, 15 (1997) 2274.

68. S. Y. Chou and P. R. Krauss, *J. Magn. Magn. Mater.*, 155 (1996) 151.
69. C. Haginoya, K. Koike, Y. Hirayama, J. Yamamoto, M. Ishibashi, O. Kitakami, and Y. Shimada, *Appl. Phys. Lett.*, 75 (1999) 3159.
70. G. A. Prinz, *Science*, 282 (1998) 1660.
71. S. D. Sarma, *Am. Sci.*, 89 (2001) 516.
72. S. A. Wolf, D. D. Awschalom, R. A. Buhrman, J. M. Daughton, S. von Molnar, M. L. Roukes, A. Y. Chtchelkanova, and D. M. Treger, *Science*, 294 (2001) 1488.
73. C. Chen, M. H. Yu, Mo. Zhu, and W. L. Zhou, unpublished work.
74. C. Montemagno and G. Bachand, *Nanotechnology*, 10 (1999) 225.
75. R. H. Austin, J. O. Tegenfeldt, H. Cao, S. Y. Chou, E. C. Cox, *IEEE Trans. Nanotechnol.*, 1 (2002) 12.
76. R. Bunk, J. Klinth, J. Rosengren, I. Nicholls, S. Tagerud, P. Omling, A. Mansson, and L. Montelius, *Microelectron. Eng.*, 67 (2003) 899.

6
Scanning Transmission Electron Microscopy for Nanostructure Characterization

S. J. Pennycook, A. R. Lupini, M. Varela, A. Y. Borisevich, Y. Peng, M. P. Oxley, K. van Benthem, M. F. Chisholm

1. Introduction

The scanning transmission electron microscope (STEM) is an invaluable tool for the characterization of nanostructures, providing a range of different imaging modes with the ability to provide information on elemental composition and electronic structure at the ultimate sensitivity, that of a single atom. The STEM works on the same principle as the normal scanning electron microscope (SEM), by forming a focused beam of electrons that is scanned over the sample while some desired signal is collected to form an image [1]. The difference with SEM is that thin specimens are used so that transmission modes of imaging are also available. Although the need to thin bulk materials down to electron transparency can be a major task, it is often unnecessary for nanostructured materials, with sample preparation requiring nothing more than simply sprinkling or distributing the nanostructures onto a commercially available thin holey carbon support film. No long and involved grinding, polishing, or ion milling is required, making the STEM a rapid means for nanostructure characterization.

As in the SEM, secondary or backscattered electrons can be used for imaging in STEM; but higher signal levels and better spatial resolution are available by detecting transmitted electrons. A bright field (BF) detector includes the transmitted beam and so the holes appear bright, whereas a dark field detector excludes the transmitted beam and holes appear dark. Each detector provides a different and complementary view of the specimen. It is one of the key advantages of the STEM to have multiple detectors operating simultaneously to collect the maximum possible information from each scan. Although transmitted electron detectors are usefully fitted to conventional SEM instruments working at relatively low voltages, there are major advantages in increasing the accelerating voltage. Increased specimen penetration means that thicker specimens can be tolerated; but more importantly, the decreasing electron wavelength leads to higher spatial resolution and the ability to see the actual atomic configurations within the nanostructure.

Thus the STEM can take many forms: a simple add-on detector to a standard low-voltage SEM; a dedicated, easy-to-use, intermediate voltage STEM with rapid throughput; or an instrument more comparable to a high-resolution transmission electron microscope (TEM), which is able to provide the ultimate spatial resolution and analytical sensitivity. All have important and complementary roles in nanostructure characterization. Rapid feedback is critical to synthesis, and commercially available SEMs with subnanometer resolution at 30 kV have the ability to image tens of samples within a few hours. Similar throughput is also available with dedicated STEMs giving at best around 0.2-nm resolution at 200-kV accelerating voltage. Such instruments can be used to guide the synthesis on a day-to-day basis and represent an invaluable first step in characterization. An example of the imaging of a gold nanocatalyst supported in mesoporous silica is shown in Fig. 6.1.

These microscopes are ideal for determining size distributions of nanoparticles at the level of 1 nm and above, but they lack sensitivity at the atomic level. For understanding the functionality of nanostructures, it is highly desirable to examine individual nanostructures with atomic level sensitivity. This requires a more sophisticated instrument at the forefront of what is technologically achievable in electron optics, electronics, and environmental stability. STEMs of this type are generally based on high-resolution TEM columns, operating at 200–300 kV. Ideally, we would like to see every atom in its 3D location, a dream that Feynman first laid forth in his famous lecture "There's Plenty of Room at the Bottom," where he not only forecast the nanotechnology era but also explicitly called for 100-fold improvement in the resolution of the electron microscope [3]. If we could see the atoms clearly, then surely we would be able to see how the nanostructure functioned.

Today we are well on the way along this path. We are able to see individual atoms of high atomic number (Z), either on surfaces or inside bulk materials [4,5]. We can even identify single atoms spectroscopically and analyze their local

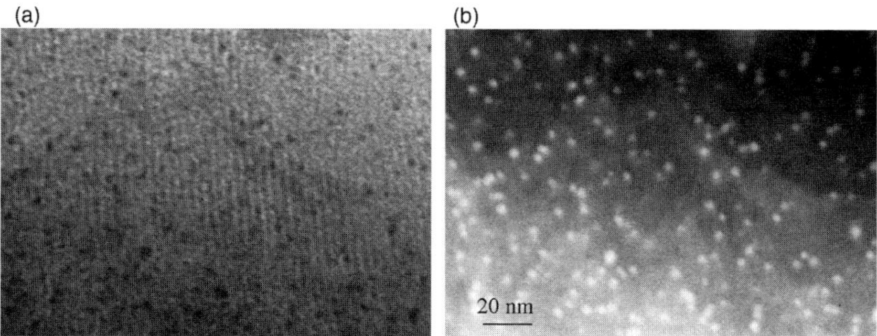

FIGURE. 6.1. (a) Bright field and (b) annular dark field (Z-contrast) transmitted electron images of Au nanoparticles (2.2 wt %) supported in mesoporous silica functionalized with a diethylenetriamine ligand. The bright field image shows the aligned mesoporous channels as vertical stripes, with the Au particles showing as dark spots. In the Z-contrast image the Au particles are bright, and smaller particles are revealed more clearly. Images recorded with a Hitachi HD2000 dedicated STEM by Lee [adapted from ref. 2].

electronic environment by electron energy loss spectroscopy (EELS) [6]. The advances in STEM capability in the past few years have been quite dramatic, comparable to progress in the previous two decades, through the successful realization of another innovation that Feynman called for in his lecture, the correction of lens aberrations. Feynman explicitly called for the incorporation of nonround lens elements to improve the resolution beyond the limit imposed by the unavoidable spherical aberration of the traditional round microscope objective lens. These revolutions in nanoscience and in electron microscopy are going on at the same time, and indeed one is fuelling the other. The ability of aberration-corrected microscopes to image nanostructures more clearly than ever is significantly increasing the demand for these instruments.

It is often said that a picture is worth a thousand words, but in the area of nanoscience a picture can sometimes be a revelation, showing up structures or phenomena that were totally unexpected. Such is the importance of feedback to synthesis, but at the same time a picture can just be a starting point into a quantitative insight into atomistic processes. Catalysis provides a perfect example where images can furnish information on the size and shape of nanoparticles, and how they change with processing conditions. From such knowledge, it becomes possible to perform theoretical modeling of the atomistic processes themselves, calculations of the binding energy of molecules onto the clusters observed, calculation of migration energies and diffusion pathways. It is also possible to calculate actual reaction pathways, processes that can never be observed directly in any microscope. Of course, it is always possible to calculate such things without any image, but then all possible configurations of a nanostructure need to be calculated, which is a vast number of trial structures. Without an image for guidance, it is certain that large numbers of irrelevant structures will be calculated, and there is the possibility that the right structure might be missed.

Poised between synthesis and atomistic processes, the STEM fills an exciting and central area of nanotechnology. In this chapter we will cover the basics of the technique, including probe formation, image resolution and contrast in different modes, and analytical techniques. More details on imaging theory have been given elsewhere [7–11]. Here, we will illustrate the discussion with a number of examples in different areas of nanotechnology, including nanocatalysis, nanocrystals, nanotubes, nanostructured magnetic materials, and nanoscale phase separation in complex oxides, pointing out how this level of characterization can provide new insights into the functionality at the nanoscale [12,13]. Other applications in material science have recently been reviewed by Varela et al. [14]. In addition, we present a new possibility that is opened up by the new aberration-corrected STEMs. Correcting the lens aberrations allows the objective aperture to be opened up, thereby obtaining higher resolution. At the same time, as in an optical instrument like a camera, the depth of field is reduced. Present-day aberration-corrected STEMs have a depth of field of only a few nanometers, and so it becomes possible to effectively depth slice through a sample and to reconstruct the set of images into a 3D representation of the structure. The technique is comparable to confocal optical microscopy, but provides a resolution that is on the nanoscale.

2. Imaging in the Scanning Transmission Electron Microscope

Figure 6.2 shows the main components comprising the aberration-corrected STEM. Electrons are accelerated from a source and focused into a point on the specimen by a set of condenser lenses and an objective lens. An objective aperture limits the maximum angle of illumination included in the incident probe, which is scanned across the sample by a set of scan coils. The output of a variety of possible detectors can then be used to form an image. In fact, multiple detectors can be used simultaneously to give different views of the sample, providing different but complementary information. The usual detectors include a BF detector that intercepts the transmitted beam and an annular dark field (ADF) detector that surrounds the transmitted beam to collect scattered electrons. The inner angle of this detector can be changed with postspecimen lenses from just outside the incident beam cone, which gives maximum efficiency for collecting scattered electrons, to several times this angle, that enhances the atomic number (Z) dependence of the image contrast. This latter configuration is often referred to as a Z-contrast or high-angle ADF (HAADF) image. Also, normally, part of the STEM is an EELS system, comprising spectrometer and parallel detection system using a charge-coupled device (CCD). Coupling lenses may be required to provide sufficient collection efficiency into the spectrometer.

Other detectors are also possible; an energy-dispersive x-ray detector is common; also secondary electrons, cathodoluminescence, or electron beam-induced

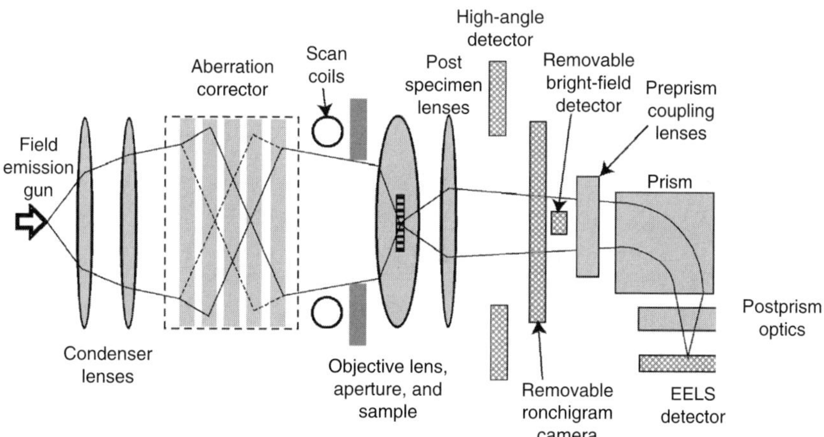

FIGURE 6.2. Schematic showing the main components of a high-resolution dedicated STEM [adapted from ref.14].

current can all be used to form an image. However, these signals tend to be lower, possibly very much lower, in intensity and therefore may be insufficient to form a clear, high-resolution image of a nanostructure in a reasonable exposure time. Typically, one would image a nanostructure using the high-intensity transmitted electron detectors, then stop the beam on the nanostructure for detection of x-rays or EELS data. Alternatively, one may be able to scan with a larger, higher current beam to obtain sufficient signal-to-noise ratio in some weaker signal of interest, sacrificing spatial resolution. Low-loss features in the energy loss spectrum or low-energy core loss edges are particularly useful in this regard.

In the past few years, the achievable resolution in STEM has more than doubled, and the first direct image of a crystal showing sub-Ångstrom resolution has been achieved with a 300-kV STEM (see Fig. 6.3) [15]. The reason for this impressive progress is that it has become possible to correct the major geometrical aberrations of the probe-forming lenses [16–18]. The traditional round magnetic lens has unavoidable aberrations that have limited the useful aperture that could be used, thus limiting resolution.

Although it has long been known that multipole lenses could in principle be used to correct these aberrations, successful implementations of aberration-correcting systems have only appeared in the past few years. The reason for this is the need to tune all the low-order aberrations simultaneously for the best focused spot. Optimization of the first-order aberrations astigmatism and focus

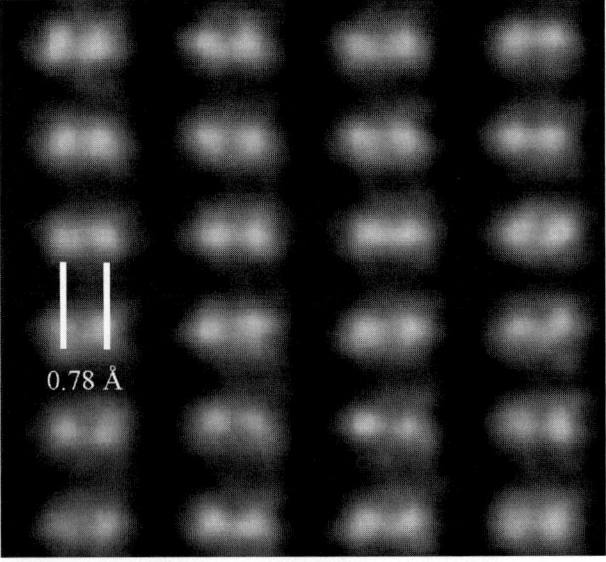

FIGURE 6.3. Z-contrast image of silicon taken along the ⟨112⟩ zone axis, resolving columns of atoms just 0.78 Å apart [adapted from ref. 15]. Image recorded with the ORNL 300 kV VG Microscopes HB603U STEM equipped with Nion aberration corrector. Image has been filtered to remove noise and scan distortion.

can be done by hand, but it is impossible to focus by hand in the much higher dimensional space needed to correct all nine first- and third-order aberrations. Computer autotuning procedures are essential, where the computer takes over the measurement of the aberrations, adjusts the multipole lenses, and iterates to the optimum settings. Commercial systems are now available from Nion and CEOS based on a quadrupole, octupole, or hexapole system, respectively [19,20]. It is also possible to insert a postspecimen aberration corrector, to cure the spectrometer aberrations and provide efficient collection of EELS data with the highest possible resolution.

2.1. Probe Formation

The STEM probe is a demagnified image of the source, as seen clearly from the ray diagram in Fig. 6.2. For high resolution a small probe is required, and source brightness then becomes an important limiting characteristic. The commonly used sources are a heated tungsten filament, a LaB_6 pointed filament, a Schottky or thermal-assisted field emission source, and a cold field emission source. Brightness depends both on the current density per unit area and on the angular range filled by the beam. It is defined as

$$B = \frac{I}{\pi A \theta^2} \tag{6.1}$$

where I is the beam current, A is the area of the beam, θ is the illumination semi-angle. The four sources are compared in Table 6.1, in the order of increasing brightness, monochromaticity, and vacuum requirements.

It may be noted that brightness scales with beam energy, hence values for 200- or 300-kV operation are 2× or 3× higher than that shown in Table 6.1. Brightness is conserved in an optical system that is free of aberrations. Given the source sizes in Table 6.1, it is clear that high demagnifications are needed to achieve probes of atomic dimensions. If the source of radius r_s is demagnified by a factor M to form the probe, giving a geometric probe radius of r_s/M in the absence of aberrations, the angular divergence will be increased by a factor M. To avoid broadening the probe due to lens aberrations, the angular aperture therefore has to be restricted at some point in the optical column. In practice, this means that only a small fraction of the emitted beam current will end up in the probe, and the choice of probe

TABLE 6.1. Comparison of the characteristics of various sources (values are approximate)

	Source size	Energy spread FWHM (eV)	Brightness (Asr^{-1} cm^{-2}) at 100 kV	Total emission current (μA)
Tungsten filament	30 μm	2	5×10^5	100
LaB_6	10 μm	1	5×10^6	50
Thermal field emission	100 nm	0.6	5×10^8	100
Cold field emission	5 nm	0.3	2×10^9	5

size becomes a trade-off between resolution and signal-to-noise ratio. The ultimate resolution can be achieved only with zero beam current.

Lens aberrations fall into two main classes: (1) geometric aberrations due to errors in the optical path lengths and (2) chromatic aberration due to a spread in energy of the beam. Geometric aberrations mean that rays traveling at an angle to the optic axis are focused at a different point to rays traveling almost parallel to the optic axis, which define the Gaussian focus point. Spherical aberration is the most well-known geometric aberration because it was the dominant aberration before correction became possible. A schematic depiction of spherical and chromatic aberration is presented in Fig. 6.4.

The aberration is the error in optical path length between the actual wave front and the perfect sphere, which is conventionally expressed as a power series in angle θ. If, for simplicity, we ignore all nonrotationally symmetric aberrations, the aberration function becomes

$$\chi(\theta) = \frac{1}{2} \Delta f \theta^2 + \frac{1}{4} C_S \theta^4 + C_5 \theta^6 + \frac{1}{8} C_7 \theta^8 + \ldots +, \qquad (6.2)$$

where Δf is the defocus, C_S is the coefficient of third-order spherical aberration, the dominant geometric aberration in an uncorrected microscope, and C_5 and C_7 are the coefficients of fifth- and seventh-order spherical aberrations, respectively. For round magnetic lenses, these are all positive coefficients and have dimensions of length. The aberration in radians is just $\gamma = 2\pi\chi/\lambda$, where λ is the electron wavelength. Before aberration correction, it was normal to partially compensate for the C_S term by a small amount of negative defocus, weakening the lens slightly to give less aberration at high angles. In Fig. 6.4b, the effect is seen by

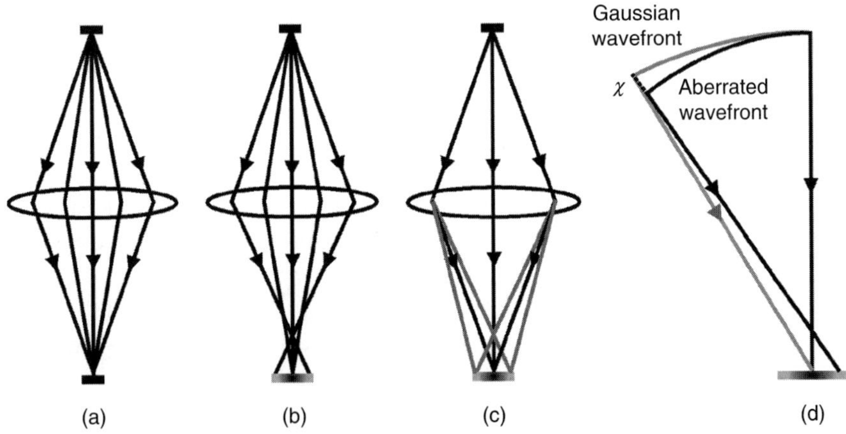

FIGURE 6.4. Schematic showing the action of a perfect lens (a), the effect of geometrical aberrations that bring rays at different angles to a different focus (b), the chromatic aberration which blurs any single ray path (c), and the definition of the aberration length (d), the path length from the Gaussian sphere to the true wavefront.

imagining the specimen raised to a point where the rays cross closest to the axis forming a disk of least confusion. However, compensation can only be achieved over a restricted range of angles because spherical aberration and defocus have a different angular dependence.

Geometric aberrations bring the rays at different angles to different focus points along the optic axis, meaning that they are displaced laterally in the Gaussian focus plane. The amount of sideways displacement δ is related to the gradient of the aberration function [21]

$$\delta(\theta) = \Delta f\,\theta + C_\mathrm{S}\theta^3 + C_5\theta^5 + C_7\theta^7 +\dots +, \tag{6.3}$$

which is the reason why the C_S term is referred to as third-order spherical aberration. An approximate idea of the magnitude of these different terms can be seen by plotting their separate contributions, as shown in Fig. 6.5.

This illustrates well how the optimum probe represents a balance between diffraction broadening at the aperture, which is less for a *large* aperture, and the geometrical aberrations, which are reduced for a *small* aperture. The diffraction pattern of the circular aperture is an Airy disk with first minimum at a radius 0.61 λ / θ, so we have taken 0.3 λ / θ as a measure of the average sideways displacement due to diffraction. For an uncorrected system, the conditions around point A represent the optimum probe size; if third-order spherical aberration is corrected,

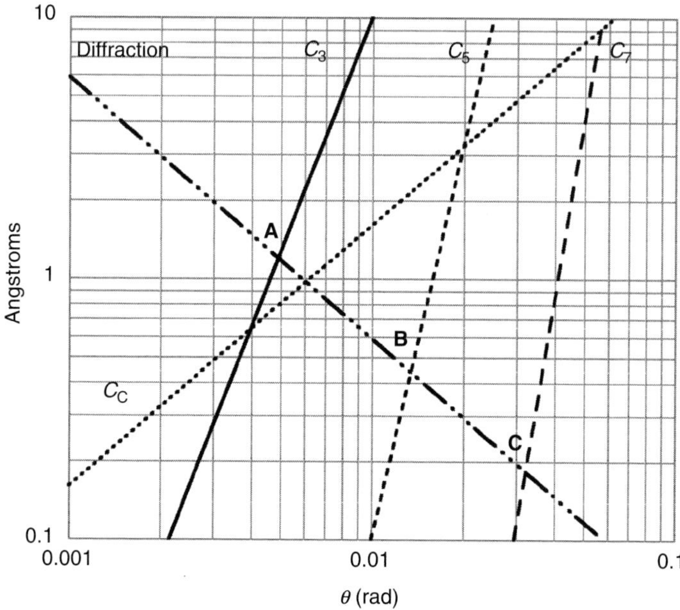

FIGURE 6.5. Sideways image spread introduced by the third-, fifth-, and seventh-order geometric aberrations, by chromatic aberration and by diffraction by the aperture of semiangle θ.

then the probe will be limited by the fifth-order term and the optimum probe will be achieved around point B, approximately a factor of two smaller. This is the reason for the factor of two gain in resolution with a third-order aberration-corrected STEM [8,12,13,16,22]. More advanced correctors are currently being constructed that will correct fifth-order aberrations, when the limiting geometric aberration will become seventh order, and another factor of two reduction in probe size is anticipated [22]. Also shown on the plot is the effect of chromatic aberration, which introduces a focal spread given by

$$\Delta = C_c \sqrt{\left(\frac{\Delta V}{V}\right)^2 + \left(\frac{2\Delta I}{I}\right)^2}, \tag{6.4}$$

where ΔV is the standard deviation of the energy spread of the beam, assumed Gaussian, V is the accelerating voltage, and ΔI is the standard deviation of fluctuations in objective lens current I. In this case the sideways spread is a little misleading, because we will see that at the limit of resolution the STEM image is insensitive to chromatic aberration effects. They do not limit the maximum resolution; instead, they reduce the contrast at lower spatial frequencies.

Although these considerations give a good feel for the magnitude of the various aberrations, calculation of the probe intensity profile and optimization of the conditions require a wave-optical formulation of the problem. As seen from Eq. (6.2) the geometrical aberrations are not independent terms to be added in quadrature, but one aberration can be balanced against another to produce an optimum result. As already noted, in the case of a C_S limited system, the optimum conditions were established a long time ago [23,24]. The optimum aperture is chosen to allow one wavelength of third-order spherical aberration at its perimeter, and is therefore given by

$$\alpha_{opt} = \left(\frac{4\lambda}{C_S}\right)^{1/4}. \tag{6.5}$$

Optimum defocus is chosen to provide an exactly compensating contribution at the aperture rim and is given by

$$\Delta f_{opt} = -\sqrt{C_S \lambda}, \tag{6.6}$$

and the two contributions give a maximum aberration of $-\lambda/2$ at an angle of $\alpha_{opt}/\sqrt{2}$. The probe profile is then quite similar to an Airy disk distribution and the resolution due to aberrations alone d_a is normally taken as the Rayleigh criterion for a circular aperture, the radius of the first zero, $0.61\,\lambda/\alpha_{opt}$, giving

$$d_a = 0.43\lambda^{3/4}C_S^{1/4} \tag{6.7}$$

This is also approximately the full-width-half-maximum (FWHM) of the probe intensity profile, which we shall use as the definition of probe size. Similar considerations apply to aberration-corrected probe-forming systems. The optimum resolution for a C_5 limited system is [25]

$$d_a = 0.43\lambda^{5/6}C_5^{1/6} \tag{6.8}$$

and for a C_7 limited system

$$d_a = 0.43\lambda^{7/8}C_7^{1/8} \tag{6.9}$$

although by this time nonround aberrations are likely to be limiting the resolution quite significantly [25,26].

These limiting values for resolution do not include any contribution due to the geometric source size. Clearly, for the highest possible resolution, the size of the geometrical image of the source, d_s, should be arranged to be significantly less than the size of the spot determined by aberrations, d_a; otherwise, the image will be blurred and the resolution degraded. Sources are usually considered to be incoherent emitters, with each point on the source emitting independently. Then the overall probe size d is approximately given by adding the two contributions in quadrature,

$$d = \sqrt{d_s^2 + d_a^2}. \tag{6.10}$$

For $d_s \ll d_a$, we achieve the electron-optical resolution limit. Under these conditions the probe will therefore be highly coherent, that is, each point on the objective aperture and each point in the probe profile will have strong and constant phase relationships with each other. At the opposite extreme, if d_s is comparable or larger than d_a, we have an essentially unaberrated image of the source, an incoherent probe. Lattice imaging, both BF and Z-contrast, relies on interference to form the image; the incoherent portion of the probe reduces the visibility of the interference fringes or may eliminate them altogether. The lattice image is generated by the coherent part of the probe d_a and blurred by the incoherent part d_s. The effect can be modeled by convoluting a simulated image with a Gaussian of FWHM d_s. To see a spacing d with high contrast, the source contribution d_s must be small compared with d.

Calculation of the actual probe profile must be done wave-optically, by integrating the contributions of partial waves within the objective aperture, including their phase aberration term $\exp(i\,\gamma)$. For convenience, we separate the electron wave vector into one longitudinal and two transverse components, i.e., $\mathbf{k} = (\mathbf{K}, k_z)$, where we have defined $|\mathbf{k}| = 1/\lambda$, and similarly the object space $\mathbf{r} = (\mathbf{R}, z)$, where $\mathbf{R}$ is a vector representing the transverse coordinates and z is the coordinate along the optic axis. Gaussian focus is at $z = 0$. The phase error from the Gaussian sphere to a transverse point $(\mathbf{R}, 0)$ is then $\exp(2\pi i \mathbf{K} \cdot \mathbf{R})$, so the probe amplitude at point $\mathbf{R}$ in object space is obtained by integrating the aberrated partial waves over the objective aperture,

$$P(\mathbf{R}) = \int e^{2\pi i \mathbf{k} \cdot \mathbf{R}}\, e^{i\gamma(\mathbf{k})}\, d\mathbf{K}, \tag{6.11}$$

which is simply the Fourier transform of the aberrated wave over the objective aperture. The probe intensity profile is then

$$P^2(\mathbf{R}) = \left| \int e^{2\pi i \mathbf{k} \cdot \mathbf{R}} \, e^{i\gamma(\mathbf{k})} \, d\mathbf{K} \right|^2, \tag{6.12}$$

Figure 6.6 shows probe intensity profiles as a function of defocus for a 300-kV STEM with and without the correction of third-order spherical aberration. Note that the defocus is inserted into the aberration function to give the probe profile in the plane, $z = \Delta f$. The profiles are shown on the same axes and for the same total current in the probe. It is immediately clear that aberration correction not only gives a smaller probe, but also much higher peak intensity. This is very important for the imaging of individual atoms or nanostructures, because it results in a higher signal-to-noise ratio.

Figure 6.7 compares an image of Pt_3 trimers on a γ-Al_2O_3 substrate taken before [27] and after [28] aberration correction. The improved image quality is striking, both in terms of resolution and signal-to-noise ratio. It is significant that the three Pt atoms do not form an equilateral triangle. This is visible in the uncorrected image, but the corrected image allows a more precise measure of the atomic configuration. The explanation for the distorted shape comes through density functional total energy calculations. Placing bare Pt_3 trimers on a γ-Al_2O_3 $\langle 110 \rangle$ surface and relaxing the structure to equilibrium results in an almost equilateral triangle with bond lengths of 2.59, 2.65, and 2.73 Å, close to the interatomic spacing in metallic Pt. The longer bonds can only be explained by adding an OH group to the top of the trimer, when two of the bonds lengthen to 3.1 and 3.6 Å, in excellent agreement with observation. The addition of the OH group also changes the electron density on the Pt, from slightly electron-rich to electron-poor, which explains the observed chemical nature of the Pt [28].

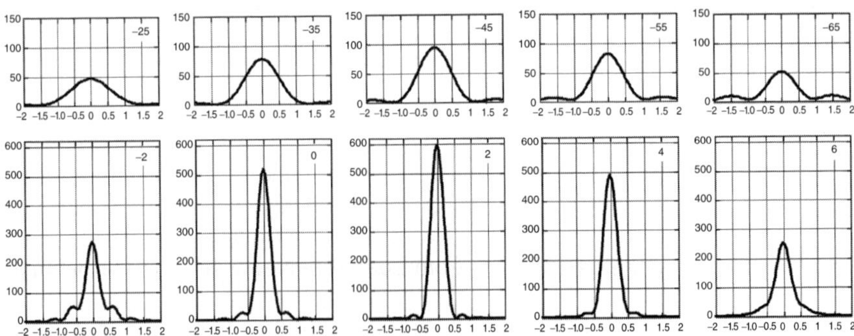

FIGURE 6.6. Through-focal series of probe intensity profiles for the uncorrected VG Microscopes HB603U STEM (*upper*) and after correction of third-order spherical aberration (*lower*), for the same incident beam current and shown on the same scale. Numbers in the top right of each plot denote defocus in nm. Parameters are: accelerating voltage 300 kV, energy spread 0.3 eV FWHM, $C_c = 1.6$ mm; uncorrected probes are with $C_S = 1$ mm and $\alpha_{opt} = 9.5$ mrad, corrected probes have $C_S = -37$ μm, $C_5 = 100$ mm, and $\alpha_{opt} = 22$ mrad. Horizontal scales are in Å.

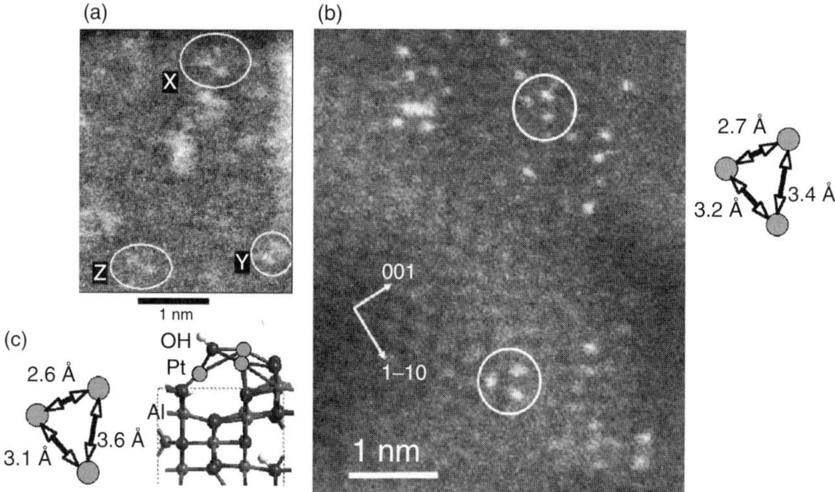

FIGURE 6.7. Images of Pt_3 trimers on γ-Al_2O_3 $\langle 110 \rangle$ surfaces obtained (a) before [27], and (b) after aberration correction, showing the improved resolution and contrast. Measured spacings from (b) are shown on the right, and closely match with first-principles simulations (c) if the Pt_3 is capped with an OH group [adapted from ref. 28].

2.2. Image Contrast

The BF and ADF detectors in STEM can be arranged to give very different and complementary images. Figure 6.8 shows a through-focal series of images of gold nanoparticles. At each focus value images were recorded simultaneously with the two detectors, although only a few representative values of defocus are shown from each series. Linescan A shows how the dark field image is sensitive to individual gold atoms; isolated gold atoms can be seen away from the nanocrystal out on the carbon film. These atoms provide an absolute intensity calibration; peaks of the same height that are seen at the edge of the nanocrystal can therefore be identified as single gold atoms. As the gold atoms move into the nanocrystal, the intensity increases according to the number of atoms in the column [29,30]. Notice that the image does not reverse contrast at any defocus value. These are the characteristics of an incoherent image, as familiar from normal photography, and the optimum focus is therefore that which reveals the nanocrystal with the best contrast. In an incoherent image the intensity is a simple convolution of a (positive) function representing the scattering cross section of the object, $O(\mathbf{R})$, and a resolution function, which for weakly scattering objects is the incident probe intensity profile, $P^2(\mathbf{R})$ i.e.,

$$I(\mathbf{R}) = O(\mathrm{R}) \otimes P^2(\mathbf{R}), \qquad (6.13)$$

The BF image, on the other hand, gives an image equivalent to that obtained with an aberration-corrected TEM, a coherent phase contrast image. The gold nanocrystal is seen with high contrast and minimal Fresnel fringes blurring the

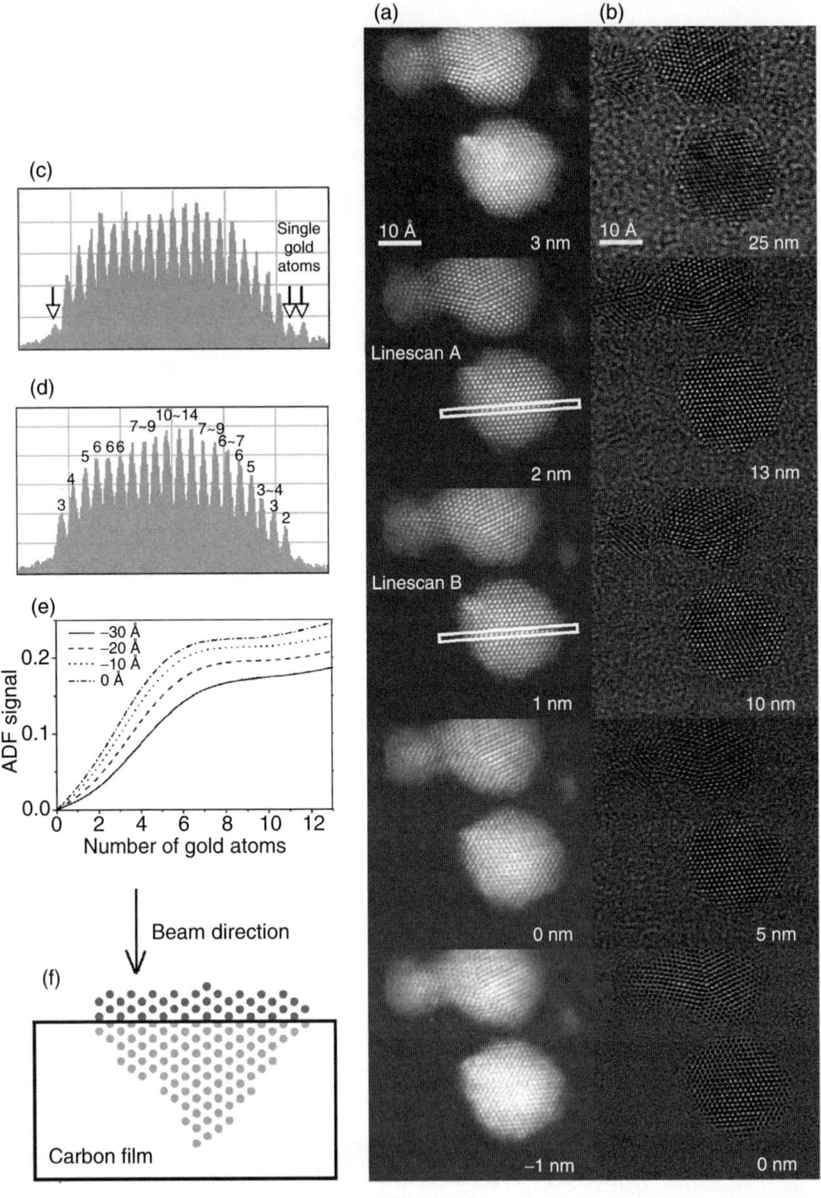

FIGURE 6.8. Selected images of gold nanoparticles supported on a thin carbon film taken from a through-focal series of (a) ADF Z-contrast images and (b) simultaneously collected BF phase contrast images, showing their very different characteristics. Defocus is shown in the lower right corner of each frame. Linescan A across the Z-contrast image shows single gold atoms (c). Linescan B can be quantified to give the number of gold atoms per column (d), based on image simulations (e), and allow the approximate 3D shape of the nanocrystal to be estimated (f) [adapted from ref. 30].

edge of the particle (which is often referred to as image delocalization). However, the image changes form with defocus, reversing contrast from 0 to 5 nm, and does not detect the individual gold atoms. The reason that individual gold atoms cannot be located from the BF image alone is because of the speckle pattern contrast from the thin carbon film used to support the sample. The signal of a single gold atom cannot be distinguished from the speckle of the support, whereas in a Z-contrast image the single gold atom ($Z = 79$) scatters about as many electrons as the whole thickness of the carbon support ($Z = 6$). The advantage of the phase contrast image is that it is more sensitive to light atoms. Figure 6.9 compares the images of multiwall carbon nanotubes in Z-contrast and phase contrast modes. The multiwall structure can just be seen in the Z-contrast image, but shows stronger contrast in the phase contrast image.

The reason for the very different forms of the BF and ADF images is quite simple to understand. It is related to the different angular sizes of the detectors. In Section 2.1, we have seen that for high-resolution imaging the incident probe is a coherent spherical wave (slightly aberrated) that converges onto the specimen. It generates some scattered waves which propagate out of the specimen and onto the detectors. Most of these scattered waves remain coherent with the unscattered beam, and will interfere with it. The important length scale here is set by the atomic spacings d, which are around 1–3 Å in all materials and lead to diffraction at angles $n\lambda/d$ where n is an integer. Typical first-order diffraction angles are around 10–20 mrad. Detectors that are *small* on this scale are sensitive to the details of the interference pattern, and phase contrast results. On the other hand, detectors that are *large* with respect to typical diffraction angles will not be sensitive to any fine details of the interference pattern, but only to the overall intensity integrated over a large number of diffraction peaks. A large detector, therefore, gives an image whose contrast is based on *intensities*, which is the definition of an incoherent image. This is illustrated schematically in Fig. 6.10, which shows the diffraction pattern in the detector plane for a simple cubic crystal.

The central disk is just the transmitted probe falling on the detector, and all the other disks are caused by diffraction. The BF detector is small in angular extent compared to the diffraction angles, while the ADF detector is large. Note that if

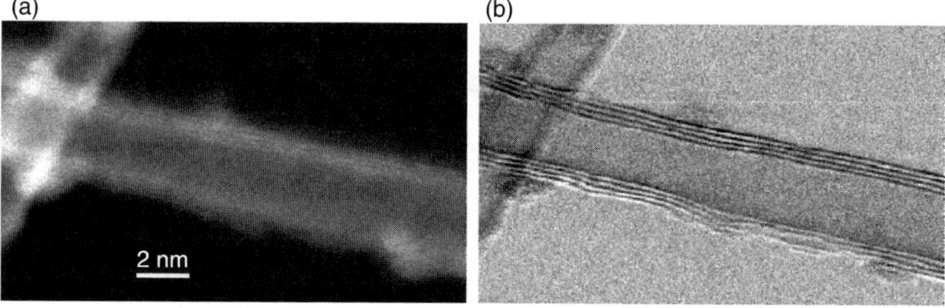

FIGURE 6.9. Images of multiwall carbon nanotubes in (a) ADF Z-contrast and (b) BF phase contrast modes.

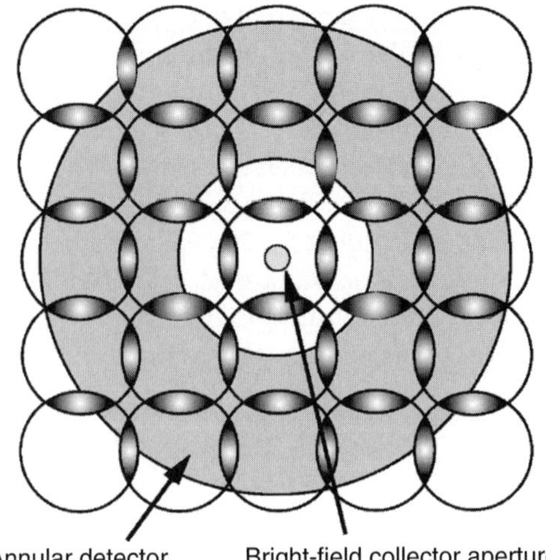

FIGURE 6.10. Schematic showing the diffraction pattern in the detector plane for a cubic crystal. Each diffracted beam has an angular extent defined by the incident probe, so the pattern takes the form of overlapping disks. The region of overlap controls the image contrast. For the spacing shown, the BF image shows no lattice resolution but the ADF detector does.

you were to look at an actual diffraction pattern, then it would not show spots or disks at high angles because of the influence of thermal vibrations. Each successive electron can see the atoms in a slightly different position because of their thermal motion, and the details of the pattern are different at high angles. The actual pattern from many scattering events looks diffuse, and the scattering is referred to as thermal diffuse scattering. Thermal effects actually assist in breaking the coherence of the ADF image.

For the same reason, the overall form of the ADF image is relatively insensitive to the thickness. Dynamical diffraction shifts the amplitude between different diffracted beams, which causes changes in the details of the diffraction pattern. The phase contrast image changes form and can reverse contrast as the thickness increases. However, the Z-contrast image does not show any dramatic changes in form, as seen from Fig. 6.8. The reason again is that changes in the interference pattern are on a local scale with respect to the ADF detector and are integrated by it. Only the changes in the total intensity falling on the detector affect the image. The thickness dependence of the Z-contrast image is therefore much simpler than that of a BF image. The signal increases initially as the number of atoms increases, but then reaches a point where much of the probe has been scattered onto the detector and the signal saturates. This leads to characteristic curves that have approximately the same shape for all materials, scaled by the overall Z^2 factor of the scattering cross section. The simulated thickness dependence for gold is shown in Fig. 6.8e for various values of defocus, and all curves show this characteristic thickness behavior.

These simulations can be compared with the experimental data to estimate the number of atoms in each column of the nanocrystal. The intensity scale is calibrated by the intensity of the single gold atom in linescan A. Short columns at the edge of the nanocrystal are in the linear range of the thickness dependence, and the precision in column length is just a single atom. In the thicker region of the nanocrystal, where saturation sets in, the precision is lower. The defocus scale is set by the image where the single atoms show the best contrast, linescan A. Optimum defocus for these conditions is 2 nm. The maximum contrast in the center of the nanocrystal is obtained at a defocus of 1 nm (linescan B), showing that the entrance face of the nanocrystal is flat to about 1 nm. Hence the approximate 3D shape of the nanocrystal can be determined, as shown in Fig. 6.8f.

Figure 6.10 also illustrates another important difference between BF and ADF imaging, the achievable resolution. Image contrast requires interference, and the example of a crystal specimen shows clearly that for this crystal spacing the BF detector sees no interference because all the diffracted disks fall outside the detector. The atoms are too close to be resolved. The ADF detector, however, does see interference effects, and will show an image. In practice, the situation may be more complicated because the BF detector can see contrast formed as a result of multiple scattering. Sum and difference frequencies can appear giving spurious image features, and image simulations are important to interpret the complicated contrast effects that result.

It seems obvious that in an incoherent image, the smaller the probe the better the resolution, or the sharper the atoms. This is true at present-day resolutions of around 0.6–1 Å, because the probe is still large compared to the region of the atom that scatters to high angles, which is just a few tenths of an angstrom. The images of Pt atoms seen in Fig. 6.7 are therefore the images of the probe rather than the images of the atoms. The same applies to crystals. As the probe is sharpened, it picks out the individual columns of atoms with more contrast; more of the probe strikes the target column and less is spread over neighboring columns. This can also be seen from the diffraction perspective in Fig. 6.10. Making the probe smaller requires enlarging the probe-forming aperture, and the diffraction disks will overlap to a larger extent. As the probe scans a crystal, the overlap region experiences alternating constructive and destructive interference. The intensity changes from white to black to white as the probe scans from one column to the next. Since it is the overlap region that gives the contrast, larger overlaps mean larger contrast. Quantitatively, this is usually expressed in terms of a transfer function in reciprocal space,

$$I(\mathbf{K}) = O(\mathbf{K}) \cdot T(\mathbf{K}), \qquad (6.14)$$

where $O(\mathbf{K})$ is the spatial frequency spectrum of the object and $T(\mathbf{K})$ is an optical transfer function (OTF) or modulation transfer function of the imaging system. Equation 6.14 is the Fourier transform of Eq. 6.13. Therefore, for a thin object, the OTF is the Fourier transform of the probe intensity. If the probe is much sharper than any details in the object, its Fourier transform is very broad in reciprocal space, so all spatial frequencies in the object are transferred to the image and a true representation of the nanostructure is seen. However, even the aberration-corrected probe is still significantly larger than the screened nuclear potential that scatters the beam to the ADF detector. High spatial frequencies in the object are therefore seen with reduced contrast in the image. This is how the probe controls the transfer of spatial frequencies to the image.

Figure 6.11a shows transfer functions for the ORNL VG Microscopes HB603U 300-kV STEM, before and after aberration corrections up to third-order, and the expected transfer function for a fifth-order corrected 200-kV Nion UltraSTEM. All have a generally triangular transfer function, with slowly decreasing transfer at higher spatial frequencies as the diffraction disks overlap progressively less. There is about a factor two improvement in resolution for each order of aberration correction, in agreement with the rough estimates from Fig. 6.5. Note that the information transfer cutoff for the third-order corrected microscope is near 0.5 Å, whereas the fifth-order corrected microscope still shows good contrast in this regime despite an accelerating voltage of only 200 kV.

Also shown is the transfer function for the third-order corrected 300-kV STEM with an unrealistically high energy spread of 1 eV, showing how the information limit and therefore resolution in the STEM is robust to chromatic aberration effects [31]. The explanation is that near the limit of resolution the diffraction disks are overlapping only at the edge of each disk. The beams responsible for the image contrast are passing almost symmetrically on either side of the optic axis. The relative phase of beams at equal angles to the optic axis is insensitive to a change in focus, and so there is an achromatic line along the center of each overlap region. Only away from this line is the interference sensitive to defocus, and hence as the disks overlap more, the image contrast becomes more sensitive to chromatic instabilities. Thus it is the mid-range spatial frequencies that suffer reduced contrast. This is an important advantage compared with the BF phase contrast image, in which chromatic instabilities have always been and remain an important limiting factor for achieving the highest resolution.

Equation 6.14 applies only if the specimen is sufficiently thin that the probe profile inside the specimen is the same as the incident probe profile; in other words, there has been no significant dynamical scattering or beam broadening. Figure 6.11b compares the Fourier transform of a simulated image of Si $\langle 112 \rangle$ just 1-nm thick with the Fourier transform of the incident probe. The probe indeed acts as the envelope of the frequencies seen in the image. The simulation was performed using Bloch wave calculations [32,33], and the probe for this comparison has been convoluted with a Gaussian source of size $d_s = 0.31$ Å, reducing the high-frequency transfer significantly. As the specimen becomes thicker, the incident probe is no longer the effective transfer function because dynamical scattering and beam broadening take place. The general effect is to add a uniform background to the image, reducing the magnitude of the contrast over the whole frequency range. The best defocus is no longer the electron optical optimum defocus, but one that pushes the focus into the specimen so that more of the specimen thickness is close to optimum focus. However, the image retains its incoherent characteristics and the optimum focus can still be chosen by eye to be the focus that gives the sharpest image.

The final resolution is determined not by the aperture cutoff, which is 0.45 Å for the 22-mrad aperture used, but by the signal-to-noise ratio of the data. In the Fourier transform of the experimental data, the $(1\bar{7}3)$ reflection at 0.71 Å had a signal-to-noise ratio of about 2.5, giving a 99% confidence that it is real. However, the $(55\bar{5})$

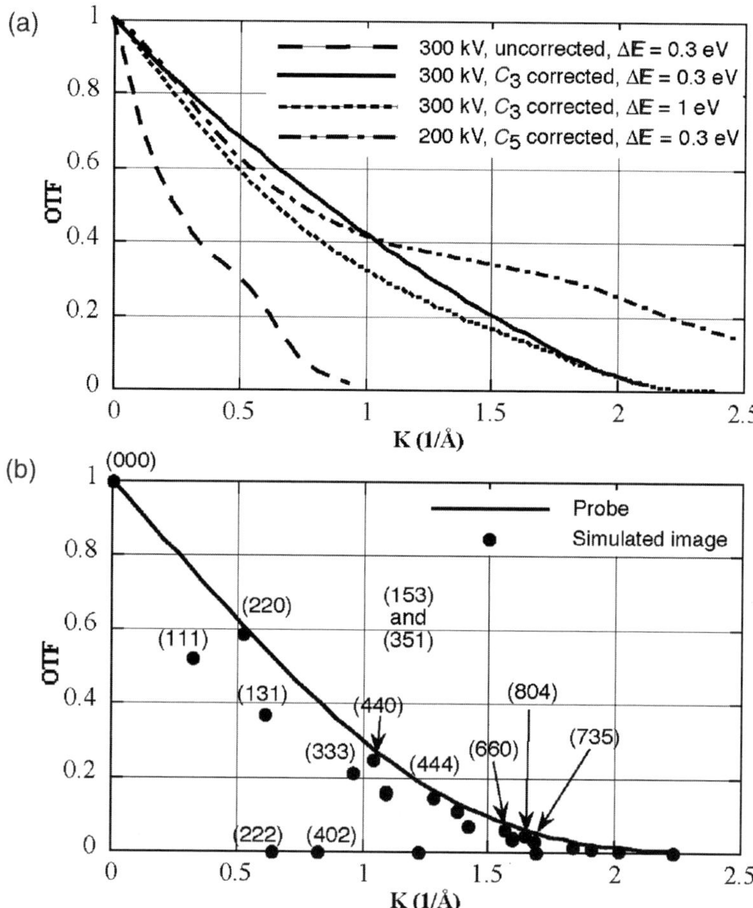

FIGURE 6.11. (a) Transfer functions for incoherent imaging in the ORNL 300 kV STEM before and after correction of third-order aberrations showing the greatly extended information transfer. Parameters are C_s = 1.0 mm, C_c = 1.6 mm, α_{opt} = 9.5 mrad, Δf = −45 nm for the uncorrected case and C_s = −37 µm. C_5 = 100 mm, α_{opt} = 22 mrad, Δf = 2 nm after correction, energy spread 0.3 eV in both cases. The dotted line shows the effect of increasing the energy spread to 1 eV, giving reduced midrange transfer but no reduction in resolution limit. Also shown is the transfer expected for a 200-kV STEM with full fifth-order aberration correction (C_s = 23 µm, C_5 = −2 mm, C_7 = 100 mm, C_c = 1.5 mm, energy spread 0.3 eV). (b) Comparison of the Fourier transform of a simulated image of Si $\langle 112 \rangle$ with the Fourier transform of the probe showing how the latter is acting as the transfer function for the image (C_s = −37 µm, C_5 = 100 mm, C_c = 1.6 mm, energy spread = 0.3 eV, α_{opt} = 22 mrad, Δf = 2 nm, thickness = 1 nm, source size = 0.31 Å, detector angles 90–200 mrad).

reflection at 0.63 Å had a signal-to-noise ratio of only 1.4, giving a confidence level of only 84%. It can be seen that the Fourier transform method is a valuable means for accurate determination of information limit. However, a word of warning is needed. Great care must be taken to ensure that the recorded data is a true representation of the intensity. No black level should be used to enhance contrast, because any clipping of the signal will introduce spurious high frequencies. Similarly, spurious spots can be introduced from instabilities in the scans or selection of a section of image whose dimensions are not exact numbers of lattice spacings. Information transfer, although important, does not define resolution, which should always be determined from a real space image as shown in Fig. 6.3 [15, M. A. O'Keefe, L. F. Allard, and D. A. Blom, J. Electron Microsc. 54 (2005) 169.].

It may seem surprising at first sight that the STEM BF image is equivalent to that obtained in an aberration-corrected TEM, but this is solidly based on time-reversal symmetry and is known as the reciprocity principle. To make the equivalence clear, Fig. 6.12 compares the essential optics for BF imaging in the fixed beam TEM and the STEM. The ray diagrams are shown for one image pixel, and it is seen that for equivalent apertures the only difference is the direction of electron propagation. Elastic scattering is the basis of phase contrast imaging, and depends only on scattering angle and not on propagation direction. Therefore, the two optical arrangements will give identical results; the TEM condenser aperture is equivalent to the STEM BF collector aperture and the objective aperture remains the same. Note, however, that the terminology can be confusing: in a TEM/STEM microscope, the STEM objective (probe-forming) aperture is usually the same

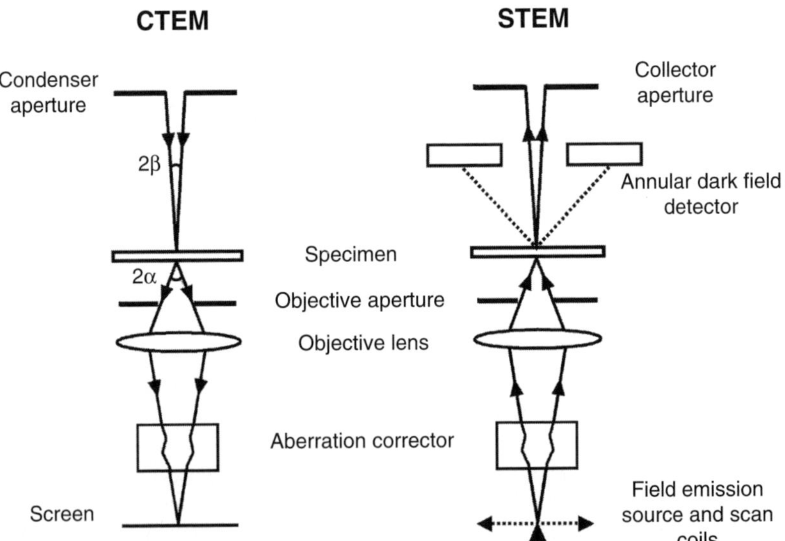

FIGURE 6.12. Ray diagrams for TEM and STEM to illustrate the reciprocal nature of the bright field images. The BF image in aberration-corrected STEM gives an image optically equivalent to that obtained with aberration-corrected TEM.

aperture used as the condenser aperture in TEM mode. The difference between the two arrangements is that in the TEM all pixels in the image are obtained in parallel, whereas in the STEM they are obtained one at a time by scanning the probe.

The BF image amplitude is now given by a convolution of the exit face wave function ψ_e (**R**) with the complex impulse response function of the lens P (**R**) given by Eq. 6.11. The image intensity becomes the square of the convolution,

$$I(\mathbf{R}) = |\psi_e(\mathbf{R}) \otimes P(\mathbf{R})|^2 \tag{6.15}$$

This is much more complicated than the corresponding expression for an incoherent image (Eq. 6.13), which is a convolution of intensities. In phase contrast imaging it is the amplitudes that are transferred through the lens system, and the intensity is taken later. As noted earlier, interference in the detector plane is important for coherent imaging, and we cannot approximate the image to any simple incoherent form as we can for the ADF image. This means that coherent image contrast can be varied from black to white through a change in one of the terms of the aberration function, for example, the defocus. It also means that information on the phases is lost on taking the image intensity, and unique structure inversion becomes difficult.

Therefore, resolution of a BF phase contrast image cannot be defined on the basis of image intensities but on the amplitude transmission characteristics of the objective lens. It is conventionally defined in terms of a weak phase object. The specimen is considered to be vanishingly thin, so it acts as a weak phase grating, refracting the incident beam but not changing its amplitude. It is simplest to consider the TEM geometry, when the exit-face wave function is given by

$$\psi_e(\mathbf{R}) = \exp\{-i\sigma\phi(\mathbf{R})\} \approx 1 - i\sigma\phi(\mathbf{R}), \tag{6.16}$$

where $\sigma = 2\pi me\lambda/h^2$ is the interaction constant and $\phi(\mathbf{R}) = \int \phi(\mathbf{r})dz$ is the projected potential. Phase changes in the exit wave therefore map the projected potential of the specimen. These phase changes can be represented vectorially by the generation of a scattered wave $\psi_s(\mathbf{R}) = -i\sigma\phi(\mathbf{R})$, which is oriented at $\pi/2$ to the incident wave as shown in Fig. 6.13a. The principle of phase contrast imaging is to rotate the phase of the scattered beams by an additional $\pi/2$. This converts the phase changes in the exit-face wave function to amplitude changes in the image, to give an image amplitude

$$\psi_i(\mathbf{R}) \approx 1 - \sigma\phi(\mathbf{R}). \tag{6.17}$$

In light optics this can be done with a phase plate, but in electron optics we must use the lens aberration function itself. We therefore try to optimize the aberration function γ to be as close to $\pi/2$ as possible over as large a range of angles as possible. Expressing the aberration phase change as $e^{i\gamma} = \cos\gamma + i\sin\gamma$, it is clear that it is the sin term that we need to maximize, and $\sin\gamma$ is defined as the phase contrast transfer function. Figure 6.13 compares phase contrast transfer functions for a 300-kV STEM before and after aberration correction. Since it is the lens aberrations that are used to give the image contrast, both begin at zero contrast for low spatial frequencies where the aberration function is zero.

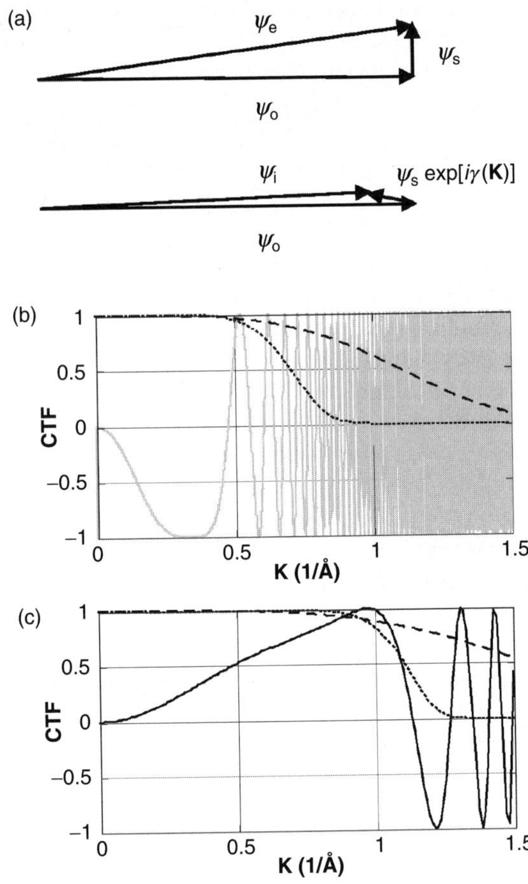

FIGURE 6.13. (a) Vector diagram for a weak phase object showing the incident beam, ψ_0, a small scattered beam, ψ_s, at $\pi/2$ to the incident beam, and the resultant exit face wave function, ψ_e. Lens aberrations are used to rotate the phase of the scattered beam an additional $\pi/2$ to create amplitude contrast from the phase changes. (b) Contrast transfer functions (CTF) for an uncorrected 300-kV microscope, gray line ($C_s = 1.0$ mm, $C_c = 1.6$ mm, $\Delta f = -44$ nm), with the damping envelopes introduced by a beam divergence of 0.25 mrad (dotted line) and an energy spread of 0.6 eV FWHM typical of a Schottky (thermal) field emission source (dashed line), (c) transfer for a corrected 300-kV microscope, solid line ($C_s = -37$ μm, $C_5 = 100$ mm, $C_c = 1.6$ mm, $\Delta f = 5$ nm), with the damping envelopes introduced by a beam divergence of 2.5 mrad (dotted line) and an energy spread of 0.3 eV typical of a cold field emission source (dashed line).

With increasing angle to the axis (increasing spatial frequency), the aberration function increases quickly for the uncorrected case, oscillating rapidly for spacings below 2 Å. This rapid oscillation means different frequencies are seen with different contrast, but also sets stringent conditions on the maximum beam divergence that can be used (the incident beam divergence or condenser aperture size in TEM, the collector aperture size in STEM). The angular aperture must be kept small on the scale of the oscillations or the aperture will average over the oscillating transfer function and the image contrast will be reduced. The effect can be simulated with an exponential damping function

$$D_\alpha = \exp\left[-\pi^2\,\alpha^2\,K^2\,(\Delta f + \lambda^2\,K^2 C_S + \lambda^4 K^4 C_5)^2\right] \qquad (6.18)$$

which multiplies the transfer function. It is shown in Fig. 6.13b for an aperture semiangle α of 0.25 mrad (modeled as a Gaussian with standard deviation 0.25 mrad [34]). After aberration correction the oscillations are pushed to higher spatial frequencies, resulting in an increased pass band before contrast reversals set in. Another consequence of the reduced aberrations is that the damping factor is also much reduced, which has important implications for the STEM. It allows the STEM collector aperture to be increased by a factor 10 to 2.5 mrad, as shown in Fig. 6.13c. The collection efficiency increases by two orders of magnitude, and BF STEM becomes a viable means for obtaining phase contrast images with the advantage that ADF images are available simultaneously and EELS can be performed with the same probe used for imaging.

Chromatic aberration effects due to a spread in beam energy or objective lens current lead to a focus spread, given by Eq. 6.4, and will again decrease the transfer. In phase contrast imaging, the effect is also described by an exponential damping factor [35]

$$D_E = exp\left[-0.5\,\pi^2\,\lambda^2\,K^4\,\Delta^2\right] \qquad (6.19)$$

and applies irrespective of whether the geometric aberrations are corrected or not. It is shown in Fig. 6.13b for an energy spread Δ of 0.6 eV as appropriate for a thermal field emission gun, and in Fig. 6.13c for a spread of 0.3 eV appropriate for a cold field emission gun.

3. Spectroscopic Imaging

One of the key advantages of the STEM is the availability of the EELS signal, which can be obtained simultaneously with an atomic-resolution ADF image [36–40]. It is therefore straightforward to place the probe on an atomic column seen in the image and obtain a spectroscopic analysis. EELS reveals elemental features through characteristic core loss edges corresponding to inner shell excitations into the first available unoccupied states. The fine structure on such edges therefore provides information on the density of states seen in the vicinity of the excited species. EELS is similar to x-ray absorption spectroscopy, except that it can provide atomic-scale spatial resolution. The brightness of the STEM probe substantially exceeds that of third-generation synchrotron sources, making the STEM a powerful means for analyzing electronic structure and identifying impurity species or dopants within nanostructures. Spectroscopic imaging can also be achieved in the TEM through use of an energy filter; images are obtained in parallel and the energy is scanned. In the STEM the spectrum is obtained in parallel and the image is scanned. The key difference between the two arrangements is that in the TEM the energy loss electrons need to be brought to a focus on the image plane, and different energies will need a different focus. For an atomic-resolution image, the spread of energies allowed to contribute to the image needs to be a fraction of an eV to avoid chromatic damping, as discussed earlier. Therefore, given the small cross section for inner shell excitations, very little signal would be detected, and so far no atomic resolution spectroscopic images have been obtained in TEM mode. In the STEM arrangement, the focusing is done before the specimen; although there may be coupling lenses into the spectrometer, the focus precision required is only that needed to maintain the energy resolution on the CCD chip, which is orders of magnitude less than needed to maintain atomic resolution in the TEM image.

With the successful correction of aberrations in the STEM, the probe is now smaller and brighter, bringing major gains for EELS analysis as well as imaging. The smaller probe brings not only better spatial resolution but also more current on an individual column, resulting in a better signal-to-noise ratio and a higher sensitivity analysis. The highest sensitivity so far achieved has been the spectroscopic identification of an individual atom in a single atomic column of a thin crystal [6]. The sample was specially made by molecular beam epitaxy to contain various known concentration of La dopants in specific layers in a $CaTiO_3$ matrix. The lowest concentration layer was $La_{0.002}Ca_{0.998}TiO_3$, grown just one-unit-cell thick. A cross-section sample revealed bright spots as shown in Fig. 6.14. The expected frequency in the thin regions of the sample was consistent with the concentration and sample thickness, confirming that individual columns contained mostly individual atoms. Placing the probe over any one of the bright columns,

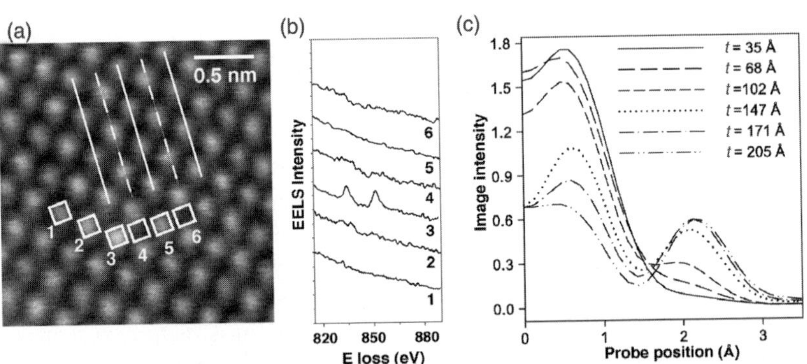

FIGURE 6.14. Spectroscopic identification of an individual atom in its bulk environment by EELS. (a) Z-contrast image of $CaTiO_3$ showing traces of the CaO and TiO_2 {100} planes as solid and dashed lines, respectively. A single La dopant atom in column 3 causes this column to show slightly brighter than other Ca columns, and EELS from it shows a clear La $M_{4,5}$ signal (b). Moving the probe to adjacent columns gives reduced or undetectable signals. (c) Dynamical simulation of the La $M_{4,5}$ signal, as the probe is scanned from column 3 through column 4, calculated for the La atom at different depths below the surface [adapted from ref. 6].

the EELS spectrum showed two distinct peaks corresponding to the La $M_{4,5}$ lines at 832 and 849 eV. This demonstrates the first spectroscopic identification of a single atom in its bulk environment. In addition, placing the probe at the neighboring TiO or O column positions that are located 2.8 and 1.9 Å away, respectively, the majority of the La $M_{4,5}$ signal disappeared, leaving a residual of about 10% and 20%, respectively. These distances are large compared to the probe size, and the residual intensity is due to dynamical diffraction and beam broadening that occur between the probe entering the crystal and reaching the depth of the atom.

It is now possible to perform full quantum-mechanical simulations for EELS core loss images, using accurate atomic wave functions and a full treatment of dynamical diffraction [6,41]. Simulated linescans are shown in Fig. 6.14 running from the Ca column through the adjacent O column for the La atom located at different depths. With the La atom close to the probe entrance surface, maximum signal occurs near the Ca column reducing monotonically toward the O column. With the La atom located deeper in the crystal, more of the probe intensity is scattered off the column before it reaches the atom, resulting in less signal when the probe is over the Ca column and correspondingly more signal when the beam is over the O column. The ratio of the two, therefore, indicates roughly the depth of the atom within the crystal and matches the experimental data for the La at around 100 Å in depth.

Figure 6.14 predicts that the peak La $M_{4,5}$ signal should appear just off the Ca column. This reflects the slightly less localized nature of the inelastic interaction with respect to the elastic image. The spatial resolution of a core loss EELS image is not therefore expected to be quite as high as that of the corresponding ADF image. With the probe centrally located over the Ca column, there is maximum scattering into the ADF detector and minimum collected by the spectrometer. As the probe moves off the column the ADF scattering reduces faster than the EELS image, and the result is an increase in EELS signal.

There have been a large number of simplified treatments of EELS localization in the past, with conflicting conclusions. These calculations now provide a definitive answer. Figure 6.15 shows plots of the FWHM across EELS images of single atoms, thus removing any contribution of channeling effects and leaving only the probe itself and any ionization delocalization effect [42]. Results are shown for two aberration-free probes with 10 mrad and 20 mrad objective semiangles, resulting in FWHM intensities of 1.0 and 0.5 Å, respectively. The calculations reveal that the extent of the ionization delocalization can be of the order of 1 Å for the lighter elements.

Another potential pitfall using core loss excitations arises because they have a much lower cross section than the elastic (or quasielastic) scattering used to form the ADF image. To achieve a usable signal, beam exposure per pixel is often increased from microseconds for an ADF image to seconds for a core loss such as the O K edge near 532 eV. Consequently, there is a much-increased likelihood of beam damage occurring. A time sequence of spectra is useful to test for beam-induced damage artifacts. If the first and last spectrum look identical, then beam damage is not likely to be a problem. Various techniques can be used to reduce

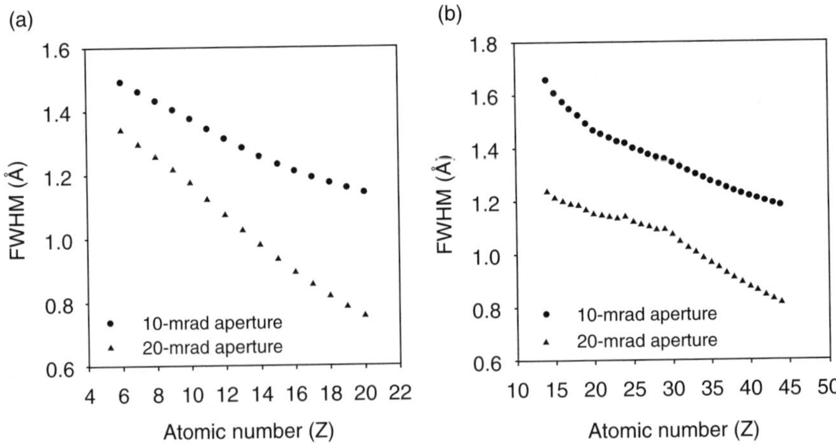

FIGURE 6.15. Calculated plots of the FWHM intensity across EELS images of single atoms using (a) K-shell (b) L-shell edges for two different probes. Calculations assumed an EELS collection semiangle of 20 mrad and an energy window of 40 eV [adapted from ref. 42].

exposure, such as rastering the beam in a 1 Å by 1 Å image or scanning a 2D image but collecting the EELS data only along the slow scan direction, but all reduce the spatial resolution of the data collected. Another useful reference, if possible, is to compare the spectra obtained by scanning a large area with minimal exposure per pixel. With nanostructures, one spectrum from one nanostructure can be compared to the sum of ten spectra from ten identical nanostructures each taken with one tenth the exposure. Such methods are necessary to ensure that data are representative of the specimen in its original state. For details on the quantification of the data and interpretation of the spectral features, see for example the book by Egerton [43].

4. 3D Imaging

In recent years, major progress has been made in obtaining 3D information from nanostructured materials through the use of tilt series tomography. A series of views in different projections can be reconstructed into a single 3D data set, which can then be viewed in any desired orientation. Spatial resolution has been demonstrated at the nanometer level [44]. The ADF image is very suitable for this procedure because it shows much reduced diffraction contrast effects compared with a BF image. This is important since the back-projection reconstruction procedure relies on the images being good representations of the projected mass thickness, so the incoherent characteristics of the ADF image are required. An additional constraint with present reconstruction algorithms is that they assume a perfect projection, that is, they assume the incident beam to be parallel. This works in practice as long as the depth of focus is long compared to the sample thickness. However, as the desired spatial resolution increases, this requirement can no longer be met; a small probe is necessarily convergent and the projection becomes sensitive to the focus, with different parts of the sample in focus at different settings.

Aberration correction provides the increased spatial resolution by allowing the objective lens aperture to be increased. Just as in a camera, this results in a reduced depth of field. In fact, the depth of field d_z decreases as the square of the aperture angle, whereas the lateral resolution, which has been the main motivation for aberration correction, increases only linearly with aperture angle. Depth resolution on the 300-kV STEM today is on the nanometer scale, and we can optically slice through a sample simply by changing the objective lens focus. With aberration correction, we find a natural changeover from the conditions appropriate for a tilt-series reconstruction to those for optical sectioning. A through-focal series now becomes a through-depth series, which can be recombined into a 3D data set [45] in a similar manner as used for confocal optical microscopy. However, the electron technique retains single-atom sensitivity in each image if the specimen is sufficiently thin, as shown by the recent location of individual Hf atoms in a subnanometer-wide region of SiO_2 in a high-K device structure [46].

Figure 6.16 shows selected frames of a through-depth image series taken from a Pt_2Ru_4 cluster-derived catalyst on a γ-alumina support. At -8 nm defocus, the lower right-hand corner of the alumina comes into focus and shows a lattice image. Individual Pt atoms are visible as bright spots. As the defocus increases, the focused region moves toward the upper left of the field of view. Profiles across the image frames show the probe FWHM of 0.07 nm. At -16 nm defocus, a rather brighter nanocrystal raft is seen. Intensity profiles across this suggest that it is two or three monolayers in thickness. Figure 6.17 shows a Z-contrast and simultaneously recorded BF phase contrast image of Pt_2Au_4 nanoparticles supported on TiO_2. The cluster is seen best in the Z-contrast image, whereas the support is seen more clearly in the phase contrast image. The 3D rendering is shown in Fig. 6.18.

5. Recent Applications to Nanostructure Characterization

5.1. Nanotubes

The Z-contrast image is particularly suited to imaging catalyst particles with high Z that are often used to grow nanotubes. Figure 6.19a shows Co–Ni catalyst particles in a network of single-wall nanotubes grown by laser ablation. The smallest particles observed were of 2-nm diameter, showing that in this case the diameter of the nanotubes was not dictated by the particle diameter.

Figure 6.19b shows a 2-nm diameter catalyst particle with a composition profile obtained by EELS. The Co/Ni ratio is constant across the nanoparticle,

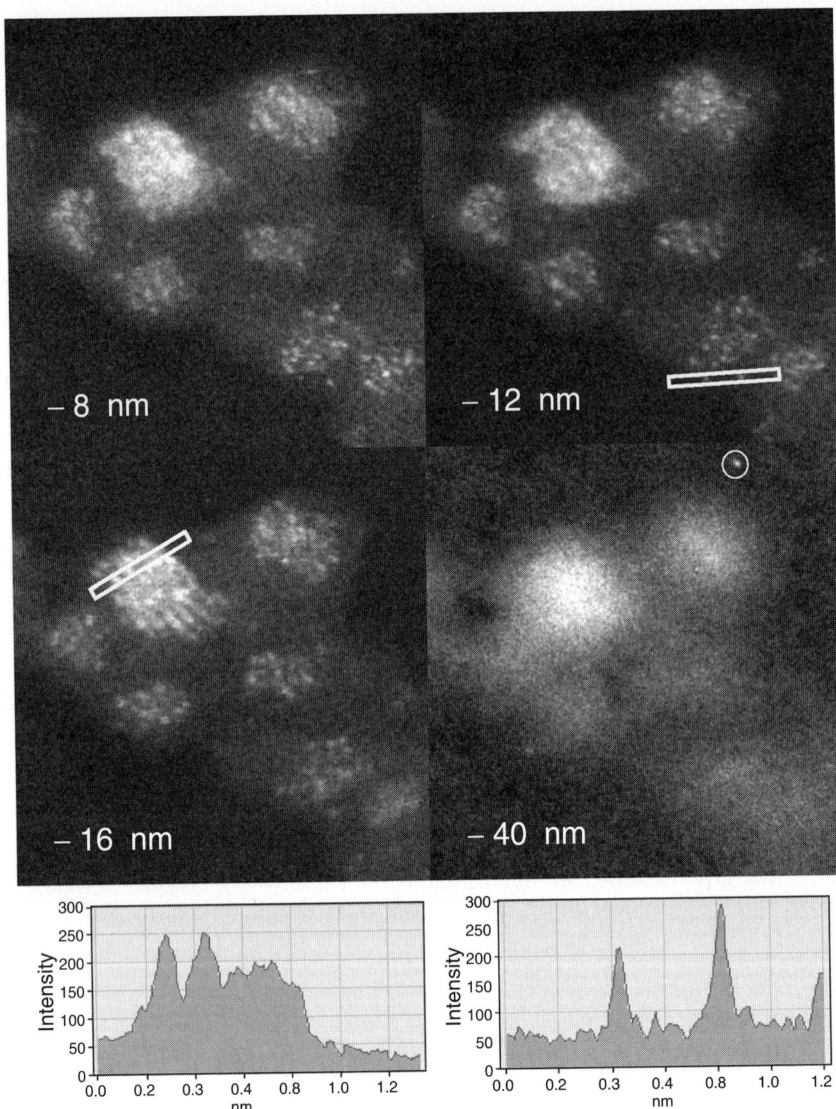

FIGURE 6.16. Four frames from a through-depth series of images of a Pt$_2$Ru$_4$ catalyst supported on γ-alumina. The alumina is 3D with thin, raft-like Pt-Ru clusters on its surface. Different clusters are resolved at different depths, and at −40-nm defocus, the carbon support film is reached and a single Pt atom comes into focus (circled). Line profiles are intensity traces along the rectangles shown in the images [adapted from ref. 45].

indicating a uniform alloy composition. The intensity of the EELS signal is highest in the center of the nanoparticle, indicating a 3D particle shape.

This is another example where density functional calculations proved highly illuminating. Microscopy could never image the nucleation processes occurring within the laser ablation plume, but theory can investigate candidate processes to

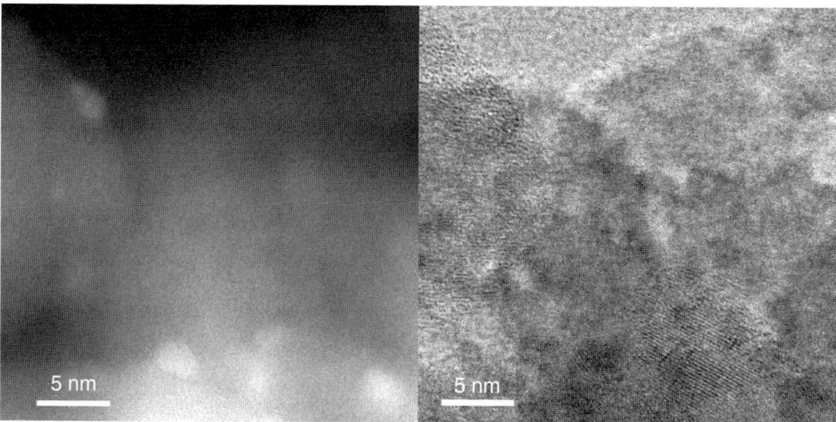

FIGURE 6.17. Z-contrast (*left*) and bright field (*right*) images of Pt_2Au_4 nanoparticles supported on TiO_2, part of a 3D data set used to reconstruct the morphology of the support and location of the nanoparticles, as seen in Fig. 6.18.

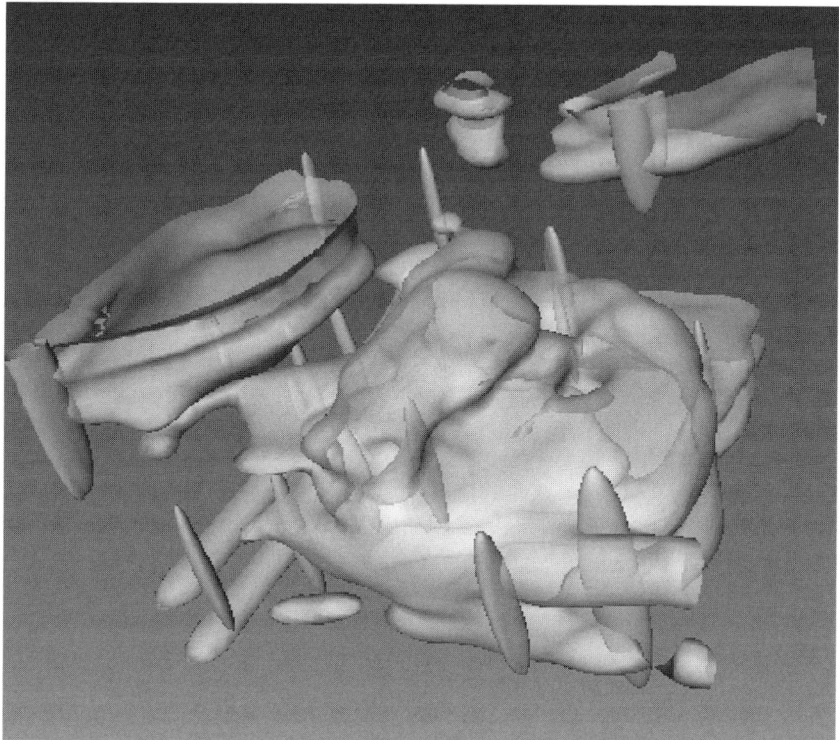

FIGURE 6.18. A 3D rendering of the data set shown in Fig. 6.17 showing the TiO_2 support morphology and the metal nanoclusters as elongated ellipsoids. This reflects the lower depth resolution of the STEM image compared with its lateral resolution (a few nanometers in depth compared with a subangstrom lateral resolution). [adapted from A. Y. Borisevich, A. R. Lupini, and S. J. Pennycook, Proc. Natl. Acad. Sci. U. S. A. 103 (2006) 3044].

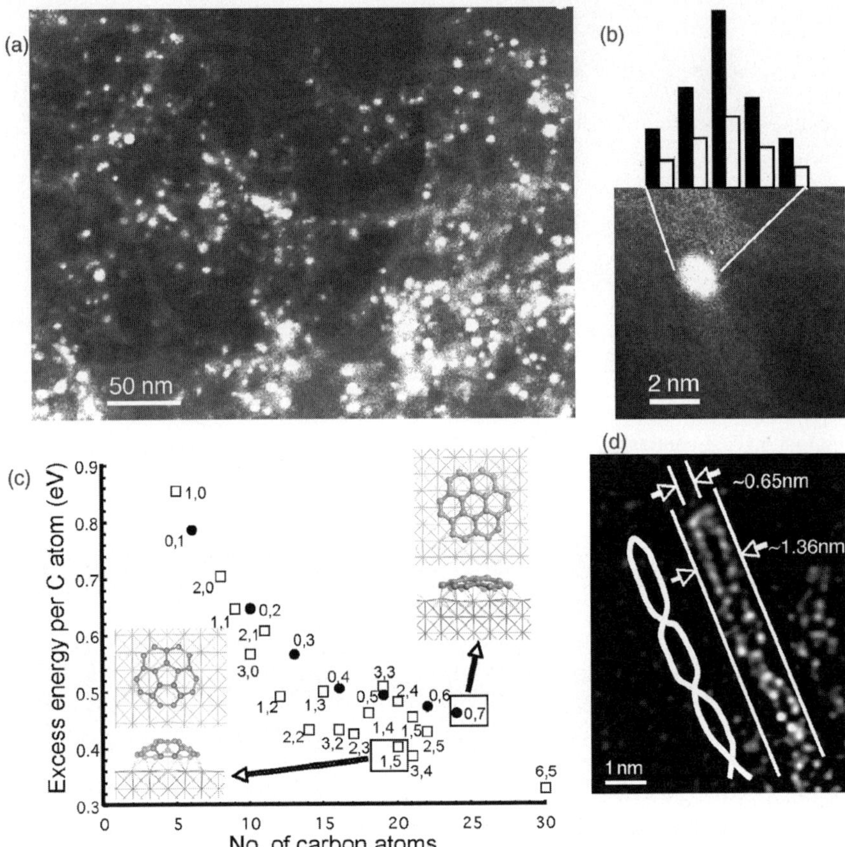

FIGURE 6.19. (a) A Z-contrast image of laser-ablation-grown single-wall carbon nanotubes bundles with Co–Ni catalyst nanoparticles. (b) A close-up image of a 2-nm nanoparticle, with EELS composition profile across the nanoparticle (black bars indicate Co, white bars Ni). (c) Total energy per carbon atom for various nuclei shapes on a nickel surface. Solid circles denote structures with all hexagons; open squares include pentagons, which are lower in energy [47]. (d) A Z-contrast image (low pass filtered) of iodine-intercalated single-wall carbon nanotubes showing the iodine atoms to form a double-helix arrangement inside the tube [adapted from ref. 48].

determine which are energetically favorable. The tube grows out from the cooling metal-carbon droplet, as the temperature falls below the eutectic point. Theory was used to calculate the energy of various trial structures, including graphite flakes, fullerene caps, and spheres, either on a metal surface or free standing. The key driving force is the very high energy of the carbon dangling bond. Without metal the fullerene has the lowest energy, but in the presence of metal it becomes favorable to bury dangling bonds in the metal and the cap or closed nanotube becomes energetically preferred. The calculations revealed that it is favorable to insert pentagon units into the graphitic flake from the very early stages of growth, because it allows the flake to bend and bury the high-energy dangling bonds at its perimeter. The nucleation pathway will incorporate pentagons from the earliest stages of growth to allow the graphite flake to spontaneously bulge out into a cap shape. Then growth can proceed easily around the perimeter of the cap, and a capped nanotube is extruded [47].

Figure 6.19d shows the imaging of iodine atoms intercalated into single-wall nanotubes. Although taken before aberration correction, the individual iodine atoms could be detected within the tube walls (shown outlined in white in the figure). The shape appeared consistent with a spiral pattern, as indicated schematically, and again density functional calculations explained the spiral form. Iodine atoms like to form 1D chains, and the spiral allows the interatomic spacing of the iodine to match the carbon lattice. The potential energy minimum is very shallow, explaining why the arrangements are not perfect [48]. Electron microscopy of nanotubes is a very active field, see for example the book by Wang and Hui [49].

5.2. Nanocatalysis

Gold is not a good catalyst in bulk or in the form of large particles, but when prepared as nanoparticles on an oxide support, it becomes one of the most active catalysts for the oxidation of CO to CO_2. The cause of this activity has been a mystery for many years, with several explanations proposed. Early work suggested that there is a correlation between particle size and activity [50,51] although it is difficult to see the smallest nanoparticles by conventional TEM, hindering the identification of likely sites and mechanisms. We have used aberration-corrected STEM to image the clusters of an Au catalyst prepared by deposition/precipitation onto nanocrystalline anatase [52]. The catalyst showed 50% conversion at 235 K, which is comparable to the highest reported in the literature at this temperature. As seen in Fig. 6.20, most of the nanoparticles are 1–2 nm in diameter, and quantifying the thicknesses by comparing to image simulations revealed that they are just 1- or 2-layers thick. More recent work on model systems has shown that the thickness may be more important than the lateral extent, with bilayer structures having the highest activity [53]. Furthermore, the contrast seen in the image is in fact that of the TiO_2 substrate showing right through the Au nanoparticles. Since we certainly have the resolution to resolve an Au crystal, and none were seen on this sample, it seems likely that the nanoparticles may be in a liquid state. This is consistent with the known dependence of Au melting point on particle diameter, which extrapolates to room temperature at the size range of 1–2 nm [54,55].

XANES studies show that the Au particles are not reoxidized by exposure to air at room temperature, or even at 573 K, and so are unlikely to reoxidize under reaction conditions or during transfer into the STEM [56]. Therefore, we believe that the images shown in Fig. 6.20 are representative of the active state of the Au nanocatalysts. Based on these images, it was then possible to perform first-principles calculations to investigate the adhesion energies of Au_N clusters on defect-free TiO_2 surfaces and also on the same surfaces with an O vacancy. Single

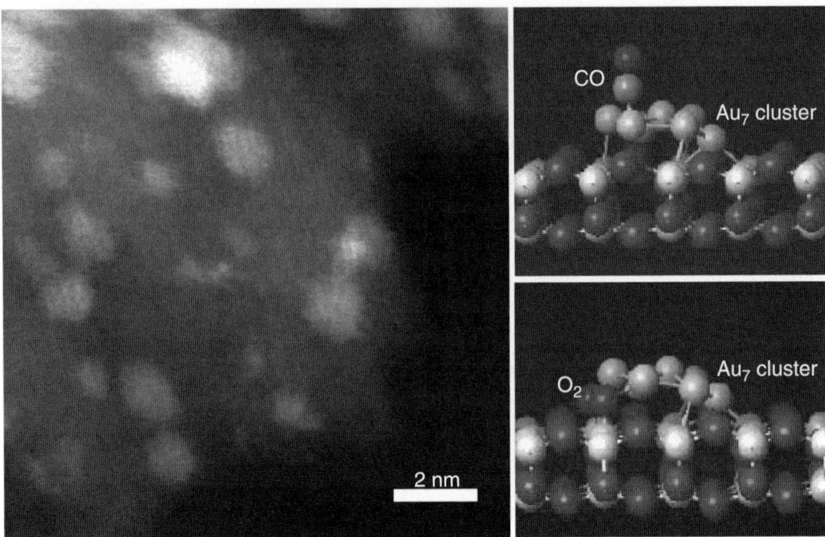

FIGURE 6.20. (*Left*) Z-contrast STEM image showing Au nanoparticles on a titanium flake. Most of the nanoparticles are between 1 and 2 nm in diameter and are 1- to 2-layers thick. First-principles calculations for the nanoparticles observed show that they are capable of bonding both CO (*right top*) and O_2 (*right bottom*), enabling the high catalytic activity.

Au atoms adsorb preferentially at O vacancy sites with a binding energy about 0.9 eV larger on the O vacancy than on the stoichiometric (001) surface. The binding energy of Au_N clusters at an O vacancy site is approximately *constant* with the number N of Au atoms, suggesting that only the Au atom on the O vacancy site forms a strong bond with the substrate. Further evidence for this conclusion is provided by the observation of the relaxed cluster structure, which shows one Au atom displaced into the O vacancy site. We conclude that O vacancies act as "anchors" to suppress coalescence, consistent with earlier calculations [57]. Such anchors facilitate the high areal density of small clusters.

Further calculations were carried out for the binding energies of CO and O_2 molecules. Although a single Au atom binds CO and O_2 only weakly, small Au clusters, such as Au_7 and Au_{10}, can adsorb *both* CO and O_2, unlike gold steps or surfaces. Thus it seems that the ability to bind both of the reactants is the key to the high activity of this unusual catalyst. For more examples of the electron microscopy of catalysts, see the book by Gai and Boyes [58].

5.3. La-Stabilization of Supports

γ-Al_2O_3 is used extensively as a catalytic support material because of its high porosity and large surface area, but at temperatures in the range of 1,000–1,200°C it transforms rapidly into the thermodynamically stable α-Al_2O_3 phase (corundum), drastically reducing the surface area and suppressing the catalytic activity of the system. The phase transformation can be shifted to higher temperatures by doping with elements such as La, but previously it was not possible to establish if dopants enter the bulk, adsorb on surfaces as single atoms or clusters, or form surface compounds.

A Z-contrast image of a flake of La-doped γ-Al_2O_3 in the [100] orientation is shown in Fig. 6.21. The square arrangement of Al–O columns is clearly resolved. Single La atoms are visible in the form of brighter spots on the background of thicker but considerably lighter γ-Al_2O_3 support. Most of the La atoms are located directly over Al–O columns (site A), but a small fraction also occupies a position shifted from the Al–O column (site B). The images reveal clearly that there is no correlation in the distribution of dopant atoms, and a through-focal series shows they are located on the surfaces of the flake. Density functional theory calculations have demonstrated that La atoms are very strongly bound to the γ-Al_2O_3 surfaces (binding energy 7–8 eV), considerably stronger than to the α-alumina surface (binding energy 4.3 eV). Thus it became apparent that the stabilization is achieved by single La atoms adsorbed on the γ-Al_2O_3 surface, which improve its stability with respect to phase transition and make sintering highly unfavorable [5].

5.4. Semiconductor Nanocrystals

Due to the quantum confinement of electrons and holes, semiconductor nanocrystals offer the potential for sensitive tuning of optical emission or absorption wavelength via control of particle size, along with the possibility of 100% quantum yields through use of a suitable passivating surface layer. Often referred to as quantum dots, these nanostructured materials are finding major applications in photovoltaics, photocatalysts, electronics, and biomedical imaging. At present there is little detailed understanding of structure–property relationships at the level of individual nanostructures. Z-contrast STEM can provide detailed

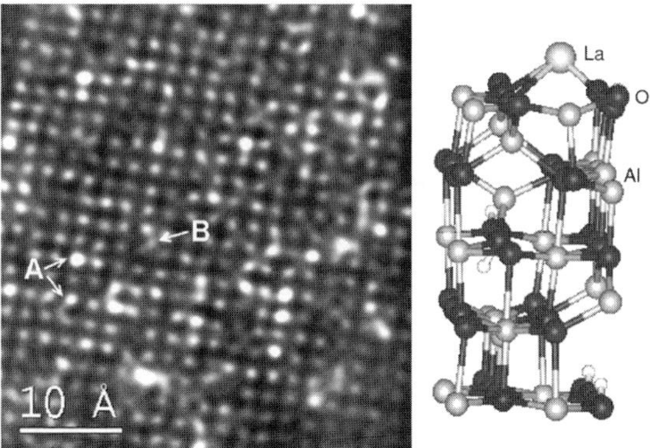

FIGURE 6.21. (*Left*) Z-contrast image of La atoms on a γ-Al_2O_3 flake in [100] orientation. (Fourier filtered to reduce noise) (*Right*) Schematic of the configuration for the La atom determined by first-principles calculations [adapted from ref. 5].

information on nanocrystal size, sublattice polarity, surface facets, defect content, and 3D shape [59,60].

Figure 6.22 shows an image of a Quantum Dot Corp. CdSe/CdS/ZnS core/shell nanorod viewed along the [010] direction. The sublattice polarity is directly observable in the raw data though it can be enhanced by application of a band-pass Fourier filter.

Knowing the beam direction to be [010] and the polarity of the hexagonal wurtzite structure, all the facets of the nanocrystal can be indexed. Furthermore, from the intensity trace the thickness of the nanocrystal can be seen. The entire 3D shape of the nanocrystal can therefore be determined from a single image. As shown in the schematic below the figure, in this particular configuration the nanocrystal is sitting on the carbon support film on a facet junction and rotated out of view before a second scan could be obtained. Images of these structures

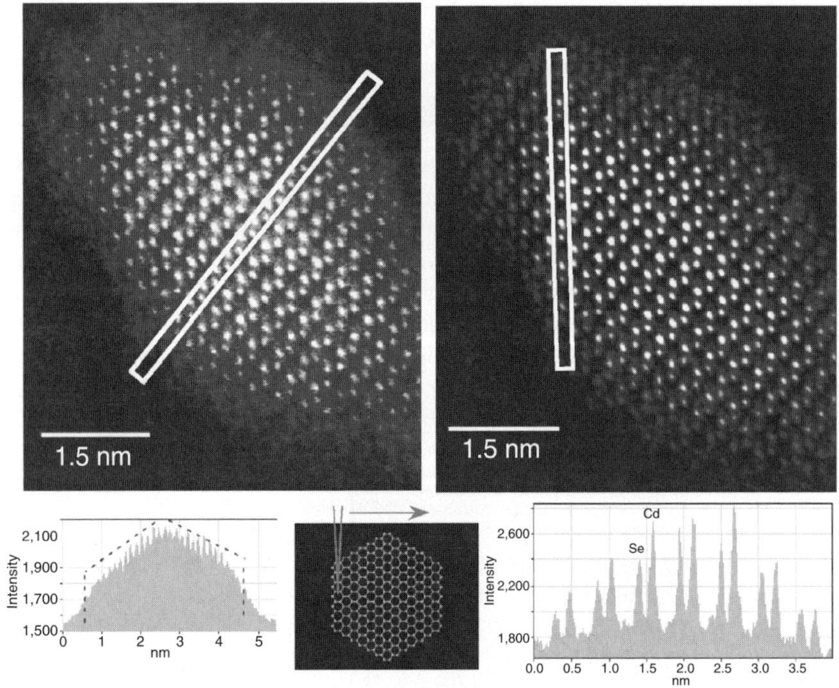

FIGURE 6.22. Z-contrast image of a CdSe/CdS/ZnS core/shell nanocrystal viewed along the [010] direction as shown in the ball and stick model below. The left-hand image shows raw data, the right-hand image has been band-pass filtered. A line trace across the raw image (*left*) taken across the long axis of the nanocrystal reveals the thickness profile of the core. The sublattice polarity is also directly visible, but is more clearly seen from a line trace across the filtered image (*right*). This nanocrystal is oriented as shown in the ball and stick model, the beam direction being vertical, and the scan direction for the thickness profile being shown by the grey arrow [adapted from ref. 60].

show the 3D shape of the core and shell, and show the anisotropy of growth. They confirm predictions in the literature that the Se-rich face is the primary growth face for the CdSe core [61].

5.5. *Magnetic Nanoparticles*

When embedded in a nonmagnetic matrix, magnetic nanoparticles, possibly with a core/shell structure, are of significant scientific and technological interest. As in the case of the optical quantum dots, magnetic nanoparticles show size-tunable magnetic properties of interest for electronic devices and magnetic recording. In addition, engineering the interparticle spacing of nanoparticle arrays allows the interparticle magnetic coupling to be tuned. The Z-contrast STEM again offers the possibility to determine the size, 3D shape, and composition of these metal nanoparticles, even after embedding into a light matrix material such as Al_2O_3. For this application, however, the ability of the STEM to provide simultaneous EELS is a particular advantage. Figure 6.23 shows an example of such an analysis on a Ni nanoparticle embedded in an Al_2O_3 film [62]. The data are collected with a 100-kV STEM in which the resolution is lower, but the multiple-twinned structure of the Ni nanoparticle is clearly seen. The resolution is sufficient to place the probe down the edge of the nanoparticle and analyze those atoms in contact with Al_2O_3.

The EELS data, both from the center of the particle and from the edge, show an $L_{2/3}$ ratio characteristic of Ni metal. Taking account of the noise level in the spectra, it was determined that less than 5% of the surface atoms were oxidized. Therefore, these nanoparticles would not be expected to exhibit any magnetic dead layer and, indeed, the magnetic size determined from hysteresis is the same as that seen in the images.

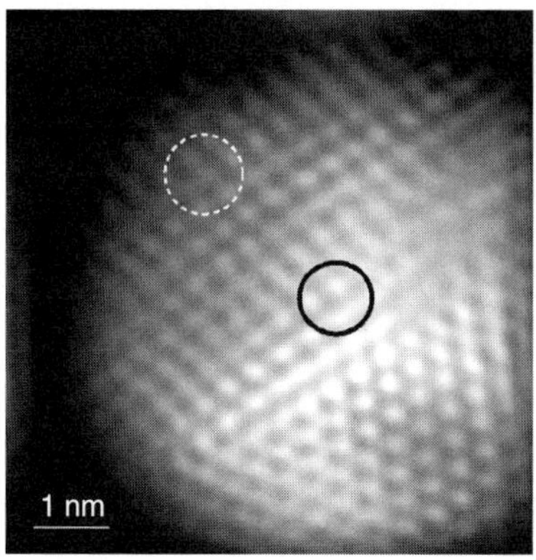

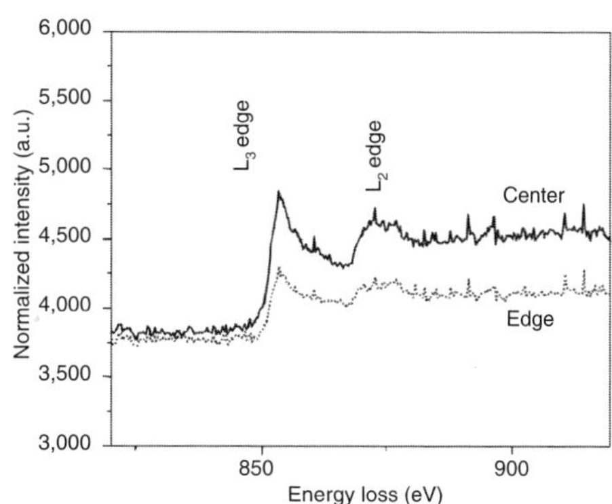

FIGURE 6.23. Z-contrast image of a Ni nanoparticle embedded in an Al_2O_3 film, with EELS spectra taken from the center and edge showing fine structure characteristics of Ni metal [adapted from ref. 62].

5.6. *ZnO Nanorods*

ZnO nanorods are of major interest as optical components due to their ability to grow as 1D wires, their visible light emission, and their high exciton-binding energy. Figure 6.24 shows a vertically aligned nanorod array containing multiple quantum wells on the tips of each nanorod [63,64]. The sample was grown by a catalyst-free method, using metal-organic vapor phase epitaxy. The interesting aspect of the growth procedure is that a uniform layer of ZnO is deposited initially at low temperature, but when the temperature is raised a spontaneous transition to a nanorod morphology occurs, driven presumably by the high surface energy anisotropy of ZnO. The quantum wells consisted of ten periods of $Zn_{0.8}Mg_{0.2}O/ZnO$ and exhibited a blue shift dependent on well width, the signature of quantum confinement. The STEM Z-contrast image showed the wells as slightly darker bands. The interfaces are believed to be structurally sharp, but not perfectly flat, accounting for the blurred nature of the image. Atomic-resolution images showed a surprising superlattice structure possibly resulting from ordered oxygen vacancies.

One of the advantages of a scanning microscope is that any signal can be monitored as a function of probe position and used to form an image. In optical materials, a particularly useful signal is the cathodoluminescence generated by the electron beam. Using an optical detection system, the light generated as the beam scans can be collected simultaneously with the ADF image to form a map showing the optical emission. Figure 6.25 shows some examples of ZnO nanorods where it is seen that only certain rods give optical emission. Although the reason for this behavior was not ascertained, the image shows how very different information is available in the cathodoluminescence image. The problem tends to be a low emission probability so that high beam currents are required for good-quality images; but in principle, this technique can be used to correlate optical emission

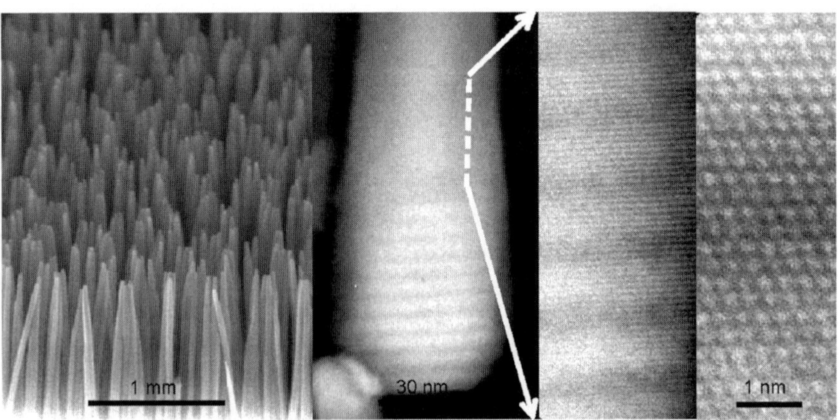

FIGURE 6.24. (*Left*) SEM image showing a vertically aligned array of ZnO nanorods, with $Zn_{0.8}Mg_{0.2}O/ZnO$ multiple quantum wells at the tip. (*Right*) Z-contrast images with increasing magnification showing the quantum well structure [adapted from ref. 63].

FIGURE 6.25. Images of ZnO nanorod bundles recorded using cathodoluminescence (total optical emission, *left*) and the ADF detector (*right*), showing that not all nanorods are efficient emitters.

characteristics, such as emission strength and peak wavelength, with the size, shape, defect, or impurity content of individual nanostructures.

5.7. Nanoscale Phase Separation in Complex Oxides

An area of intense interest in recent years has been that of nanoscale phase separation in complex oxides, often referred to as charge-ordering [65]. Recently, we have been successful in observing stripes in $Bi_{0.38}Ca_{0.62}MnO_3$ with a periodicity of ~12 Å in the [100] direction. This material is ordered at room temperature, and the stripes are detected by EELS using the Mn L_{23} ratio, as shown in Fig. 6.26. The ratio of the L_3 to L_2 peaks is sensitive to the Mn 3D occupation. Mn^{3+} stripes are seen confined to a single Mn plane. No oxygen nonstoichiometry is detected, representing perhaps the first definitive observation of stripes of different valence in the bulk. This result highlights the very different nature of the structural image, which shows the random array of Bi dopants as bright spots, and the electronic effect, which is a quasiperiodic array of 1D stripes. Note that the EELS measure-

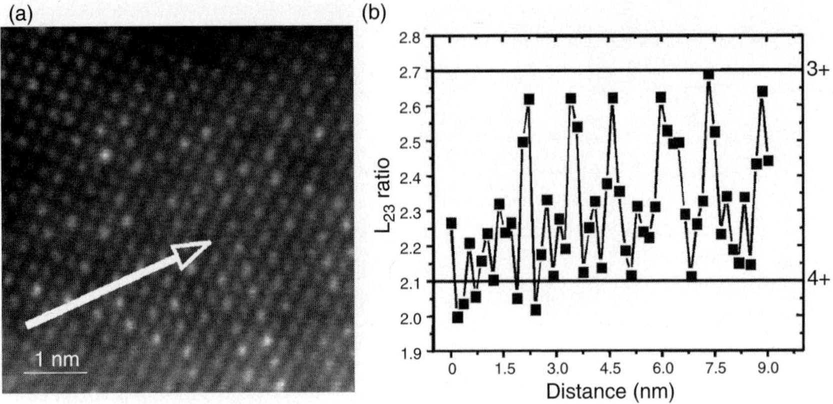

FIGURE 6.26. (a) Z-contrast image of $Bi_{0.38}Ca_{0.62}MnO_3$. The specimen is a crushed sample and does not have an amorphous surface layer. The bright columns show directly the random locations of Bi atoms on the Ca sites and appear to be quantized in this thin region. (b) L_{23} ratio on the Mn edge plotted in a 1D line along [100] showing approximately a 12-Å periodicity due to charge ordering. Horizontal lines represent the ratio observed in bulk Mn^{4+} and Mn^{3+} compounds.

ment does not measure actual charge, just orbital occupation. Theoretical calculations indicate that the charges on the different Mn sites are in fact almost equal, the increased electron density that might be expected on the Mn^{3+} site being compensated by an outward structural relaxation of the neighboring oxygen atoms to maintain approximate charge uniformity. Spatially resolved EELS, unlike diffraction techniques, is unique in its ability to measure orbital occupation independent of any structural relaxations.

6. Future Directions

Future generations of aberration correctors promise the ability to correct all aberrations up to and including fifth order, and will bring lateral resolution to the sub-0.5 Å level with nanometer-scale depth resolution. The smaller probe will further improve sensitivity, allowing lighter atoms to be imaged in Z-contrast mode and improved detection limits for EELS. The higher current density available in these probes will enable data to be taken much faster. It will be possible to make movies of dynamical processes. With the depth resolution of the Z-contrast image, it will be possible to directly image atoms diffusing inside a host material or on its surface. We can expect to be able to section through nanostructured materials and reconstruct their 3D morphology, with a precision at the single-atom level. We will have the sensitivity to image single dopant atoms inside nanocrystals or along dislocation lines inside a bulk material and to probe their effect on the host electronic structure by EELS.

Along with such exciting new imaging techniques comes renewed interest in *in situ* capabilities. If it were possible to maintain the resolution while also heating or cooling the sample, it would greatly facilitate studies of atomic diffusion, nanostructure nucleation, and growth processes and phase transformations. If a gas environment could also be maintained, it would be possible to image a catalyst nanocluster in operation or a nanotube growing out of a metal nanoparticle. We are just at the beginning of this revolution in nanostructure imaging. With such improved sensitivity we can look forward to more insights into the origin of the unique properties of the nanoscale, aided by first-principles electronic structure and total energy calculations. We still have a way to go to realize Feynman's dream to just look at the atoms, but we are progressing toward that goal at a pace more rapid than ever before. The aberration-corrected STEM is allowing us to probe the atomic-level details of the nanoworld with sensitivity and versatility that is unprecedented in history.

Acknowledgment. The authors are grateful to many colleagues for the collaborations mentioned in this article, including P. D. Nellist, O. L. Krivanek, N. Dellby, M. F. Murfitt, Z. S. Szilagyi, S. D. Findlay, L. J. Allen, E. C. Cosgriff, N. Shibata, E. Abe, S. H. Overbury, M. D. Amiridis, S. Dai, R. D. Adams, D. Kumar, G. Duscher, W. I. Park, L. G. Wang, R. Buczko, X. Fan, M. Kim, H. S. Baik, A. A. Puretzky, D. B. Geohegan, E. C. Dickey, A. V. Kadavanich, T. C. Kippeny, M. M. Erwin, S. J. Rosenthal, S. N. Rashkeev, M. V. Glazoff, L. G. Wang, K. Sohlberg, S. T. Pantelides, W. H. Sides, and J. T. Luck. This research was supported in part by the Laboratory Directed Research and Development Program of ORNL, managed by UT-Battelle, LLC, for the US Department of Energy under Contract No. DE-AC05-00OR22725, by appointments to the ORNL postdoctoral program and distinguished visiting scientist program administered jointly by ORNL and Oak Ridge Institute for Science and Education, and by a fellowship from the Alexander-von-Humboldt Foundation (K. v. B.).

References

1. A. V. Crewe, J. Wall, and J. Langmore, *Science*, 168 (1970) 1338.
2. S. H. Overbury, L. Ortiz-Soto, H. G. Zhu, B. Lee, M. D. Amiridis, and S. Dai, *Catal. Lett.*, 95 (2004) 99.
3. R. P. Feynman, http://www.zyvex.com/nanotech/feynman.html
4. P. M. Voyles, D. A. Muller, J. L. Grazul, P. H. Citrin, and H. J. L. Gossmann, *Nature*, 416 (2002) 826.
5. S. W. Wang, A. Y. Borisevich, S. N. Rashkeev, M. V. Glazoff, K. Sohlberg, S. J. Pennycook, and S. T. Pantelides, *Nat. Mater.*, 3 (2004) 143.
6. M. Varela, S. D. Findlay, A. R. Lupini, H. M. Christen, A. Y. Borisevich, N. Dellby, O. L. Krivanek, P. D. Nellist, M. P. Oxley, L. J. Allen, and S. J. Pennycook, *Phys. Rev. Lett.*, 92 (2004) 095502.
7. P. D. Nellist and S. J. Pennycook, in: *Advances in Imaging and Electron Physics*, Vol. 113, B. Kazan, T. Mulvey, and P. Hawkes (Eds.), Elsevier, New York (2000), pp. 147–203.

8. S. J. Pennycook, in: *Advances in Imaging and Electron Physics*, Vol. 123, P. G. Merli, G. Calestani, and M. Vittori-Antisari (Eds.), Elsevier Science, New York (2002), pp. 173–206.

9. S. J. Pennycook and P. D. Nellist, in: *Impact of Electron and Scanning Probe Microscopy on Materials Research*, D. G. Rickerby, U. Valdré, and G. Valdré (Eds.), Kluwer Academic Publishers, Dordrecht, The Netherlands (1999), pp. 161–207.

10. E. J. Kirkland, *Advanced Computing in Electron Microscopy*, Plenum Press, New York (1998).

11. J. C. H. Spence, *High Resolution Electron Microscopy*, Oxford University Press, Oxford (2003).

12. S. J. Pennycook, A. R. Lupini, A. Kadavanich, J. R. McBride, S. J. Rosenthal, R. C. Puetter, A. Yahil, O. L. Krivanek, N. Dellby, P. D. L. Nellist, G. Duscher, L. G. Wang, and S. T. Pantelides, *Z. Metallkunde*, 94 (2003) 350.

13. S. J. Pennycook, A. R. Lupini, M. Varela, A. Borisevich, M. F. Chisholm, E. Abe, N. Dellby, O. L. Krivanek, P. D. Nellist, L. G. Wang, R. Buczko, X. Fan, and S. T. Pantelides, in: *Spatially Resolved Characterization of Local Phenomena in Materials and Nanostructures*, Vol. 738, D. A. Bonnell, A. J. Piqueras, P. Shreve, and F. Zypman (Eds.), Materials Research Society, Warrendale, Pennsylvania (2003), p. G1.1.

14. M. Varela, A. R. Lupini, K. van Benthem, A. Borisevich, M. F. Chisholm, N. Shibata, E. Abe, and S. J. Pennycook, in: *Annu. Rev. Mater. Res.*, 35 (2005) 539–569.

15. P. D. Nellist, M. F. Chisholm, N. Dellby, O. L. Krivanek, M. F. Murfitt, Z. S. Szilagyi, A. R. Lupini, A. Borisevich, W. H. Sides, and S. J. Pennycook, *Science*, 305 (2004) 1741.

16. P. E. Batson, N. Dellby, and O. L. Krivanek, *Nature*, 418 (2002) 617.

17. N. Dellby, O. L. Krivanek, P. D. Nellist, P. E. Batson, and A. R. Lupini, *J. Electron Microsc.*, 50 (2001) 177.

18. O. L. Krivanek, N. Dellby, and A. R. Lupini, *Ultramicroscopy*, 78 (1999) 1.

19. http://www.ceos-gmbh.de/

20. http://www.nion.com/

21. M. Born and E. Wolf, *Principles of Optics*, Pergamon Press, Oxford (1980).

22. O. L. Krivanek, P. D. Nellist, N. Dellby, and et al., *Ultramicroscopy*, 96 (2003) 229.

23. A. V. Crewe and D. B. Salzman, *Ultramicroscopy*, 9 (1982) 373.

24. O. Scherzer, *J. Appl. Phys.*, 20 (1949) 20.

25. O. L. Krivanek, P. D. Nellist, N. Dellby, M. F. Murfitt, and Z. Szilagyi, *Ultramicroscopy*, 96 (2003) 229.

26. M. Haider, S. Uhlemann, and J. Zach, *Ultramicroscopy*, 81 (2000) 163.

27. P. D. Nellist and S. J. Pennycook, *Science*, 274 (1996) 413.

28. K. Sohlberg, S. Rashkeev, A. Y. Borisevich, S. J. Pennycook, and S. T. Pantelides, *Chemphyschem*, 5 (2004) 1893.

29. Y. Peng, A. R. Lupini, A. Borisevich, S. M. Travaglini, and S. J. Pennycook, *Microsc Microanal.*, 10 (Suppl. 2) (2004) 1200.

30. K. van Benthem, Y. Peng, and S. J. Pennycook, in: *Electron Microscopy of Molecular and Atom-Scale Mechanical Behavior, Chemistry, and Structure*, Vol. 839, Materials Research Society, Warrendale, Pennsylvania (2005), pp. 3–8.

31. P. D. Nellist and S. J. Pennycook, *Phys. Rev. Lett.*, 81 (1998) 4156.

32. L. J. Allen, S. D. Findlay, M. P. Oxley, and C. J. Rossouw, *Ultramicroscopy*, 96 (2003) 47.

33. S. D. Findlay, L. J. Allen, M. P. Oxley, and C. J. Rossouw, *Ultramicroscopy*, 96 (2003) 65.

34. J. Frank, *Optik*, 38 (1973) 519.

35. P. L. Fejes, *Acta. Crystallogr. A*, 33 (1977) 109.

36. N. D. Browning, M. F. Chisholm, and S. J. Pennycook, *Nature*, 366 (1993) 143.
37. P. E. Batson, *Nature*, 366 (1993) 727.
38. D. A. Muller, Y. Tzou, R. Raj, and J. Silcox, *Nature*, 366 (1993) 725.
39. G. Duscher, N. D. Browning, and S. J. Pennycook, *Phys. Status Solidi A*, 166 (1998) 327.
40. A. C. Diebold, B. Foran, C. Kisielowski, D. A. Muller, S. J. Pennycook, E. Principe, and S. Stemmer, *Microsc. Microanal.*, 9 (2003) 493.
41. L. J. Allen, S. D. Findlay, A. R. Lupini, M. P. Oxley, and S. J. Pennycook, *Phys. Rev. Lett.*, 91 (2003) 105503.
42. E. C. Cosgriff, M. P. Oxley, L. J. Allen, and S. J. Pennycook, *Ultramicroscopy*, 102 (2005) 317.
43. R. F. Egerton, *Electron Energy-Loss Spectroscopy in the Electron Microscope*, Plenum Press, New York (1996).
44. P. A. Midgley and M. Weyland, *Ultramicroscopy*, 96 (2003) 413.
45. S. J. Pennycook, A. R. Lupini, A. Borisevich, Y. Peng, and N. Shibata, *Microsc. Microanal.*, 10 (Suppl.2) (2004) 1172.
46. K. van Benthem, A. R. Lupini, M. Kim, H. S. Baik, S. Doh, J.-H. Lee, M. P. Oxley, S. D. Findlay, L. J. Allen, and S. J. Pennycook, *Appl. Phys. Lett.*, 87 (2005) 034104.
47. X. Fan, R. Buczko, A. A. Puretzky, D. B. Geohegan, S. T. Pantelides, and S. J. Pennycook, *Phys. Rev. Lett.*, 90 (2003) 145501.
48. X. Fan, E. C. Dickey, P. C. Eklund, K. A. Williams, L. Grigorian, R. Buczko, S. T. Pantelides, and S. J. Pennycook, *Phys. Rev. Lett.*, 84 (2000) 4621.
49. Z. L. Wang and C. Hui, *Electron Microscopy of Nanotubes*, Kluwer Academic Press, Dordrecht, The Netherlands (2003).
50. M. Haruta, *Catalysis Today*, 36 (1997) 153.
51. M. Valden, X. Lai, and D. W. Goodman, *Science*, 281 (1998) 1647.
52. H. G. Zhu, Z. W. Pan, B. Chen, B. Lee, S. M. Mahurin, S. H. Overbury, and S. Dai, *J. Phys. Chem. B*, 108 (2004) 20038.
53. M. S. Chen and D. W. Goodman, *Science*, 306 (2004) 252.
54. P. Buffat and J.-P. Borel, *Phys. Rev. A*, 13 (1976) 2287.
55. K. Dick, T. Dhanasekaran, Z. Y. Zhang, and D. Meisel, *J. Am. Chem. Soc.*, 124 (202) 2312.
56. V. Schwartz, D. R. Mullins, W. F. Yan, B. Chen, S. Dai, and S. H. Overbury, *J. Phys. Chem. B*, 108 (2004) 15782.
57. J. A. Rodriguez, G. Liu, T. Jirsak, J. Hrbek, Z. P. Chang, J. Dvorak, and A. Maiti, *J. Am. Chem. Soc.*, 124 (2002) 5242.
58. P. L. Gai and E. D. Boyes, *Electron Microscopy in Heterogeneous Catalysis*, Institute of Physics, London (2003).
59. A. V. Kadavanich, T. C. Kippeny, M. M. Erwin, S. J. Pennycook, and S. J. Rosenthal, *J. Phys. Chem. B*, 105 (2001) 361.
60. J. R. McBride, T. C. Kippeny, S. J. Pennycook, and S. J. Rosenthal, *Nano Lett.*, 4 (2004) 1279.
61. D. V. Talapin, R. Koeppe, S. Gotzinger, A. Kornowski, J. M. Lupton, A. L. Rogach, O. Benson, J. Feldmann, and H. Weller, *Nano Lett.*, 3 (2003) 1677.
62. D. Kumar, S. J. Pennycook, A. Lupini, G. Duscher, A. Tiwari, and J. Narayan, *Appl. Phys. Lett.*, 81 (2002) 4204.
63. W. I. Park, G. C. Yi, M. Kim, and S. J. Pennycook, *Adv. Mater.*, 15 (2003) 526.
64. W. I. Park, G. C. Yi, M. Y. Kim, and S. J. Pennycook, *Adv. Mater.*, 14 (2002) 1841.
65. C. Renner, G. Aeppli, B. G. Kim, Y. A. Soh, and S. W. Cheong, *Nature*, 416 (2002) 518.

7
Introduction to In Situ Nanomanipulation for Nanomaterials Engineering

Rishi Gupta and Richard E. Stallcup II

1. Introduction

Nanomanipulation represents the logical next step in applications of electron microscopy: why simply image when one can image and manipulate in real time simultaneously with no loss of resolution? The incorporation of robotic manipulators in scanning electron microscopes (SEM) began as a way to characterize the mechanical properties of novel nanostructures [1]. However, the applications for nanomanipulators have expanded to include electrical characterization of nanostructures [2], as well as contact level integrated circuit (IC) probing, and the manipulation of virus nanoblocks [3].

The term nanomanipulator can be loosely defined as any kind of electromechanical device used for controlled placement of an end effector with better than 100 nm resolution. The restriction on the resolution of the tool arises from the industry standard length scale definition of nanotechnology: structures with one or more dimensions at sub-100 nm. Obviously, the positioning resolution of the nanomanipulator must conform to the length scales of the materials to be characterized and/or manipulated. Constructing a nanomanipulator requires conforming to several other restrictions that will be discussed later.

Before the introduction of nanomanipulators, scientists were forced to resort to arduous and time-consuming techniques in order to characterize nanostructures. For example, to measure the electrical properties of a nanotube, one had to deposit nanotubes onto a surface, locate an isolated nanotube using atomic force or SEM, deposit metal contacts with lithographical techniques, and finally perform the electrical characterization using microprobes. Although this technique has produced excellent data [4], the overhead in equipment and personnel is prohibitive.

The nanomanipulator operated inside an SEM allows the scientist to locate a nanostructure using SEM imaging, connect electrodes or other end effectors to it using the manipulator, and perform mechanical and/or electrical measurements all in a single experiment. Essentially, an SEM with a nanomanipulator is a direct analog of an optical light microscope with a microprober station. The SEM can image nanometer-scale structures in real time, and the vacuum pressures allow measurements to be made in a clean, dry environment.

Additional advantages over previous techniques come from the ability to make dynamic measurements on nanostructures, such as measuring the electrical response of a multiwalled carbon nanotube (MWNT) to a mechanical deformation [5]. Since the end effectors serve as the electrodes, moving them to deform a nanotube does not affect the metal–nanotube contact or contact resistance. Performing similar experiments using AFM resulted in strained contacts, which obviously affect the outcome of the experiment [6].

Nanomanipulators are also compatible with focused ion beam (FIB) systems. FIB systems are used for milling sections out of semiconductor devices for failure analysis (FA). Materials and structures can be extracted with high precision using nanomanipulators for transmission electron microscopy (TEM) characterization [7]. However, they can also be used to deposit metal to modify probe tips and create electrodes on surfaces. FIB systems coupled with nanomanipulators form the tool of choice for nanotechnology applications in physical characterization of nanostructures and nanomaterials.

Nanomanipulators have also been adapted for use in transmission electron microscopes. Nanostructures can be analyzed and even altered with the added functionality of manipulation systems in situ to the TEM [8,9]. TEM manipulation of multiwalled nanotubes yielded some of the first insights into the engineering of nanoelectromechanical systems (NEMS).

2. SEM Contamination

Before discussing in situ applications, it is vital to discuss contamination issues associated with the SEM vacuum system. Attempts at performing electrical measurements in situ has brought to light how contamination problems associated with electron microscopy affect the ability to perform low-noise, low-current measurements. Specimen contamination has plagued electron microscopy as far back as the 1950s. Each time resolution advancements in electron microscopy were made, specimen contamination would require additional advancements in vacuum technology. Essentially, the electron beam used for imaging will dissociate ubiquitous hydrocarbons (HC), depositing carbonaceous material onto the region of interest. This process, know as electron beam-induced deposition (EBID) or contamination writing, results in a hard material that is a poor conductor. It creates darkening of the scanned area, loss of resolution, changes in conductivity, and other artifacts. The EBID buildup can rapidly become thicker than the resolvable features of the specimen. Figure 7.1 shows controlled growth of EBID nanostructures on a tungsten probe. Deposits created by the interaction of the probe beam with the surface specimen also may interfere with the probe beam itself, or with emitted electrons and x-rays, thus adversely affecting analysis accuracy. Deposits add uncertainty to SEM-measured line widths for semiconductor device-critical dimension metrology [10].

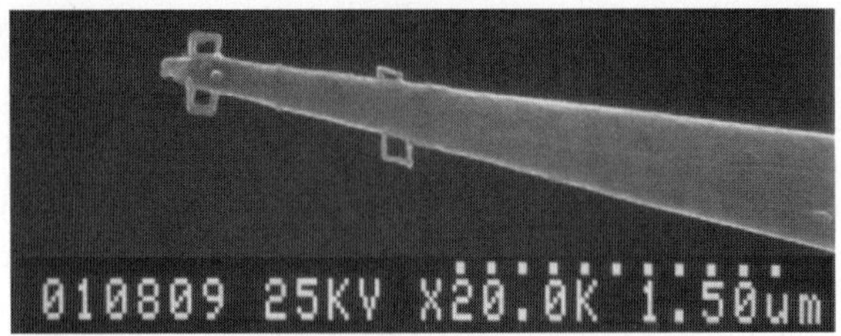

FIGURE 7.1. Controlled growth of electron beam-induced deposition (EBID) structures on a tungsten probe tip. The EBID structures are the four rectangular "handlebars" on either side of the probe tip. (Adapted from [22].)

Contaminants are typically introduced into the instrument by one of four ways: (1) the specimen can contain HCs, inherently or otherwise; (2) the specimen stage and its lubrication greases or other introduced surfaces can be contaminated; (3) the pump system might use oil that can diffuse into the system; (4) and/or the instrument could have been contaminated during its initial assembly. Contamination introduced from the vacuum system can be reduced by cold trapping, purging with dry nitrogen, or using cleaner (dry) pumps. Electron microscope and FIB manufacturers have resorted to dry pumping systems with turbo molecular pumps to stop these HCs from entering and have installed liquid nitrogen traps or anticontaminators to immobilize them. However, even with these steps, residual HCs can enter and/or remain in the chamber.

HCs are present at trace levels in ordinary room air and come from living organisms and man-made material. All surfaces exposed to air at atmospheric pressure accumulate HCs. Surfaces are further contaminated by handling them without gloves, the use of low vapor pressure materials in the vacuum system, or poor vacuum practices in general. A film of HC deposit will accumulate on a surface if left exposed to ordinary room air for any length of time. The sources of these HCs are most living things, plastics, other organic objects, or other sources of HC vapors in the vicinity. Reducing and controlling atmospheric molecular contamination (AMC) is an active area of concern for semiconductor manufacturers, as device dimensions get smaller.

All specimens can carry contaminants into the chamber. These contaminants might be part of the specimen, residues from sample preparation or mounting techniques, created by production process, or caused by improper sample handling or storage techniques. The specimens are also subject to AMC. While the part of the films created in these processes dissipates under vacuum conditions, a small amount generally remains on surfaces and is sufficient to cause problems when the specimen is examined. Once present inside the chamber, these contaminants will reside on the chamber surfaces.

These residues are widely distributed and are generally at low concentrations on the various surfaces. Some of the contaminant molecules become mobile in the vacuum environment. At high vacuum, the mean free path of vaporized molecules is comparable with or longer than the dimensions of the vacuum chamber of these instruments. The contaminants move in the vapor phase from surface to surface in the vacuum environment and are attracted to any focused electron probe beam, forming deposits through an ionization and deposition process. Since these contaminants can travel large distances within the vacuum chamber and over the surface of a specimen, it is important to remove or immobilize these contaminants as much as possible prior to an analysis without disturbing the microstructure of the specimen.

Carbon deposition comes about because of the interaction of the electron beam with free HC molecule vapors in the vacuum chamber or on the surface of the wafers. The HC molecules are ionized and follow the electron beam to the scanned area. Surface diffusion of the hydrocarbon molecules to the scanned area is also an important mechanism over short distances. Low energy FE-SEM work is very sensitive to contamination because the electron beam can image these deposits rather than blast through them. Simple imaging of structures often takes only one or two scans, allowing the microscopist a way to avoid contamination artifacts. However, nanomanipulation takes longer. With four separate probes involved in doing electrical measurements, for example, and the small distances involved, this intricate work may take half an hour or longer in any particular spot. This is a lengthy scan time and requires the complete absence of HC to prevent carbon buildup. Carbon contamination will create a dielectric layer on the probes, thereby making them less sharp and less conducting. This contamination distorts the electrical measurements. The buildup appears as a translucent coating on the probe tips, as shown in Fig. 7.2.

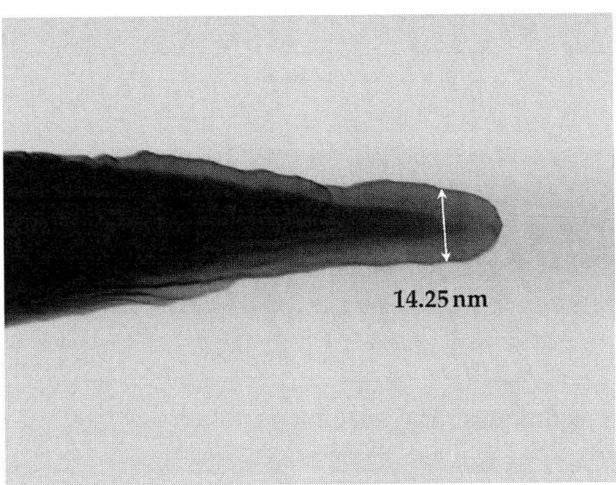

FIGURE 7.2. SEM micrograph of an electrochemically etched tungsten wire. The semitransparent edge is electron beam-induced deposition (EBID). Although the probe is still quite sharp in spite of the EBID growth, the true sharpness is overshadowed by contamination. This SEM image was taken at 25 keV at a magnification of 200,000×. (Adapted from [22].)

2.1. Preventing Contamination

The best method of cleaning is by using a downstream ash process. Oxygen radicals created by an RF plasma bind with HCs creating water, carbon monoxide, and carbon dioxide. These molecules are then pumped out of the SEM chamber. XEI Scientific in Redwood City, California, manufactures the Evactron System, which is a production model downstream ash system.

Figure 7.3a shows a tungsten probe imaged in a chamber treated with the downstream ash process. Figure 7.3b shows the same probe imaged in an SEM chamber that has not undergone cleaning. The thin layer of EBID is seen encasing the probe. The amount of HC removed depends on the length of time the device is operated and the nature of the carbon compound. Light, volatile species are quickly removed and cleaned (taking 5–10 min), but heavy polymerized or carbonized deposits may take hours for the initial cleaning. Because the process works by a chemical etch mechanism, rather than sputter etching, damage to the instrument and specimens is avoided. Once the initial cleaning of an instrument is done, then most Evactron cleaning times take less than 5 min to achieve satisfactory cleaning results. Note, however, that a chamber must always be cleaned before imaging and nanomanipulation. Contaminants can diffuse into the chamber over time, and poor sample preparation can introduce HCs during each operation.

2.2. Removing Contamination

Removing EBID is a relatively new concept and it should be considered a last resort. There are several techniques currently being investigated. One such technique is ion

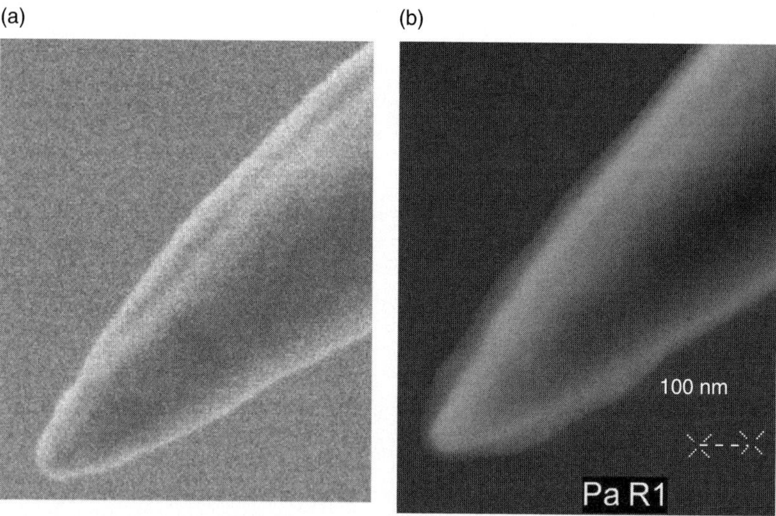

FIGURE 7.3. (a) A tungsten probe imaged in a chamber cleaned of hydrocarbons using the downstream ash technique. (b) The same probe, but imaged later in the same chamber not treated for contamination. Note the layer of growth on Fig. 15.3b. (Adapted from [22].)

sputtering of the deposited carbon material. A low energy ion source can be implemented into the SEM chamber to introduce energetic ions into the chamber. These ions (either focused or not) will impact the carbon material and sputter it away. When working in an FIB system, a focused gallium ion beam is readily available for the task of quickly removing the EBID layer. Ion sputtering can cause damage to the nanostructures that are under study, so care has to be taken when using ion sputtering. Ashing the sample with an oxygen plasma is one way to chemically react the EBID layer away, but oxidation of the specimen is a potential side effect of this process. Mechanically polishing the specimen is an effective way to remove the EBID layer and can be used for cleaning the tungsten contacts of an IC. Chemical mechanical polishing (CMP) with colloidal silica slurry for only a few seconds is usually all that is needed to strip the EBID from an IC at the contact level.

3. Types of Nanomanipulators

Several companies currently sell production quality manipulation systems, each with its own set of competitive advantages. Zyvex Corporation based in Richardson, Texas, sells the S100 Nanomanipulator System (and its variants) for SEM, FIB, and OLM applications. The S100 is based on a four-quadrant positioner utilizing true Cartesian coordinate motion. Kleindieck based in Reutlingen, Germany, sells the micromanipulator MM3A, a high quality system compatible with the various microscopy techniques. Micromanipulator is technically a misnomer: the MM3A has nanometer resolution, which qualifies it as a nanomanipulator based on the aforementioned definition. It uses rotational-based motion (polar cylindrical), which is equivalent to linear motion over small distances (nanometers). Figure 7.4 shows the Zyvex and the Kleindieck head units.

Omniprobe, Inc., based in Dallas, Texas, manufactures a nanomanipulator system for TEM sample lift-out used primarily in FIB microscopes for semiconductor FA, as well as nanomechanical and electrical testing. Hitachi, Ltd., in Tokyo, Japan, manufactures an FIB microscope with a built-in Omniprobe nanomanipulator for in situ lift-out.

Nanofactory based out of Gothenburg, Sweden, manufactures nanomanipulators for TEM applications. Danaher Precision Systems, Dynamic Structures and Materials, Inc., Omicron, and Physik Instruments, Polytec PI all make systems as well.

3.1. Homebuilt Units

The construction of a nanomanipulator requires a multitude of engineering expertise, including mechanical engineering, vacuum technology, electrical engineering, and control systems. Many research groups have developed their own nanomanipulators to fit their specific needs. A team at Northwestern University led by Professor Rod Ruoff uses a manipulator built in-house; as does Professor Min-Feng Yu of the University of Illinois, Urbana-Champaign.

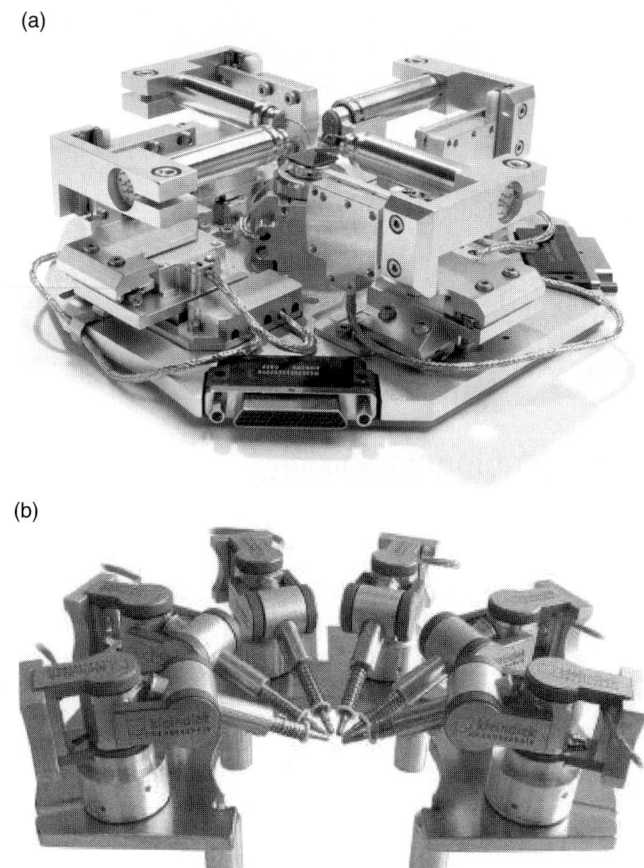

FIGURE 7.4. (a) The Zyvex S100. (Adapted from [22]). (b) The Kleindieck MM3A. (Adapted from [23].)

When building a manipulation system, there are many constraints that have to be kept in mind. Primarily, when building a system for in situ manipulation, the system must be vacuum compatible. Certain materials will outgas, air can be trapped in small crevasses and tapped holes, and long exposure times to the atmosphere can lengthen pump downtime, and hence, time to experiment.

An example of an in-house unit built by engineers at Zyvex Corporation is shown in Fig. 7.5. Even though typical SEM base vacuum is on the order of 1×10^{-6} Torr, the materials that go into the construction of a nanomanipulator for SEM need to have vapor pressures much lower than the SEM base pressure. The partial pressure of each component in the construction of a nanomanipulator adds to the total of the SEM base pressure. If these gases are volatile HCs, the contamination level of the SEM vacuum will increase and that will then increase EBID on the sample.

FIGURE 7.5. An in-house nanomanipulation unit with three X-Y-Z positioners and one rotational positioner. (Adapted from [22].)

Nanomanipulation depends not only on positioning resolution, but also on the ability to maintain an end effector at a desired location. Thus, thermal drift of the actuators, piezo creep, and other mechanical noise must be minimized. Thermal drift of the actuators can cause a probe tip to move hundreds of nanometers over the period of an experiment. When performing electrical or mechanical measurements, this is undesirable.

If the system utilizes piezoelectric elements for actuation, piezo creep must be minimized. Piezo creep usually occurs immediately after initial actuation of an element. Generally, once a "cold" piezo has been actuated, it may take several minutes for creep to dissipate. That is, no external factors are required to reduce it. However, proper control circuitry can optimize piezo motion. One way to minimize creep is to have piezos operated in closed loop with position sensors to feedback into the loop.

Mechanical stability of the entire nanomanipulator is also vital for accurate placement of probes. Too much vibration during probe translation will cause the end effector to oscillate uncontrollably. When working near surfaces or sample holders, unwanted vibrations can severely damage the probe and ruin the experiment. Mechanical vibrations can also affect the SEM signal; it can also alter the location or damage other probes that have already been placed. Mechanical stability is crucial to performing nanomanipulation experiments.

In order to perform electrical measurements on nanoscale structures and materials, and to power the actuators, electrical signals must be received and sent into the SEM chamber. This requires a number of items. First is a feedthrough flange. The flange must conform to the SEM specifications in order to maintain the proper operating vacuum pressure. Also, input–output terminals into which probes and other devices, such as preamplifiers, can be connected are needed. Nanomanipulators in production have as few as one and as many as five electrically isolated I/O terminals. Finally, an in-vacuum cable system is required to connect the nanomanipulator to the laboratory environment. Vacuum compatibility of the cable is implied. However, space in the chamber and tension on the cable are also of importance. The cable should not interfere with the motion of the actuators or the SEM specimen stage. The cable should also be properly shielded. Shielding will allow for low current, low noise measurements in addition to preventing electrical shorts. The electronics to drive piezo actuators tend to utilize high voltage and some are pulse-driven, so proper electromagnetic shielding is necessary.

4. End Effectors

The business end of the nanomanipulator is the end effector. An end effector can be any type of device used for the actual manipulation. A very basic example of an end effector is a sharpened metal wire, which is usually referred to in the vernacular as a probe. This and other examples are described below.

4.1. Probes

Sharpened metal wires, referred to as probes or nanoprobes, are the most common and often the most useful nanomanipulation end effector (Fig. 7.6). They are relatively simple to obtain and are useful for both electrical and mechanical operations. Probes can be purchased from a number of places, including Veeco Instruments in Santa Barbara, California, Micromanipulator in Carson City, Nevada and Zyvex Corporation.

The metal chosen for in situ applications depends on the experiment. For example, tungsten probes are great for performing IC device testing because the contacts used are tungsten as well. This reduces contact resistance associated with Fermi level dissociations from different metals. However, palladium is the current metal of choice for nanotube electrical measurements. Etching palladium, gold, and platinum wires into a sharp point is not trivial, but can be done with etching systems like those supplied by Obbligato Objectives in Toronto, Ontario, and Canada. These metals are much softer than tungsten and have little spring or memory. Metals can be deposited onto preetched tungsten probes by physical vapor deposition, sputter coating, or FIB deposition if the spring constant of tungsten and the electrical properties of a noble metal are needed.

Colloids of many varieties and sizes can be obtained from chemistry supply houses and can be deposited onto a probe. Figure 7.7 shows a tungsten probe that was prepared first by chemically removing the native tungsten oxide and then dipping it into a solution of 40 nm gold colloids. The colloid-functionalized probe can then be chemically modified or can act as a sample holder for manipulating gold colloids in situ.

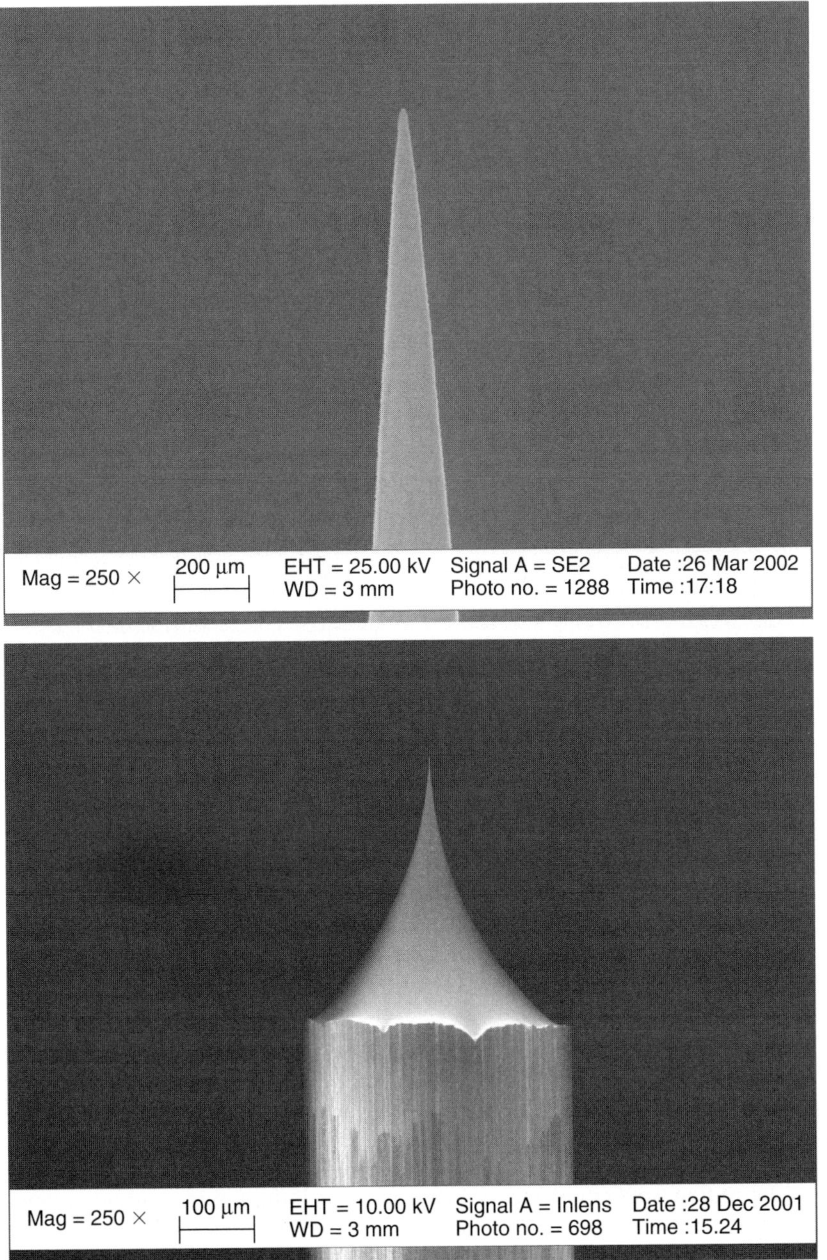

FIGURE 7.6. SEM micrograph of tungsten probes. The sharpness and geometry are dependent upon the etch process used in its fabrication. (Adapted from [22].)

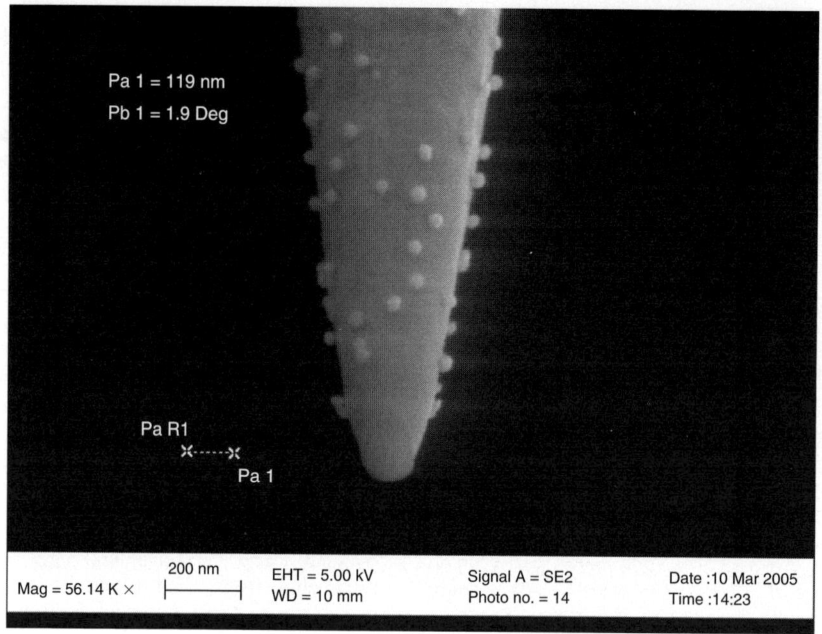

FIGURE 7.7. SEM micrograph of a tungsten probe prepared with 40 nm gold colloids.

As stated above, tungsten is a good choice for making probes, and the standard electrochemical etch techniques used to sharpen tungsten wires for STM applications [11] are also good techniques for fabricating probes intended for nanomanipulation. The sharpness of the probe depends on the quality of the etch. However, whether purchased or homemade, tungsten probes must go through a cleaning process to remove native oxide layers before they can be used for electrical measurements.

4.1.1. Tungsten Oxide

As mentioned above, tungsten is a good choice for electrochemically etching probe tips. The degree of sharpness obtained by using tungsten as the probe material far surpasses any other material. However, tungsten does react in air, quickly creating an oxide layer. Bulk tungsten oxide can be removed through a number of ways. Usually, potassium hydroxide (KOH) and hydrofluoric acid (HF) rinse are sufficient to remove most of the native oxides.

Figure 7.8 shows a TEM image of a tungsten probe that was in storage and exposed to air for about 1 month. The oxide layer is a significant portion of the cross section of the probe tip diameter. Actually the tungsten metal has been oxidized to the point that the metal tip is several hundred nanometers recessed. In general, the activation energy for oxidation is smaller for nanoparticles than it is for the same bulk material. Thus, the sharpened probe tip will oxidize more rapidly than its bulk wire surface.

FIGURE 2.18. Ni superalloy showing grain boundaries and all-Euler coloring.

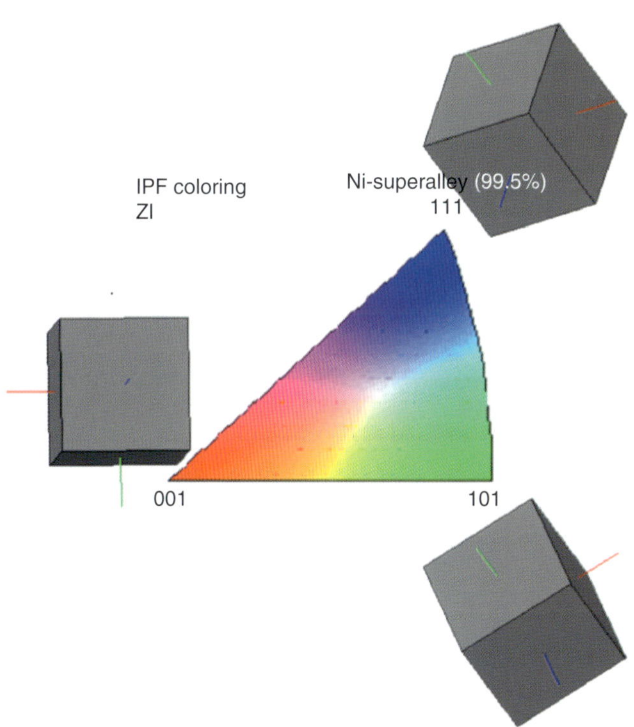

FIGURE 2.19. Inverse pole figure colored map.

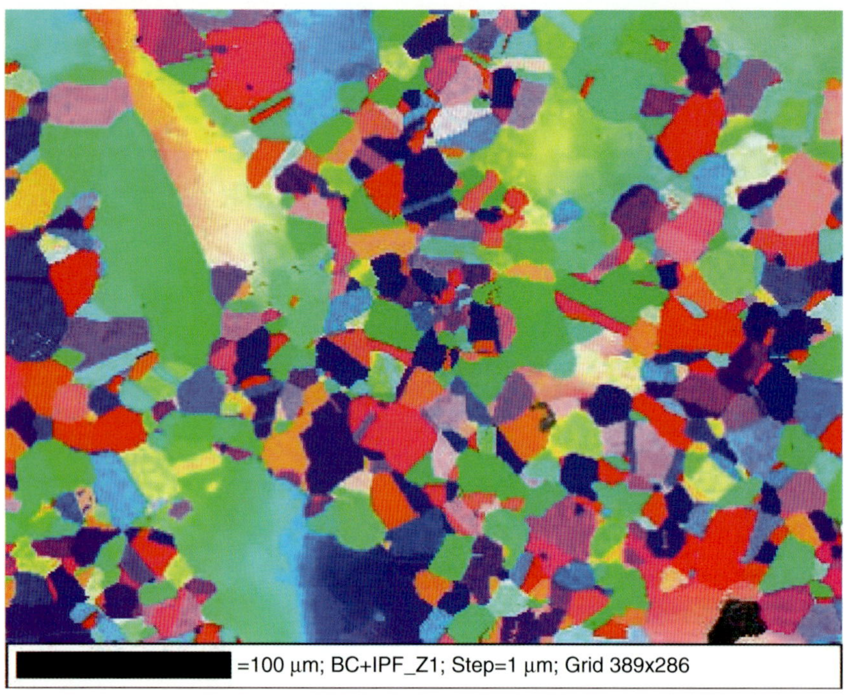

FIGURE 2.20. Surface normal-projected inverse pole figure orientation map.

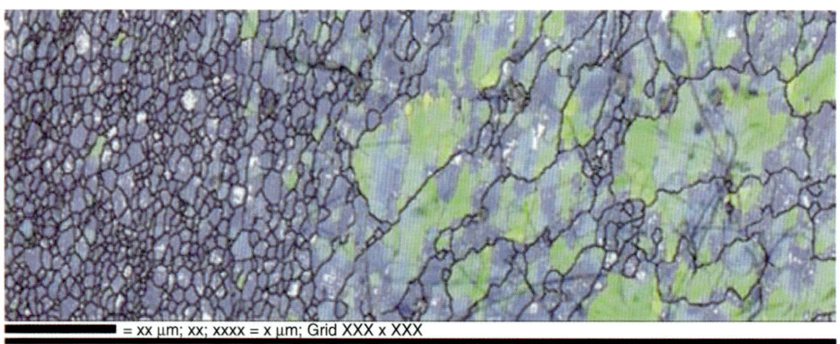

FIGURE 2.23. Grain internal deformation map, based on intragranular lattice rotation relative to a reference point in each grain. Coloring scheme follows a rainbow scale, where blue = smallest rotation from reference orientation, red = 10° of rotation. Scale bar = 200 μm.

FIGURE 2.24. Pattern quality (band contrast) map, high angle boundaries in black, low angle grain boundaries in red, very low angle in yellow. Vertical black scale bar on right = 200 μm.

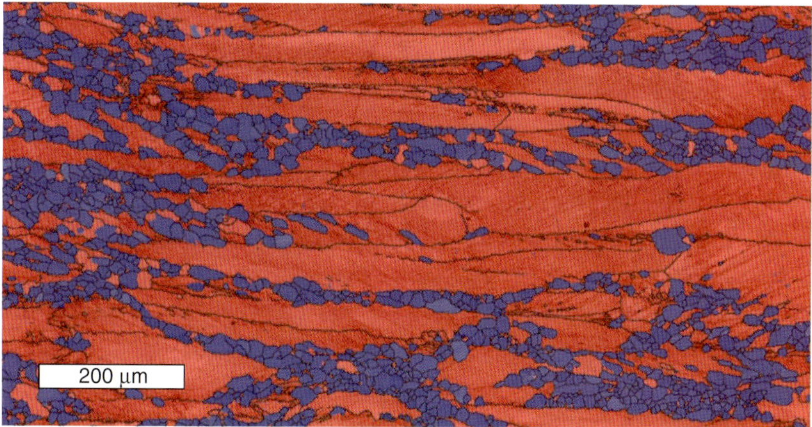

FIGURE 2.25. Map with two classifications of grains: red = deformed, blue = recrystallized, based on degree of internal lattice rotation. High angle grain boundaries are also shown in black. (Adapted from [14].)

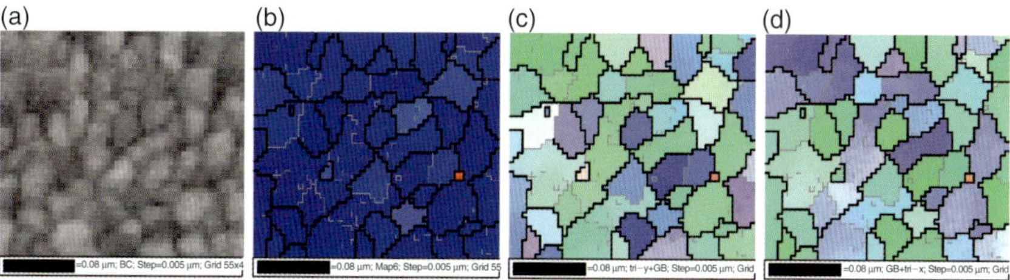

FIGURE 2.28. EBSD maps of Pt thin film. (a) Band contrast (pattern quality) map showing basic grain structure independent of indexing; (b) surface normal-projected IPF-based orientation map, blue color indicating strong <111> ∥ surface normal texture; (c) and (d) are surface plane horizontal and vertical-projected IPF maps, respectively, and show no strong texture in these directions. Scale bar = 80 nm.

FIGURE 2.32. All-Euler orientation map with grain boundaries. Most grains are heavily twinned and span the length of interconnect lines. Drawn box shows location of higher-resolution run. Scale bar = 2 µm.

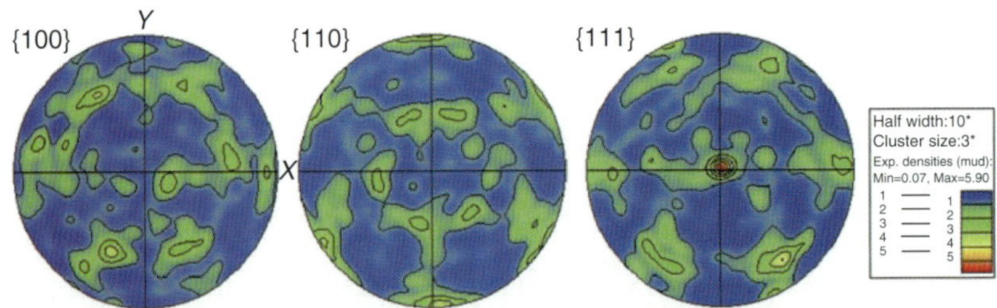

FIGURE 2.33. Contoured pole figure set, displaying strong <111> ‖ surface normal texture.

FIGURE 2.34. Grain size (equivalent-circle diameter) map, with all Σ3 twins as grain-delimiting boundaries. Scale bar = 2 μm.

FIGURE 2.35. Grain size (equivalent circle diameter) map, disregarding Σ3 twins as grain-delimiting boundaries. Scale bar = 2 µm.

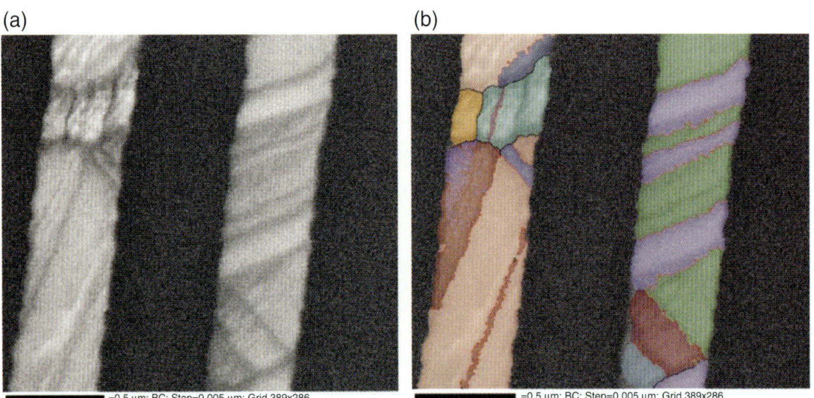

FIGURE 2.36. (a) Pattern quality (band contrast) map; (b) All-Euler orientation map. Wavyness due to instability during acquisition. Note <30 nm twin domain at bottom of left interconnect. Scale bar = 500 nm.

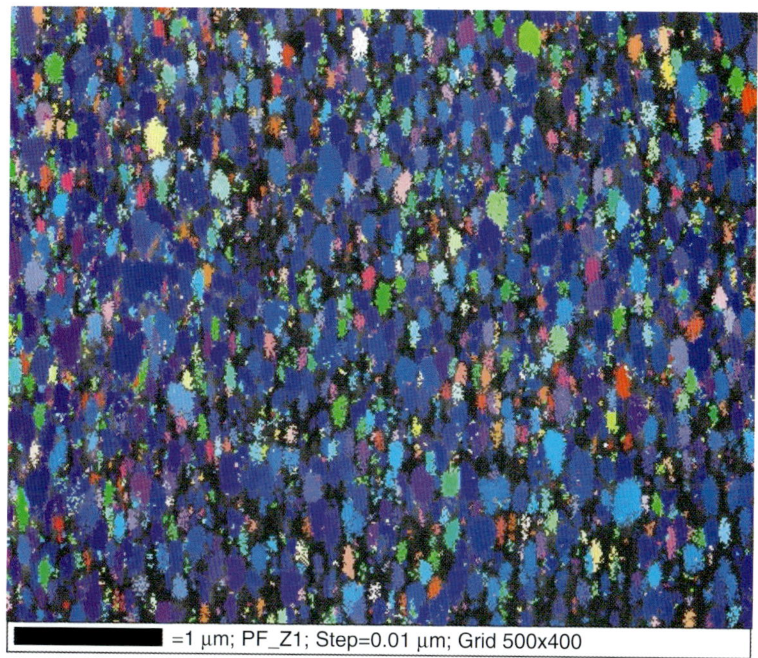

FIGURE 2.38. Surface normal-projected IPF-based orientation map, legend in Fig. 2.17. Predominance of blue grains implies a strong <111> ∥ surface normal texture.

(a)

(b)

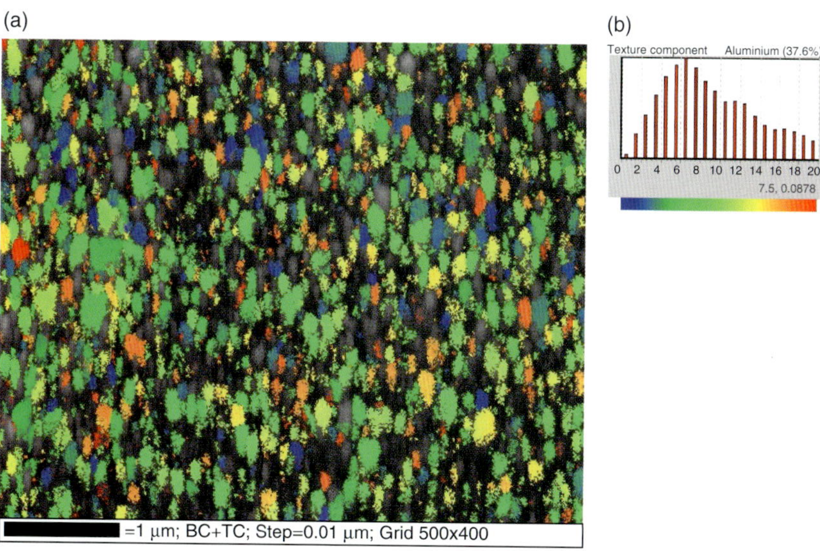

FIGURE 2.40. (a) Orientation map with grain coloring by degrees of misorientation relative to a perfect <111> ∥ surface normal parallelism, with blue = closest to parallel, red = 20° from parallel. (b) Legend showing rainbow scale fitted to 0–20° range.

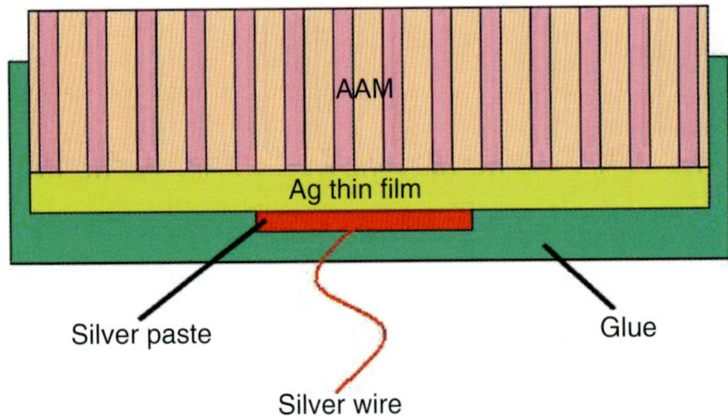

FIGURE 12.3. The structure of microelectrode.

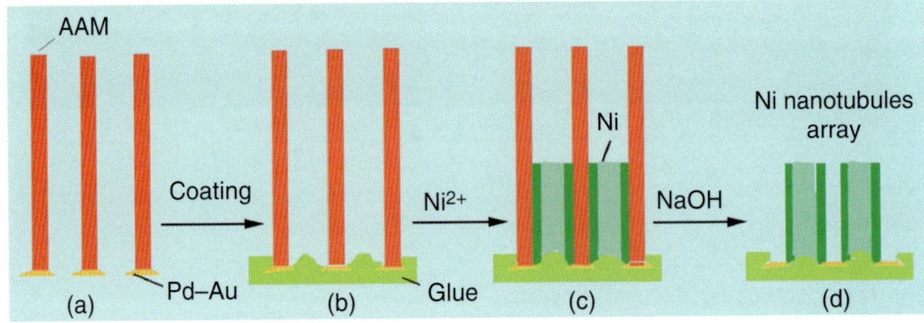

FIGURE 12.5. Processing scheme for fabrication of Ni nanotubes from glue nanowires–modified template. After partially coating the anodic alumina membrane (AAM) with Pd/Au (a) and attaching a metal wire contact (not shown), the metal surface is coated with glue that diffuses into the open pores of the membrane to produce glue wires (b). The glue wires help to direct the electrodeposition of nickel along the wall of the pores producing nickel nanotubes (c). Ni nanotube array after removal of AAM and glue wires.

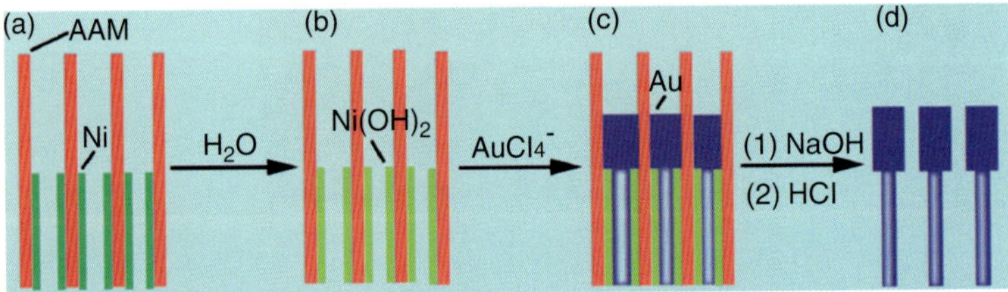

FIGURE 12.8. Scheme of directed growth of nanowires with structured tips.

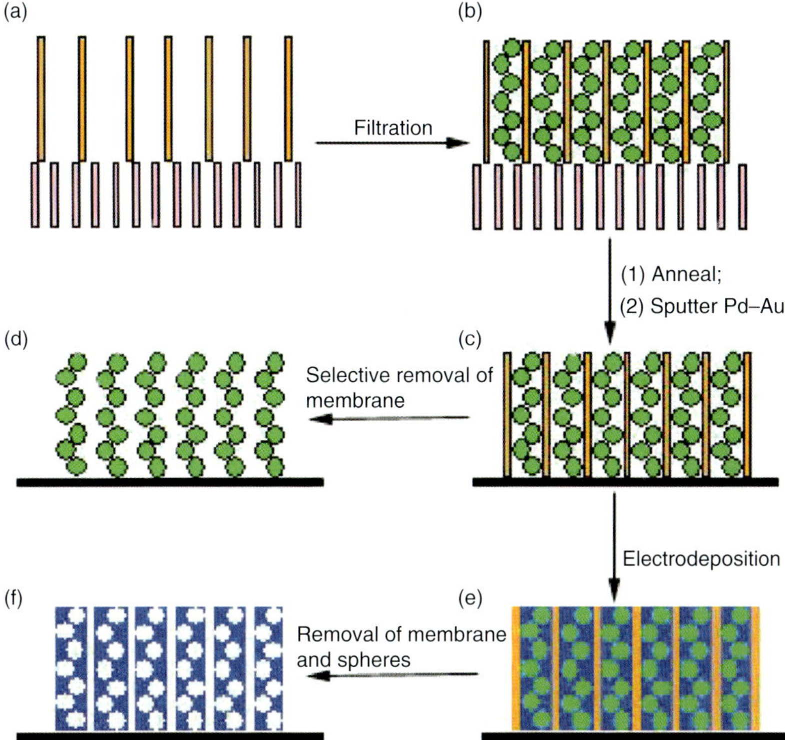

FIGURE 12.11. Fabrication procedure of colloidal crystal wires and porous wires: (a) two porous membranes are placed flush with each other; (b) colloid spheres are infiltrated into top membrane; (c) metal deposition to convert the composite into electrode and obtain metal base; (d) remove template to get colloidal crystal wires on metal base; (e) electrodeposition in colloid spheres-modified template; and (f) removal of templates to release porous wires.

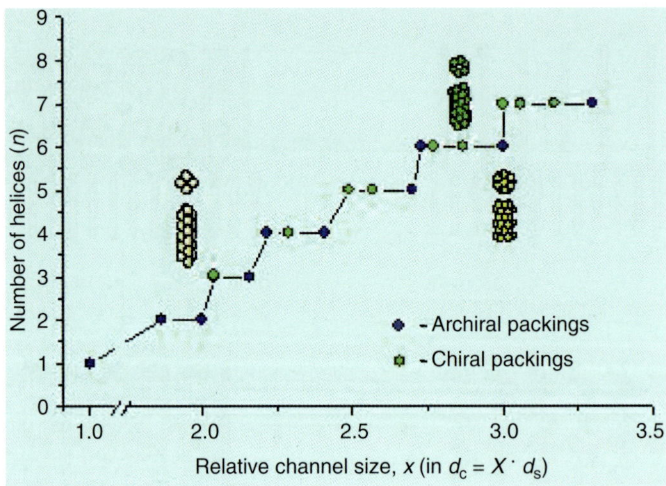

FIGURE 12.16. Number of helices vs. relative channel size for the packing of spheres within channels. Both chiral and achiral packings are indicated.

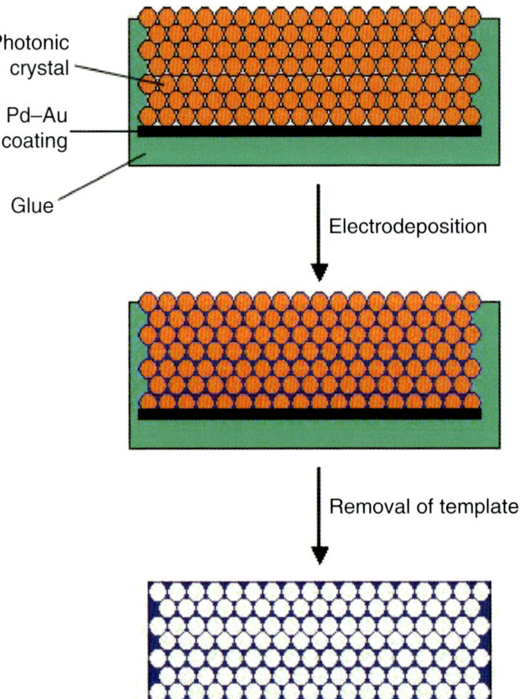

FIGURE 12.17. Procedure scheme of 3D metal inverse photonic crystals.

FIGURE 12.22. Color changes of photonic-crystal metal-mesh composites as a function of angle. The thin composite piece is displayed on green paper with a line of text to help indicate changes in orientation. The observed colors are green, brown, yellow, and blue and black, for the relative orientation angle of 30°, 60°, 90°, 120°, and 150°, respectively. Red, purple, and other colors (not shown) can also be observed at other angles.

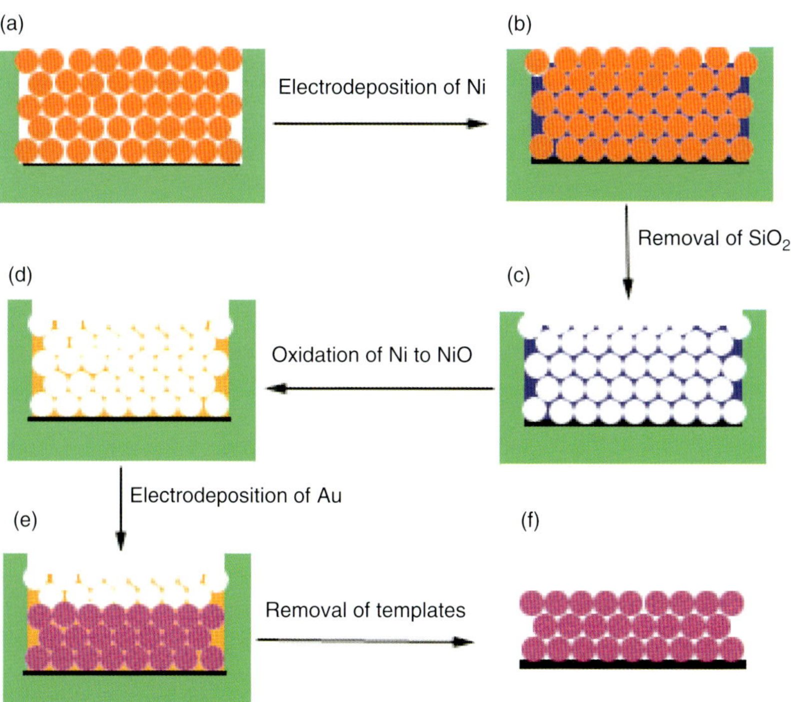

FIGURE 12.23. Procedure for the preparation of Au nanosphere arrays in a nonconductive matrix.

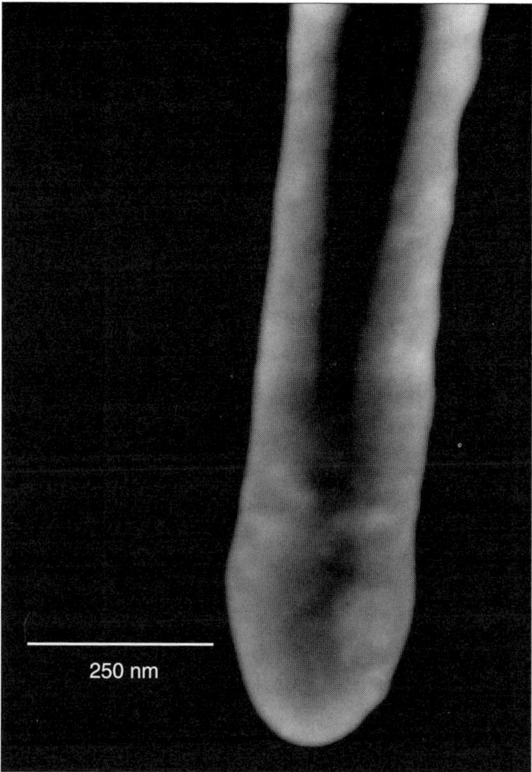

FIGURE 7.8. TEM micrograph of an oxide-coated tungsten probe. The tungsten is the darker material encased in the lighter colored oxide. The sharpness of the probe will increase on the order of 80% with the complete removal of oxide. (Adapted from [22].)

Although KOH and HF rinse is required for removing the bulk of oxide on a tungsten probe, it is not usually sufficient in removing all the oxide from the probe tip. A thin layer of oxide can quickly accumulate on the previously cleaned probe tip before it can be put in the vacuum system of the SEM. An in situ cleaning process can be performed in order to further clean the probe tips. Making the probe tips short together and passing a relatively large current through the tips can cause local heating that can result in cleaning of the tips. If the heating current is sufficiently high, usually a few milliamperes, the tungsten oxide will sublimate from the tip and the clean tungsten surface will then be available for probing. If local melting of the probe tip is achieved in a controlled manner, a clean tungsten sphere will form at the end as shown in Fig. 7.9.

Oxide and carbonaceous SEM contamination are killers of nanoscale electrical measurements. A clean chamber and clean probes are vital to perform such experiments. Thus, ensure that the probes and the nanomanipulator itself are included when performing downstream ash cleaning of the SEM chamber and that the local tip heating in vacuum follows soon after.

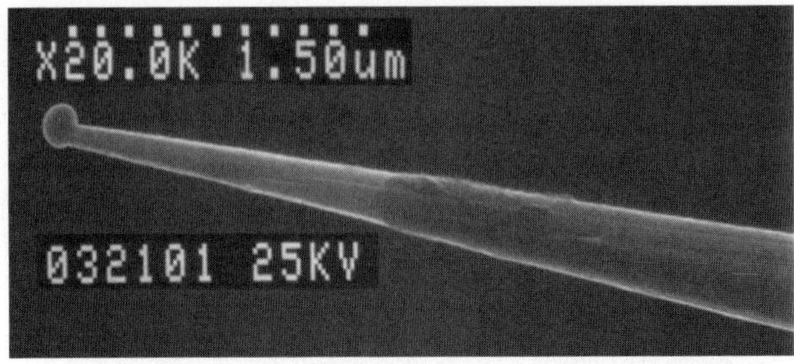

FIGURE 7.9. An SEM micrograph of a tungsten probe tip after melting the end. (Adapted from [22].)

4.2. Cantilevered Probes

Atomic force microscopy tips also make useful end effectors. An AFM tip (Fig. 7.10) is typically made out of single crystal silicon with a regular geometry and a sharp tip (radius of curvature in the order of 10 nm). Because of its regular shape and material properties, the mechanical properties of the cantilever can be calculated. Thus, AFM tips can be used to perform tensile tests of nanometer-scale tubes, wires, coils, and ropes [12–14]. AFM cantilevers can be purchased from the many companies that sell AFMs.

4.3. MEMS Grippers

Microelectromechanical systems (MEMS)-based grippers are also a valuable tool for manipulation. It has been demonstrated that they can be used for TEM lift-out [7]. They can also be used to characterize nanostructures in a similar way that an AFM cantilever is used. The mechanical properties of the gripper beams can be calculated or modeled. Thus, the mechanical properties of a nanotube suspended

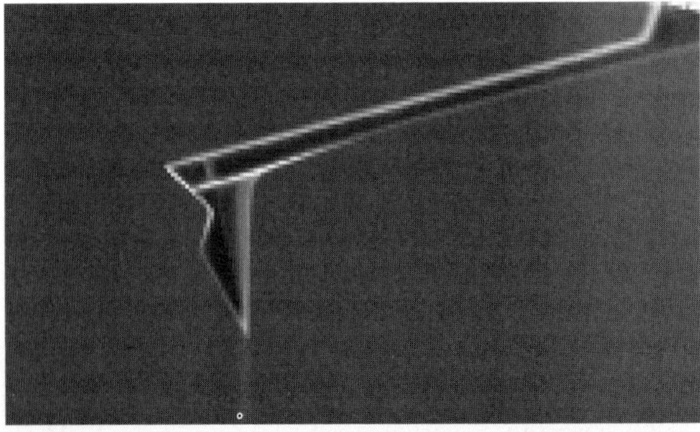

FIGURE 7.10. AFM style cantilevered probe.

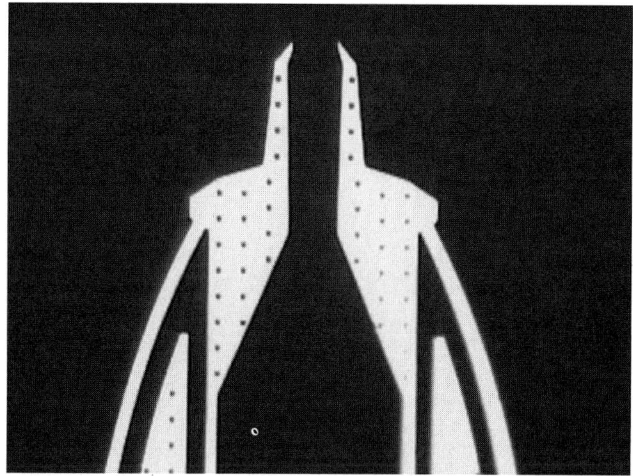

FIGURE 7.11. High-resolution photograph of a thermally actuated MEMS gripper. (Adapted from [22].)

between two gripper arms can be extrapolated. Figure 7.11 shows one of many examples of thermally actuated MEMS grippers.

MEMS gripper requires up to 10 V to operate, which can be delivered through the I/O terminals in situ. Special holders are required to attach the grippers to the nanomanipulator and to deliver power to them.

5. Applications of Nanomanipulators

The significant increase of applications for nanomanipulators over the past several years was caused by both innovation in nanomaterials as well as from attempts to expand the marketplace of production-level manipulation systems. The first applications were for the characterization of MWNTs. The advent of other types of nanowires of various materials such as boron and gallium makes the nanomanipulator a mainstay apparatus in nanotechnology research. There are also applications in IC device probing at the contact level. As devices become smaller, it becomes increasingly important to perform failure analyses at the device level, which requires SEM imaging and in situ probing. Nanomanipulators are well suited for this application because of their high-resolution positioning capabilities. These and other applications are described below.

5.1. Nanopositioning

The most straightforward application of nanomanipulation is positioning of nanostructures in situ. This can be done using a variety or combination of end effectors. Figure 7.12 shows a sharpened tungsten probe making contact to a

(a)

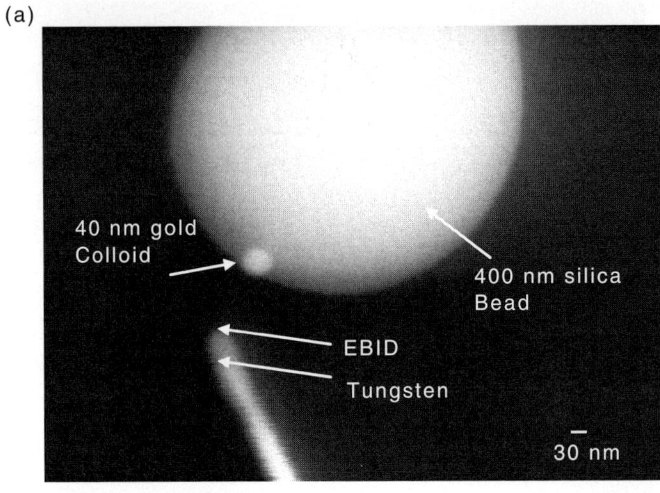

(b)

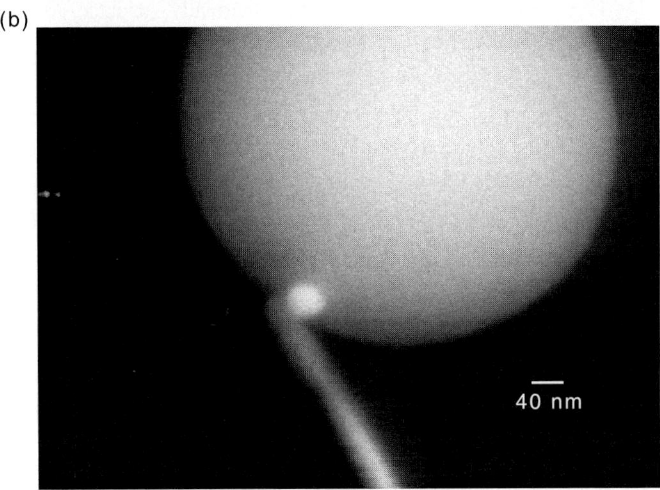

FIGURE 7.12. (a) A tungsten probe on approach to a 40 nm gold colloid affixed to a 400 nm silica bead. (b) The same probe in contact with the colloid. It is bent slightly after a force is applied, indicating a strong bond between the substrate and the silica bead (i.e., the bead does not move upon an application of a force). Also noteworthy is the layer of EBID growth enshrouding the tungsten probe. (Adapted from [22].)

40 nm gold colloid affixed to a 400 nm silica bead. The image demonstrates the positioning resolution of nanomanipulators.

A clean system is once again vital when performing nanometer-scale positioning. This is because nanoparticles will get bound to the surface by EBID. There is already a strong bond to the surface through van der Waals forces, and that bond can be overcome by the nanopositioner. EBID bonds, however, are strong, and attempts to perform nanomanipulation of particles that are cemented to a surface by EBID and can in fact damage probe tips and the particles under study. Figure 7.12b shows a probe tip in contact with a gold colloid. In comparison with Fig. 7.12a, the probe is bent slightly after application of a force.

The nanoparticles under study can be picked and placed, or pushed into place if they are not cemented to the surface. A sharp probe tip can be directed to gently scoop up a nanoparticle, move to a new location, and gently deposit the particle. The electron beam of the microscope may assist in this by adding additional charge, thus creating local electric fields. Particles have been observed jumping from probe to surface or from probe to probe. Figure 7.13 shows picking a 40 nm gold colloid from a surface.

5.2. *Mechanical Probing of Nanostructures*

Carbon nanotubes have attracted much attention for their remarkable physical properties, including high tensile strength and electrical conductivity. Theoretical predictions of elastic properties suggest that carbon nanotubes are extremely strong [15]. However, theoretical predictions must always be corroborated by experimentation.

Carbon nanotubes can be mechanically tested using a nanomanipulator. Yu et al. [12,13] used a cantilevered atomic force microscopy probe mounted onto

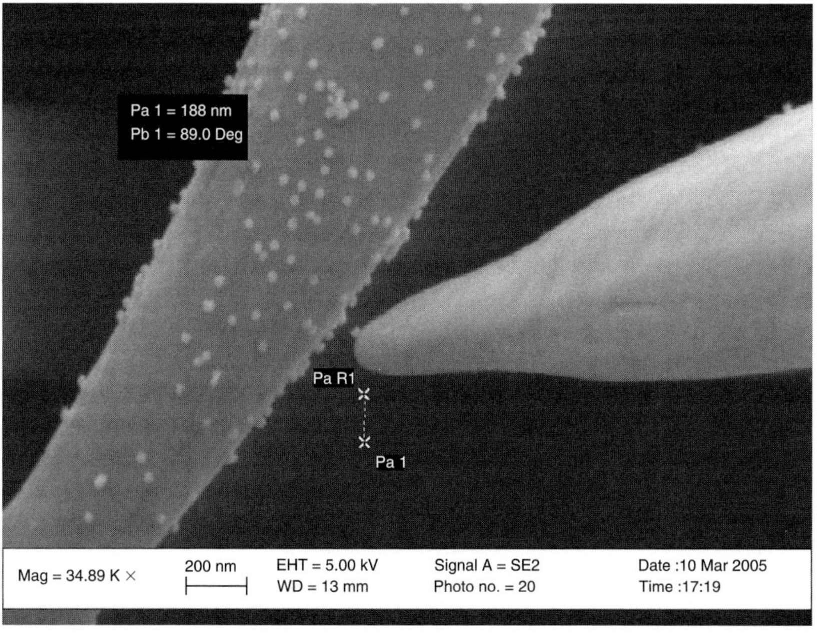

FIGURE 7.13. SEM micrograph of a 40 nm gold colloid picked from a surface by a tungsten probe. (Adapted from [22].)

a nanomanipulator to acquire and test both multiwalled and single-walled carbon nanotubes.

A nanotube can be attached to a probe tip using EBID. A probe is made to contact the nanotube and the electron beam is raster-scanned over a very small area overlapping the tube and a portion of the probe. The resulting carbonaceous layer (in fact, an EBID layer) acts to secure the nanotube almost like an adhesive. Once a nanotube is suspended as such between two AFM probes, a stress can be applied to the nanotube by pulling apart the AFM tips using the nanomanipulator. Figure 7.14 shows this process. The nanotube can be stressed until it breaks, and stress/strain curves can be extrapolated from data collected using the deflection of the AFM cantilevers [13]. Figure 7.15 shows how AFM cantilevers can be used to measure the mechanical strength of carbon nanocoils [14].

Further mechanical tests were performed by controllable placement of a carbon nanotube onto an MEMS-based device [16]. MEMS devices, like AFM cantilevers, can be accurately gauged for tensile strength and can be actuated with a high degree of control, making them excellent tools for performing nanoscale mechanical measurements.

Nanostructures of other materials besides carbon can also be characterized. Tungsten nanowires, for example, can be driven to vibrate to resonance by an applied electric field, as shown in Fig. 7.16. The Young's modulus of the nanotube or nanowire can be deduced by determining the fundamental frequency of the nanowire in resonance and its effective length [17].

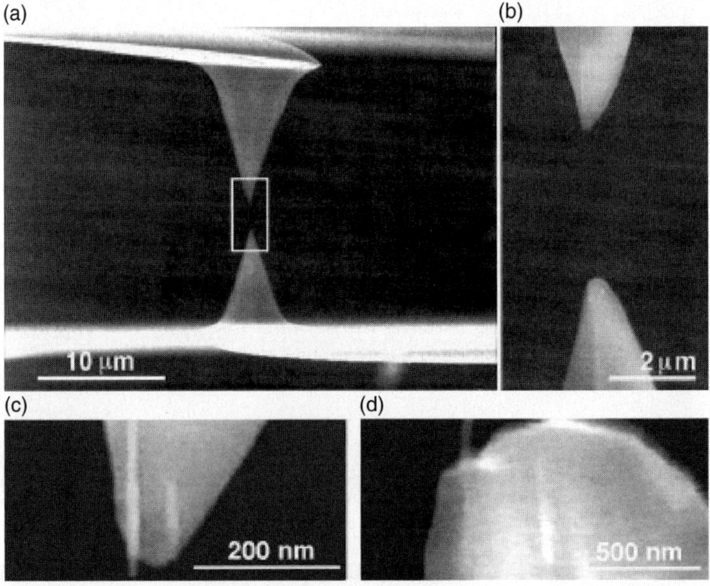

FIGURE 7.14. SEM micrographs showing the process by which carbon nanotubes can be gauged for tensile strength using in situ nanomanipulation. (Adapted from [12].)

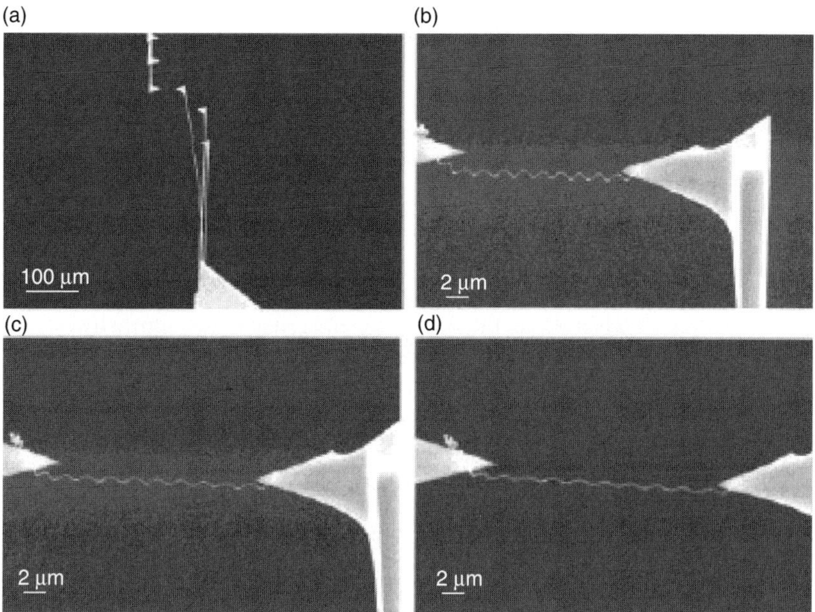

FIGURE 7.15. SEM micrographs of AFM cantilevered probes being used to measure the mechanical properties of a carbon nanocoil in situ. (Adapted from [14].)

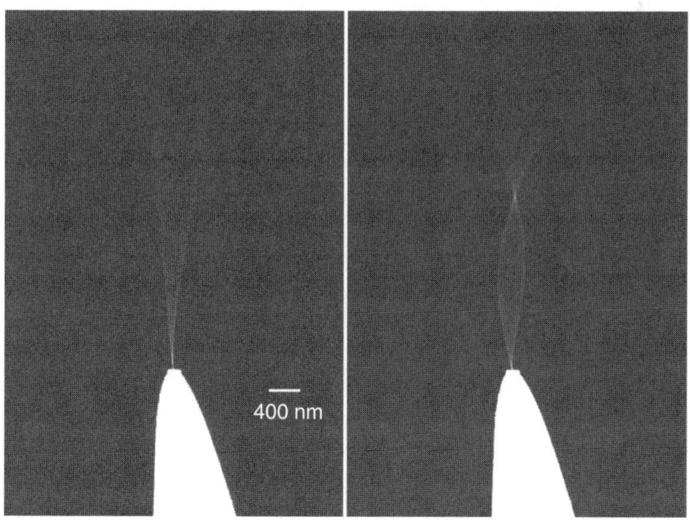

FIGURE 7.16. First and second harmonic mode of a tungsten nanowire. (Adapted from [22].)

5.3. Electrical Probing

Electrical probing in situ using nanomanipulators has proven a challenge to scientists for several reasons. EBID is one reason. The carbonaceous contamination will form on top of nanotubes as well as on probe tips. When performing electrical measurements, this layer serves as a dielectric barrier as does even a small amount of native oxide on probe tips. A KOH and HF rinse can remove much of the oxide, but not all. Tungsten oxide, even in very small amounts, can affect low current measurements.

Advances in nanoprobing have helped to resolve these issues. The removal of excessive levels of HCs from the SEM chamber using the downstream ash process allows scientists to examine an object for a longer period of time without the worry of contaminating it. For a nanomanipulation experiment, this translates into being able to image a nanostructure, bring one or several probes into contact with it, and perform the measurement before any contamination can form.

When electrically probing nanostructures, additional factors come into play. For example, contact resistance associated with the metal probe–nanostructure junction creates difficulties when making low current measurements [18]. When making electrical measurements while sweeping voltage, tests must be run slowly to overcome the high RC (resistance–capacitance) of the complete circuit. That is, high capacitance associated with the probe cables and contact resistance serves to increase the time constant, RC, of the circuit. Longer measurement times with slower sweeps can reduce the effects of the high RC. Ways of reducing the RC include clean probe junctions, the choice of metal for the electrode, and shorter cables or preamps in vacuum near the probes. Figure 7.17 shows a two-probe configuration; Fig. 7.18 shows a four-probe configuration for performing electrical measurements.

5.3.1. Probing Carbon Nanotubes

Single-walled carbon nanotubes (SWNT) were grown across trenches using a chemical vapor deposition technique. Four-point probe measurements were taken on the nanotubes using a Zyvex S100 nanomanipulator and clean tungsten probes. Figure 7.19 shows the probe configuration for the experiment. The nanotube was removed from the left side of the trench and contacted with four probes as shown. The tube is supported by the four probes and the right side of the trench, where it is still connected but electrically isolated. Figure 7.20a shows the source–sink measurement, which is a two-probe measurement between the two outer probes. The nonlinearity of the curves is typical of the contact barrier of the electrode–sample junction. Figure 7.20b shows the I–V trace between the two inner probes. This measurement is made as confirmation of contact between the inner probes. Figure 7.20c is the sense voltage measured at each of the two inner probes plotted against the sweep voltage of the source probe. The voltage drop or difference between the two sense probes is calculated and also plotted and shows a similar nonlinearity of the source current. In Fig. 7.20d, this voltage drop curve is used as the x-axis to plot the source current of Fig. 7.20a. The results show the

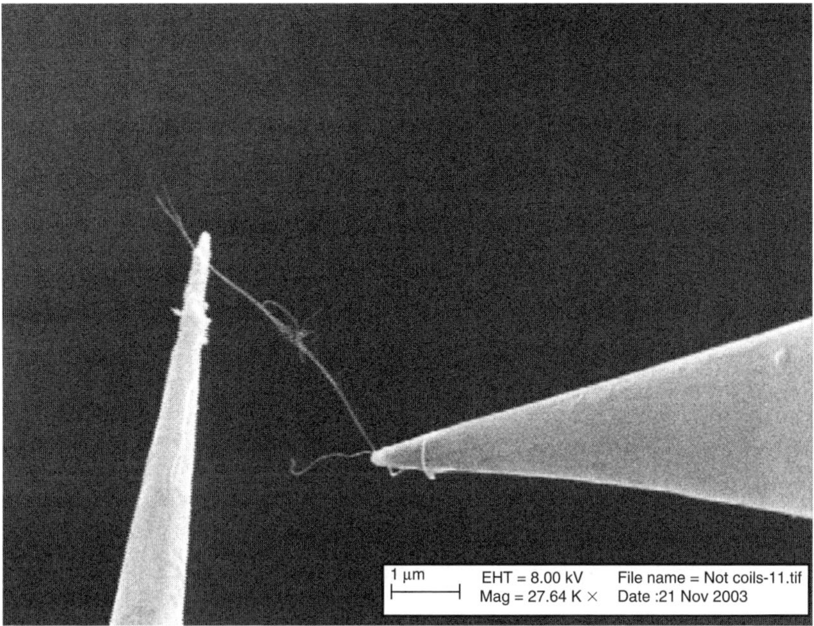

FIGURE 7.17. Two probe electrical measurement configuration of a group (rope) of multi-walled carbon nanotubes. (Adapted from [22].)

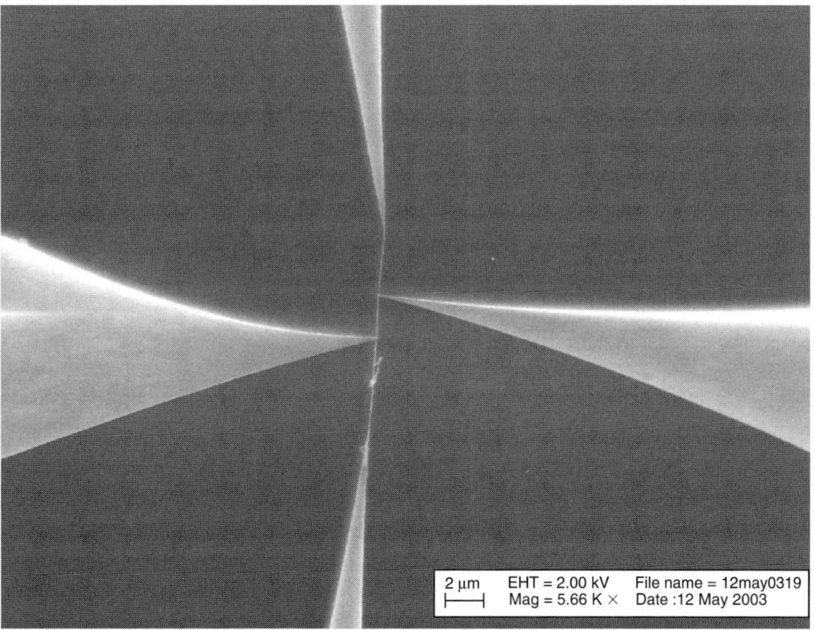

FIGURE 7.18. Four-probe configuration for performing electrical measurements. The four-probe configuration is considered to be sufficient for extrapolating electrical data in the absence of contact resistance values. (Adapted from [22].)

FIGURE 7.19. Four-probe configuration used to collect electrical data on a single-walled carbon nanotube. The nanotube is attached to a trench, but pulled away from it in order to perform the experiment. (Adapted from [22].)

true electrical behavior of the nanotube's region between the two voltage sense probes. This carbon nanotube's behavior is relatively linear and Fig. 7.20d shows the four-probe measurement curve, from which the resistance of the nanotube can be extrapolated. The four-point probe measurement eliminates some of the problems associated with high contact resistance.

The nanomanipulator must be stable enough for multiple measurements to be made. If the nanotube cannot be held fixed for at least several minutes, then subsequent data sets will not reflect accurate electrical characteristics. Each of the data sets presented in Fig. 7.20 correspond to the image in Fig. 7.19. The RC of the complete electrical circuit combined with the integration time for the high impedance voltage sense units (10^{16} Ω) was considerable and required 20 min to collect the I–V curves.

5.3.2. Probing ZnO Nanowires with Mn-Coated Tips

A bunch of zincoxide (ZnO) nanowires was placed into an SEM, and individual nanowires were probed in situ using the Kleindieck system [19]. Figure 7.21 shows the experimental setup. The probes used were coated with manganese (Mn). The ZnO nanowires were not removed from the substrate; rather, a single

(a)

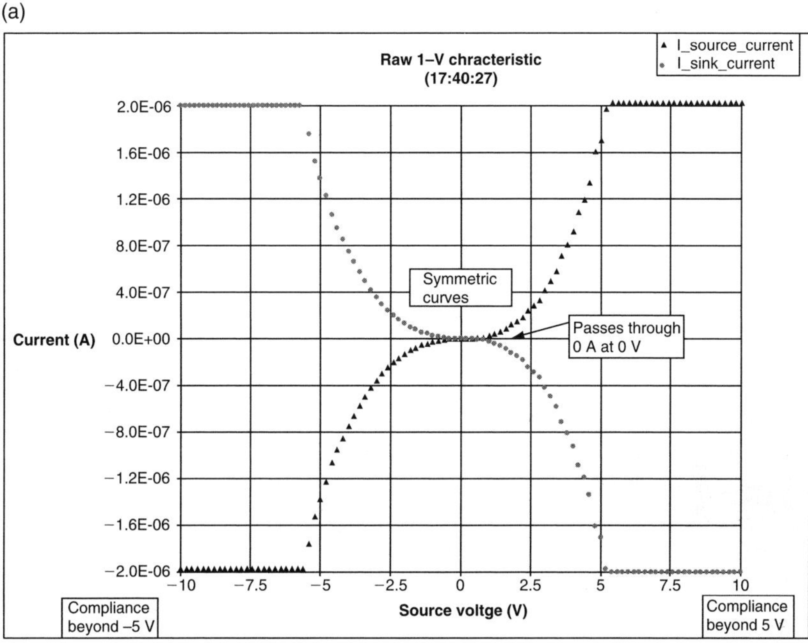

(b)

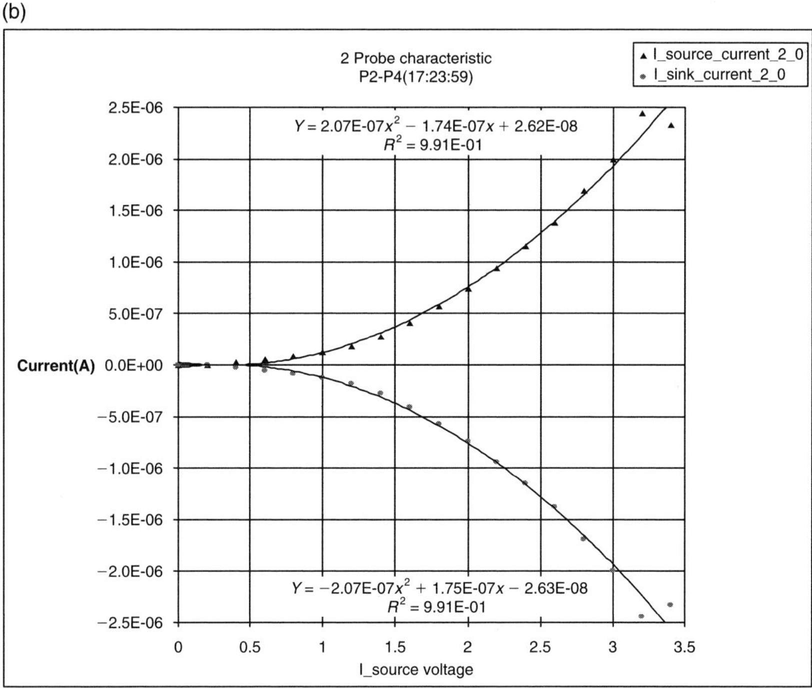

FIGURE 7.20. (a) The I–V measurement of the single-walled carbon nanotube (SWNT) made between the source probe and the sink. (b) The I–V measurement made between the inner (measurement) probes. This is made to verify contact to the nanotube.

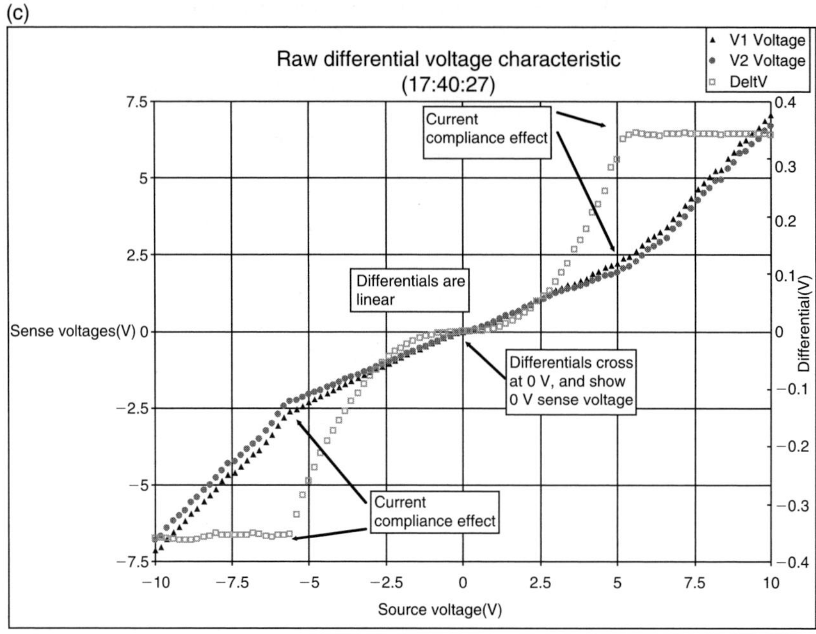

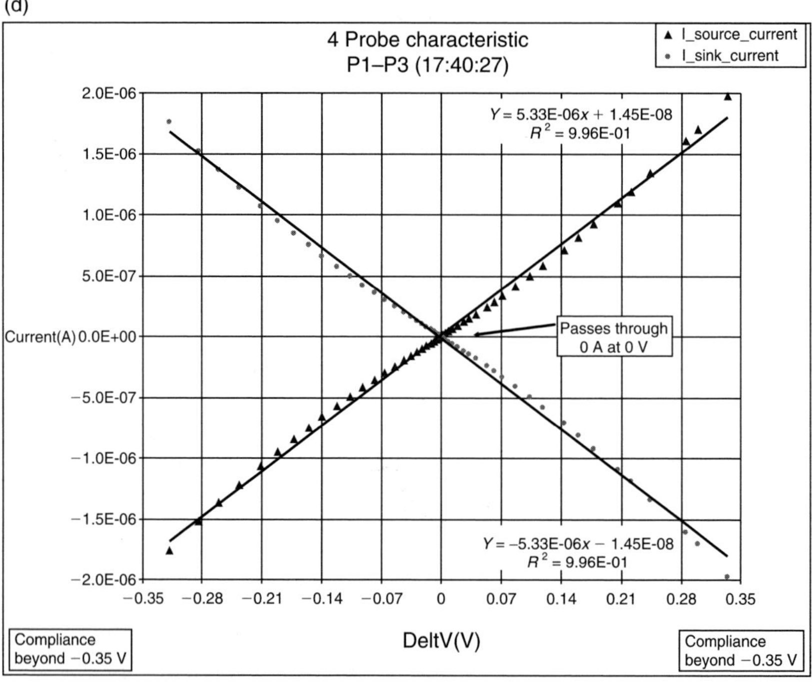

FIGURE 7.20. (*Continued*) (c) The differential voltage plots showing the difference in voltage between the inner probes measured against the driving voltage. (d) The characteristic four-point probe measurement of a single-walled carbon nanotube (SWNT). The linear response is typical of a metallic tube. Each measurement was made in succession and corresponds to the image in Fig. 7.19. The nanomanipulator has the capability to maintain a fixed position long enough for multiple measurements to be made (each measurement took 20 min). (Adapted from [22].)

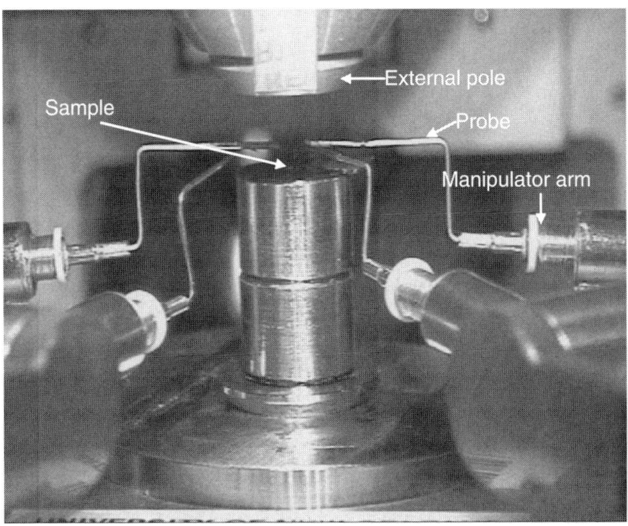

FIGURE 7.21. The Kleindieck micromanipulator in an FE-SEM for nanowire electrical characterization. (Adapted from [19].)

nanowire was contacted from near its base and from the top, as shown in Fig. 7.22a. Initial contact, as shown in Fig. 7.22b resulted in a Schottky type electrical response. Changes in the current were observed when the ZnO nanowire was mechanically deformed during the electrical test. Figure 7.23a shows the probe configuration forcing a 5° bend in the ZnO nanowire; Fig. 7.23b shows the same nanowire with a 20° bend. The measured current was observed to increase upon a higher degree of deflection, as shown in Fig. 7.23c. Changes in electrical response were also observed as the SEM chamber pressure was increased. As the graph shown in Fig. 7.24, the conductivity decreased with increasing pressure. This feature has applications in sensor technology at room temperature.

5.4. IC Probing

Nanomanipulators can be used in semiconductor FA. Determining failure mechanisms and localizing defect sites in IC has proven challenging, even when considering the litany of powerful techniques analysts utilize throughout the FA process. The complexities of failure verification and identification are compounded because of the lack of a standardized FA process flow; each procedure is unique to the device under test.

Any process flow will include failure verification followed by a visual inspection. Visual inspections can include anything from an external visual inspection, such as an examination of the packaging, to optical and electron microscopy. Other techniques include infrared analysis (i.e., hot spot detection), photon emission, and scanning probe techniques. When visual inspections do not yield

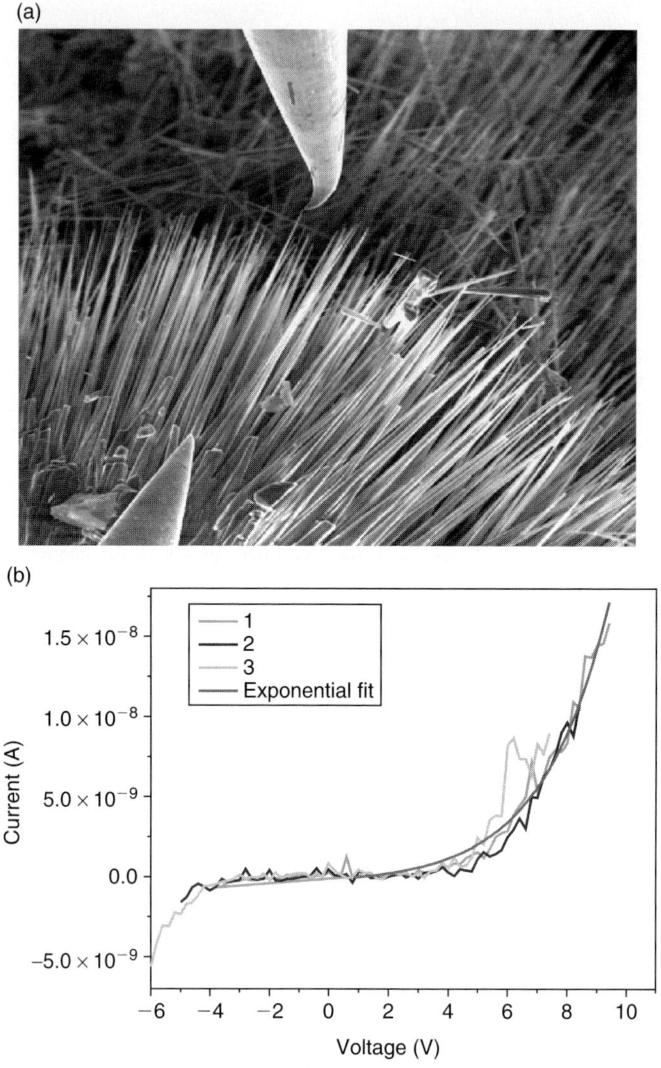

FIGURE 7.22. (a) Manganese-coated probes making contact to a ZnO nanowire. (b) Electrical response upon contact showing the Schottky barrier. (Adapted from [19].)

enough or any information to assist the analyst, techniques to evaluate nonvisual failures must be employed. Microprobing has been used extensively to access contact points on the IC while analyzing the electrical behavior of various parts of the circuit, bit, or transistor.

Probing ICs with sharpened metal wires under an optical microscope has been the primary means of characterizing an IC's electrical performance for many years. Since advances in semiconductor technology follow Moore's Law of

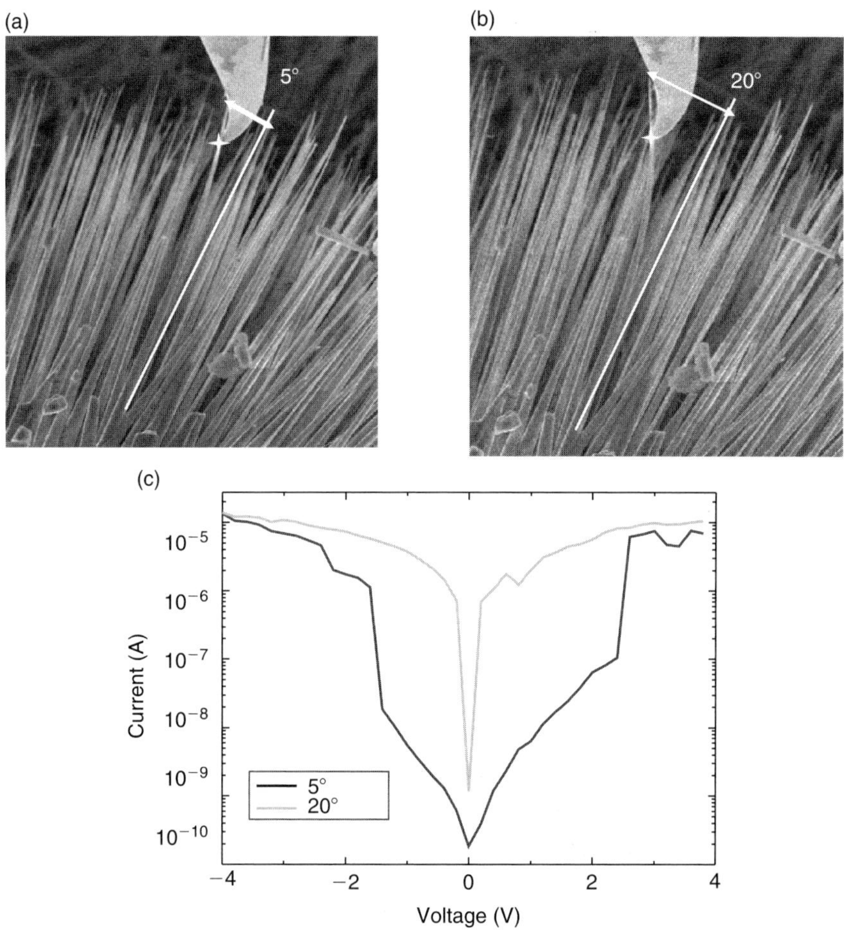

FIGURE 7.23. (a) Manganese-coated probe making contact to a ZnO nanowire creating a 5° bend. (b) The same nanowire, but with a 20° bend. (c). The corresponding I–V response. (Adapted from [19].)

decreasing scale, the complexity of producing and testing advanced IC's has increased. More powerful microscopes and high precision probe placement is needed in order to test these next-generation ICs.

SEM and FIB systems have emerged as the tools of choice for imaging and measuring the results of different IC fabrication steps. SEMs are commonly used in fabrication laboratories and are workhorse instruments for metrology and FA. The Zyvex KZ100 IC Nanoprober System and the Kleindieck MM3A, which are compatible with both imaging platforms, greatly enhance the utility of these platforms by enabling IC electrical performance measurements at the nanoscale. Probing in-die transistors, which are actual devices at the contact level has proven to be an invaluable technique. Figure 7.25 shows an SEM micrograph of the

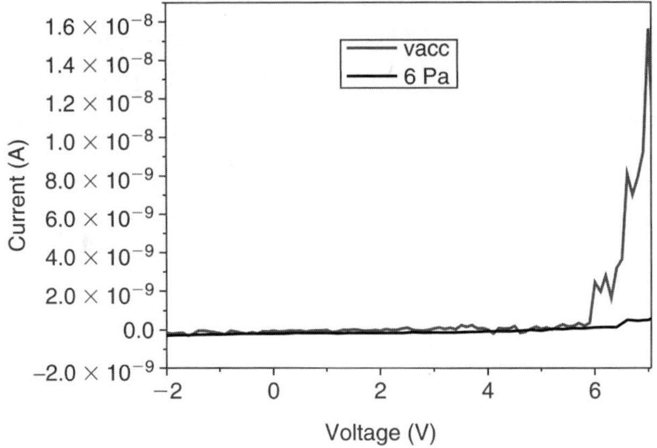

FIGURE 7.24. A decrease in current is observed as the chamber pressure is increased. This feature can be used to develop gas sensors at room temperature. (Adapted from [19].)

four-probe configuration during a particular device test. Every wafer brought in for probing should be freshly cleaned to remove AMC. With data collected from individual on-chip transistors using nanomanipulator, IC design engineers can feed device data into design models to improve modeling accuracy.

5.5. Semiconductor Coupon Extraction

Part of the IC FA cycle is to perform TEM analyses on FIB-milled, semiconductor coupons. These coupons must be transferred from a sample in situ to a TEM sample grid. Various instruments exist to perform this transfer, including the Omniprobe AutoProbe 200 and the Zyvex S100. The AutoProbe 200 is a chamber-mounted nanomanipulator for TEM lift-out and nanomechanical and electrical testing. It uses an etched W probe tip and a metal CVD process to lift coupons from samples and place them onto TEM grids for analysis (Fig. 7.26). The Zyvex S100 is a four-positioner system that utilizes MEMS grippers or metal probes to remove the FIB coupons. Figure 7.27 shows a sequence of images where MEMS grippers are being utilized for TEM coupon extraction [7].

5.6. In Situ TEM Manipulation

The Nanofactory TEM manipulator (STM-Holder) has been used to deduce electrical transport mechanisms and mechanical characterization of nanotubes and nanowires by several research groups. For example, Cummings et al. [8] showed that carbon nanotubes can be structurally altered using applied voltages. This structural change can be monitored in situ to the TEM simultaneously, as shown in Fig. 7.28. The TEM offers increased magnification such that individual layers of an MWNT can be imaged.

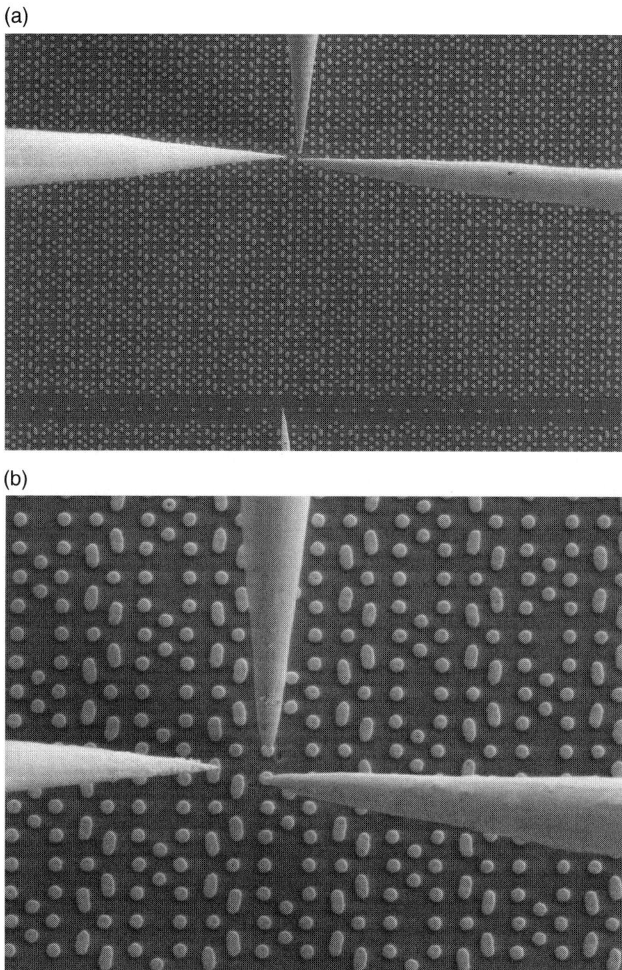

FIGURE 7.25. (a) Figure showing SEM micrographs of IC device probing using a four-probe nanomanipulator in situ. (b) High magnification image of (a). (Adapted from [22].)

The TEM manipulator can also be used to measure single-atom nanowires and point contacts. The force interaction between nanoparticles was studied by Erts et al. in Latvia and Sweden. An AFM cantilever was used to measure force deflection upon contact with gold-coated probe tips [20]. Another study by Erts et al. utilized the TEM manipulator to characterize electrical properties of gold point contacts in situ. It has been shown that single-atom wires can be created by pushing gold-coated STM style probes into surfaces, running a current, and retracting the probe. The softening of the gold at the nanometer scale creates a small wire connecting the probe to the surface. The limit between ballistic and diffusive transport was studied using the TEM manipulator [21].

(a)

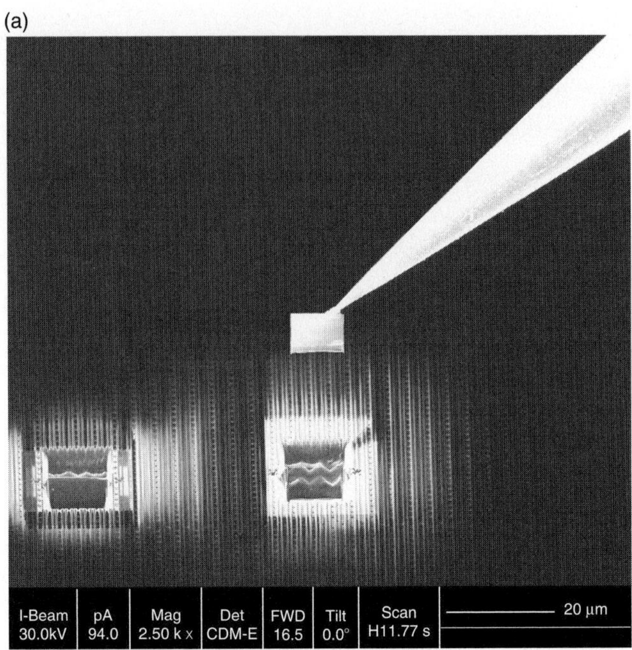

(b)

FIGURE 7.26. (a) SEM micrograph of an Omniprobe nanomanipulator extracted an FIB-milled coupon. (b) SEM micrograph of an Omniprobe nanomanpulator probe placing and securing an FIB-milled coupon onto a TEM grid for analysis. (Adapted from [24].)

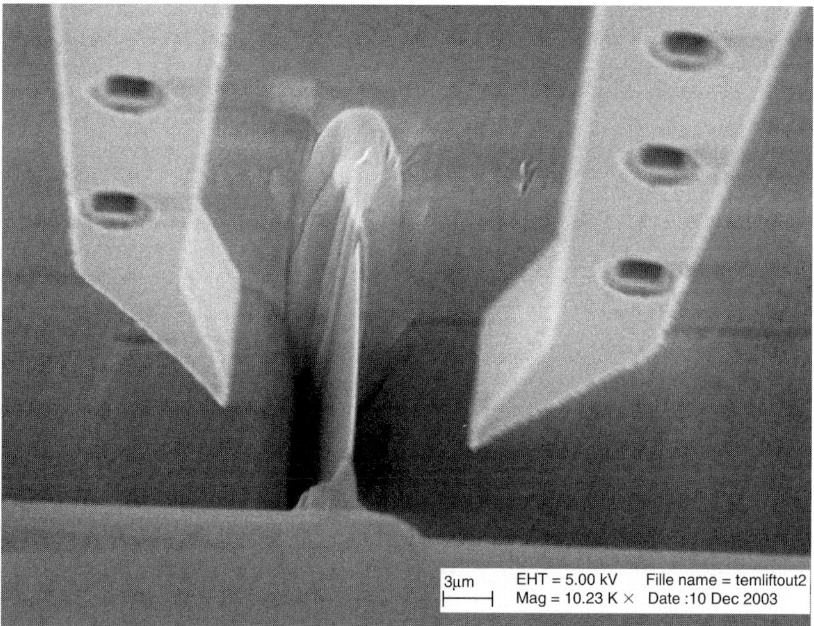

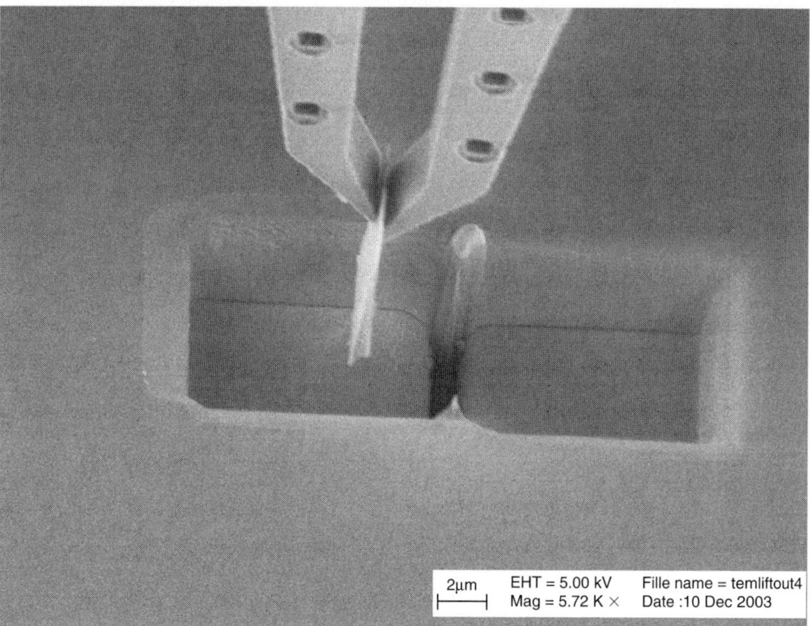

FIGURE 7.27. SEM micrographs of an MEMS gripper removing and placing an FIB-milled coupon for TEM analysis. (Adapted from [7].)

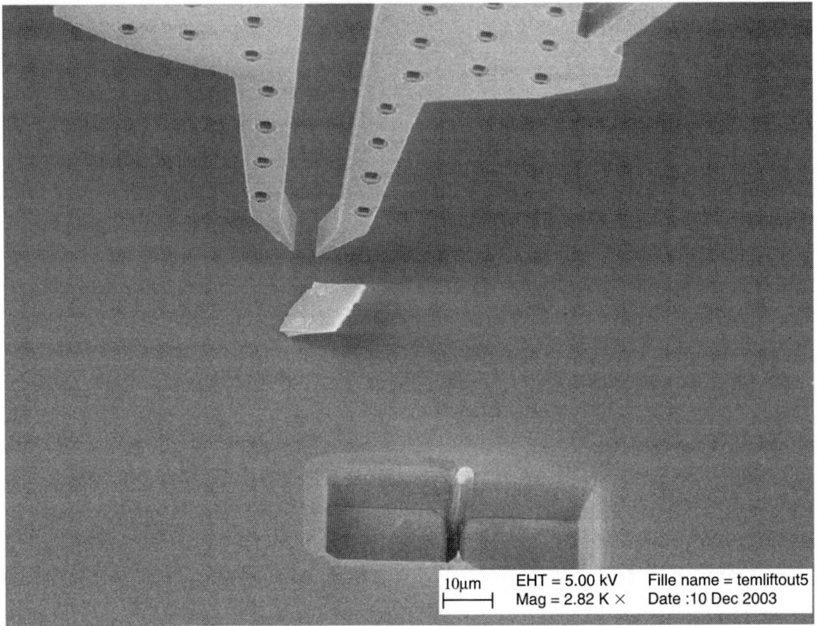

FIGURE 7.27. (*Continued*)

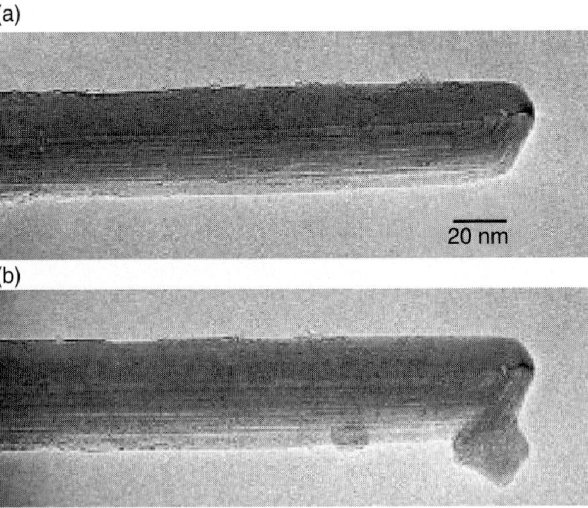

FIGURE 7.28. TEM images of a multiwalled carbon nanotube being structurally altered using the Nanofactory TEM nanomanipulator in situ. (Adapted from [8].)

(c)

(d)

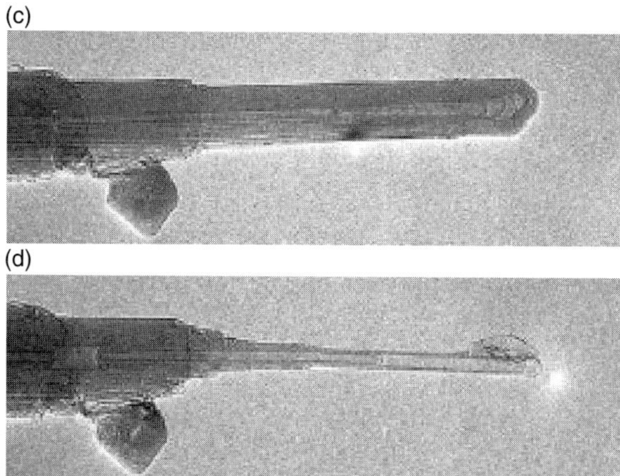

FIGURE 7.28. (*Continued*)

6. Summary

The discovery of novel nanostructures and the decreasing length scales of semi-conductor devices have created the need for high precision, mechanical manipulators for in situ probing and positioning. There are several types that already exist as production models, though in-house systems are also used. The various systems have different advantages, which depend entirely on the application. Nanomanipulators have been used for years to study the mechanical and electrical properties of nanostructures and for semiconductor device FA. They are compatible with SEM, FIB, and even TEM systems, and provide an excellent means by which real-time imaging and analysis can be enhanced.

References

1. M. F. Yu, M. J. Dyer, G. D. Skidmore, H. W. Rohrs, X. K. Lu, K. D. Ausman, J. R. VonEhr, and R. S. Ruoff, *Nanotechnology*, 10 (1999) 244.
2. R. Gupta, T. M. Cavanah, and M. in het Panhuis, *Microsc. Microanal.*, 10 (Suppl. 2) (2004) 962.
3. M. J. Kim, M. in het Panhuis, R. Gupta, A. S. Blum, B. R. Ratna, B. E. Gnade, and R. M. Wallace, *Microsc. Microanal.*, 10 (Suppl. 2) (2004) 26.
4. P. G. Collins and Ph. Avouris, *Appl. Phys. A*, 74 (2002) 329.
5. R. Gupta, R. Stallcup, and M. in het Panhuis, *Nanotechnology*, (2005) (in press).
6. S. Paulson, M. R. Falvo, N. Snider, A. Helser, T. Hudson, R. M. Taylor II, R. Superfine, and S. Washburn, *Appl. Phys. Lett.*, 75 (1999) 2936.
7. K. Tuck, M. Ellis, A. Geisberger, G. Skidmore, and P. Foster, *Microsc. Microanal.*, 10 (Suppl. 2) (2004) 1144.
8. J. Cummings, P. G. Collins, and A. Zettl, *Nature*, 406 (2000) 586.
9. J. Cummings and A. Zettl, *Science*, 289 (2000) 602.

10. R. Vane and R. E. Stallcup II, *Scanning*, 27(2) (2005) 106.
11. I. Ekvall, E. Wahlström, D. Claesson, H. Olin, and E. Olsson, *Meas. Sci. Technol.*, 10 (1999) 11.
12. M. F. Yu, O. Lourie, M. J. Dyer, K. Moloni, T. F. Kelly, and R. S. Ruoff, *Science*, 287 (2000) 637.
13. M. F. Yu, B. S. Files, S. Arepalli, and R. S. Ruoff, *Phys. Rev. Lett.*, 84(24) (2000) 5552.
14. X. Chen, S. Zhang, D. A. Dikin, W. Ding, and R. S. Ruoff, *Nano Lett.*, 3(9) (2003) 1299.
15. R. S. Ruoff, D. Qian, and W. K. Liu, *CR Physique*, 4 (2002) 993.
16. P. A. Williams, S. J. Papadakis, M. R. Falvo, A. M. Patel, M. Sinclair, A. Seeger, A. Helser, R. M. Taylor II, S. Washburn, and R. Superfine, *Appl. Phys. Lett.*, 80(14) (2002) 2574.
17. M. F. Yu, G. J. Wagner, R. S. Ruoff, and M. J. Dyer, *Phys. Rev. B*, 66 (2002) 073406.
18. J. Tersoff, *Appl. Phys. Lett.*, 74(15) (1999) 2122.
19. J. J. Liu, J. J. Chen, M. Zhu, and W. L. Zhou (2005) (to be submitted).
20. D. Erts, A. Lõhmus, R. Lõhmus, H. Olin, A. V. Pokropivny, L. Ryen, and K. Svensson, *Appl. Surf. Sci.*, 188 (2003) 460.
21. D. Erts, H. Olin, L. Ryen, E. Olsson, and A. Thölén, *Phys. Rev. B*, 61(19) (2000) 12725.
22. Image courtesy of Zyvex Corporation, Richardson, Texas.
23. Image courtesy of Kleindieck, Reutlingen, Germany.
24. Image courtesy of Omniprobe, Dallas, Texas.

8
Applications of Focused Ion Beam and DualBeam for Nanofabrication

Brandon Van Leer, Lucille A. Giannuzzi, and Paul Anzalone

1. Introduction

The use of focused ion beam (FIB) technology in the area of nanoprototyping and nanofabrication is becoming increasingly important as dimensions of emphasis continue to shrink from the micrometer to the nanometer level. The characterization of materials and devices using FIB and/or DualBeam technology (i.e., an FIB and a scanning electron beam on the same platform as shown in Fig. 8.1) has proven to be critical for the research, development, and failure analysis research laboratory. FIB/SEM technology is also being utilized in novel ways to engineer nanostructures and devices employing ion-and/or electron-beam deposition of metals, organic materials or insulators, and milling of materials with the ion beam.

The interaction of ions and electrons with target materials is a bit different. The signals generated from interaction of electrons and target materials have been addressed in Chapter 1. Ion–solid interactions produce secondary ions, x-rays, backsputtered ions, neutral atoms, secondary ions, and clusters from target materials, and the penetration depth is only about 10–20 nm, quite different from that of electrons as shown in Fig. 8.2.

The fabrication of micrometer scale structures with the FIB [1] has easily transformed into nanoscale fabrication. For example, fabrication at the nanoscale with FIB technology has been utilized to make sensors and electrical devices [2], to serve as nucleation sites for precise growth of either carbon nanotubes [3] or quantum dots [4], and in the fabrication of nanostructures such as photonic gratings [5]. The ability to ion beam deposit 3D free-standing structures [6] has enabled a wide range of structures to be directly processed [7]. FIB nanofabrication has also been performed in the processing of existing structures and materials such as probe tip modification for atomic force microscopy [8].

The use of FIB and DualBeam instrumentation for the nanometer precision of transmission electron microscopy specimen preparation is well known [9,10]. FIB based TEM specimen preparation techniques have been directly transferable to other analytical methods [11]. As recently summarized [12], the use of FIB/DualBeam instrumentation as shown in Fig. 8.3 has become quite popular to

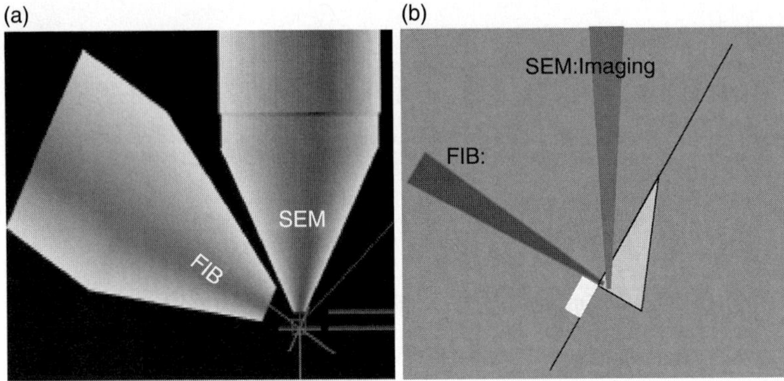

FIGURE 8.1. A FIB and a scanning electron beam on the same platform.

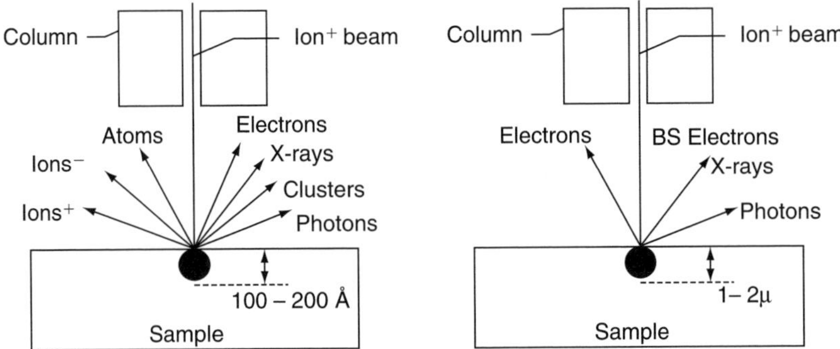

FIGURE 8.2. Ion interaction (a) and electron interaction (b) with targeted materials, respectively.

directly thin sample tips for field ion atom probe microscopy. Nanofabrication with FIB and DualBeam technology has been enhanced with the addition of either onboard or external pattern generator engines and the use of scripting for individual pixel control of beam parameters such as beam overlap and dwell time. In the section below, we will discuss these beam variables as they pertain to the nanofabrication of structures and devices, and show examples of DualBeam use for nanofabrication.

2. Onboard Digital Patterning with the Ion Beam

FIB based processes to remove or deposit material are dependent on several parameters that include the ion beam current, beam dwell time, raster refresh time, and if using a chemical gas precursor, the gas flux. Historically, beam overlap was fixed in either the x- or y-direction to ensure uniform exposure of the surface by the FIB [13]. Recently, leading FIB instrument manufacturers have begun providing digital pattern generators, which allow for milling and deposition of

FIGURE 8.3. FEI FIB/SEM DualBeam instrumentation.

complex structures employing user defined inputs such as geometrical patterns like circles, rectangles, polygons and/or direct import of graphical bitmap files. For example, the flexibility of the FEI onboard pattern generator allows the user to vary up to 30 parameters to achieve structures for nanotechnology applications, and can be used for milling or in conjunction with either ion-beam assisted or electron-beam assisted chemical vapor deposition (CVD) [14].

Critical beam parameters that control the time the beam resides in one spot (i.e., the dwell time) or the relative distance between beam position (e.g., defined either by a percentage overlap or by an actual distance or "pitch") can be key to achieving optimum results in machining or deposition. The beam overlap (OL) is defined by the beam diameter and the step size of the beam movement as shown in Fig. 8.4.

Figure 8.5 shows schematic diagrams that define positive, zero, and negative beam overlap conditions. A positive overlap is generally used for milling and imaging, while zero or a negative overlap is typically used when the ion beam is used with gases such as in CVD or enhanced etching.

Other important beam parameters in FIB applications are the beam dwell time and raster refresh time. As previously stated, the dwell time is defined as the time the beam rests in one position. These dwell time values can typically vary between 100 ns and 4 ms. The number of pixels and the dwell time per pixel determines the time required to complete one raster across the pattern and is aptly

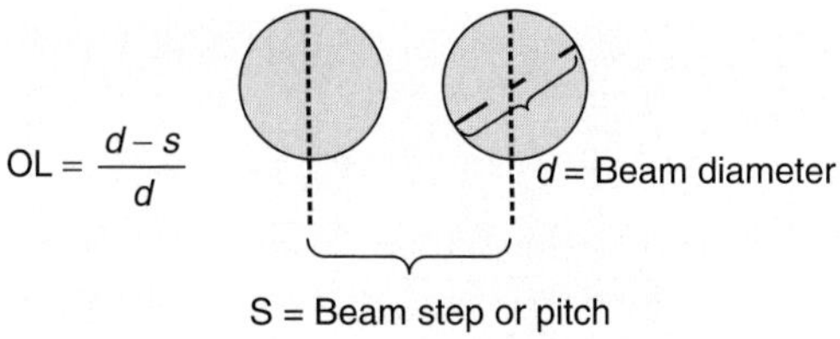

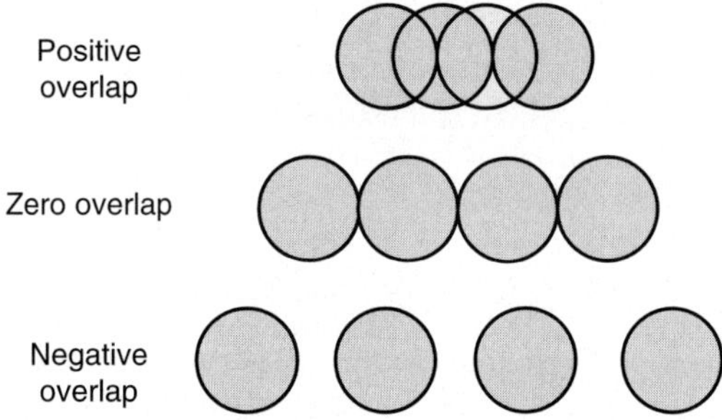

FIGURE 8.4. A schematic diagram defining beam overlap.

FIGURE 8.5. A schematic diagram showing beam overlap conditions.

named raster refresh time. Figure 8.6 shows how the beam dwell time influences the effective FIB milled line width. A beam of 1 pA was used to FIB mill into Si using by varying the dwell time from 250 ns to 10 μs. A constant 0% beam overlap and the same number of passes were used for all lines. As shown in Fig. 8.6, as the dwell time decreases, the depth of the cut will decrease, and the effective FIB mill line width will also decrease. The actual FIB milled dimensions that can be achieved will always be somewhat larger than the ultimate beam diameter and will vary depending on the collision cascade defined by the ion–solid interactions for any given target. Note that with a FIB resolution of ~5 nm, 10 nm wide FIB milled line is possible in Si (see Fig. 8.6), but that a factor of 3× larger line width is observed if the beam conditions are not optimized.

One may use ion–solid interaction theory to alter the geometry and aspect ratio of FIB milled nanostructures. Figure 8.7 shows SEM images of FIB milled cross-sections of FIB milled lines performed at 52° incidence angle (left) and 0° incidence angle (i.e., with the ion beam perpendicular to the original sample surface, right). The lines were milled with all beam parameters identical. The only differences in the milled lines were the angle of incidence of the beam with respect to the sample surface. The differences in the aspect ratios of the FIB milled lines are evident in the SEM images in Fig. 8.7, where the 52° incidence angle cut shows a deeper cut with an overall improvement in the aspect ratio of the cut from ~2:1 to 3:1. Since different materials exhibit different collision cascade characteristics which also vary with incidence angle [15], it is expected that different FIB milled aspect ratios will vary with material as well as incidence angle that are consistent with ion–solid interaction theory. In addition, differences in FIB milled lines are

FIGURE 8.6. Nanometer scale FIB milled lines in Si with varying dwell times.

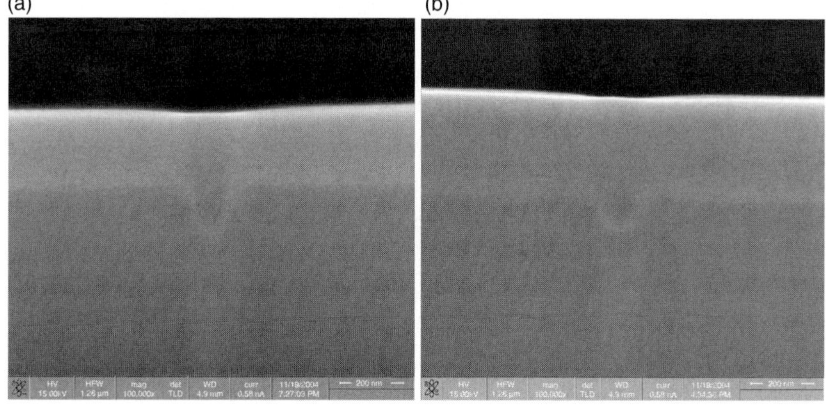

FIGURE 8.7. SEM images of FIB milled lines milled at 52° (*left*) and 0° (*right*).

also observed when milled in a single beam pass down, up, or across, an inclined slope, which is also consistent with ion–solid interaction theory [13].

3. FIB Milling or CVD Deposition with Bitmap Files

Pattern generation via milling or deposition in conjunction with graphic bitmap files allows the user to mill complex 3D structures [14]. The pixel spacing of each bitmap defines the beam location and the overlap is then defined by controlling the beam size. The color value of each pixel in the bitmap can be delineated to define the beam dwell time and beam blanking. Figure 8.8 shows an example of FIB milling a 3D structure from a bitmap image. The inset in Fig. 8.8 shows a grayscale bitmap image and the corresponding FIB milled SEM image in Fig. 8.8 shows how the grayscale levels defined in the bitmap allows for a 3D structure to be milled directly from the 2D bitmap by varying the dwell times for each pixel. Thus, pixels exhibiting a longer dwell time yield deeper FIB milled regions. As shown in Fig. 8.8, pixels with a color value of 255 (white) provide the longest dwell time and pixels with a color value of 0 (black) have zero dwell time and are therefore regions which are not FIB milled.

The deposition of 3D structures can also be achieved with bitmap files. Fundamentally, ion beam assisted CVD occurs when the precursor gas is cracked and decomposed directly by the ion beam as well as by the secondary electrons that are generated as a result of ion–solid interactions. Electron beam CVD can also be accomplished, where the size of the deposit is limited by the primary beam as well as the electron–solid interactions. The precursor is emitted from a

FIGURE 8.8. FIB milling of a 3D structure from a 2D bitmap image (*inset*).

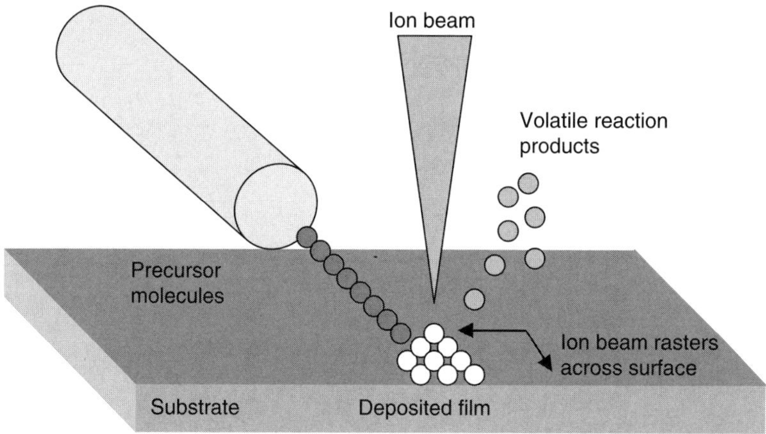

FIGURE 8.9. A schematic diagram of the ion beam assisted CVD process. Electron beam CVD process is performed in a similar fashion.

nozzle at a predefined height from the sample surface (typically ~50–200 μm depending on the gas) with a predetermined flow rate and adsorbs onto the sample surface. Next, the impinging beam and subsequent ion–solid interactions react with the adsorbed molecules decomposing the organic constituents. The volatile reaction products are removed by the vacuum system and the remaining material deposits onto the substrate surface. The schematic diagram in Fig. 8.9 depicts the CVD deposition process.

Three-dimensional features via bitmap FIB deposition can be achieved using a similar process to FIB milling. Figure 8.10 shows how the subtlety in utilizing 2D grayscale control allows the user to precisely deposit 3D films with the ion beam. The bitmap image that was used to create the 3D Pt structure is inset in the SEM image. Note that the SEM image was obtained with the sample surface tilted to emphasize the 3D nature of the deposited Pt film, and hence does not correspond to the exact geometrical dimensions as depicted in the bitmap image.

4. Onboard Digital Patterning with the Electron Beam

Changing variables with the onboard pattern generator can also be used to directly fabricate structures [14]. As an example, Fig. 8.11 shows use of the onboard pattern generator applied to the electron beam of the SEM. Figures 8.11a and 8.11b show helical-shaped electron beam deposited nanoscale (<100 nm) Pt lines that were deposited by scanning with one pass of the beam and independently changing the beam pitch and overall scan dimensions to form the different features shown.

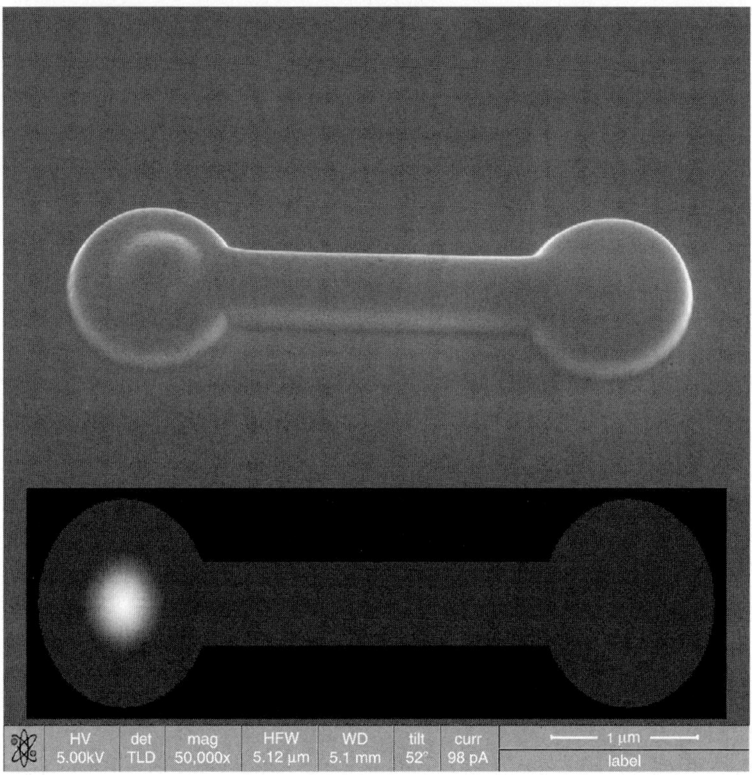

FIGURE 8.10. An SEM image of a 3D Pt structure fabricated using the 2D bitmap image shown in the *inset*.

(a)

(b)

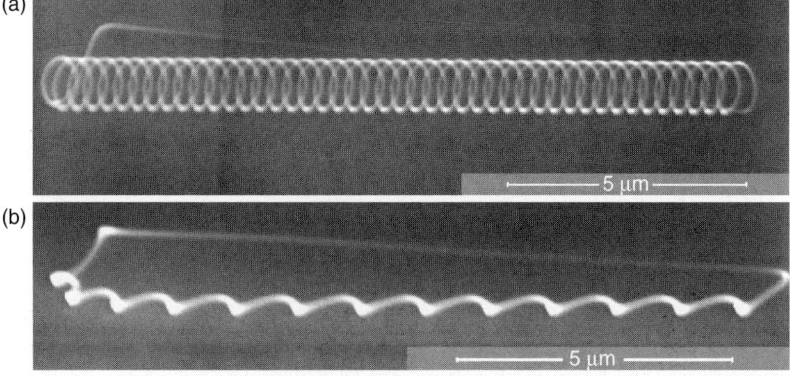

FIGURE 8.11. (a) and (b) are digitally patterned electron beam deposited Pt nanolines.

5. Automation for Nanometer Control

The use of scripting with DualBeam systems provides a powerful tool for automating process steps for deposition and milling. It also provides precise control of the FIB parameters such as scan direction, beam position and a host of other variables. A script is just a set of software instructions that control the DualBeam instrument. Figure 8.12 is an SEM image showing 3D spiral growth of ion beam assisted CVD Pt deposited performed by scripting. In this example, the speed and location of the beam is managed with precise control of the dwell and overlap such that the Pt grows continuously in a 3D free-standing spiral shape.

Scripting may also be used to control the DualBeam system to perform various milling, imaging and/or deposition tasks without user intervention. For example, automation is available for site-specific transmission electron microscopy specimen preparation [16] as well as cross-section preparation to capture either a single SEM image, or for sectioning multiple images serially for subsequent 3D reconstruction and tomography via AutoSlice and View™ [17,18]. Figure 8.13 below demonstrates the ability to perform site specific TEM specimen preparation (i.e., to within ~20 nm) where the TEM lamella thickness of 100 nm is easily achievable.

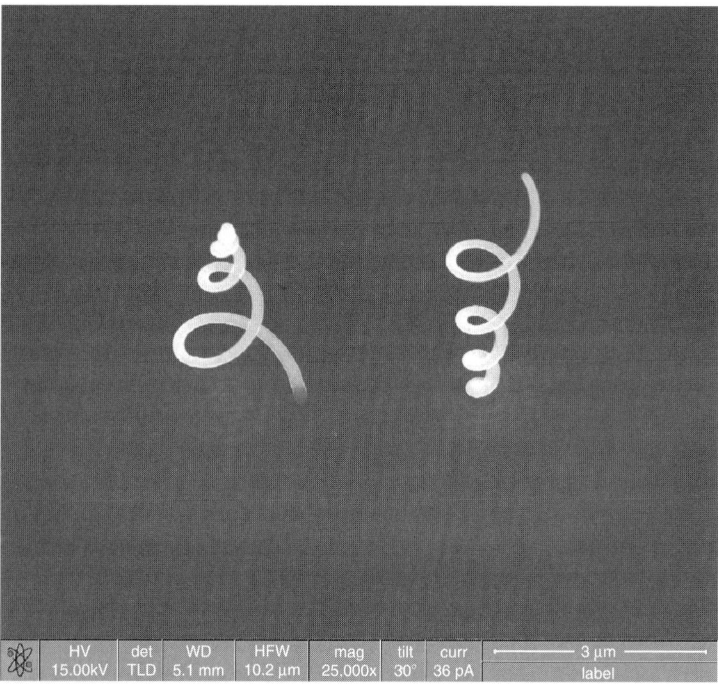

FIGURE 8.12. SEM image of 3D free-standing Pt FIB deposited growth. (Image courtesy of S. Reyntjens.)

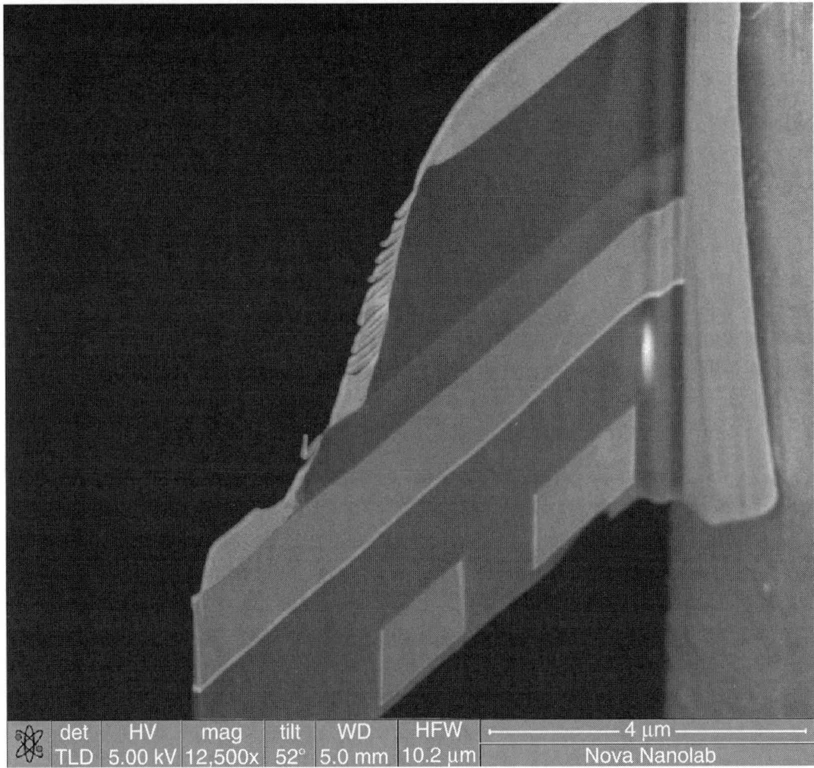

FIGURE 8.13. FIB prepared TEM specimen having a final thickness ~75 nm.

6. Direct Fabrication of Nanoscale Structures

Post- and fine-processing of structures and materials is a powerful FIB application for users in the nanoelectronics and nanoresearch communities. As already shown, examples include device modification or edit, AFM tip or atom probe tip creation, and TEM specimen preparation. Figure 8.14 is an SEM image of a probe tip consisting of ion beam assisted CVD Pt deposited on silicon. Note that the effective tip radius is smaller than 45 nm.

7. Summary

System advances in SEM/FIB dual platform instruments that allow precise control of beam parameters like dwell time and overlap have allowed users to scale 3D prototyping from the micrometer to the nanometer scale. Just as opportunities exist for *in situ* nanocharacterization, the same can be said for nanoprototyping. As developments in software applications, resolution, beam control and chemistry continue

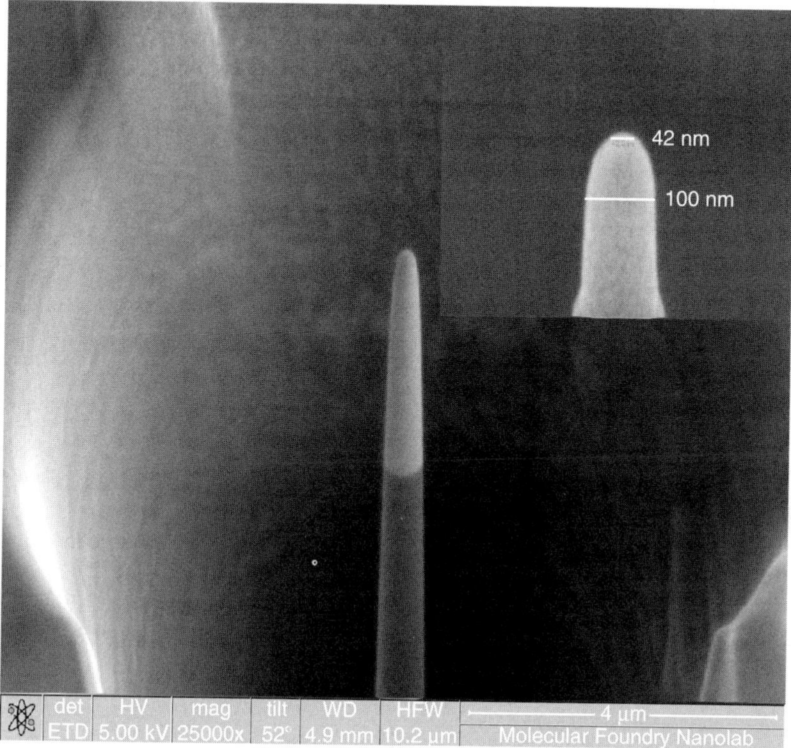

FIGURE 8.14. Nanoprobe tip creation of FIB deposited Pt on single crystal Si.

to grow with the use of FIBs, the user will have a larger tool box to draw from for 3D nanofabrication and nanomanipulation. Additionally, postprocessing of existing micro- and nanostructures with the SEM/FIB will be useful to repair damaged structures or modify existing assemblies for novel purposes.

Acknowledgments. We thank Steve Reyntjens and Daniel Phifer for their contributions.

References

1. J. H. Daniel, D. F. Moore, and J. F. Walker, *Eng. Sci. Educ. J.*, 7 (1998) 53.
2. G. Ben Assayag, J. Gierak, J. F. Hamet, C. Prouteau, S. Flament, C. Dolabdjian, F. Gire, E. Lesquey, G. Gunther, C. Dubuc, D. Bloyet, and D. Robbes, *J. Vac. Sci. Technol.*, B13 (1995) 2772.
3. L. Chow, D. Zhou, A. Hussain, S. Kleckley, K. Zollinger, A. Schulte, and H. Wang, *Thin Solid Films*, 368 (2000) 193.
4. A. J. Kubis, T. E. Vandervelde, J. C. Bean, D. N. Dunn, and R. Hull, *Mat. Res. Soc. Symp. Proc.*, 818 (2004) M14.6.1.

5. S. Reyntjens and R. Puers, *J. Micromech. Microeng.*, 10 (2000) 181.

6. B. A. Ferguson and R. A. Young, *Proc. SPIE Int. Soc. Opt. Eng.*, 3180 (1997) 73.

7. R. Kometani, T. Morita, K. Watanabe, T. Hoshino, K. Kondo, K. Kanda, Y. Haruyama, T. Kaito, J. -I. Fujita, M. Ishida, Y. Ochiai, and S. Matsui, *J. Vac. Sci. Technol.*, B22 (2004) 257.

8. A. Lugstein, E. Bertagnoli, C. Kranz, A. Kueng, and B. Mizaikoff, *Appl. Phys. Lett.*, 81 (2002) 349.

9. L. A. Giannuzzi, P. Anzalone, and D. Phifer, *Technical Proceedings of the 2005 NSTI Nanotechnology Conference and Trade Show, Nanotech*, Vol. 2 (2005) 683.

10. L. A. Giannuzzi and F. A. Stevie, *Micron*, 30 (1999) 197.

11. F. A. Stevie, C. B. Vartuli, L. A. Giannuzzi, T. L. Shofner, S. R. Brown, B. Rossie, F. Hillion, R. H. Mills, M. Antonelli, R. B. Irwin, and B. M. Purcell, *Surf. Interface Sci.*, 31 (2001) 345.

12. M. K. Miller, K. F. Russell, G. B. Thompson, *Ultramicroscopy*, 102 (2005) 287.

13. L. A. Giannuzzi and F. A. Stevie (Eds.), *Introduction to Focused Ion Beams*, Springer, New York (2005).

14. P. A. Anzalone, J. F. Mansfield, and L. A. Giannuzzi, *Micros. Microanal.*, 10(Suppl. 2) (2004) 1154CD.

15. B. I. Prenitzer, L. A. Giannuzzi, S. R. Brown, T. L. Shofner, R. B. Irwin, F. A. Stevie, *Micros. Microanal.*, 9 (2003) 216.

16. R. J. Young, P. D. Carleson, X. Da, and T. Hunt, *Proceedings of the 24th International Symposium for Testing and Failure Analysis*, ASM (1998), pp. 329–336.

17. R. J. Young, *AVS 47th International Symposium* (2000).

18. M. D. Uchic, M. Groeber, R. Wheeler, F. Scheltens, and D. M. Dimiduk, *Micros. Microanal.*, 10(Suppl. 2) (2004) 1136.

9
Nanowires and Carbon Nanotubes

Jianye Li and Jie Liu

1. Introduction

Due to their unique properties and novel applications, 1D structures such as nanowires and carbon nanotubes (CNTs) attracted a great deal of attention in the past few years. A variety of inorganic materials have been prepared in the form of nanowires by both vapor-growth and solution-growth processes [1]. CNTs including multiwalled CNTs and single-walled CNTs have been grown by many methods.

As one of the most powerful and maneuverable tools in nanotechnology, scanning electron microscopy (SEM) has played an important role in the research of nanowires and CNTs. It has been extensively used in the study of 1D nanomaterials including observing their morphologies in low and high magnifications, confirming their orientation, and for other purposes.

In this chapter, we describe the applications of SEM in the studies of nanowires and CNTs.

2. III–V Compound Semiconductors Nanowires

Wurtzite structure gallium nitride (hexagonal GaN), an important III–V semiconductor with a direct band gap of 3.4 eV, is an ideal material for use as ultraviolet (UV) or blue photon emitters, photodetectors, high-speed field effect transistors, and high-temperature/high-power electronic devices [2,3]. In the past few years, GaN nanowires have received considerable attention because of their great potential for realizing photonic and biological nanoscale devices such as blue light–emitting diodes, short-wavelength UV nanolasers, and biochemical sensors [4–9]. The reported synthetic methods for GaN nanowires include template growth [4], laser ablation [5], sublimation [6], metal-organic chemical vapor deposition (CVD) [7], hydride vapor epitaxy [8], and CVD [9].

SEM technique was extensively used in the characterization of GaN nanowires. GaN nanowires were grown on catalysts-patterned substrates by CVD method in our group [9] and the morphologies of the nanowires were observed by field emission SEM (FESEM, Philips FEI XL30SFEG).

Figure 9.1 is a FESEM image of GaN nanowires grown on a polished silicon wafer and the wafer was photolithographically patterned with iron (III) acetylacetonate. Iron (III) acetylacetonate was used as the precursor of catalyst iron oxide. From the image, the overall morphology of the controllably grown nanowires can be seen.

Figure 9.2a and b are FESEM images of GaN nanowires grown from "dip-pen" nanolithography (DPN)-patterned $Ni(NO_3)_2$ sites on polished silicon with 1-μm thick thermal oxide. Figure 9.2a is a low-magnification image and the controllable growth of the nanowires on the DPN-patterned catalyst sites can be seen. Figure 9.2b is a high-magnification image of the nanowires and the smooth surface and uniform diameters of the individual GaN nanowires were observed.

Figure 9.3 is a FESEM image of the GaN nanowires grown on a silicon wafer, and the wafer was patterned with nickel nanoparticles. The GaN nanowires had controlled growth.

Usually, the SEM is used only to characterize the morphologies and orientations of nanowires. The structure and composition of nanowires need to be characterized by means of x-ray powder diffraction (XRD), transmission electron microscopy (TEM), selected area electron diffraction (SAED), Raman, energy dispersive x-ray (EDX), electron energy loss spectroscopy (EELS), x-ray photoelectron spectroscopy (XPS), etc.

Boron nitride (BN) is a III–V wide-gap semiconductor with high melting point, high mechanical strength, hardness, corrosion resistance, oxidation resistance, and outstanding thermal and electrical properties [10].

Deepak et al. synthesized BN nanowires, and they characterized the assynthesized BN nanowires by SEM (Leica S-440) and TEM [11]. Figure 9.4a shows an SEM image of the BN nanowires obtained by reacting a mixture of H_3BO_3 and activated carbon with NH_3 at 1,300°C. The high-resolution TEM

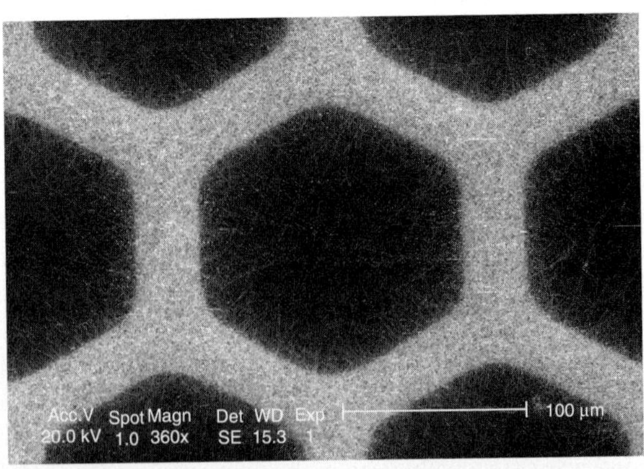

FIGURE 9.1. FESEM image of GaN nanowires grown on a polished silicon wafer and the wafer was photolithographically patterned with iron (III) acetylacetonate.

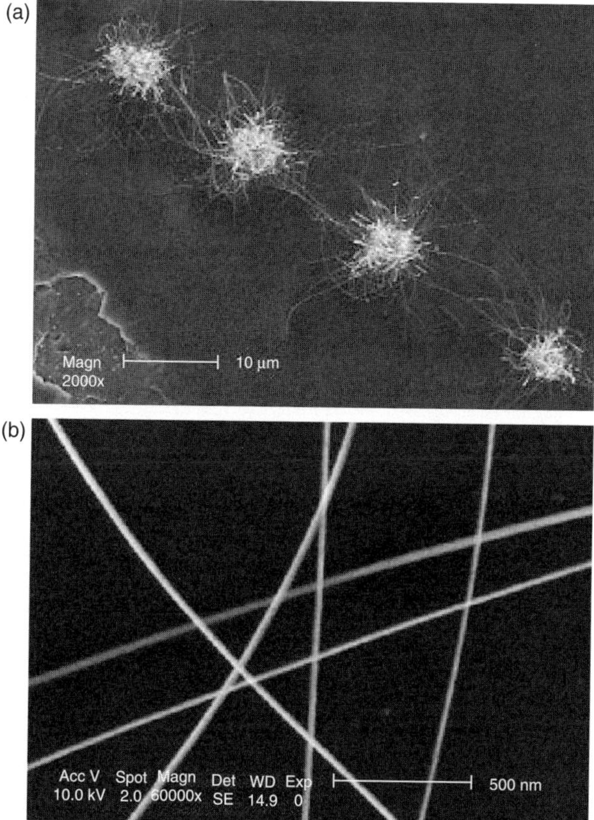

FIGURE 9.2. FESEM images of GaN nanowires grown from DPN-patterned $Ni(NO_3)_2$ sites on polished silicon with 1-μm thick thermal oxide. (a) Low-magnification image and (b) high-magnification image of the nanowires.

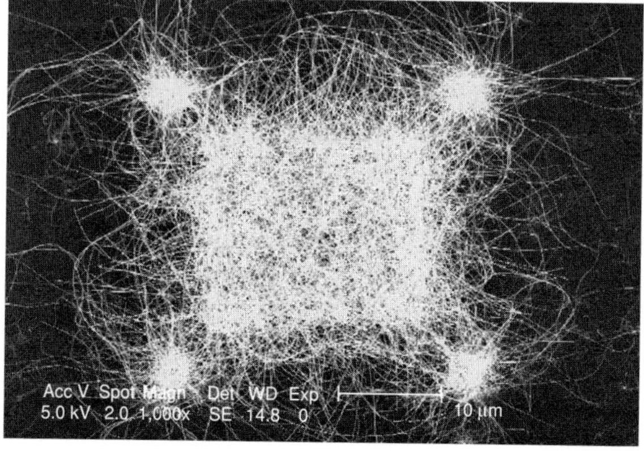

FIGURE 9.3. FESEM image of GaN nanowires grown on a silicon wafer patterned with nickel nanoparticles.

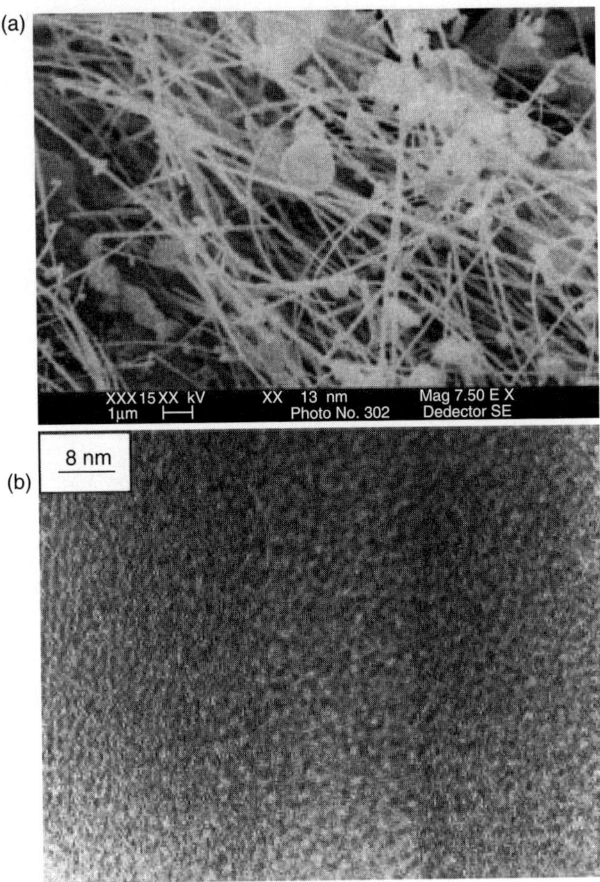

FIGURE 9.4. (a) SEM image of BN nanowires obtained by the reaction of H_3BO_3 and activated carbon with NH_3 by procedure. (b) High-resolution TEM image indicating the discontinuous lattice planes. (Adapted from [11] with permission. Copyright (2002) Elsevier.)

(HRTEM) image in Fig. 9.4b shows discontinuous lattice planes with a spacing of 0.33 nm corresponding to the interplanar distance between (002) planes (0.333 nm in bulk hexagonal BN) and indicates the hexagonal structure of a BN nanowire.

As a wide-gap III–V semiconductor, aluminum nitride (AlN) has many attractive properties, including high thermal conductivity, low coefficient of thermal expansion that closely matches that of silicon, high electrical resistivity, good mechanical strength, and excellent chemical stability [12]. Thus AlN nanowires have attracted extensive interest [12–14].

Wu et al. synthesized hexagonal AlN nanowire arrays through the direct reaction of Al and NH_3/N_2 under the confinement of anodic porous alumina template (APAT) and they characterized the as-synthesized AlN nanowires by SEM (JEOL JSM-6300), TEM, XRD, and XPS [13].

Figure 9.5a is a typical SEM image of the APAT used for growing AlN nanowires, and the pore size of the APAT is approximately 75 nm. Figure 9.5b is a SEM image of the as-prepared AlN nanowires array and the aligned hexagonal AlN nanowire array grown under the confinement of APAT. Figure 9.5c shows the top-view SEM image of the AlN nanowire array. Due to van der Waals interaction, the nanowires are aggregated together after growing out of the channels of the APAT.

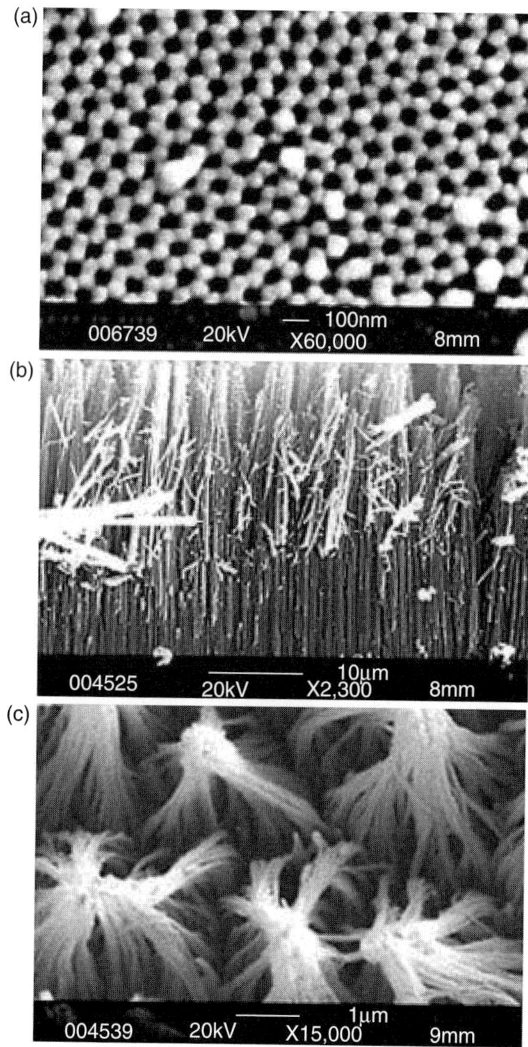

FIGURE 9.5. (a) SEM image of the APAT fabricated by the two-step anodization. (b) Side-view SEM image of the as-prepared hexagonal AlN nanowire array with length >30 µm. (c) Top-view SEM image of hexagonal AlN nanowire array. (Adapted from [13] with permission. Copyright (2004) Elsevier.)

Figure 9.6a shows the XRD pattern of the as-grown AlN nanowires. The peaks marked with "h" and "κ" are assigned to hexagonal AlN and κ-Al_2O_3, respectively, and it reveals that hexagonal AlN nanowires are synthesized. The relatively stronger (002) diffraction peak for hexagonal AlN here implies that the as-prepared AlN nanowires have grown preferentially along the c-axis parallel to the channels of the template. Figure 9.6b is a TEM image of the AlN nanowires. The nanowires are very straight and the diameters are approximately 80 nm. The nanowire surface looks rather rough, which should result from the coarse channel

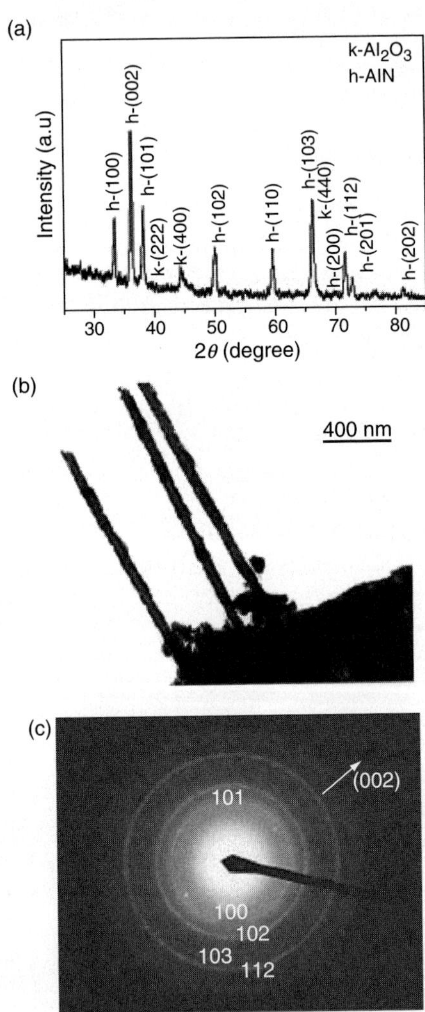

FIGURE 9.6. (a) XRD pattern of as-prepared AlN nanowires with Cu Kα radiation. (b) TEM image of the AlN nanowires grown within APAT and the diameters are ~80 nm. (c) SAED pattern of the hexagonal nanowires. (Adapted from [13] with permission. Copyright (2004) Elsevier.)

wall of APAT for confinement. The SAED pattern shown in Fig. 9.6c indicates the polycrystalline structure of the nanowires. The diffraction rings marked with corresponding indices in the figure can be assigned to those of hexagonal AlN. It could be inferred from Fig. 9.6b that the nanowire has a preferential growth direction along the c-axis, in agreement with the XRD result in Fig. 9.6a.

Indium nitride (InN) is a III–V semiconductor with promising transport and optical properties, and its large drift velocity at room temperature could render it better than GaAs and GaN as field effect transistors [15].

Zhang et al. synthesized InN nanowires through a gas reaction, in which a mixture of indium metal and In_2O_3 powders reacted with flowing ammonia at 700°C via a vapor–solid process [15]. They observed the morphologies of the as-synthesized InN nanowires by SEM (JEOL JSM-6300) and analyzed the structure and composition of the nanowires by means of XRD, TEM, EDX, and XPS.

Figure 9.7 is a typical SEM image of the InN nanowires and it shows that the nanowires are straight and smooth on their surfaces. The nanowires have diameters of 10–100 nm and a maximum length of several hundred micrometers.

Figure 9.8a shows an XRD pattern of the product and all peaks in the figure can be indexed to a pure hexagonal phase InN with the lattice parameters $a_0 = 3.52$ Å and $c_0 = 5.71$ Å. Figure 9.8b is an HRTEM image of an InN nanowire and it indicates a single-crystalline structure of the nanowire. The interplanar distance of (002) planes is about 0.285 nm. The inset of Fig. 9.8b is an SAED pattern of the nanowire. The electron diffraction patterns could be indexed to the single-crystalline hexagonal and consistent with the XRD result. The SAED pattern also indicates that the nanowire grows along [001] direction.

Figure 9.9a shows the EDX spectrum of an InN nanowire and it indicates that the nanowire consisted of indium and nitrogen with a molecular ratio of about 1:1. Figure 9.9b is an XPS spectra of the InN nanowires and the In (3d) and N (1s) core level regions were examined. The two strong peaks at 396.1 and 443.6 eV correspond to the N (1s) and In (3d) binding energy for the InN, respectively. The ratio of the quantification of peaks N:In is close to 1:1 and is consistent with the EDX result.

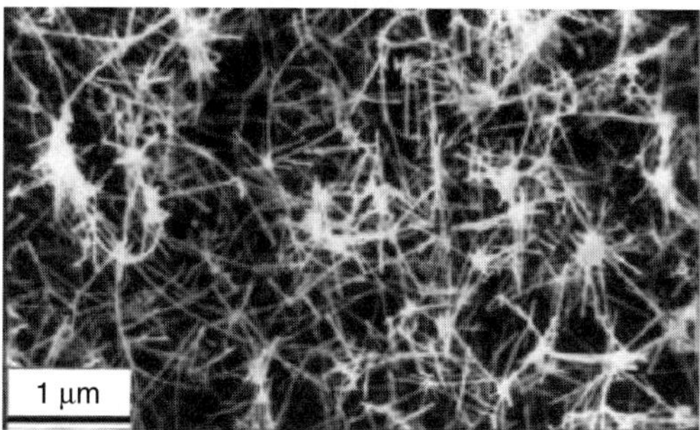

FIGURE 9.7. Typical SEM image of the as-grown InN nanowires. (Adapted from [15] with permission. Copyright (2002) Royal Society of Chemistry.)

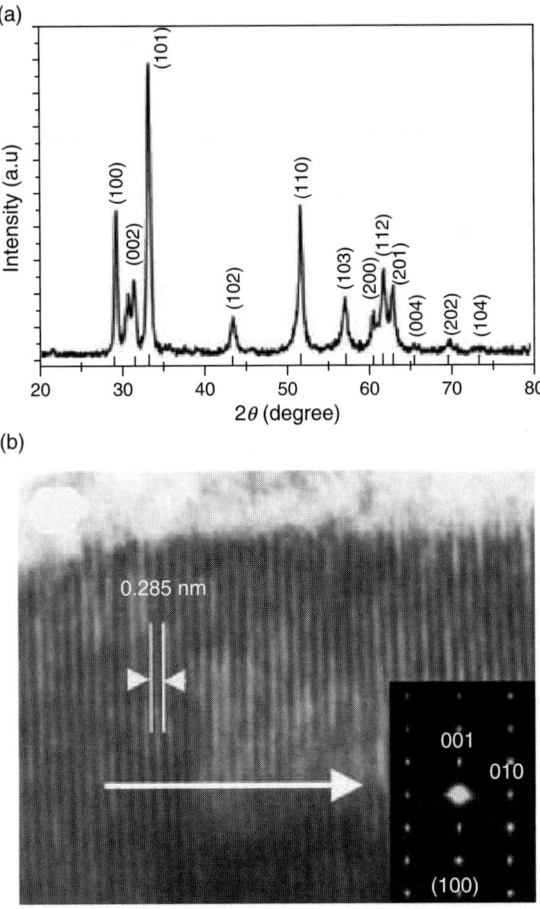

FIGURE 9.8. (a) XRD pattern taken on bulk InN nanowires. (b) HRTEM image of an InN nanowire and the nanowire grown along [001] direction. The space distance between (002) planes is about 0.285 nm. The *inset* is the SEAD patterns of the nanowire. The nanowire grows along [001] direction and is enclosed by ±(001) and ±(010) crystallographic facets. (Adapted from [15] with permission. Copyright (2002) Royal Society of Chemistry.)

Gallium phosphide (GaP) is a III–V semiconductor with a band gap of 2.26 eV, and it can be used to fabricate light-emission devices in the visible range [16]. There are several reports about the growth and characterization of GaP nanowires. Duan and Lieber synthesized GaP nanowires by a laser-assisted catalytic growth method [17], Shi et al. reported the synthesis of GaP nanowires using laser-ablation method [18], and Lyu et al. synthesized GaP nanowires by directly vaporizing a mixture of gallium and GaP powder in argon ambient [19].

Figure 9.10 shows the SEM (Hitachi S-4700) images of the GaP nanowires synthesized with NiO nanoparticles as catalyst by Lyu et al. [19]. The nanowires have

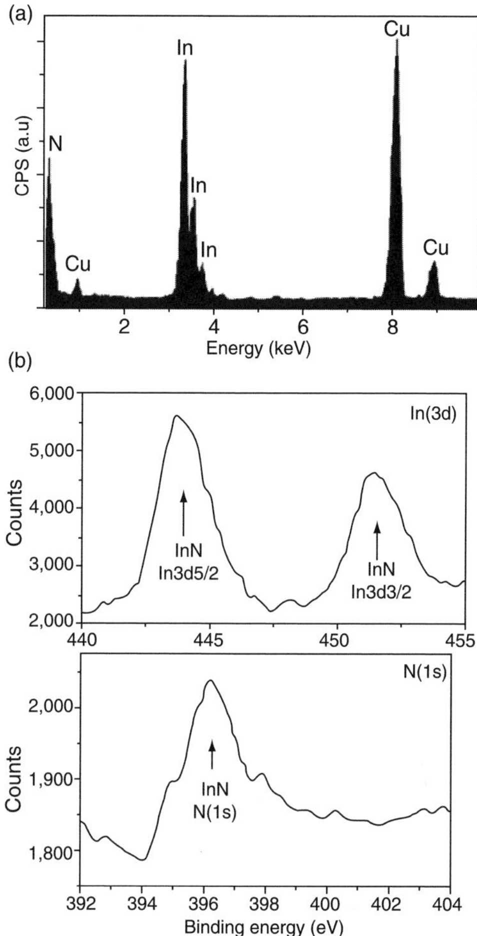

FIGURE 9.9. (a) EDX of an InN nanowire. The Cu peaks are generated from the grid that supports the nanowires. (b) XPS spectra of the InN nanowires. (Adapted from [15] with permission. Copyright (2002) Royal Society of Chemistry.)

diameters in the range of 38–105 nm and lengths up to several hundreds of micrometers. SEM images show that the nanowires have a very clean surface without any nanoparticles.

They also characterized the structures of the as-grown GaP nanowires by HRTEM and XRD. Figure 9.11a is a HRTEM image of a GaP nanowire and it indicates that the nanowire consisted of a core-shell structure with a single-crystalline GaP core and $GaPO_4$ and Ga_2O_3 as outer shell layers. The SAED pattern (inset) clearly demonstrates that the core is zinc blende–structured GaP and the outer layers are orthorhombic-structured $GaPO_4$ and amorphous Ga_2O_3. Figure 9.11b is an XRD pattern of the GaP nanowires synthesized on the alumina

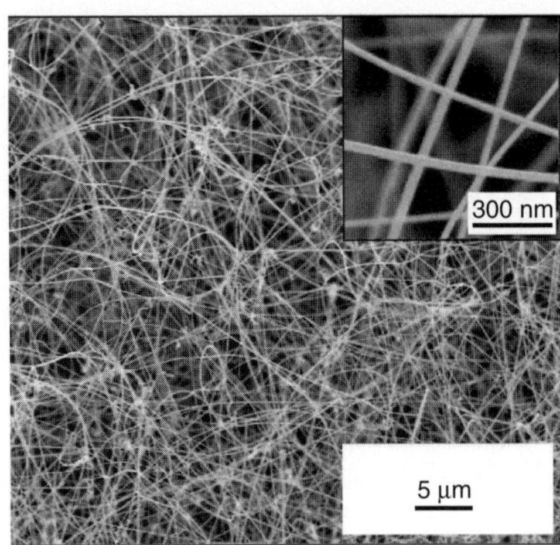

FIGURE 9.10. Low-magnification SEM image of the GaP nanowires synthesized with NiO catalyst. (*Inset*) High-magnification SEM image of the GaP nanowires. (Adapted from [19] with permission. Copyright (2003) Elsevier.)

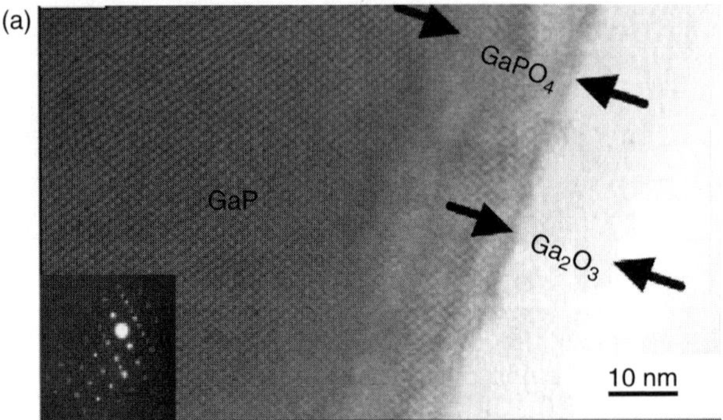

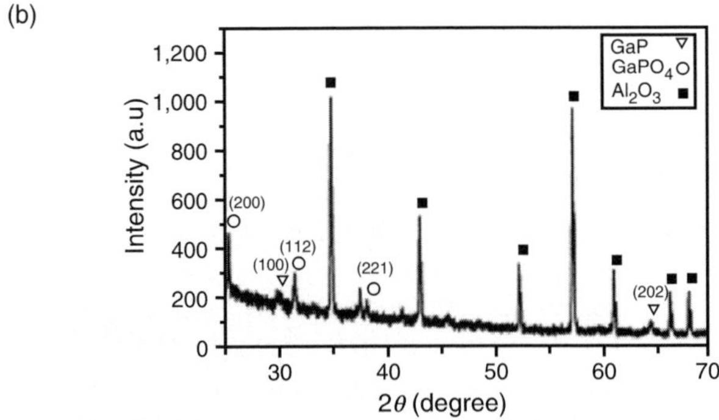

FIGURE 9.11. (a) An HRTEM image of a GaP nanowire with GaPO$_4$ and Ga$_2$O$_3$ outer-shell layers. *Inset* is a SAED pattern. (b) XRD pattern of the GaP nanowires synthesized on the alumina substrate. (Adapted from [19] with permission. Copyright (2003) Elsevier.)

substrate and it reveals that the nanowires consisted of single-crystalline zinc blende GaP and orthorhombic $GaPO_4$.

Gallium arsenide (GaAs) is a direct-band-gap III–V semiconductor with high electron mobility and it has been widely used for the fabrication of laser diodes, full-color flat-panel displays, and high-speed transistors. Recently, the remarkable properties of nanowires have stimulated intensive interest in the synthesis, characterization, and applications of GaAs nanowires [20–23].

The methods applied to synthesize GaAs nanowires include laser-assisted catalytic growth [20], oxide-assisted growth [21,22], and chemical beam epitaxy [23].

Shi et al. grew GaAs nanowires by the oxide-assisted growth method [22]. They observed morphologies of the GaAs nanowires by SEM (Philips XL 30 FEG). Figure 9.12a is a typical SEM image of the GaAs nanowires and they have diameters of about 50 nm and lengths up to 10 μm. They further characterized the GaAs

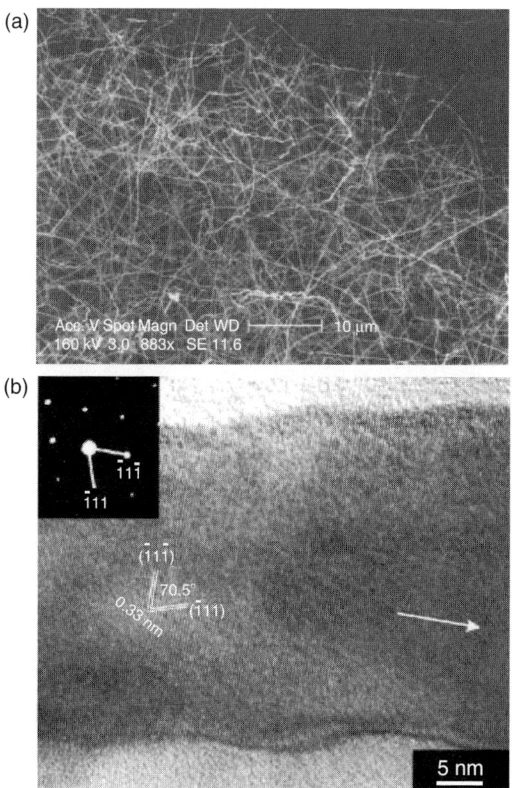

FIGURE 9.12. (a) A typical SEM image of GaAs nanowires synthesized by an oxide-assisted growth method. (b) HRTEM image of a GaAs nanowire and the growth axis is close to the [$\bar{1}1\bar{1}$] direction. The *inset* is the corresponding SAED pattern recorded along the [110] zone axis. (Adapted from [22] with permission. Copyright (2001) John Wiley & Sons.)

nanowires by TEM. Figure 9.12b is a HRTEM image of a GaAs nanowire and it corresponds to the {111} lattice planes of the crystalline GaAs. The growth axis of the nanowire is close to the [−11−1] direction. The inset of Fig. 9.12b is the corresponding SAED pattern recorded along the [110] zone axis and it can be indexed as the single-crystalline GaAs with a zinc blende structure.

Indium phosphide (InP) is a III–V semiconductor with a band gap of about 1.35 eV, and the InP nanowires have been synthesized by laser catalytic growth process [24] and metal-organic vapor-phase epitaxy [25,26], etc.

Bhunia et al. grew InP nanowires by metal-organic vapor-phase epitaxy method [25,26]. Figure 9.13 is an SEM (Hitachi S-5000) image of the InP nanowires grown on 20- and 10-nm Au nanoparticles, respectively, and they were

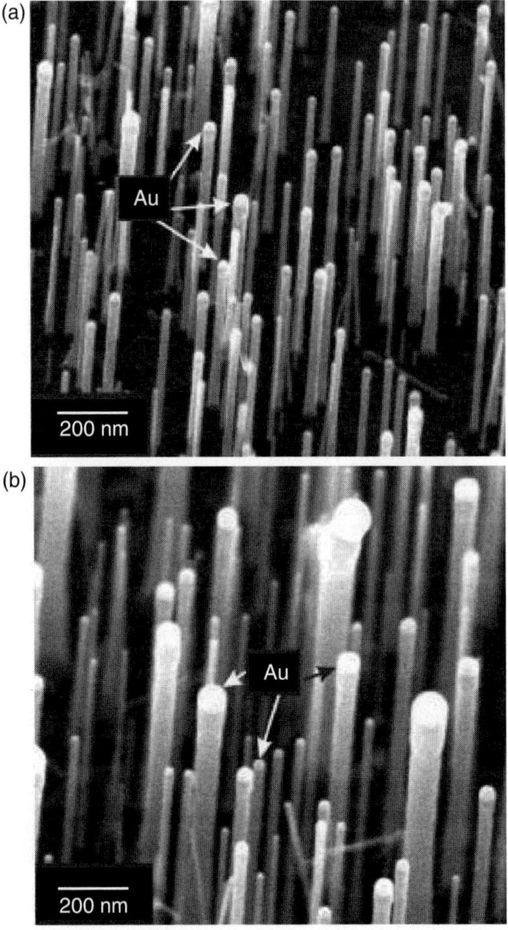

FIGURE 9.13. SEM images of InP nanowires grown on (a) 20 and (b) 10 nm Au nanoparticles at annealing and growth temperatures of 500°C and 430°C, respectively. (Adapted from [25] with permission. Copyright (2004) Elsevier.)

with annealing and growth temperatures of 500°C and 430°C [25]. The individual nanowires were uniform in cross section throughout their length, vertically aligned and isolated from each other for both images. A characteristic Au ball was clearly visible on the top of each nanowire.

Figure 9.14a shows an SEM image of the InP nanowires viewed along a direction that is perpendicular to the growth direction of the nanowires [26]. It can be clearly seen that the nanowires were mounted on the substrate, aligned perpendicular to it, and densely spaced. The nanowires with thicker diameter were quite rigid and the thinner ones were bent under the electric field and current during the SEM observation process. The Au nanoparticles were clearly observed on the top of each nanowire. Figure 9.14b is an SEM image of the InP nanowires with two adjacent nanowires to join to become a single nanowire, and a single nanowire to bifurcate to make branching during their growth. The node was clearly seen on both kinds of the nanowires [26].

To further investigate the structure and impurity of the nanowires, TEM studies were carried out. Figure 9.15 shows a typical lattice-resolved HRTEM image that includes the tip of an individual nanowire grown on 20-nm Au particles with

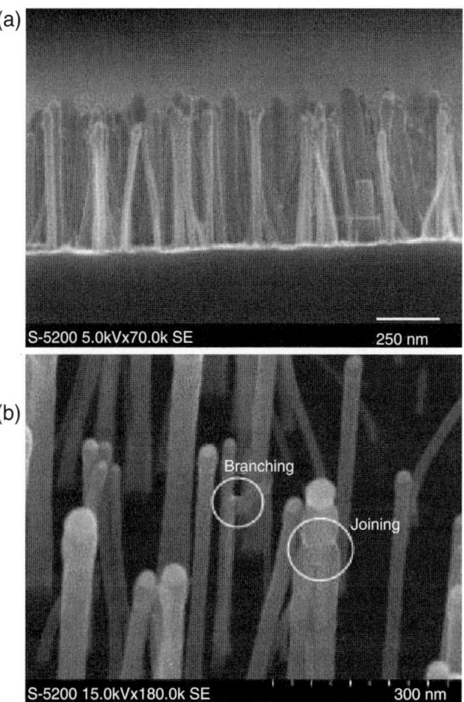

FIGURE 9.14. (a) SEM image of the InP nanowires taken in the direction perpendicular to the axis of the nanowires. (b) SEM image of the InP nanowires showing the branching and joining during the growth. (Adapted from [26] with permission. Copyright (2004) Elsevier.)

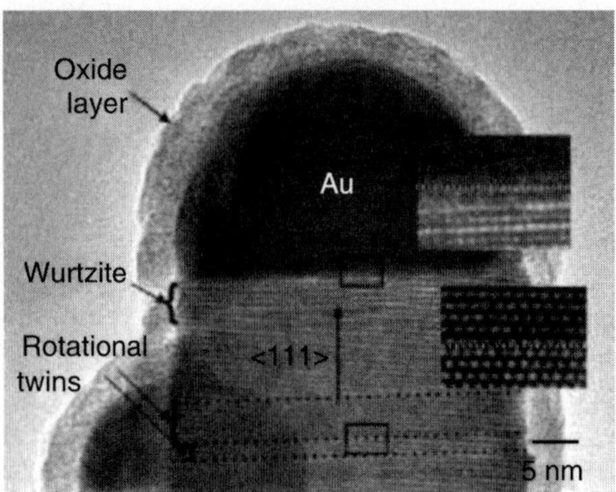

FIGURE 9.15. HRTEM lattice-resolved image of a top part of an InP nanowire grown on 20-nm Au particle with annealing and growth temperatures of 500°C and 430°C, respectively. The *insets* show the magnified views of the marked areas in the figure. (Adapted from [25] with permission. Copyright (2004) Elsevier.)

self-explanatory insets [25]. The Au particle was separated by an atomically flat interface from InP and thus confirms the vapor-liquid-solid (VLS) growth mechanism. The HRTEM image reveals the [111] growth direction and zinc blende structure of InP. However, the nanowire contained periodic blocks of rotational twin structures along its length, which appeared as alternate black contrasts (inset of Fig. 9.15), and the rotational axis being [111]. Few monolayers of InP just below the Au tip with wurtzite structure, which was believed to form during the cooling process after the growth, were observed. The native oxide layer with an average thickness of about 5 nm formed after exposing the nanowires to the atmosphere, and it was clearly observed on the entire surface.

3. II–VI Compound Semiconductors Nanowires

Wurtzite structure zinc oxide (ZnO) is an important II-VI group semiconductor with a direct band gap of 3.37 eV at room temperature [27–29]. It has large exciton binding energy (60 meV for ZnO vs. 28 meV for GaN) and high optical gain (300 cm^{-1} for ZnO vs. 100 cm^{-1} GaN) at room temperature [28,29]. Wurtzitic ZnO is a very interesting material for its applications for low-voltage and short-wavelength (green or green/blue) electro-optical devices such as light-emitting diodes and laser diodes. It can also be widely used as transparent UV-protection films, transparent conducting oxide materials, piezoelectric materials, electron-transport medium for solar cells, chemical sensors, photocatalysts, and so on [27–31].

Recently, a great deal of attention has been focused on the study of ZnO nanowires for their great prospects in fundamental physical science, novel nanotechnological applications, and significant potential for nanoscale optoelectronics [27,32].

Yuan et al grew ZnO crystal nanowires by thermal evaporation of ZnS powders and they characterized the nanowires by means of SEM (Hitachi S-4200) and TEM [32].

Figure 9.16a is an SEM image of the ZnO nanowires and it can be seen that the obtained products consist of nanowires with diameters of about 20–60 nm and lengths up to tens of micrometers. Figure 9.16b is an HRTEM image of a ZnO nanowire with diameter of about 30 nm. The (001) atomic planes with an inter-planar spacing of 0.513 nm are clearly seen. The inset is the corresponding SAED from the nanowire. The SAED pattern and HRTEM image reveal that the ZnO nanowires are structurally uniform and single-crystalline. They also suggest that the nanowire grew along [0 0 1] direction.

We obtained vertically aligned ZnO nanowires on a wafer scale by thermal evaporation of a mixture of ZnS and graphite carbon powders. Figure 9.17a is a low-magnification SEM (Philips FEI XL30SFEG) image of the ZnO nanowires,

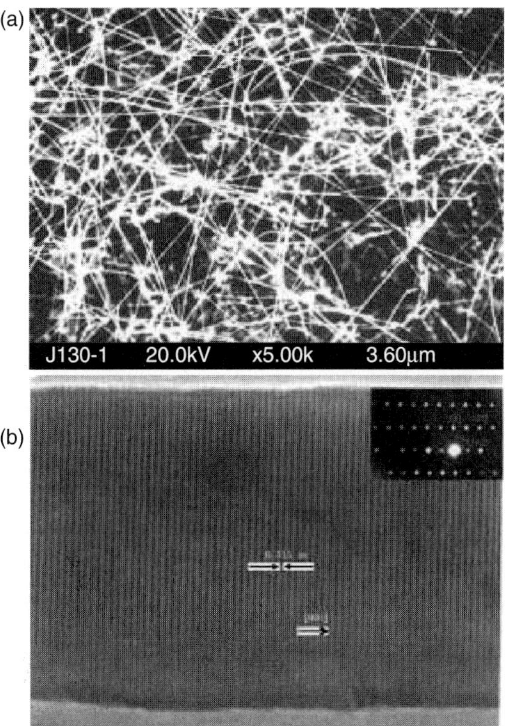

FIGURE 9.16. (a) SEM image of the ZnO nanowires. (b) high-resolution TEM image of a ~30-nm diameter ZnO nanowire. The (001) lattice planes (separation = 0.513 nm) are clearly visible and perpendicular to the axis of the nanowire. The *inset* is the corresponding selected area electron diffraction from the nanowire. (Adapted from [32] with permission. Copyright (2003) Elsevier.)

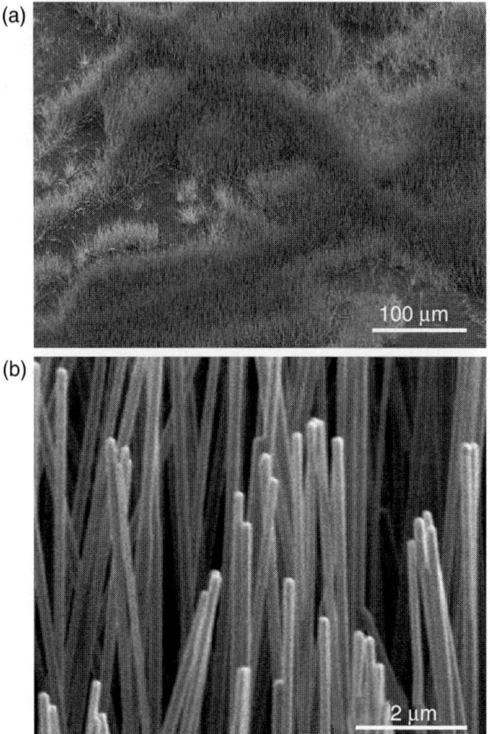

FIGURE 9.17. (a) Low-magnification SEM image of wafer-scale vertically aligned ZnO nanowires. (b) high-magnification SEM of the vertically aligned ZnO nanowires.

which shows the wafer-scale vertically aligned nanowires. The high-magnification SEM image in Fig. 9.17b shows that the vertically aligned ZnO nanowires have uniform diameters and smooth surfaces.

Hexagonal zinc sulfide (ZnS) is a II–VI group semiconductor with a direct wide band gap of 3.7 eV at room temperature. It has been extensively studied due to its wide applications as excellent phosphors and photocatalysts. [33–37]. As an important luminescence material, ZnS shows various luminescence properties [33–37]. It has been widely used in the fields of UV light-emitting diodes, injection lasers, flat-panel displays, cathodes-ray tube luminescence, thin-film electroluminescent devices, infrared (IR) windows, sensors, solar cells, etc. [33–39].

Nanocrystalline ZnS has been reported to have some characteristics different from the bulk crystal and this may extend its applications [34,40] Attention has recently been focused on the study of ZnS nanowires for their great prospects in fundamental physical science research, novel nanotechnological applications, and significant potential for optoelectronics [33–39].

Wang et al. obtained ZnS nanowires by thermal evaporation of ZnS powders at 900°C and they characterized them by SEM (JEOL JSM-6300) and TEM [38].

SEM observation (Fig. 9.18a) indicates that the synthesized products consist of a large quantity of nanowires with lengths in the range of several tens of micrometers. The HRTEM image (Fig. 9.18b) reveals that the ZnS nanowires are structurally uniform and the (110) atomic planes are with an interplanar spacing of 0.1921 nm. Both the HRTEM image and the SAED pattern (inset of Fig. 9.18b) confirm the single-crystalline wurtzite structure of the ZnS nanowires.

Lin et al. [39] prepared dense and uniform hexagonal ZnS nanowires on Au-coated Si wafers. Figure 9.19 is a typical SEM (JEOL-JSM 6700F) image of the ZnS nanowires. It shows the large quantities of densely packed nanowires obtained, and the diameters and lengths of the nanowires are about 80–100 nm and 10 μm, respectively. The EDX pattern (inset of Fig. 9.19) indicates that the nanowires are composed mainly of Zn and S.

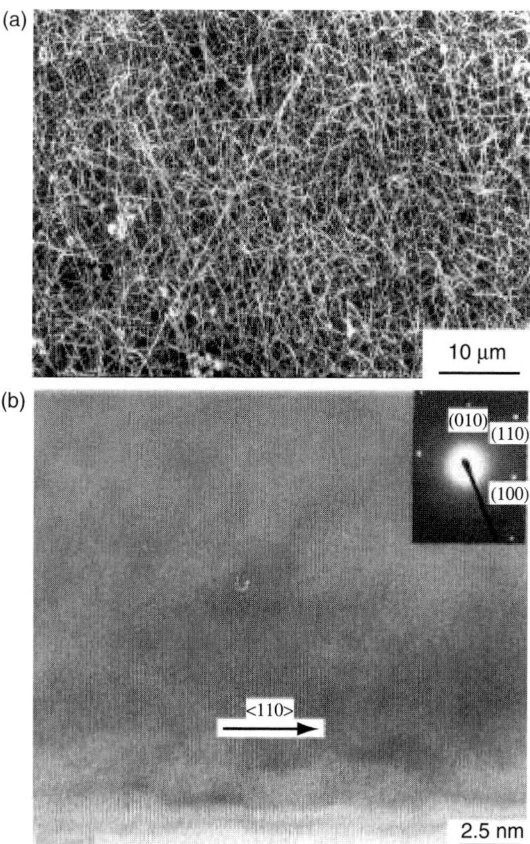

FIGURE 9.18. (a) Typical SEM image of the as-synthesized ZnS nanowires. (b) HRTEM image of a nanowire with a diameter of 40 nm. The *inset* is an SAED pattern of the nanowire and it is recorded perpendicular to the nanowire long axis. (Adapted from [38] with permission. Copyright (2002) Elsevier.)

FIGURE 9.19. SEM image of ZnS nanowires and the *inset* shows an EDX spectrum of a ZnS nanowire. (Adapted from [39] with permission. Copyright (2004) Elsevier.)

Zinc selenide (ZnSe) is a II–VI group semiconductor with a direct band gap of 2.8 eV at room temperature. It could be used for the fabrication of blue- and green-light-emitting devices [41]. Jiang et al. synthesized ZnSe nanowires by laser ablation method and characterized these materials by SEM (Philips XL 30 FEG) and TEM.

Figure 9.20a is a low-magnification SEM image of the products and it shows that the product consists of a large quantity of wirelike structures with lengths of 10–20 µm. Figure 9.20b is a typical high-magnification SEM image and it reveals that both nanowires and nanobelts exist in the product.

A typical bright-field TEM image of a ZnSe nanowire is shown in Fig. 9.21a and a gold spherical particle at one end can be observed. The inset is the corresponding [100] zone axis SAED pattern of the nanowire. Figure 9.21b and c show the HRTEM images of the nanowire taken in a region near the gold particle and in the wire body, respectively. The white arrows in them indicate the length direction of the nanowire. Both the SAED pattern and the HRTEM images confirm the single-crystal wurtzite-2H structure of the ZnSe nanowire, and the nanowire is grown along the [001] direction. Figure 9.21b also shows that the boundary between ZnSe and gold is very sharp and there are some stacking faults near the gold tip.

Cadmium sulfide (CdS) is a typical wide band gap II–VI semiconductor with a band gap of 2.42 eV at room temperature. It has many commercial or potential applications in light-emitting diodes, solar cells, and other photoelectric devices [42]. The growth of CdS nanowires has been reported by several groups to date [42–45].

Xu et al. prepared aligned CdS nanowires by direct current (dc) electrodeposition in porous anodic aluminum oxide template from dimethylsulfoxide solution containing cadmium chloride and elemental sulfur [43]. They characterized the

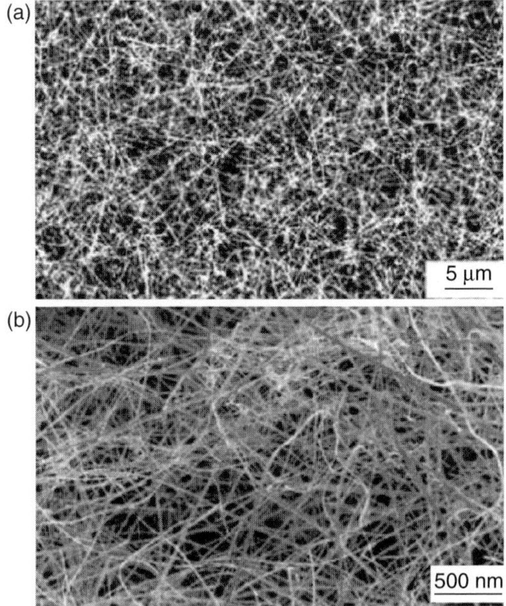

FIGURE 9.20. SEM images of the as-synthesized nanowires: (a) low magnification and (b) high magnification. (Adapted from [41] with permission. Copyright (2004) American Chemical Society.)

as-prepared CdS nanowires by means of SEM (AMRAY 1910FE), TEM, EDX, and Raman.

Figure 9.22a is an SEM image of the product, and aligned CdS nanowire arrays were observed. The diameters of the nanowires are about 100 nm and the lengths are up to 30 µm. The inset is an SAED pattern and the diffraction spots correspond to the (002), (101), and (102) diffraction planes of hexagonal CdS. The diffraction spots were dispersed and elongated in some sort and this implies that the crystals in the nanowires should have a similar orientation, that is, the nanowires should grow along similar direction.

The EDX spectrum of the CdS nanowires is shown in Fig. 9.22b and it reveals that the atomic composition of S and Cd is very close to a 1:1 stoichiometry. Figure 9.22c gives the Raman spectrum of bulk nanowires. Due to surface enhancement by Ag substrates, the Raman modes are very strong. All Raman modes are in good agreement with those of pure hexagonal CdS.

Figure 9.23a is an HRTEM image of a CdS nanowire and the interplanar spacing is about 0.32 nm, which corresponds to the {101} plane of the hexagonal CdS. The HRTEM lattice image reveals the single-crystalline hexagonal structure of the CdS nanowire. Figure 9.23b indicates that the continuous {101} lattice fringes are approximately vertical to the nanowire axis and the growth plane may be the {101} plane.

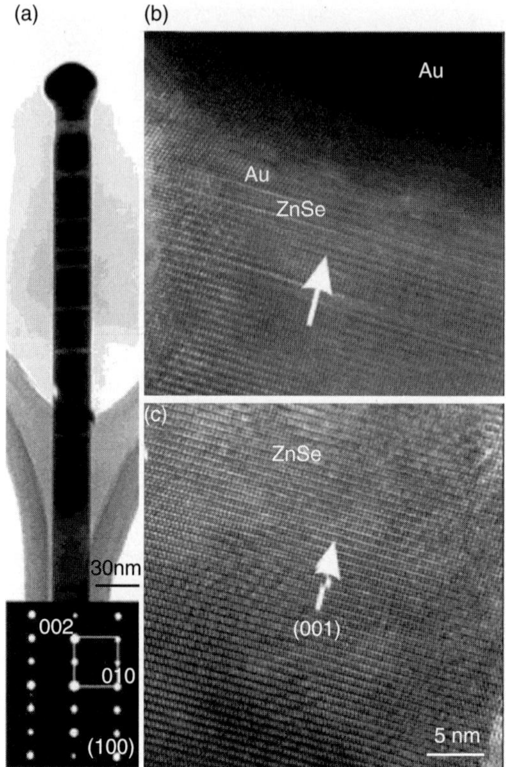

FIGURE 9.21. (a) TEM image of single ZnSe nanowire with a gold tip. *Inset* is the [100] zone axis SAD pattern. (b) Corresponding HRTEM image of the tip region of the ZnSe nanowire. And (c) corresponding HRTEM image of the ZnSe nanowire body. (Adapted from [41] with permission. Copyright (2004) American Chemical Society.)

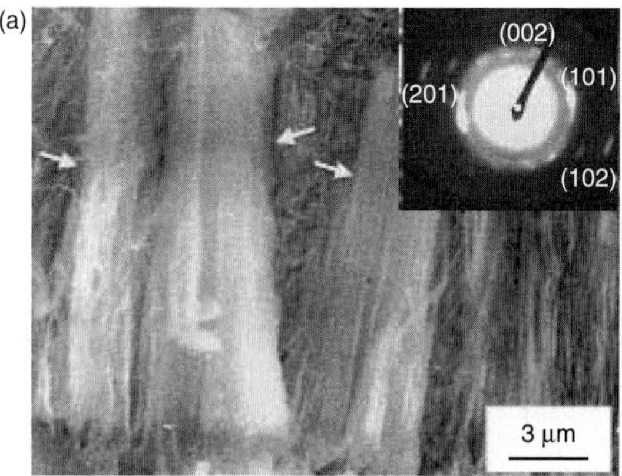

FIGURE 9.22. (a) SEM image of the CdS nanowires. The *inset* is an SAED pattern.

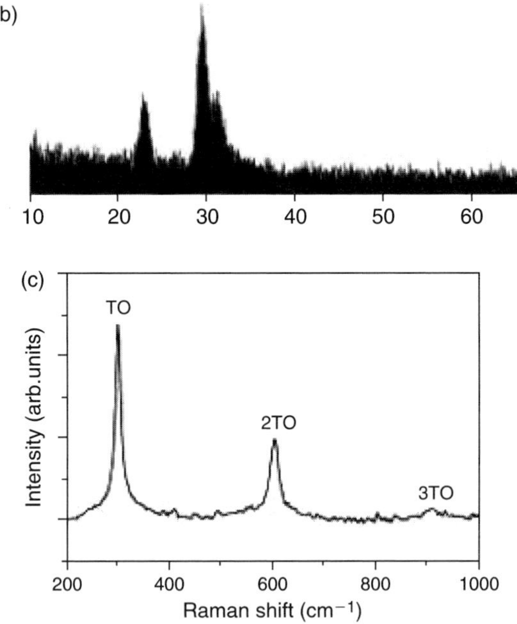

FIGURE 9.22. (*Continued*) (b) EDX spectrum and (c) Raman spectrum of the nanowires with a diameter of about 20 nm. (Adapted from [43] with permission. Copyright (2000) Elsevier.)

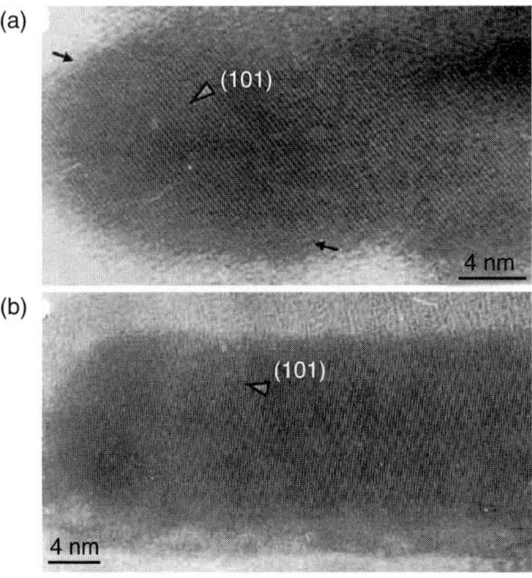

FIGURE 9.23. HRTEM images of individual nanowires with different diameters: (a) 15 nm and (b) 25 nm. (Adapted from [43] with permission. Copyright (2000) Elsevier.)

CdSe is a II-VI group semiconductor, and CdSe nanowires have been fabricated by dc [46,47] or by alternating current (ac) [48] electrodeposition with anodic alumina membrane (AAM) templates.

Peng et al. synthesized ordered polycrystalline CdSe nanowire arrays by dc electrodeposition in AAM from ammonia alkaline solution containing $CdCl_2$ and SeO_2 [47].

Figure 9.24a is a typical cross-sectional view SEM (JEOL JSM-6300) image of the CdSe nanowires and it shows that the CdSe nanowires have uniform diameters of about 60 nm, close to the pore sizes of the used AAM. The nanowires have a length of about 5 μm and highly ordered. Figure 9.24b is a top-view SEM image of the CdSe nanowires.

Figure 9.25a is an XRD pattern of the CdSe nanowire arrays and it indicates that the nanowires are in the hexagonal crystalline phase of CdSe. The chemical composition of the CdSe nanowires was determined by XPS and the XPS results that are shown in Fig. 9.25b, which indicates that the stoichiometric CdSe is formed. The SAED pattern taken from a CdSe nanowire is shown in Fig. 9.25 (c) and it reveals the hexagonal polycrystalline structure of the CdSe nanowires, which agrees well with the XRD result.

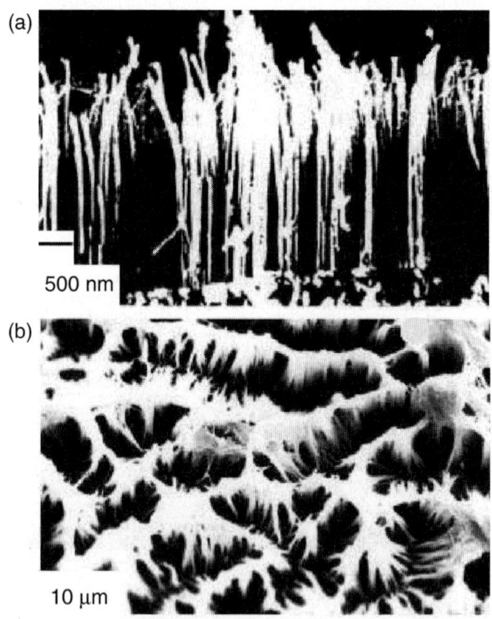

FIGURE 9.24. SEM images of the ordered CdSe nanowire arrays with diameters about 60 nm. (a) A cross-sectional-view image; and (b) a top-view image. (Adapted from [47] with permission. Copyright (2001) Elsevier.)

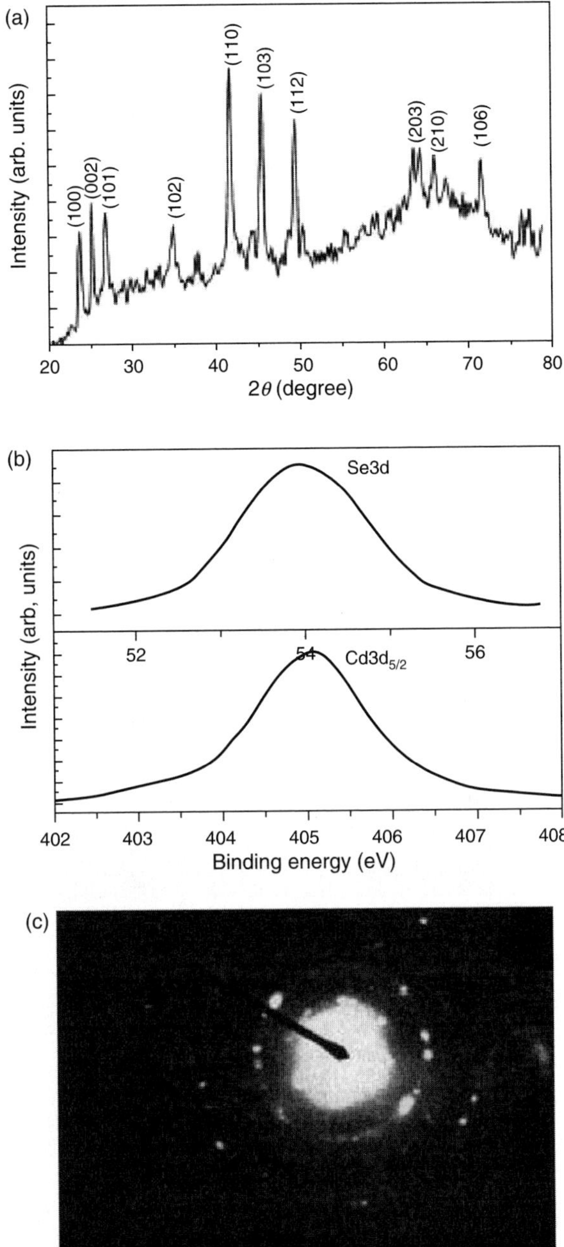

FIGURE 9.25. (a) XRD pattern of the CdSe nanowire arrays. (b) Se3d and Cd3d$_{5/2}$ XPS spectrum of the CdSe nanowires. The peaks binding energies are 53.8 eV and 405.05 eV, respectively. The C$_{1s}$ ($E = 284.5$ eV) level served as the internal standard. (c) SAED pattern of the CdSe nanowires. (Adapted from [47] with permission. Copyright (2001) Elsevier.)

4. Elemental Nanowires

Silicon is a very important semiconductor, and silicon nanowires (SiNWs) have attracted much attention because of their possibility of testing a fundamental concept of quantum physics and their potential application in nanoelectronics [49,50].

Yan et al. synthesized amorphous SiNWs with an average diameter of 20 nm [49]. Figure 9.26a is an SEM (AMRAY FEG-1910) image and shows the general morphology of the SiNWs. Inset is an EDX spectrum, which reveals that the nanowires mainly consist of silicon. The TEM image in Fig. 9.26b shows that the SiNWs have a smooth morphology and average diameter of 40 nm. The inset is an SAED pattern of the SiNWs. The highly diffusive ring pattern reveals that the nanowires are completely amorphous.

Bulk Ge has a much larger excitonic Bohr radius (24.3 nm) than that of Si (4.9 nm). Therefore, the quantum size effects will be more prominent in Ge nanowires [51].

Wu and Yang synthesized single-crystalline Ge nanowires with diameters less than 30 nm through a vapor transport process [51]. Figure 9.27a shows a typical

FIGURE 9.26. (a) SEM image of the SiNWs. *Inset* is an EDX spectrum with the peak corresponding to Si. (b) TEM image of the SiNWs with a smooth morphology and average diameter of 40 nm. The *inset* is an SAED pattern with a characteristic diffusive ring pattern, showing that the nanowires are completely amorphous. (Adapted from [49] with permission. Copyright (2000) Elsevier.)

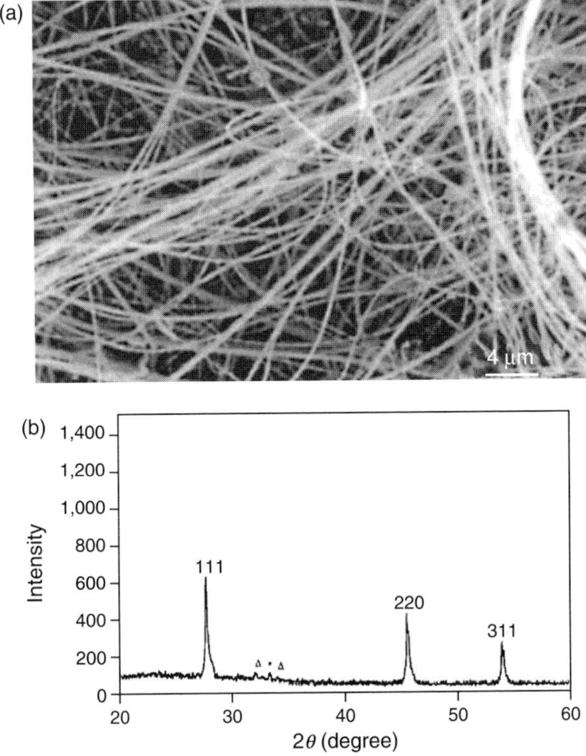

FIGURE 9.27. (a) SEM image of as-made germanium nanowires; and (b) XRD pattern recorded on the germanium nanowires. (Adapted from [51] with permission. Copyright (2000) American Chemical Society.)

SEM (JEOL JSM-6400) image of the as-grown nanowires. Pure nanowires with diameters ranging from 5 to 300 nm were found. These wires generally are several hundreds of micrometers long. The average size of these nanowires can be controlled by the thickness of the gold thin film deposited on the substrate. The purity and structure of these nanowires were examined by XRD and Fig. 9.27b is the diffraction pattern collected on the as-grown nanowires and it can be indexed as crystalline diamond structure germanium. A small amount of I_2 and Au/Ge alloy was also detected in the product.

Boron is a trivalent element and exhibits some of the most interesting chemistries of all the elements in the periodic table. Boron and boron-rich materials have important technological applications [52].

Meng et al. synthesized boron nanowires by laser ablation [52]. The as-synthesized boron nanowires were characterized by means of SEM (Philips XL 30 FEG), HRTEM, SAED, etc. SEM image shown in Fig. 9.28a gives a general overview of the morphology of the product. It consists of smooth wires that are several tens of micrometers in length and 30–60 nm in diameter. Figure 9.28b is

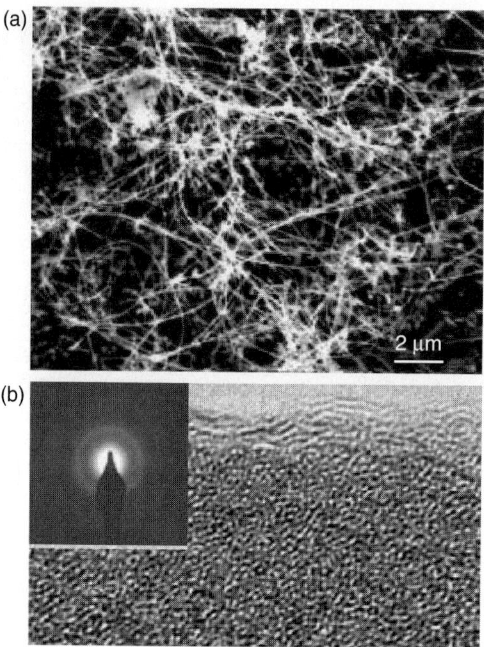

FIGURE 9.28. (a) SEM image of the boron nanowires. (b) HRTEM image of a boron nanowire. The *inset* is an SAED pattern of a nanowire. (Adapted from [52] with permission. Copyright (2003) Elsevier.)

an HRTEM image of a boron nanowire and it indicates an amorphous structure. The inset is an SAED pattern of a nanowire and shows some amorphous halo rings and no diffraction spots, confirming the HRTEM result.

In the past few years, great attention has been focused on the study of gold nanowires for their potential applications, and highly ordered Au nanowire arrays were electrochemically fabricated by Zhang et al. [53].

Figure 9.29a is an SEM (JEOL JSM-6300) image of the highly ordered Au nanowire arrays prepared in the porous nanochannel alumina templates with a diameter of 70 nm [53]. It can be clearly seen that the Au nanowires have an equal height and form a highly ordered tip array. The structure of the Au nanowires was characterized by HRTEM, and an HRTEM image of a nanowire is shown in Fig. 9.29b. The single-crystalline structure of the Au nanowire can be clearly observed. The interplanar spacings of $(\bar{1}11)$ and (002) lattice fringes are around 0.24 nm and 0.20 nm, respectively.

Silver nanowires are interesting material for their optical and electronic properties, and great efforts have been focused on the synthesis and study of silver nanowires [54].

Choi et al. obtained monodisperse silver nanowires with a high aspect ratio via electrochemical plating into monodomain porous alumina templates [55]. Figure 9.30a is an SEM image of a cross-section view of the silver nanowires in the

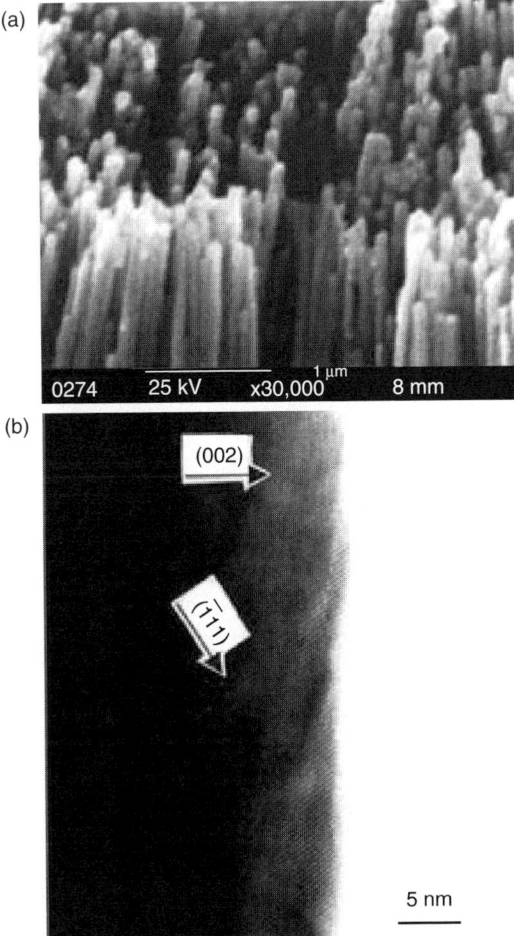

FIGURE 9.29. (a) SEM images of the highly ordered array of Au nanowires. (b) HRTEM image of a single-crystalline Au nanowire. (Adapted from [53] with permission. Copyright (2001) Royal Society of Chemistry.)

porous alumina template with a pore diameter of 180 nm. The bright strips are silver wires. The nanowires with smooth surfaces are very straight and have a length of 30 µm or more. Figure 9.30b shows a typical SEM image of the silver nanowires with several small branches and one large nanowire grow out between two and five small nanowires.

The fabrication of nickel nanowires has recently attracted much attention because of their unique magnetic properties and other potential applications [56,57].

Chu et al. fabricated nickel nanowire arrays on glass substrates by a template method [56]. Figure 9.31a is an SEM image of the nickel nanowire arrays standing on tin-doped indium oxide (ITO)/glass substrate in anodic alumina films

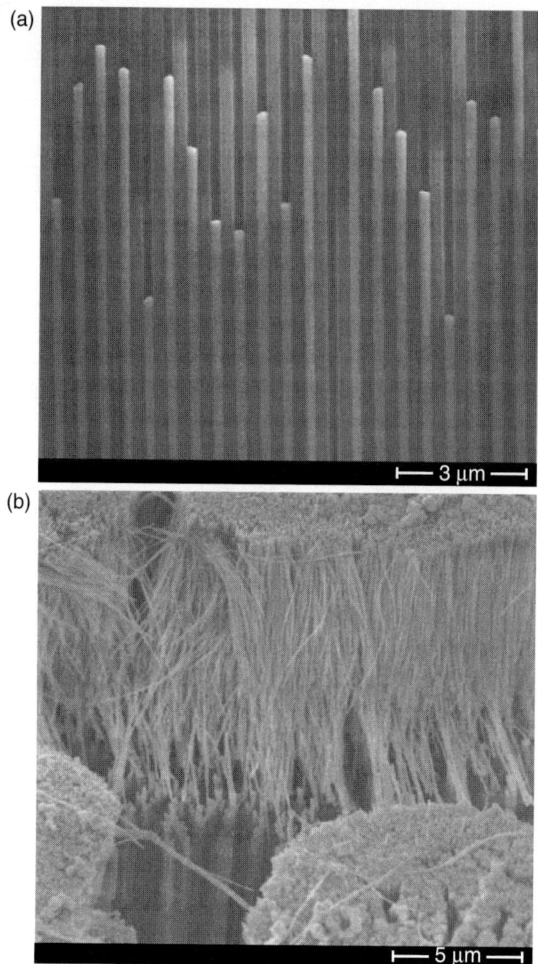

FIGURE 9.30. (a) SEM image of a cross-section view of the silver nanowires in the porous alumina template with a pore diameter of 180 nm. The bright strips are silver wires. (b) SEM image of the silver nanowires with several small branches. (Adapted from [55] with permission. Copyright (2003) American Chemical Society.)

formed in oxalic acid. The nickel nanowires are uniform in diameters and are straight. They are dense with a smooth surface and have average diameters of 50 nm. Figure 9.31b shows a TEM image of the free Ni nanowires formed from the anodic alumina films formed in sulfuric acid films. The Ni nanowires are composed of tetrahedral crystalline grains with dimensions equivalent to the pore sizes of the anodic alumina templates and this is due to the growth of Ni nanowires confined by the pore walls. Inset of Fig. 9.31a is an SAED pattern

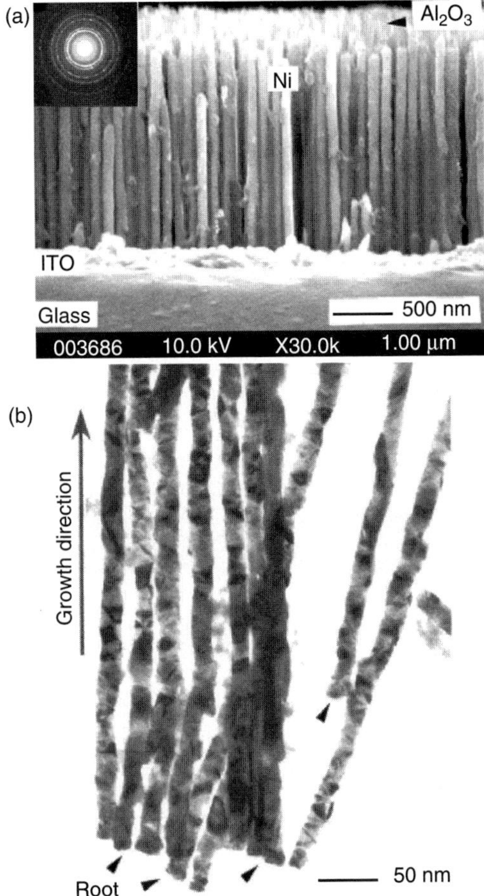

FIGURE 9.31. (a) SEM image of nickel nanowire arrays standing on ITO/glass substrate in anodic alumina films formed in oxalic acid. *Inset* is a SAED pattern. (b) TEM image of the free Ni nanowires fabricated from the anodic alumina films formed in sulfuric acid solution. (Adapted from [56] with permission. Copyright (2002) American Chemical Society.)

taken from the bundle of the Ni nanowires in Fig. 9.31b. It shows several continuous rings and indicates the polycrystalline structure of the Ni nanowires.

Recently, great efforts have been devoted to the development of magnetic metal Fe nanowires, and Cao et al. synthesized an array of iron nanowires by electrodeposition [58].

Figure 9.32a is an SEM (Hitachi X650) image of the as-synthesized iron nanowires within polyaniline nanotubules. The nanowires are 60 μm in length, corresponding to the thickness of the template membrane. The structure of the iron nanowires was characterized by XRD. Figure 9.32b is an XRD pattern of the iron nanowires array and can be indexed as pure cubic-structure iron.

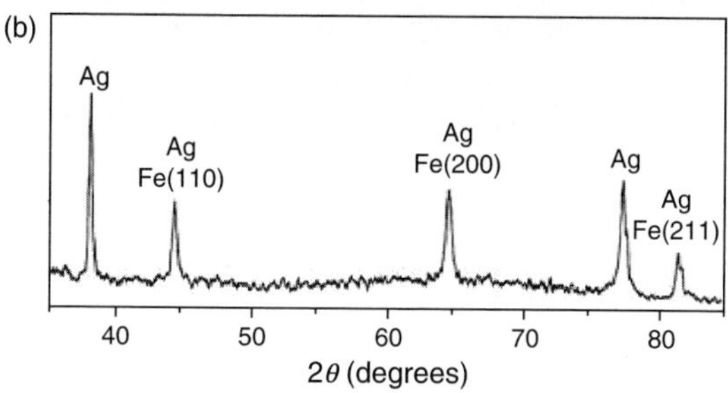

FIGURE 9.32. (a) SEM image of iron nanowires filled within polyaniline nanotubules. (b) XRD pattern of iron nanowires array within the polyaniline nanotubules. (Adapted from [58] with permission. Copyright (2001) Royal Society of Chemistry.)

Metal nanowires have attracted much attention due to their potential uses in future nanoelectronics and application possibilities for magnetic devices, nanosensors, electron emitters, and many more [59]. Among them, copper nanowires should be able to enhance the performance of the currently existing electric devices as a result of quantum-size effects [60].

Choi and Park synthesized copper nanowires by a CVD method [59]. Figure 9.33a is an SEM image of the copper nanowires and it shows that the freestanding nanowires have diameters of 70–100 nm. The SAED and XRD patterns shown in Fig. 9.33 indicate that the nanowires are pure crystalline copper and grow in the [111] orientation with a highly oriented (111) surface.

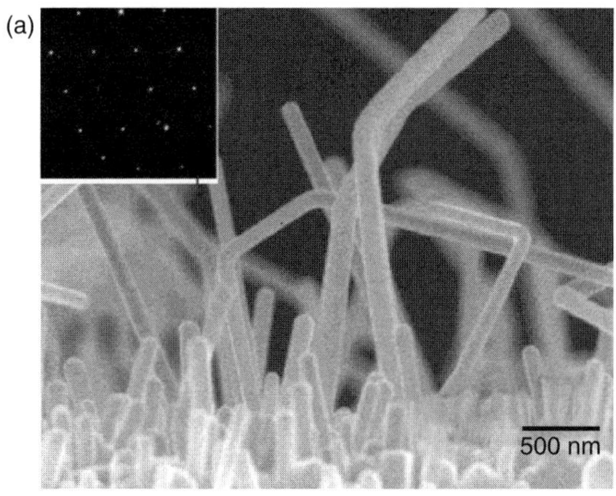

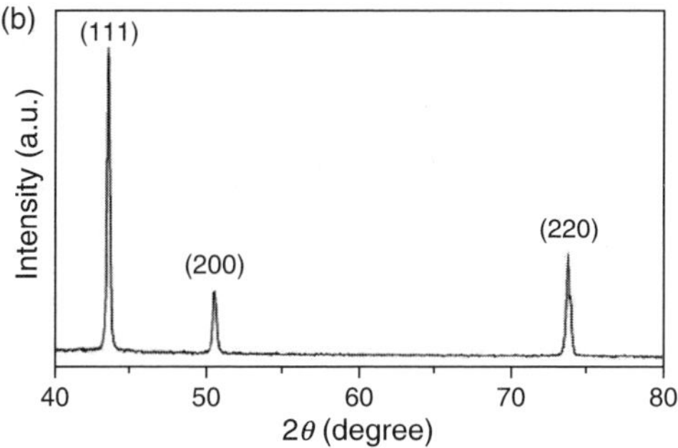

FIGURE 9.33. (a) SEM image of the copper nanowires. *Inset* is a [110] zone SAED pattern. (b) XRD pattern of the copper nanowires. (Adapted from [59] with permission. Copyright (2004) American Chemical Society.)

5. Carbon Nanotubes

CNTs are a type of materials formed by self-assembling of carbon atoms and they have hollow cylinder structures with diameters of less than 100 nm [61].

The first report of tubular filaments of carbon was by Oberlin et al. of France in 1976, and they obtained them through benzene decomposition [62]. In 1984, Gary G. Tibbetts of General Motors also reported the carbon tubular filaments [63]. However, the tubular filaments did not attract much attention from researches.

In 1991, Iijima used an HRTEM to observe the soot created in an electrical discharge between two carbon electrodes at the NEC Fundamental Research Laboratory in Japan [64]. He found that the soot contained structures that consisted of several concentric tubes of carbon and this was the discovery of multiwalled carbon nanotubes (MWNTs). In 1993, Iijima et al. [65] and Bethune et al. [66] reported the discovery of single-walled carbon nanotubes (SWNTs) independently. Since then, the synthesis and study of CNTs have drawn great deal of attention from researches in physics, chemistry, and material sciences.

5.1. Multiwalled Carbon Nanotubes

Since Iijima reported MWNTs in 1991, a large number of reports about MWNTs have appeared. As a very useful tool, SEM has been extensively used in the investigation of MWNTs.

Andrews et al. synthesized high-purity aligned MWNTs through the catalytic decomposition of a ferrocene–xylene mixture at ~675°C in a quartz tube reactor [67].

Figure 9.34a is a typical SEM image (Hitachi 3200N, 5 kV) of aligned MWNTs grown on a quartz substrate. It shows that catalyst particles are present at either ends (tips and roots) of the MWNTs and the nanotubes grow upward from the substrate. Figure 9.34b shows an SEM image of the MWNT arrays after

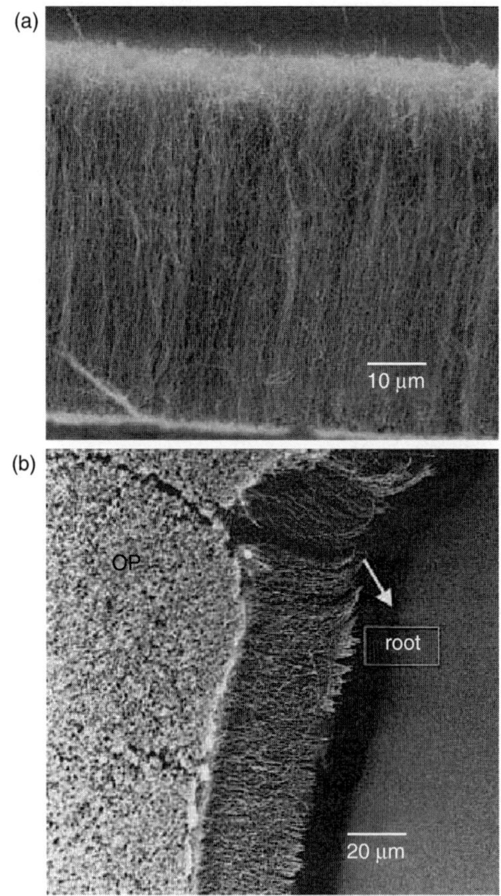

FIGURE 9.34. SEM images (Hitachi 3200N, 5 kV) of (a) as-grown MWNTs on a quartz substrate, and (b) a peeled film of as-grown MWNTs. (Adapted from [67] with permission. Copyright (1999) Elsevier.)

being peeled off from quartz substrates and the nanowires retain their parallel orientation of as-grown MWNTs.

Figure 9.35 shows the SEM images (Hitachi 3200N, 5 kV) of the expanded views of the tip and the middle section of the peeled MWNTs film, respectively. A high degree of alignment between adjacent MWNTs is shown clearly.

Figure 9.36a shows an SEM image (Hitachi 3200N, 5 kV) of expanded view of the root of the peeled MWNTs film. Figure 9.36b is a higher-magnification SEM (Hitachi 5900) image of the root region and it exhibits the open ends of the MWNTs. It suggests that in the predominant growth mode the catalyst nanoparticles detach from the substrate and travel at the head of the growing nanotube. Figure 9.36c is an HRTEM image of the multilayered structure of a single MWNT and it confirms the presence of MWNTs structure with the dominant tube diameter in the range of 20–25 nm in the sample. Inset shows typical (002) electron diffraction spots observed in a microdiffraction pattern and it reveals a high degree of structural order.

Sato et al. grew MWNTs using bimetallic particles of Ti and Co as catalysts [68].

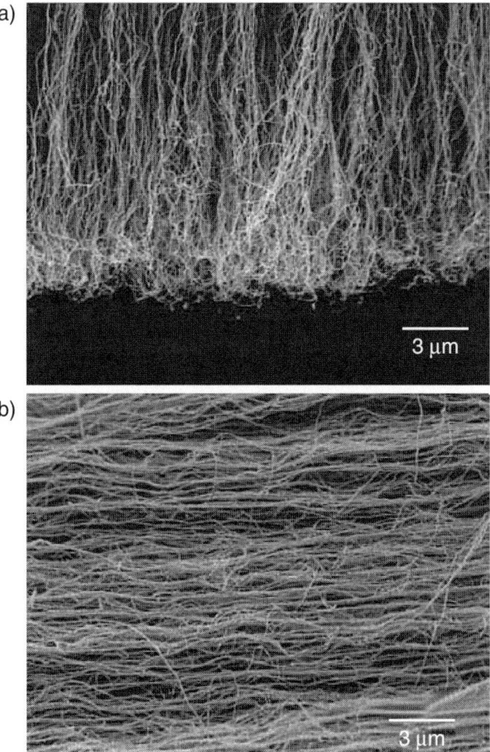

FIGURE 9.35. SEM images (Hitachi 3200N, 5 kV) of (a) expanded views of the tip and (b) the middle section of the peeled MWNTs film. (Adapted from [67] with permission. Copyright (1999) Elsevier.)

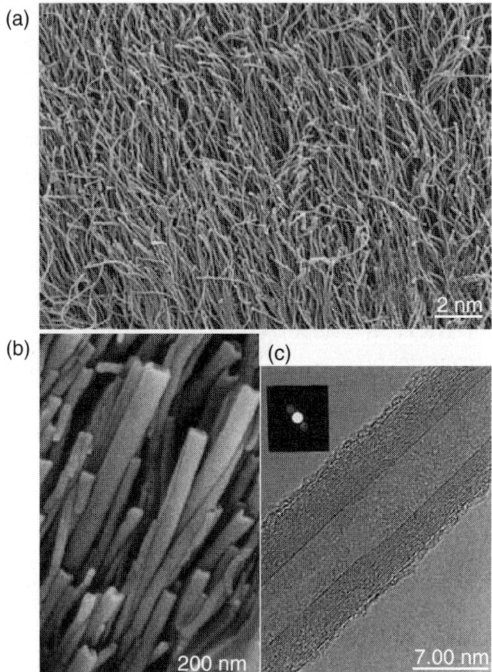

FIGURE 9.36. (a) SEM image (Hitachi 3200N, 5 kV) of expanded view of the root of the peeled MWNTs film. (b) Higher-magnification SEM (Hitachi 5900) image of the root region reveals the open ends of the MWNTs. (c) HRTEM image (Philips CM 200, 200 kV) of the multilayered structure of a single MWNT. *Inset* is a typical (002) electron diffraction spots observed in a microdiffraction pattern. (Adapted from [67] with permission. Copyright (1999) Elsevier.)

Figure 9.37a–f are SEM images of the MWNTs grown at 610°C from Ti–Co particles with Ti fractions varied from 1% to 80%. There are no aligned CNTs obtained at a Ti fraction of 1% (Fig. 9.37a). By increasing the Ti fraction from 1% to 5%, the growth of CNTs can be greatly enhanced. Aligned CNTs were obtained with Ti fractions ranging from 5% to 50% and the CNTs with a maximum height of about 30 μm were achieved at 20%. Titanium can be used to enhance the growth of MWNTs significantly.

Figure 9.38 is a TEM image of CNTs obtained from the particles with a titanium fraction of 50% at 610°C, and it reveals the multilayered structure of the MWNTs clearly and gives evidence for production of MWNTs.

5.2. Single-Walled Carbon Nanotubes

SWNTs have attracted much attention due to their unique structural, mechanical, and electrical properties. SWNTs are the basis for molecular electronics such as

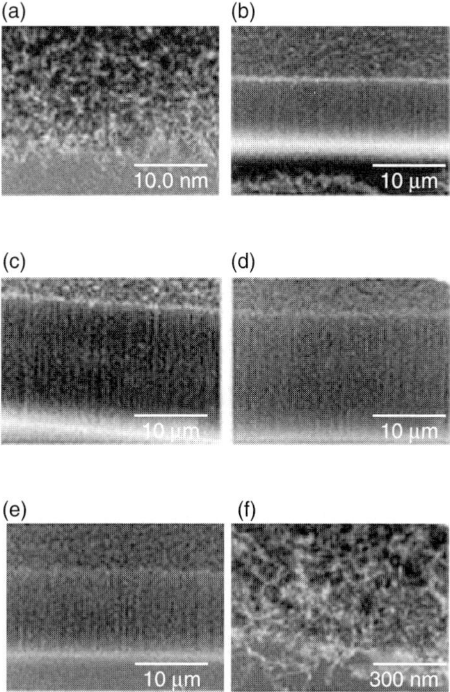

FIGURE 9.37. SEM images of MWNTs grown at 610°C from Ti–Co particles with Ti fractions varied from 1% to 80%: (a) 1:99; (b) 5:95; (c) 10:90; (d) 20:80; (e) 50:50; (f) 80:20. (Adapted from [68] with permission. Copyright (2005) Elsevier.)

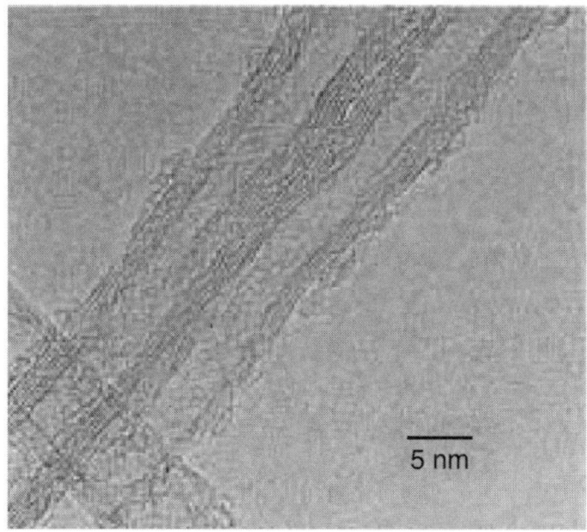

FIGURE 9.38. TEM image of the MWNTs grown from Ti–Co particles with 50% Ti. (Adapted from [68] with permission. Copyright (2005) Elsevier.)

field-effect transistors, single-electron tunneling transistors, rectifiers, simple logic devices, etc. [69].

However, to proceed to applications in molecular electronics, methods are needed to place SWNTs precisely on insulating substrates, and such methods are the subject of much current research. FESEM, a very useful and easy tool, can rapidly image the locations and electrical connectivity of SWNTs with insensitivity to surface contamination. In this mode, the FESEM scans two to three orders of magnitude quicker than a scanning probe microscope (SPM). This technique relies on the differential charging of the conducting SWNTs and insulating substrate, and therefore it is much more robust to surface defects than the SPM. This technique is generally useful in imaging SWNTs on insulators [69].

Figure 9.39a and b are compared AFM and FESEM images of the same SWNTs reported by Brintlinger et al. [69]. Figure 9.39a was taken with a Zeiss DSM982 FESEM operating at 1 kV with 20 pA beam current using in-lens secondary-electron detector. The entire $12 \times 12 \ \mu m^2$ image was obtained in ~10 s and a portion is shown. Figure 9.39b is a topograph taken from a JEOL JSPM-4210 AFM operating in an intermittent-contact mode. The entire $10 \times 10 \ \mu m^2$ image was obtained in ~700 s and a portion is shown. Two isolated CNTs, gold pads, and gold-alignment markers are shown in both images, respectively. Although the CNTs appear in the same location in both the FESEM scan and the AFM scan, their apparent diameters in the FESEM image are much larger than that indicated by the AFM height profile; in fact, it is larger than the electron beam spot size. Also, the general surface contamination appearing in the AFM image does not appear in the corresponding FESEM image. This is due to the fact that the FESEM detects contrast for conducting areas on the substrate (corresponding to

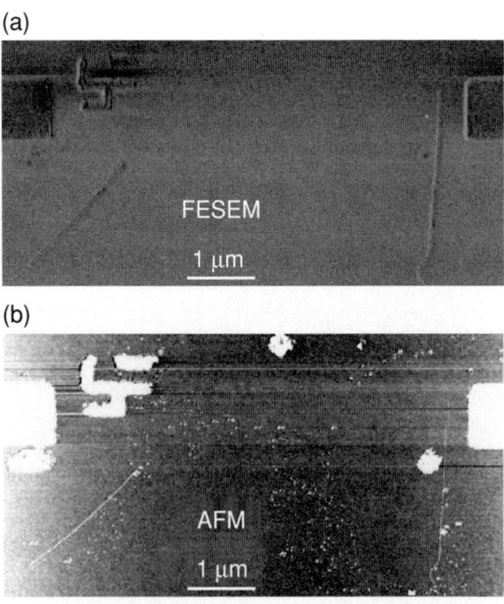

FIGURE 9.39. Compared images of the same SWNTs (a) FESEM image and (b) AFM image. (Adapted from [69] with permission. Copyright (2002) American Institute of Physics.)

the location of CNTs) but not for other nonconducting areas, even those with considerable surface roughness. The FESEM image contrast stems from the local potential difference between the nanotubes and the insulating substrate. [69].

Homma et al. proposed a different mechanism for SWNTs image formation on the insulator surface and they explained it by the electron-beam-induced current (EBIC) on the insulator surface [70]. When SWNTs are observed with SEM at a low primary-electron voltage using in-column-type secondary-electron detectors, low-energy electrons are detected. Instead of SWNTs, the insulator surface around SWNTs is imaged and provides bright, highly selective imaging of individual SWNTs [70].

Figure 9.40a–c are secondary-electron images of SWNTs on insulator SiO_2 surface reported by Homma et al., who observed the SWNTs using low-voltage SEM [70]. Figure 9.40a is an image obtained with 0.3-keV primary electrons by the LEO 1530 SEM, and Fig. 9.40b and c are images obtained using S-5000 SEM with primary electrons of 1 and 1.5 keV, respectively.

They found that SWNTs exhibited a remarkable contrast on the SiO_2 surface when imaged with a low-energy electron beam (1 keV or less) of an in-column detector. SWNTs appeared as bright and rather thick wires in the secondary-electron image (Fig. 9.40a). There are several levels of brightness in the images. With the increase in the energy of primary electron, the contrast of the images changes greatly (Fig. 9.40b and c). At a voltage of 1 keV or lower, only the nanotubes are imaged and the substrate surface topography is not visible. At a voltage of 1.5 keV or higher, both the nanotubes and the substrate surface are imaged, and the contrast difference between the nanotubes and substrate surface decreases.

To make the SWNT image formation mechanism at low voltage clearer, Homma et al. investigated secondary-electron images for various topographies of SWNTs. They compared the images taken by the in-column detector with those obtained by the outer detector in the LEO 1530 SEM. Figure 9.41a is taken on SWNTs bundles by the in-column detector and does not show as a bright contrast in the image. As shown in Fig. 9.41b, they appear almost the same when imaged by the outer detector. Figure 9.41c and d show another interesting case, where the nanotube indicated by an arrow is partially raised off the substrate surface. In Fig. 9.41c, it appears as a normal contrast and thinner than the bright images of other parts in contact with the insulating surface. The same raised part was also imaged by the outer detector and shown in Fig. 9.41d.

According to the explanation of Brintlinger et al., the contrast of CNTs on an insulator surface is due to a voltage contrast [69]. But Homma et al. thought that it should be corrected when a bias voltage is applied or when CNTs are charged on a thick insulator surface [70]. In Homma's case, there was no evidence of charging and they measured electron spectra from the same samples using 1-keV primary electrons in an Auger electron spectrometer, but no energy shift was observed. Also, the difference in the imaging of raised and contacting nanotubes cannot be explained by voltage contrast. As shown in Figs. 9.40 and 9.41, the substrate surface around the points where CNTs are connected appears bright, and the CNTs should be connected to the outside area of the primary-electron beam

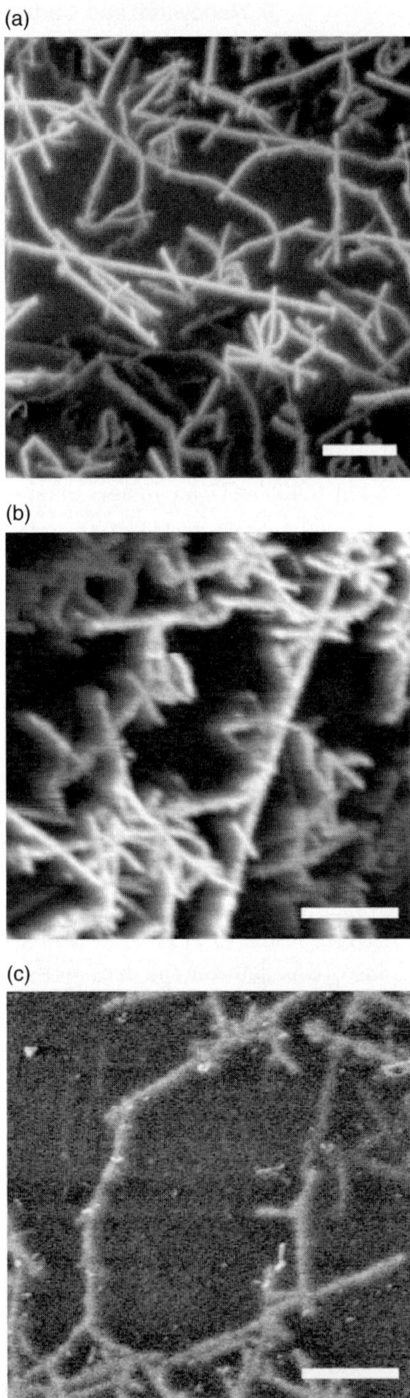

FIGURE 9.40. Secondary-electron images of SWNTs on insulating SiO_2 surface observed using low-voltage SEM. Scale bar are 1 μm, respectively. (Adapted from [70] with permission. Copyright (2004) American Institute of Physics.)

(a)

(b)

(c)

(d)

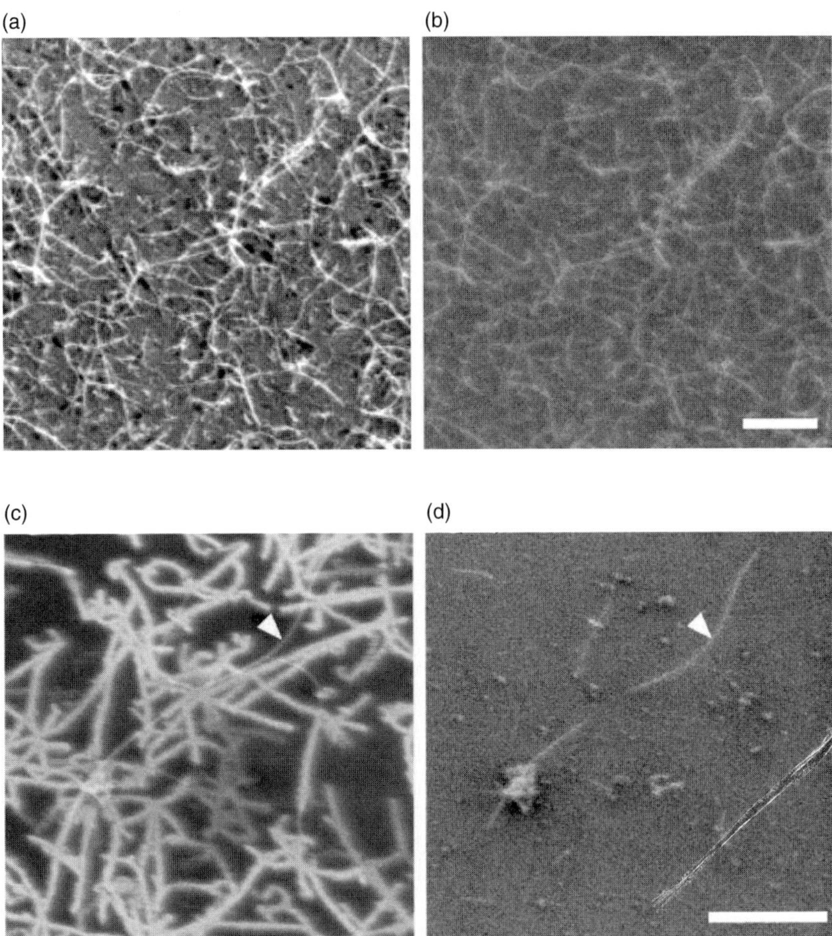

FIGURE 9.41. Secondary-electron images of SWNTs showing different image contrasts. (a) Image of SWNT bundles obtained by in-column detector. (b) Image of the same SWNT bundles obtained by outer detector simultaneously with (a). (c) Image of a SWNT raised off the surface (indicated by an *arrow*) obtained by in-column detector. (d) Image of the same SWNT obtained by outer detector simultaneously with (c). The primary electron energy was 0.5 keV with normal incidence. Scale bar, 1 μm. (Adapted from [70] with permission. Copyright (2004) American Institute of Physics.)

scanning. They explained it by an EBIC in SiO_2. The EBIC from the CNTs affects the charging state of the SiO_2 surface, thereby increasing the secondary-electron emission from the surface. So, the images reflect the EBIC range around the SWNTs. As a result, the SWNTs appear much thicker than their actual diameter (Fig. 9.42a). The EBIC range is the sum of the electron-scattering range and the electron diffusion length. The half widths of a CNT in image are about 20 and

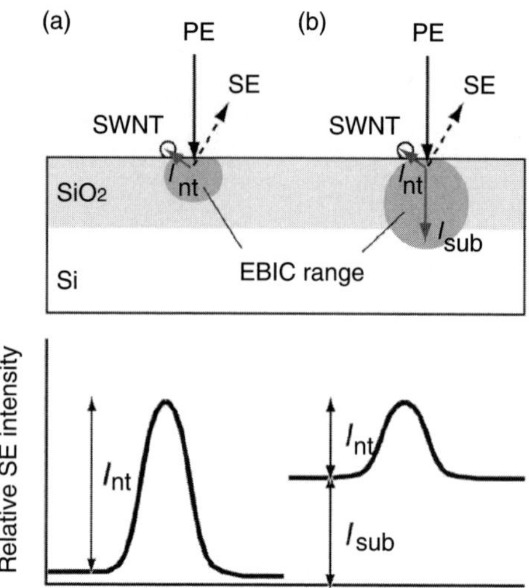

FIGURE 9.42. Model for selective image formation of SWNTs on a conductive substrate with an insulating SiO_2 layer. (a) Schematic of primary-electron penetrating range and secondary-electron yield for a low-energy primary-electron beam. (b) The same for a high-energy primary-electron beam. I_{nt} and I_{sub} are the EBIC currents to SWNTs and the substrate, respectively. (Adapted from [70] with permission. Copyright (2004) American Institute of Physics.)

15 nm for 1- and 30-keV electrons, respectively. For high-density nanotubes, bright regions may overlap each other. For SWNT bundles shown in Fig. 9.41a, there would be a large number of SWNTs in contact with the substrate surface; hence the substrate surface appears bright [70].

The bright contrast is prominent at low primary-electron energy because the primary electrons do not penetrate through the thin (100 nm) SiO_2 layer. Thus, the EBIC is supplied only through the SWNTs (Fig. 9.42a). By increasing the energy of primary-electrons, the electrons may penetrate through the SiO_2 layer and reach the Si substrate. In this case, EBIC also flows to the substrate (Fig. 9.42b), and the yield of secondary electrons from the SiO_2 surface increases uniformly. This decreases the CNT contrast. Consequently, the contrast changing at about 1.5 keV is explained by the penetration of electrons through the thin SiO_2 layer. The maximum electron energy allowed for the bright image depends on the thickness of SiO_2 layer [70].

5.3 Precision Cutting Carbon Nanotubes

At present the CNTs are relatively easy to synthesize, but at the synthesis level it is difficult to control the geometrical configurations such as length, number of walls, chirality, etc. Of great utility would be a method whereby the geometrical features of nanotubes could be altered postsynthesis, either before or after their fabrication into functional devices. A versatile method for cutting nanotubes would be particularly useful for CNTs applications [71].

In addition to the easiest tool to observe morphologies of CNTs, another powerful use for SEM was reported by Yuzvinsky et al. They presented a technique by which CNTs were precisely cut using the low-energy focused electron beam of an SEM. Their method is compatible with most device architectures and the nanotubes need only be viewable in a SEM. This method offers complete control over where the nanotube will be cut. It is relatively fast, requiring only several minutes to load, locate, and cut the nanotube in an SEM [71].

The cutting was carried out with an FEI XL30 Sirion SEM. During cutting, the SEM was operated in line scan mode at maximum magnification ($10^6\times$), with the CNT axis perpendicular to the scan line. The CNTs could be cut at a variety of acceleration voltages, beam currents, and gas pressures within the microscope chamber. Any cut can be interrupted by blanking the beam or switching the microscope out of line scan mode. Oblique cuts can also be made through rotating the scan line relative to the CNT, which may be useful for making sharpened AFM tips [71].

Figure 9.43a is a TEM image of an uncut nanotube suspended across a gap and the ends of the nanotube perpendicular to its longitudinal axis. Figure 9.43b shows a TEM image of the same nanotube after cutting. Figure 9.43c is a close-up image of a section of the nanotube before cutting and Fig. 9.43d is the same section after cutting, with the two cut sections rotated and aligned to vertically correspond with Fig. 9.43c. The cut removed about 40 nm of the nanotube and this gap is bigger than the size of the beam spot (~3 nm), most probably due to the drifts of beam position over the duration of the cutting process. Even so, the damage done by the

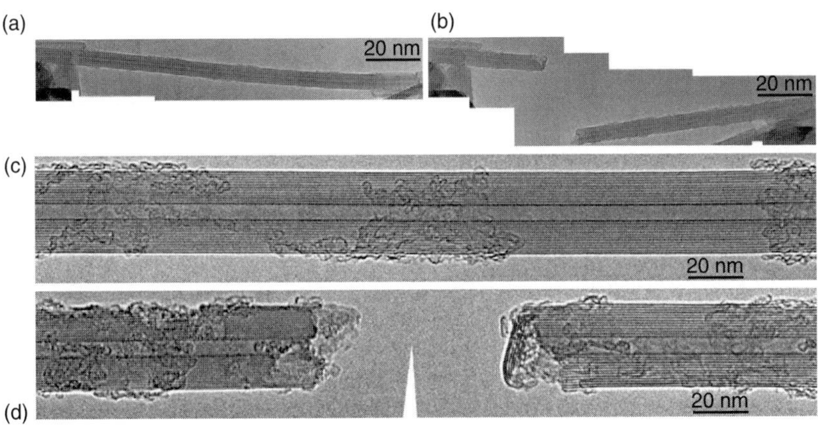

FIGURE 9.43. TEM images of (a) a CNT in its pristine state on a TEM grid; (b) the same CNT after cutting; (c) a close-up image of the same CNT; and (d) the cut segments of the CNT correspond with the same sections in (c). The scale bars are 20 nm, respectively. (Adapted from [71] with permission. Copyright (2005) American Institute of Physics.)

electron beam was confined to the immediate region of the cut, and was finished to each subsequent layer of the MWNT with equal damage [71].

6. Conclusions

In summary, SEM is one of the most powerful and useful tools in nanotechnology and it plays an important role in the investigation of nanowires and CNTs. It is the most maneuverable tool to observe the morphologies and orientations of a variety of nanowires and CNTs. It also can be used to precisely cut CNTs, and this method offers complete control over where the nanotube will be cut. This method is relatively fast and will be particularly useful for the applications of CNTs.

References

1. C. R. Rao, F. L. Deepak, G. Gundiah, and A. Govindaraj, *Prog. Solid State Chem.*, 31 (2003) 5.
2. G. Fasol, *Science*, 272 (1996) 175.
3. F. A. Ponc and D. P. Bour, *Nature*, 386 (1997) 351.
4. W. Han, S. H. Fan, Q. Li, and Y. Hu, *Science*, 277 (1997) 1287.
5. X. F. Duan and C. M. Lieber, *J. Am. Chem. Soc.*, 122 (2000) 188.
6. J. Y. Li, X. L. Chen, Z. Y. Qiao, Y. G. Cao, and Y. C. Lan, *J. Crystal Growth*, 213 (2000) 408.
7. T. Kuykendall, P. J. Pauzauskie, S. Lee, Y. F. Zhang, J. Goldberger, and P. D. Yang, *Nano Lett.*, 3 (2003) 1063.
8. H. M. Kim, D. S. Kim, Y. S. Park, D. Y. Kim, T. W. Kang, and K. S. Chung, *Adv. Mater.*, 14 (2002) 991.
9. J. Y. Li, C. G. Lu, B. Maynor, S. M. Huang, and J. Liu, *Chem. Mater.*, 16 (2004) 1633.
10. K. F. Huo, Z. Hu, F. Chen, J. J. Fu, Y. Chen, B. H. Liu, J. Ding, Z. L. Dong, and T. White, *Appl. Phys. Lett.*, 80 (2002) 3611.
11. F. L. Deepak, C. P. Vinod, K. Mukhopadhyay, A. Govindaraj, and C. N. R. Rao, *Chem. Phys. Lett.*, 353 (2002) 345.
12. Y. Zhang, J. Liu, R. He, Q. Zhang, X. Zhang, and J. Zhu, *Chem. Mater.*, 13 (2001) 3899.
13. Q. Wu, Z. Hu, X. Wang, Y. Hu, Y. Tian, and Y. Chen, *Diamond Related Mater.*, 13 (2004) 38.
14. Q. Wu, Z. Hu, X. Wang, Y. Lu, K. Huo, S. Deng, N. Xu, B. Shen, R. Zhang, and Y. Chen, *J. Mater. Chem.*, 13 (2003) 2024.
15. J. Zhang, L. Zhang, X. Peng, and X. Wang, *J. Mater. Chem.*, 12 (2002) 802.
16. B. Y. Liu, L. W. Wei, Q. M. Ding, and J. L.Yao, *Nanotechnology*, 15 (2004) 1745.
17. X. F. Duan and C. M. Lieber, *Adv. Mater.*, 12 (2000) 298.
18. W. S. Shi, Y. F. Zheng, N. Wang, C. S. Lee, and S. T. Lee, *J. Vacuum Sci. Technol. B*, 19 (2001) 1115.
19. S. Lyu, Y. Zhang, H. Ruh, H. Lee, and C. Lee, *Chem. Phys. Lett.*, 367 (2003) 717.
20. X. F. Duan, J. Wang, and C. M. Lieber, *Appl. Phys. Lett.*, 76 (2000) 1116.
21. W. S. Shi, Y. F. Zheng, N. Wang, C. S. Lee, and S. T. Lee, *Appl. Phys. Lett.*, 78 (2001) 3304.

22. W. S. Shi, Y. F. Zheng, N. Wang, C. S. Lee, and S. T. Lee, *Adv. Mater.*, 13 (2001) 591.
23. A. I. Persson, B. J. Ohlsson, S. Jeppesen, and L. Samuelson, *J. Crystal Growth*, 272 (2004) 167.
24. M. S. Gudiksen, J. F. Wang, and C. M. Lieber, *J. Phys. Chem. B*, 105 (2001) 4062.
25. S. Bhunia, T. Kawamura, S. Fujikawa, K. Tokushima, and Y. Watanabe, *Physica E: Low-dimensional Systems and Nanostructures*, 21 (2004) 583.
26. S. Bhunia, T. Kawamura, S. Fujikawa, and Y. Watanabe, *Physica E: Low-dimensional Systems and Nanostructures*, 24 (2004) 138.
27. M. H. Huang, S. Mao, H. Feick, H. Q. Yan, Y. Wu, H. Kind, E. Weber, R. Russo, and P. D. Yang, *Science*, 292 (2001) 1897.
28. E. M. Wong and P. C. Searson, *Appl. Phys. Lett.*, 74 (1999) 2939.
29. S. Choopun, R. D. Vispute, W. Noch, A. Balsamo, R. P. Sharma, T. Venkatesan, A. Iliadis, and D.C. Look, *Appl. Phys. Lett.*, 75 (1999) 3947.
30. E. A. Meulenkamp, *J. Phys. Chem. B*, 102 (1998) 5566.
31. J. Y. Lao, J. G. Wen, and Z. F. Zen, *Nano Lett.*, 2 (2002) 1287.
32. H. J. Yuan, S. S. Xie, D. F. Liu, X. Q. Yan, Z. P. Zhou, L. J. Ci, J. X. Wang, Y. Gao, L. Song, L. F. Liu, W. Y. Zhou, and G. Wang, *Chem. Phys. Lett.*, 371 (2003) 337.
33. C. Ma, D. Moore, J. Li, and Z. L. Wang, *Adv. Mater.*, 15 (2003) 228.
34. Y. Jiang, X. Meng, J. Liu, C. Xie, C. Lee, and S. T. Lee, *Adv. Mater.*, 15 (2003) 323.
35. Y. Jiang, X. Meng, J. Liu, Z. Hong, C. S. Lee and S. T. Lee, *Adv. Mater.*, *15* (2003) 1195.
36. Q. Li and C. Wang, *Appl. Phys. Lett.*, 82 (2003) 1398.
37. Q. Li and C. Wang, *Appl. Phys. Lett.*, 82 (2003) 359.
38. Y. Wang, L. Zhang, C. Liang, G. Wang, and X. Peng, *Chem. Phys. Lett.*, 357 (2002) 314.
39. M. Lin, T. Sudhiranjan, C. Boothroyd, and K. Loh, *Chem. Phys. Lett.*, 400 (2004) 175.
40. N. A. Dhas, A. Zaban, and A. Gedankan, *Chem. Mater.*, 11 (1999) 806.
41. Y. Jiang, X. Meng, W. Yiu, J. Liu, J. Ding, C. Lee, and S. T. Lee, *J. Phys. Chem. B*, 108 (2004) 2784.
42. H. Zhang, X. Ma, J. Xu, J. J. Niu, J. Sha, and D. R. Yang, *J. Crystal Growth*, 246 (2002) 108.
43. D. Xu, Y. J. Xu, D. P. Chen, G. L. Guo, L. L. Gui, and Y. Q. Tang, *Chem. Phys. Lett.*, 325 (2000) 340.
44. L. Dong, J. Jiao, M. Coulter, and L. Love, *Chem. Phys. Lett.*, 376 (2003) 253.
45. Y. Xiong, Y. Xie, J. Yang, R. Zhang, C. Wu, and G. Du, *J. Mater. Chem.*, 12 (2002) 3712.
46. D. Xu, X. Shi, G. Guo, L. Gui, and Y. Tang, *J. Phys. Chem. B*, 104 (2000) 5061.
47. X. S. Peng, J. Zhang, X. F. Wang, Y. W. Wang, L. X. Zhao, G. W. Meng, and L. D. Zhang, *Chem. Phys. Lett.*, 343 (2001) 470.
48. D. Routkoyitch, A. A. Tager, J. Haruyama, D. Almawlawi, M. Moskorvits, and J. M. Xu, *IEEE Trans. Electron Devices*, 43 (1996) 1646.
49. H. F. Yan, Y. J. Xing, Q. L. Hang, D. P. Yu, Y. P. Wang, J. Xu, Z. H. Xi, and S. Q. Feng, *Chem. Phys. Lett.*, 323 (2000) 224.
50. X. B. Zeng, Y. Y. Xu, S. B. Zhang, Z. H. Hu, H. W. Diao, Y. Q. Wang, G. L. Kong, and X. B. Liao, *J. Crystal Growth*, 247 (2003) 13.
51. Y. Wu and P. D. Yang, *Chem. Mater.*, 12 (2000) 605.
52. X. M. Meng, J. Q. Hu, Y. Jiang, C. S. Lee, and S. T. Lee, *Chem. Phys. Lett.*, 370 (2003) 825.
53. X. Y. Zhang, L. D. Zhang, Y. Lei, L. X. Zhao, and Y. Q. Mao, *J. Mater. Chem.*, 11 (2001) 1732.

54. N. R. Jana, L. Gearheart, and C. J. Murphy, *Chem. Commun.*, 7 (2001) 617.

55. J. Choi, G. Sauer, K. Nielsch, R. B. Wehrspohn, and U. Gösele, *Chem. Mater.*, 15 (2003) 776.

56. S. Chu, K. Wada, S. Inoue, and S. Todoroki, *Chem. Mater.*, 14 (2002) 4595.

57. H. Cao, C. Tie, and Z. Xu, *Appl. Phys. Lett.*, 78 (2001) 1592.

58. H. Cao, Z. Xu, D. Sheng, J. Hong, H. Sang, and Y. Du, *J. Mater. Chem.*, 11 (2001) 958.

59. H. Choi and S. Park, *J. Am. Chem. Soc.*, 126 (2004) 6248.

60. Z. Liu, Y. Yang, J. Liang, Z. Hu, S. Li, S.G. Peng, and Y. Qian, *J. Phys. Chem. B*, 107 (2002) 12658.

61. P. Avouris, *Acc. Chem. Res.*, 35 (2002) 1026.

62. A. Oberlin, M. Endo, and T. Koyama, *J. Crystal Growth*, 32 (1976) 335.

63. G. G. Tibbetts, *J. Crystal Growth*, 66 (1984) 632.

64. S. Iijima, *Nature*, 354 (1991) 56.

65. S. Iijima and T. Ichihashi, *Nature*, 363 (1993) 603.

66. D. S. Bethune, C. H. Kiang, M. S. de Vries, G. Gorman, R. Savoy, J. Vazquez, and R. Beyers, *Nature*, 363 (1993) 605.

67. R. Andrews, D. Jacques, A. M. Rao, F. Derbyshire, D. Qian, X. Fan, E. C. Dickey, and J. Chen, *Chem. Phys. Lett.*, 303 (1999) 467.

68. S. Sato, A. Kawabata, D. Kondo, M. Nihei, and Y. Awano, *Chem. Phys. Lett.*, 402 (2005) 149.

69. T. Brintlinger, Y. F. Chen, T. Dürkop, E. Cobas, M. S. Fuhrer, J. D. Barry, and J. Melngailis, *Appl. Phys. Lett.*, 81 (2002) 2454.

70. Y. Homma, S. Suzuki, Y. Kobayashi, M. Nag, and D. Takagi, *Appl. Phys. Lett.*, 84 (2004) 1750.

71. T. D. Yuzvinsky, A. M. Fennimore, W. Mickelson, C. Esquivias, and A. Zettl, *Appl. Phys. Lett.*, 86 (2005) 053109.

10
Photonic Crystals and Devices

Xudong Wang and Zhong Lin Wang

1. Introduction

During the revolution in electronic industry in the last decades, understanding and applications of electronic phenomena and materials have experienced a tremendous advancement. The technology relies on electron-operated devices and circuits, such as transistors, capacitors, inductors, interconnects, and dielectrics. In comparison, light has three main advantages over electrons, running $1,000 \times$ faster, with negligible heat dissipation and fast switching. A natural tendency is to build the devices and circuits operated by photons.

Light, which is an electromagnetic wave, is being widely utilized as a signal carrier in today's communication technology, and is likely the driver in future computer chips. However, comparing to our ability to control the electrons in conductors, semiconductors, and insulators, it is still challenging to manipulate the flow of light in the "circuit." Challenges exist in the miniaturization and integration of optical devices such as waveguide, switch, splitter, and detector. Over a decade ago, it is predicted theoretically that the propagation of light can be manipulated and controlled using a periodic structure, as electron transport in semiconductors, which is called photonic crystal (PC) [1,2]. This phenomenon explored a new approach for the manipulation of photon in a much more reduced dimensions, thus it is very promising to realize more advanced optical devices with smaller sizes and higher functionality.

1.1. Photonic Crystal: What is it?

A PC is a periodically arranged/built structure composed of alternating dielectric materials with high and low refractive indexes, the periodicity of which is controlled in the order of the wavelength of the light, so that the light can "sense" the change in refraction index as it transmits through the structure. Light striking on the structure will be refracted and reflected on each dielectric interface, and the subsequent interference can be constructive or destructive depending on the wavelength. The propagation of light in certain photon energy range can thus be prohibited inside the structure; thus, the "photonic stop band" or "photonic band gap" (PBG) is created.

The PBG is an analogy to the electron energy band gap in semiconductors. Mathematically, the motion of electrons in a semiconductor is described by the Schrödinger equation, and analogously, the traveling behavior of light in a PC is described by the Maxwell's equations. It is well known that reflection and refraction are angular and geometry-dependent in conventional optics. The forbidden energy gap in a PC also varies with the incident angle. Only when the photons can be completely reflected from the material at *any incident angles*, the PBG is described as a "full" or "complete" band gap. A PBG that only prohibits the propagation of light in some directions/orientation is referred as an "incomplete" or "pseudo" band gap.

According to the dimensions in periodic structures, PCs are classified into 1D, 2D, and 3D, in which the flow of light can be modulated in one, two or three dimensions, respectively. A typical example of 1D PCs, also known as "Bragg mirrors," is a structure with alternating dielectric layers, as shown in Fig. 10.1a. Well-developed thin film deposition technique realized the applications of these structures, e.g., dielectric mirrors, reflective coatings, filters, and distributed feedback lasers. A 2D PC is a dielectric media with 2D pattern, which is normally an ordered array of dielectric rods or airholes in a dielectric slab (Fig. 10.1b). Owing to the advancement of lithography technique, 2D PCs can be manufactured on dielectric chips, which are very attractive to engineers for optical circuits [3]. A 3D PC can control the light in all the directions, which have just been achieved in laboratory in the last several years due to the difficulty in manufacture [4]. Nature gemstone opal, which consists of a close-packed submicron-sized silica spheres, is a typical example of 3D PC with a pseudo PBG, as shown in Fig. 10.1c.

It is obvious now that the properties and applications of PCs are highly relied on their geometry and periodicity. In Section 1.2, we will discuss how the band gaps are formed by their periodic structures.

1.2. Physical Background and Band Gaps of Photonic Crystal [5]

The propagation of light in a PC is governed by the Maxwell equations (in CGS units):

$$\nabla \times \bar{E} + \frac{1}{c}\frac{\partial \bar{H}}{\partial t} = 0, \quad \nabla \times \bar{H} - \frac{1}{c}\varepsilon\frac{\partial \bar{E}}{\partial t} = \frac{4\pi}{c}\bar{J},$$

$$\nabla \times \varepsilon\bar{E} = 4\pi\rho, \qquad \nabla \times \bar{H} = 0 \tag{10.1}$$

where $\bar{E}$ and $\bar{H}$ are the electric and magnetic fields, respectively; ρ and $\bar{J}$ are the free charges and currents; and ε is the position-dependent dielectric constant. Additionally, PCs are mixed dielectric materials without any free charges or currents, so $\bar{J}$ and ρ can be set as zero. By separating the spatial component from the

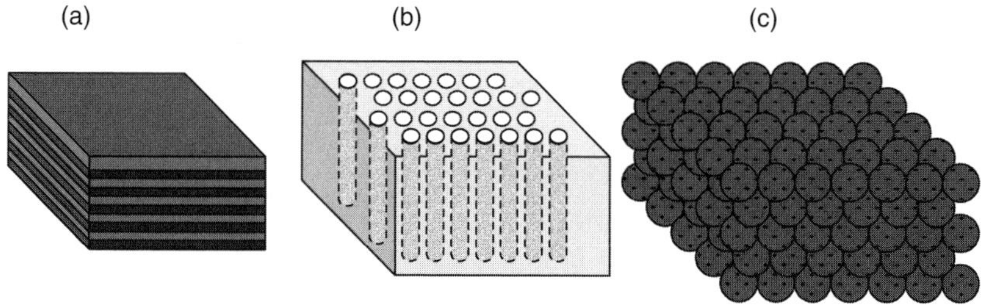

FIGURE 10.1. Schematics of (a) 1D; (b) 2D; and (c) 3D photonic crystals.

time-dependent part, such as $\vec{E}(r,t) = \vec{E}(r)\exp(i\omega t)$ and $\vec{H}(r,t) = \vec{H}(r)\exp(i\omega t)$, where ω is the angular frequency. Thus, Eq. 10.1 can be simplified as follows for PCs:

$$\nabla \times \vec{E} + \frac{i\omega}{c}\vec{H} = 0, \quad \nabla \times \vec{H} - \frac{i\omega}{c}\varepsilon\vec{E} = 0.$$

$$\nabla \times \varepsilon\vec{E} = 0, \qquad \nabla \times \vec{H} = 0.$$

(10.2)

A combination of the first two equations in Eq. 10.2 yields:

$$\nabla \times \left(\frac{1}{\varepsilon}\nabla \times \vec{H}\right) = \left(\frac{\omega}{c}\right)^2 \vec{H}.$$

(10.3)

By solving Eq. 10.3 at a given ε, a continuous spectrum of $\vec{H}$ as a function of frequency ω can be generated. To find the "band structure" of a PC, we need to introduce the case of a periodic dielectric constant.

The Bloch-Floquet theorem reveals an eigenvector of Eq. 10.3, whose operator is a periodic function of position, which can be written in the form:

$$\vec{H}_{\vec{k}}(\vec{r}) = e^{i(\vec{k}\cdot\vec{r})}\vec{u}_{\vec{k}}(\vec{r}).$$

(10.4)

where $\vec{u}_{\vec{k}}(\vec{r})$ is a periodic function of position as $\vec{u}_{\vec{k}}(\vec{r}) = \vec{u}_{\vec{k}}(\vec{r} + \vec{R})$ for all lattice vectors $\vec{R}$ and $\vec{k}$ is the wavevector. It is important to point out that the eigenvalues at $k + 2\pi/a$ are identical to those at k, where a is the periodicity. As a result, only the values in the range between $-\pi/a$ and π/a are unique and this region is called the Brillouin zone.

By inserting Eq. 10.4 into Eq. 10.3, $\vec{u}_{\vec{k}}(\vec{r})$, which is the mode profiles, can be determined within the Brillouin zone and a set of modes are expected to be found for each value of $\vec{k}$, which can be labeled by index n. Therefore, continuous functions $\omega_n(\vec{k})$, which change discretely with increasing band index number, describe the allowed modes in a PC. An ordinary 2D diagram can thus be plotted through $\omega_n(\vec{k})$ vs. $\vec{k}$ as the band structure of the PC. An example of a band structure calculated for an inverse face-centered cubic (FCC)-structured 3D PC is shown in Fig. 10.2 (e.g., FCC packed air spheres in a dielectric medium) [6]. A full PBG can be observed between the 8th and 9th bands.

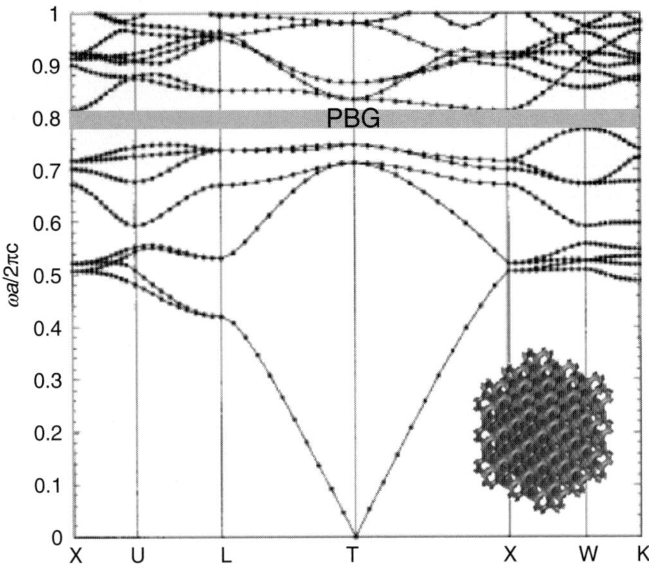

FIGURE 10.2. Calculated photonic band structure for an inversed FCC lattice of silicon ($\varepsilon \approx 11.9$). *Inset*: a schematic of the inversed FCC lattice structure.

1.3. Overview of the Applications of Photonic Crystals

A number of unique and novel optical applications can be achieved using the PBG characteristic of a PC, and they could not be realized by ordinary optical materials. Applications in many cases have been studied, including waveguides, perfect mirrors, resonators, lasers, optical fibers, prisms, optical switches, and more. Here, we will briefly introduce some state-of-the-art examples about the devices or applications that have been demonstrated using PCs as waveguides and photonic integrated circuits (IC) in optics and optoelectronics.

1.3.1. Waveguides

High-efficient light guiding is very important in optical communication and computing. Comparing with the conventional dielectric waveguide, a PC waveguide can guide light with great efficiency either along a straight path or around a small-size sharp corner and the guiding loss due to material absorption can be minimized as well.

A PC waveguide is normally fabricated by introducing a linear defect into a 2D PC slab. A typical example is shown in Fig. 10.3a [7]. Since a certain range of modes are forbidden by the PBG, light with the corresponding wavelengths is not allowed to exist inside the PC with perfect periodic structure, and it only exists in the area that has defects. Thus, it is possible to confine the light at the defect

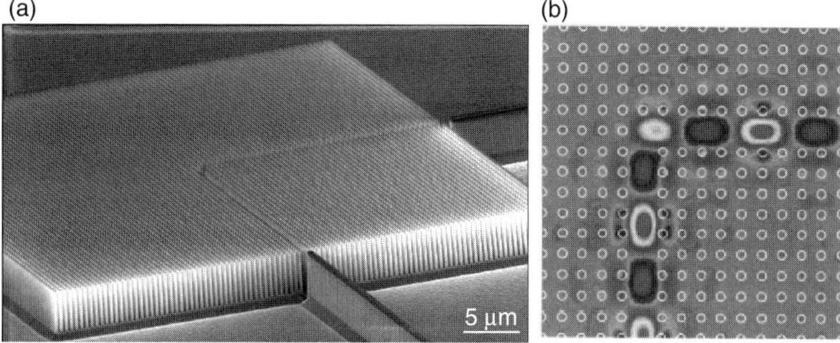

FIGURE 10.3. (a). SEM image of a waveguide in a 2D PC composed of periodic silicon pillars. (Adapted from [7] with permission.); (b) simulation of the propagation of light traveling around a 90° bent waveguide in a 2D square PC lattice. (Adapted from [3] with permission.)

region and guide it along a linear channel. Figure 10.3b shows an optical mode in traveling around a sharply bent waveguide in a 2D PC composed of a cubic array of dielectric cylinders. Moreover, zero-loss transmitting has been observed through this 90° bending channel with a bending radius less than one wavelength of the light, which is the smallest radius that can be achieved by a light without any loss [8]. Another advantage of the PC waveguide is that the only loss is due to a very small portion of light penetrating into the PC. Therefore, the signal loss due to material absorption can be reduced by a factor of ~10 comparing with normal dielectric waveguide [9].

2D PCs were proved to be a perfect waveguide for in-plane light transmitting. But the waves are not confined in the direction perpendicular to the plane, where imperfections or disorders may scatter the wave out of the PC. Creating a line defect in a 3D PC may be an alterative approach to reduce the loss. High transmittance has been achieved through a 90° bending channel within either metallic or dielectric 3D PCs [10]. Despite the complexity in the fabrication process and the high cost, the propagation of light can be confined in designed directions with minimum loss.

1.3.2. Optical Fibers

Optical fiber is a special type of waveguide, in which waves are traveling along the normal direction of the 2D PC plane (e.g., axial direction). A common structure is a bunch of parallel fine dielectric fibers that are arranged periodically around a central core, or it can be an array of air channels in a solid matrix [11], as shown in Fig. 10.4a. The core can be either a solid cladding material or an airhole. Silica is an ideal material to make a PC fiber because of its high stability, easy fabrication process, and low intrinsic optical loss. However, its relatively low refractive index limits the width of wavelengths that can be transported by the PCs.

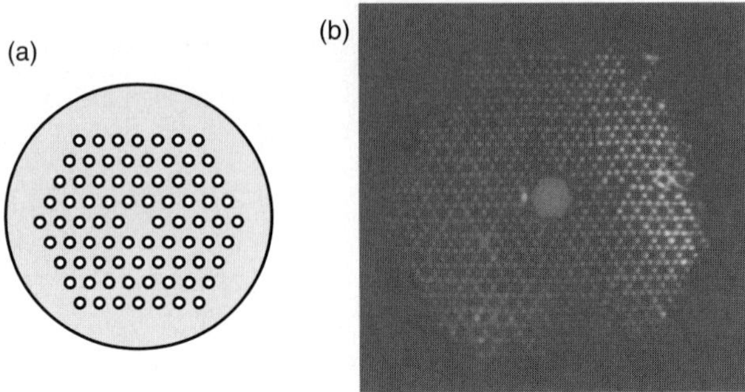

FIGURE 10.4. (a) A schematic of the cross section of a PC fiber; (b) optical image of the output pattern at the exit face of a 3-cm long 2D air-guiding PC fiber with white light excitation at the entrance face. (Adapted from [12] with permission.)

The operation wavelength of a PC cable is determined by the forbidden modes in the 2D air-channel PC, which is a function of a few parameters, including the size of air channels, the lattice type and dimensionality, and the size of the central core. Single mode waveguide can be achieved in the fiber by controlling the center core size, which is normally a couple of times larger than the periodicity of the air channels. Figure 10.4b shows an example of an air-guiding PC fiber. With a white light source inputting from its far end, only the guided mode propagates through the center core resulting in a strong red output [12]. Although the light in a wide wavelength range is introduced into the core, all of the other modes, which are allowed in the 2D airhole PC, quickly leak out as they travel through the fiber.

The PC structure requires an ordered arrangement in wavelength scale, which makes it more complex to fabricate than the conventional optical fibers. However, only a few layers of the air channels are sufficient to reduce the leakage to almost zero and a certain degree of randomness will not exhibit a significant effect on its optical property. Therefore, it is possible to commercially fabricate a PC optical fiber with infinite length. Introducing the PBG effect into optical fiber design will enhance or enable a variety of new applications for optical fibers. Their extremely low loss and mode controllability make them good candidates for generating and delivering high-power laser, quantitative fiber spectroscopy, and data transmission in optical computing. They are very likely to play an important role in future generations of optical systems.

1.3.3. Lasers

Dielectric mirrors are the first application of PC structures in laser diodes, such as distributed Bragg reflection (DBR) and distributed feedback (DFB) laser structures, where 1D PC layers serve as perfect mirrors aside an active region to achieve low threshold current and single wavelength laser emission.

One of the mostly investigated PC laser structures is the utilization of a defect state in the PBG. Similar to the doped semiconductor, the introduction of defects into a 2D or 3D PC generates a 2D or 3D PBG with defective energy/wavelength levels, in which a particular mode of light is confined and oscillated [13]. As a result, some of them can also be considered DFB type lasers. Because of the propagation confinement of PBG, spontaneous emission can be largely reduced and a single mode emission becomes easier to achieve. Thus, comparing with a conventional solid-state laser, a PC laser exhibits much lower threshold level. However, due to the small active wavelength of the PBG, it is normally difficult to achieve high-power output. Figure 10.5 shows a model of surface-emitting laser using 3D PC cavity [14]. In this structure, a diamond or nonspherical FCC-structured 3D PC is used as a mirror to form an absolute optical forbidden band. The PC media serve as both a laser mirror and the spatial filter of the spontaneous emission. In the center of the 3D PC, a plane phase shift region (λ-cavity) with a light-emitting active area is assumed in order to make a lasing level in the forbidden band region. Using the plane phase shift region makes it possible for the spontaneous emission to radiate in the lasing direction with narrow radiation angle. By tuning the emission wavelength of the active region to the phase shift level not only stimulated emission, but also the spontaneous emission are coupled in one direction with narrow radiation angle. This laser structure is predicted to have a very low threshold current and high output power owing to the large volume of the active region. In general, the ultimate aim of developing a PC laser is to achieve the thresholdless lasing property, which will be an ideal light source for optical circuits.

1.3.4. Photonic Integrated Circuit

A photonic IC is an integration of light emitters, waveguides, splitters, and detectors that generates and modulates light signal or perform logical computations.

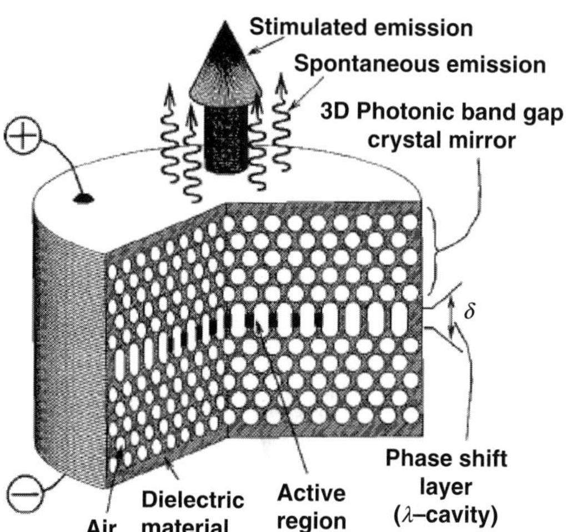

FIGURE 10.5. Schematic of a surface-emitting laser using layer defect in a 3D PC. (Adapted from [14] with permission.)

Since the performance of conventional optical devices can be highly improved, it comes out to be a very attracting aspect to combine different PC devices to achieve complex optical functions. Furthermore, the size of the optical IC can be greatly minimized because the light can be bent or split sharply without loss in a few wavelength regions inside a PC lattice.

The first example has been illustrated by Joannopoulos and his coworkers by introducing the PC waveguide and cavities to manipulate the on-chip light flow. A more complex system was proposed by Noda et al. [15]. On the basis of a 3D PC lattice, various functional optical devices are integrated, such as nanoampere laser arrays with different oscillation frequencies, optical modulator, wavelength selectors, and waveguide with very sharp turns. As shown in Fig. 10.6a, those devices are integrated in a very small area only by introducing appropriate artificial defects inside the crystal. More recently, 2D PC-based Mach-Zehnder optical switch was realized by Tanaka et al. [16]. In their structure, additional waveguides for controlling pulses were introduced, which enabled this structure to be monolithically integrated (Fig. 10.6b). 50%, 100%, and wavelength-dependent power couplings were also demonstrated. These results strongly indi-

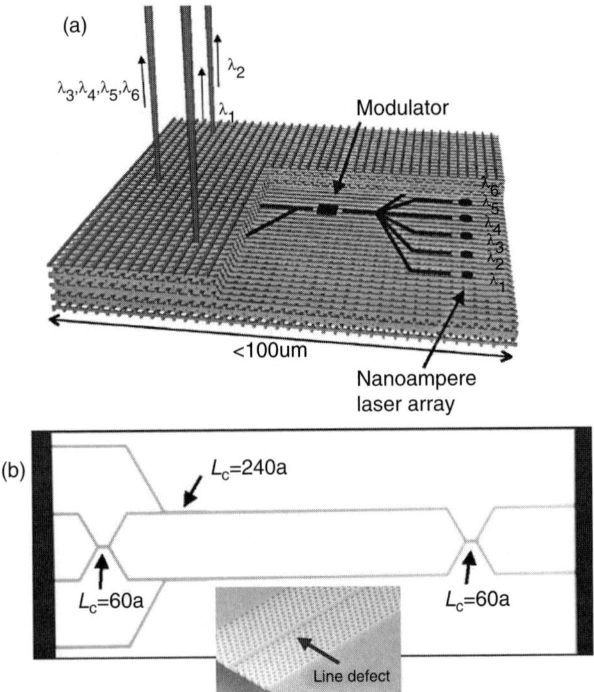

FIGURE 10.6. (a) Schematic of an example of compact quantum optical circuits based on PC slab. (Adapted from [15] with permission.); (b) optical image of a PC-SMZ structure; *inset*: SEM image of the waveguide structure in the circuit. (Adapted from [16] with permission.)

cated the feasibility to utilize PC structure for enhancing the optical performance and reducing the size of a photonic IC system.

2. SEM Imaging of Photonic Crystals

It has been demonstrated in the first part of this chapter that PBG is highly dependent on PCs' morphology and structure, including type of lattice, periodicity, size of unit cell, and the contrast of dielectric constants. The operation wavelength of PCs is normally in visible light and near–infrared (IR) region, which requires the unit cell and periodicity lay in the corresponding size range from 100 nm to a few microns. Therefore, comparing with current imaging techniques, SEM is most suitable for revealing the structural properties of PCs. In this part, different types of currently investigated PCs are categorized and their representative characterizations by SEM are illustrated.

2.1. 2D Photonic Crystals

2.1.1. PC Slabs and Waveguides

PC slab is one of the most intensively studied 2D PC structures for waveguides. The typical structure is a layer of dielectric material penetrated with ordered airholes/channels. Line defects, such as roles of missing holes are normally introduced into the structure for confining/trapping and propogating the light. Typical SEM images are shown in Fig. 10.7. Figure 10.7a is a 270-nm thick AlGaAs layer with a triangular lattice of airholes [17]. The slab was connected by a 3.0-mm wide and 0.4-mm long stripe waveguide on both sides, where the Γ-M direction of the corresponding 2D Brillouin zone is parallel to the waveguides. This configuration presents a good model for measuring the guided modes in 2D PCs. SEM images of suspending silicon PC waveguides with [18] or without [19] a strip access are shown in Fig. 10.7b and c, respectively, and a top view of a 60° bending waveguide is shown in Fig. 10.7d. These 2D PC slabs were composed of triangular lattice of airholes, which were fabricated by e-beam lithography. The operation wavelength of these PC waveguides was ~1,500 nm and both of them exhibited very low propagation losses.

2.1.2. Aligned Rods

2D ordered dielectric nano- or microrod array could be treated as an inversed structure of airhole 2D PC slab. Due to the large volume of air in the structure, their PBG are relatively narrow and light is more difficult to be confined inside the PC plane. However, light-emitting or lasing properties could be facilitated by this morphology [20]. Strong enhancement in light extraction efficiency has been observed in GaInAsP 2D microcolumns [21], whose SEM image is shown in Fig. 10.8a. The columns are ordered hexagonally with a diameter of 0.3 μm and

FIGURE 10.7. SEM images of 2D PC slab and waveguides. (a) AlGaAs slab with a triangular lattice of airholes. (Adapted from [17] with permission.); (b) silicon PC slab waveguide with a strip access. (Adapted from [18] with permission.); (c) silicon PC slab waveguide. (Adapted from [19] with permission.); and (d) 60° bending waveguide on a silicon PC slab. (Adapted from [19] with permission.)

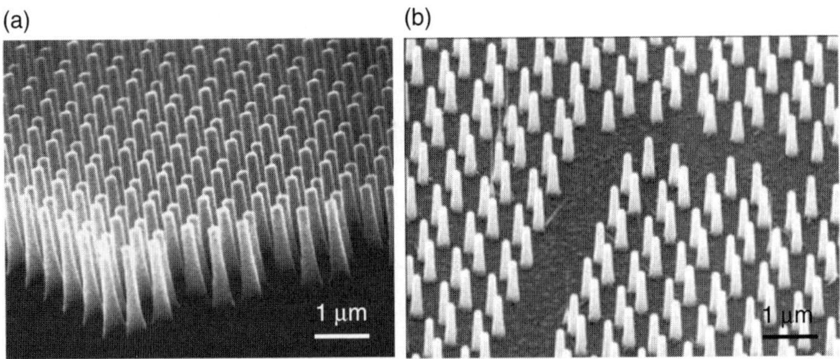

FIGURE 10.8. SEM images of 2D PC composed of aligned nanorods. (a) Hexagonal patterned GaInAsP submicrorods. (Adapted from [21] with permission.); (b) waveguide in a 2D InP nanorods PC. (Adapted from [22] with permission.)

a height of 2.5 µm. A waveguide structure has also been fabricated using the 2D pillar arrays. By using nanoimprinting lithography to pattern the catalyst, InP nanorods are grown by MOCVD [22]. A representative SEM image is shown in Fig. 10.8b. This technique presented a high throughput process with great potential for commercial applications.

2.1.3. Deep Holes

The high aspect ratio of 2D PCs are typically fabricated by wet chemical etching. An SEM image is shown in Fig. 10.9 to illustrate this type of morphology. Owing to its large aspect ratio, the transmission loss as a waveguide can be effectively reduced in the normal direction [23]. Furthermore, other dielectric materials, such as liquid crystals can be infiltrated into the holes, thus the contrast of refractive index can be changed, leading to a tunable PBG. It has been reported that the wavelength of the H-polarized air band edge was continuously tuned for over 70 nm by raising the temperature of the infiltrated liquid crystal from the nematic to the isotropic phase [24].

2.1.4. Fibers

As mentioned in the first part of this chapter, optical fiber is one of the promising applications of 2D PCs. Figure 10.10a shows an SEM image of the cross section of a typical air-guiding PC optical fiber. The fiber was formed by stacking silica capillary canes in a triangular lattice leaving a large airhole in the center for wave traveling. The detailed structure at the core area is shown in a higher resolution

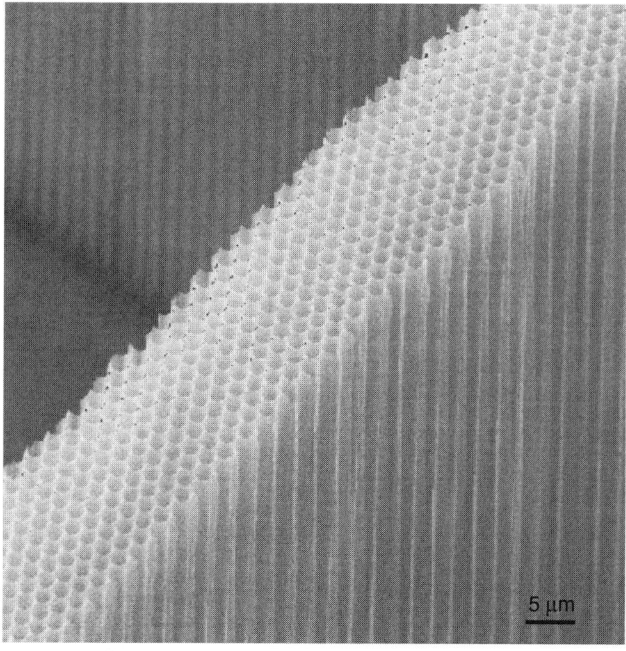

FIGURE 10.9. SEM image of a 2D silicon PC with high-aspect-ratio airholes. (Adapted from [24] with permission.)

SEM image in Fig. 10.10b, where we can see the diameter of each silica capillary cane being 4.9 μm, and the core has a diameter of 14.8 μm. This configuration provides a single-mode light transmitting and its narrow-band performance suggests that it could be used as a spectral filtering device.

2.1.5. Monolayer Spheres

Other than the masks for pattern generation, the self-assembled monolayer of micron or submicron dielectric spheres has been investigated as a type of 2D PCs. Figure 10.11 shows that the 2D triangular lattice of spheres can be formed into a particular shape by micromanipulation technique (Fig. 10.11a) [25]. The spheres can also be self-assembled into a fairly large area (Fig. 10.11b) [26]. Partial PBG has been observed in the near IR region, and the photonic band dispersion curves

FIGURE 10.10. (a) SEM image of the end surface of the air-guiding 2D PC fiber; (b) higher resolution SEM image of the fiber structure. (Adapted from [12] with permission.)

(a)

(b)

FIGURE 10.11. SEM images of ordered submicron spheres monolayer. (a) 2D triangular lattice with well-controlled number of spheres made by micromanipulation. (Adapted from [25] with permission.); (b) large-area monolayer of spheres made by self-assembly technique.

were retrieved in a finite area, which agreed to the theoretical predicted result for an infinite lattice. By measuring the transmission spectra at normal direction of the monolayer assembly of spheres with design sizes, the full photonic bands from the whispering gallery modes has also been observed as the total number of spheres in the assembly being increased.

2.2. 3D Photonic Crystals

2.2.1. Opal Structure

Gem opal is an amazing example of naturally formed 3D PC, which is composed of self-assembled spheres packed into FCC lattice. In mimicking this natural process, self-assembly techniques have been developed for organizing submicron

FIGURE 10.12. SEM image of ~300 nm silica spheres-packed FCC lattice (opal).

dielectric spheres into FCC structure. The most common materials for the opal are silica, polystyrene, and PMMA. Figure 10.12 demonstrates an SEM image of an opal structure made from close-packed 230-nm silica spheres, which is a result of self-assembly through a confinement cell process [27]. Other methods, such as electric fields-driven [28] or capillary force-driven processes [29] have also been successfully applied for fabricating high quality self-assembly opal film. Although the opal structure presented in Fig. 10.12 cannot produce a full PBG, but owing to its low cost and flexible fabrication process, this structure has received intensive investigation either for theoretical studying of 3D PCs or serving as templates for fabricating other 3D PCs.

2.2.2. Inversed Opal Structure

The inversed opal structure, which consists of FCC-packed air spheres inside a high refraction index matrix, exhibits a full PBG between the eighth and ninth bands that is corresponding to wavelength ~3.5× of the sphere radius [30]. This structure is typically fabricated by infiltrating an opal lattice followed by etching the spheres. Various dielectric materials can be filled into the vacancies between the spheres and form a 3D full PBG in visible or near IR region. For example, Fig. 10.13a shows a phenolic inverse opal, which was eventually converted to glassy carbon inverse opal by pyrolysis [31]. An SEM image of a silicon inverse opal is shown in Fig. 10.13b, where the template was 1 μm silica sphere opal and exhibited a full PBG of ~1.5 μm [32]. Figure 10.13c shows the (100) facet of a silicon FCC inverse opal lattice and the ten-layer slab is shown in Fig. 10.13d, which clearly illustrates its FCC structure [33]. Inversed opal is a complex 3D periodic structure that can be realized in large area through a low-cost colloidal assembly process, thus provides a promising sample for PBG investigation and optical device application.

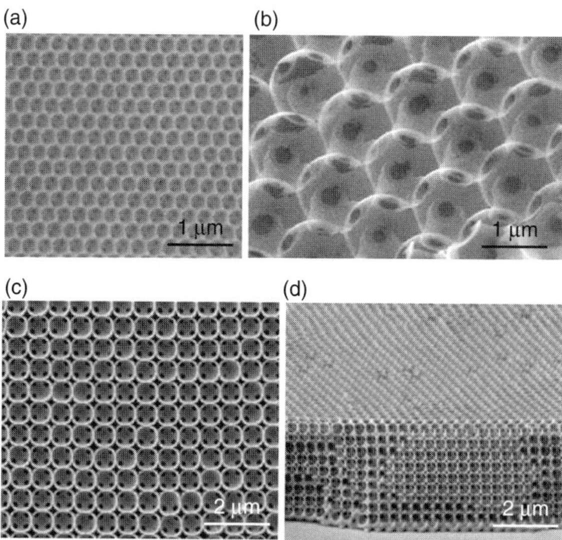

FIGURE 10.13. SEM images of inversed opal structure. (a) Phenolic inverse opal structure. (Adapted from [31] with permission.); (b) silicon inverse opal structure. (Adapted from [32] with permission.); (c) (100) facet of a silicon FCC inverse opal structure. (Adapted from [33] with permission.); and (d) ten-layer slab of the silicon FCC inverse opal. (Adapted from [33] with permission.)

2.2.3. Woodpile

The woodpile 3D PCs are composed of a layer-by-layer periodic stacking sequence of 1D dielectric rods [34]. The top-view SEM image of such a structure is shown in Fig. 10.14a. The periodicity in perpendicular direction is 4× of the thickness of the rods since it repeats itself every four layers. Within each layer, the axes of the rods are parallel to each other with a pitch of d. The orientations of the axes are rotated by 90° between adjacent layers. Between every other layer, the rods are shifted relative to each other by $0.5d$. The resulting structure exhibits a face-centered tetragonal lattice symmetry as shown in Fig. 10.14b. Because of its uniformity, this structure shows a large IR PBG from 10 to 14.5 μm, strong attenuation of light within this band (~12 dB per unit cell) and a spectral response uniform to better than 1% over the area of the 6-in. wafer.

Modifications have been applied on this structure to extend the applications. A 90° bending waveguide was created inside the woodpile lattice by removing a pair of rods, as shown in Fig. 10.14c. Very high transmission property within a wide frequency range was predicted for the waveguide [35]. Other than silicon, metallic rods such as tungsten have been successfully fabricated into this structure [36]. Figure 10.14d shows an SEM image of a woodpile PC consisting of four-layer tungsten tubes. A wide IR PBG was observed from 8 to 20 μm. Although a perfect PBG can be achieved in the woodpile structure, the costly and time-consuming

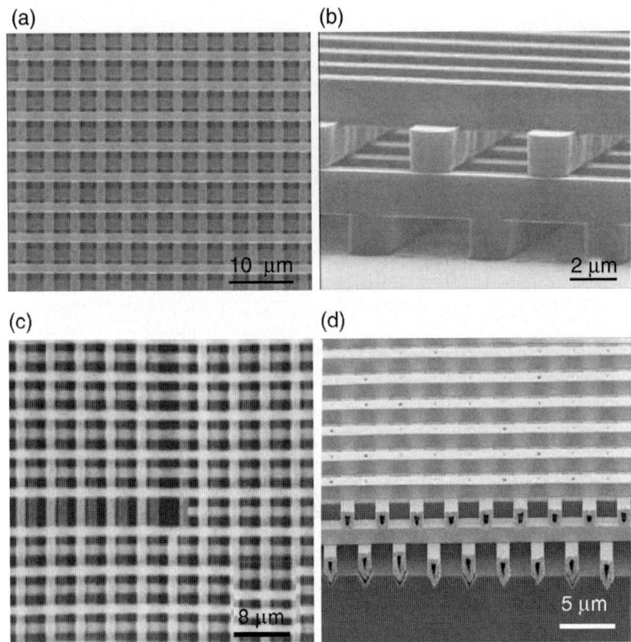

FIGURE 10.14. SEM images of woodpile 3D PCs. (a) Top-view image of the woodpile structure. (Adapted from [34] with permission.); (b) four-layer symmetric structure of a woodpile PC. (Adapted from [34] with permission.); (c) a waveguide in a woodpile 3D PC structure. (Adapted from [35] with permission.); and (d) a woodpile PC consisting of four-layer tungsten tubes. (Adapted from [36] with permission.)

lithography process hindered the further application for commercialization and the PBG is hard to be achieved in visible light region because of the resolution limit of current lithography techniques.

2.2.4. 3D Cubic Lattice

Wave beam etching has come out to be a more efficient technique for fabricating 3D PCs. A cubic periodic lattice was generated by interference of four noncoplanar laser beams in a film of photoresist. The intensity distribution in the interference pattern has 3D translational symmetry. Highly exposed photoresist is rendered insoluble; unexposed areas are dissolved away to reveal a 3D periodic structure formed of cross-linked polymer with air-filled voids. This process is called holographic lithography [37]. Figure 10.15a shows a polymeric PC generated by exposure of a 10-mm film of photoresist to the interference pattern. Its (111) surface is shown in Fig. 10.15b. This structure can also be infiltrated to form an inversed lattice as shown in Fig. 10.15c, which is the (111) surface of a titania inverse structure.

Multiple tilted x-ray beams have been used to drill interconnected holes on an x-ray sensible PMMA layer. The residual polymer exhibited a diamond-like 3D

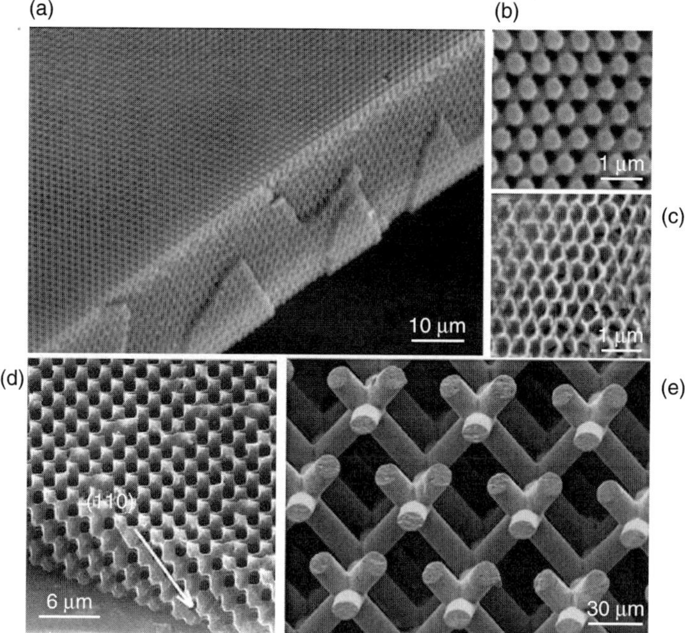

FIGURE 10.15. SEM images of 3D PC with cubic lattice. (a) A polymeric 3D PC made by holographic lithography. (Adapted from [37] with permission.); (b) (111) facet of the 3D PC. (Adapted from [37] with permission.); (c) an inversed lattice of the PC structure. (Adapted from [37] with permission.); (d) a polymeric 3D PC made by three-beam x-ray lithography. (Adapted from [38] with permission.); and (e) "three-cylinder" 3D PC formed by inversing the lattice in (d). (Adapted from [39] with permission.)

lattice after illuminated by three x-ray beams that were tilted 120° between each other [38]. An SEM image of this structure is shown in Fig. 10.15d. Additional infiltration also can inverse the lattice, forming a "three-cylinder" structure [39]. Figure 10.15e shows such a PC made from negative tone resist with a lattice constant of 114 μm. Full 3D PBG has been observed on these structures in IR region; however, the beam resolution still limits the blue shift of the PBG into visible light region.

2.2.5. 2D Ordered Helix Nanowires

It has been predicted that 2D-ordered helix silicon wires can produce a 3D PBG between the fourth and fifth bands, which can be as large as 15% of the gap center frequency [40]. Glancing angle deposition (GLAD) process provides a simple, versatile means of producing this PC structure in a single processing step on a wide variety of elementary, prepatterned wafers. The principle of this technique lays in the deposition of target materials at large incidence angles. By rotating the substrate appropriately, the deposited materials can follow the rotating angles and form amorphous helix nanowires [41]. The SEM image of a tetragonal square of the helix silicon structure is shown in Fig. 10.16. The periodicity

FIGURE 10.16. SEM image of the 2D-patterned silica helix nanowires made by glancing angle deposition technique. (Adapted from [41] with permission.)

and the size of the wires can be made within a few nanometers, thus enables this structure to produce a full PBG in visible light region.

3. Fabrication of Photonic Crystals in SEM

As we introduced in the previous part, with the increasing of research interests on PCs, various techniques, such as lithography, beam interference etching, and colloidal assembly, have been developed for fabricating different types of PCs. Generally, each technique is restricted on making particular PC structure and it is still very hard to realize a complete control over point defects inside the crystal lattice. In addressing these limitations, an SEM-based micromanipulation technique has been introduced for PC fabrication. Traded-off by the constructing efficiency, this technique can provide a precise control of the building blocks so as to achieve complex PC structures that are of great interest.

3.1. Micromanipulation System in SEM

Controlling objects in very small scale has been one of the most important targets of engineers in a variety of disciplines. Advanced tools, such as atomic force microscopy (AFM) or scanning tunneling microscopy (STM) was developed and enabled people to manipulate nano-sized species or even molecules and atoms. Considering the size range of PCs' building blocks, which is from a few hundreds nanometer of a few microns, SEM can serve as an ideal platform for in situ constructing 3D PC lattice once an appropriate manipulator is adopted inside an SEM chamber.

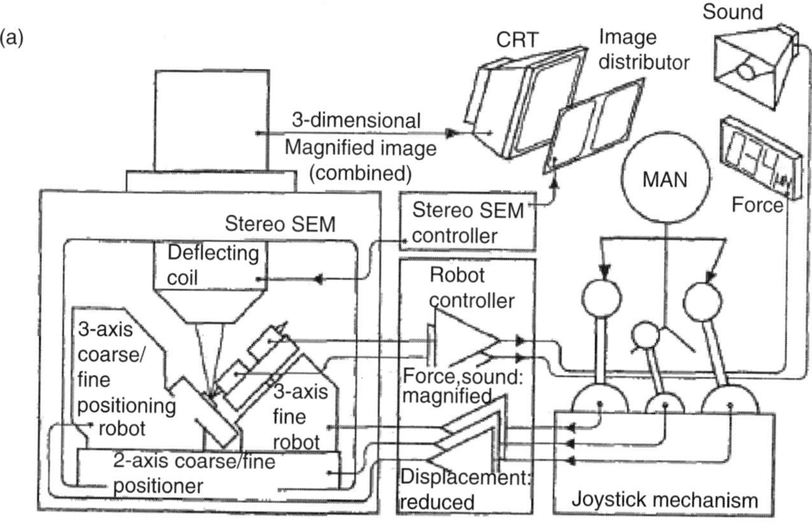

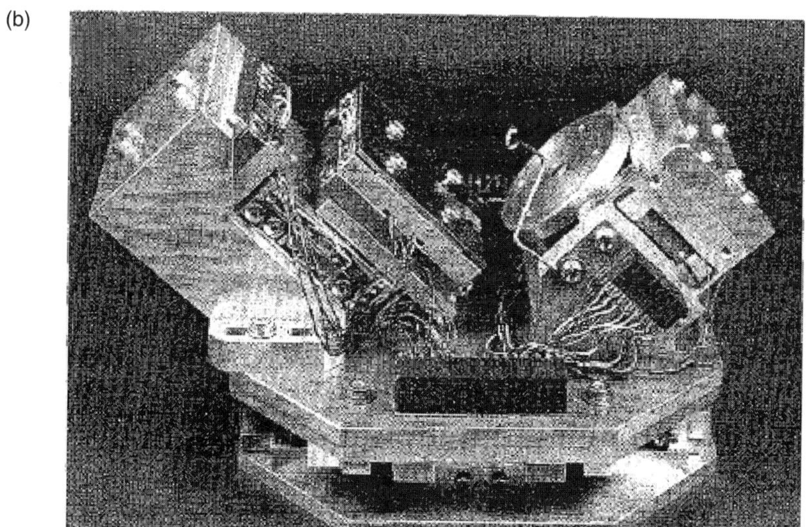

FIGURE 10.17. (a) Schematic diagram of a submicron manipulation system in SEM chamber; (b) a photo of the manipulator. (Adapted from [42] with permission.)

In order to achieve the submicron manipulation in an SEM, three key technologies are needed: 3D imaging technology in the SEM, small force sensing technology, and nanopositioning technology. In 1993, Morishita [42] realized such a system by combining these three techniques in a field emission SEM chamber with the magnification as high as 100 k. The system structure is schematically illustrated in Fig. 10.17a. On the left-hand side is a positioning robot that holds the sample stage. It can provide coarse/fine motion along 3D axes with a resolution of 10 nm in the space of $20 \times 20 \times 20$ mm^3. On the right-hand side is a 3-axis fine positioning robot with a turret at the end for changing tools. The full range of motion is 15 μm and the resolution is 10 nm. A tungsten or diamond probe was fixed on the tip of the fine positioning robot. The force applied on the probe during manipulation is measured by a force sensor at the root of the probe and can be displayed outside the SEM chamber. A photo of this micromanipulator is shown in Fig. 10.17b. This whole unit was located under the electron beam so the manipulation can be observed in situ by the SEM.

A more advanced micromanipulation system has been developed a few years later [43]. Figure 10.18 shows the system which consists of primary and secondary manipulators, a worktable, and an optical microscope equipped with a CCD camera. The primary manipulator is in charge of moving and rotating the subject; while the secondary manipulator helps in fixing the object on the worktable. The base frame of the manipulators and worktable can be rotated and horizontally moved with respect to the SEM for centering the observation point. All components are installed in the SEM vacuum chamber. With this configuration, more freedom has been applied on the stages so that manipulation can be more easily observed using the SEM and a better control can be achieved.

3.2. Photonic Crystals Fabricated by Micromanipulation

By using the system introduced above, objects in the size of a few hundreds nanometers can be freely moved and arranged into any arbitrary structure. PCs with well-controlled local arrangement can thus be constructed by laying up the building blocks one by one. Some examples of 2D and 3D PCs will be presented in this part.

3.2.1. 2D Lattices of Microspheres

This example has been previously shown as one type of 2D PC structures. In the micromanipulation system, a fine glass rod with a tip diameter of 700 nm was used as the probe, which was coated with a thin layer of gold. The building blocks are 2 μm polyvinyltoluene spheres. Silicon substrate was used for supporting, onto which a 300-nm thick Si_3N_4 membrane was coated to prevent the electrification by allowing the primary electron beam to penetrate and to prevent optical interference during optical measurements. In such a small scale, the spheres are sticky to the probe, the substrate, or other spheres because electrostatic or van der Waals forces are much stronger than the gravity force. In SEM, the acceleration voltage of the electron beam could adjust the electrostatic forces to make the adhesion force between spheres and the substrate becoming stronger than the force between spheres and the probe [44]. Experimentally, 10 kV electron beam enabled the easy picking up and depositing operations of the spheres.

Under the in situ observation of SEM, only the spheres with a diameter mismatch below ±0.5% were chosen from the sphere pool to build the 2D lattice. The hexagonal pattern has been shown in Fig. 10.11a. This precise fabrication process

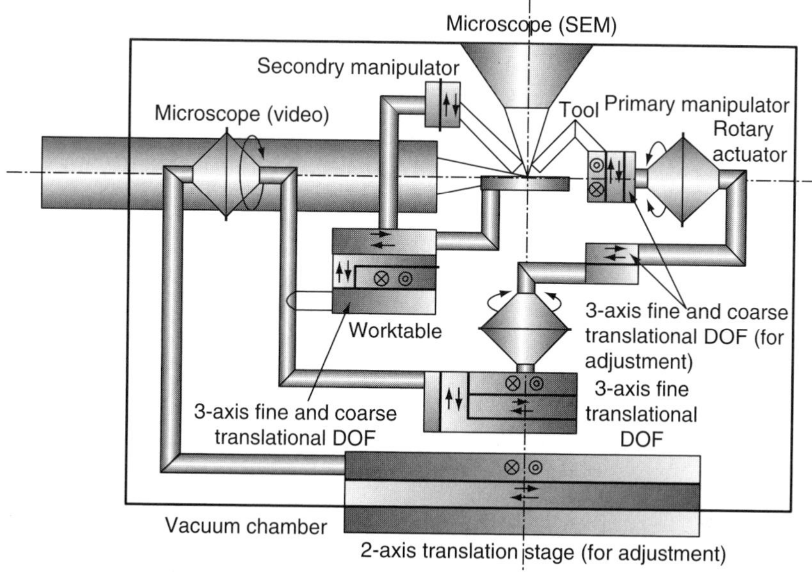

FIGURE 10.18. Schematic diagram of the second generation of the submicron manipulation system. (Adapted from [43] with permission.)

provided a new pathway to systematically investigate various photonic band effects in a finite PC system.

3.2.2. Diamond Architectures of Microspheres

The diamond lattice is a nonclose-packed structure, which is almost impossible to be fabricated through the colloidal self-assembly techniques, although a full 3D PBG has been predicted for this structure. By using the micromanipulation system, silica spheres can be built into diamond lattice utilizing latex spheres as a temporary supporting scaffold [45]. Since the body center cubic (BCC) lattice is formed by two interpenetrating diamond lattices, the diamond lattice can thus be realized by assembling a BCC lattice of mixed silica and latex spheres of equal diameter followed by removal of the latex spheres. In the assembly process, the spheres of two materials were picked one by one and placed at the predefined pattern on the substrate to form the first layer. The upper layers were stacked by putting one sphere in a stable location among four supporting spheres underneath. Figure 10.19a shows the final BCC structure composed of 165 latex spheres and 177 silica spheres. The resulting diamond lattice after plasma etching latex spheres is shown in Fig. 10.19b.

Since the SEM observation involves the formation of an amorphous carbon-rich contaminant film, silica spheres may be firmly attached after focusing the electron beam on the junctions for a few seconds [46]. Therefore, the latex supporting will no longer be required for fabricating the silica sphere diamond lattice. Figure 10.19c shows a five-layer (001) orientated diamond lattice directly

FIGURE 10.19. SEM images of a 3D PC structure made by the submicron manipulation system. (a) Body center cubic lattice composed of 165 latex spheres and 177 silica spheres; (b) diamond lattice of silica spheres after etching latex spheres; (c) five-layer (001) orientated diamond lattice directly made from silica spheres; and (d) contacting effects between two silica spheres induced by electron beam contamination. (Adapted from [45] with permission.)

made from silica spheres, where the contacting points are fixed by the electron beam contamination. The contacting situation is shown in Fig. 10.19d. This method reveals novel and complex PC lattices for experimental studying of the corresponding photonic properties.

3.2.3. Woodpile Structure

Other than lithography technique, woodpile structure has also been fabricated by micromanipulation [47], where the building blocks were prefabricated 2D photonic plates, as shown in Fig. 10.20a. Unlike the probe for assembling spheres, multiple glass needles were used in the system. The location of each plate is confined by predeposited polystyrene spheres in the fiducial holes. The picking-up and putting-down of plates were also controlled by varying the accelerating voltage of SEM beam. The final structure of the 3D woodpile lattice is shown in Fig. 10.20b. This technique created a much more accurate and maybe more efficient method for fabricating 3D PCs with a repeating planar structure.

4. Summary

This chapter introduced the basic theory and fabrication techniques of PCs. As a versatile and convenient tool, SEM is playing an important role in the characterization and fabrication of PCs. Their lattice structure and periodicity are always

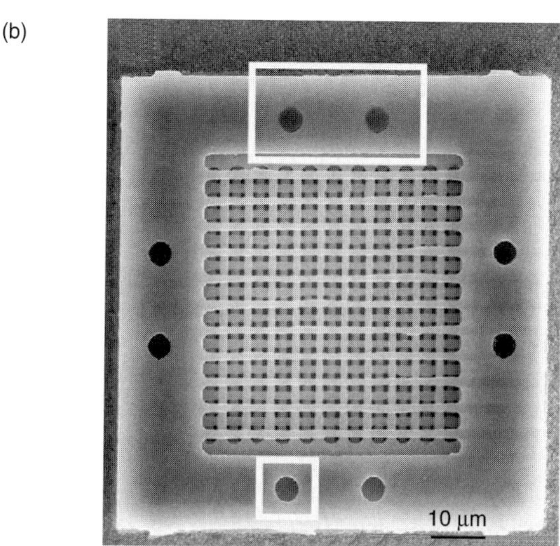

FIGURE 10.20. (a) SEM image of four 2D slabs as building blocks for woodpile 3D PCs; (b) a woodpile 3D PC formed by the 2D slabs, where their position are fixed by inserting polystyrene spheres in the fiducial holes. (Adapted from [47] with permission.)

determined by SEM investigation. The typical SEM images of different types of PCs have been illustrated and explained in this chapter. Micromanipulation technique has been improved by combining the probe stages with SEM, so the nanosized objects can be manipulated under in situ SEM observation. By this means, more precise controlled local arrangement or more complex structures of PCs such as diamond lattice have been achieved. In general, SEM is an indispensable machine for researchers to study and improve the properties of PCs.

References

1. E. Yablonovitch, *Phys. Rev. Lett.*, 58 (1987) 2059.
2. S. John, *Phys. Rev. Lett.*, 58 (1987) 2486.
3. J. D. Joannopoulos, P. R. Villeneuve, and S. Fan, *Nature*, 386 (1997) 143.
4. K. M. Ho, C. T. Chan, and C. M. Soukoulis, *Phys. Rev. Lett.*, 65 (1990) 3152.
5. J. D. Joannopoulos, R. D. Meade, and J. N. Winn, *Photonic Crystals*, Princeton University Press, Princeton, NJ (1995).

6. S. John and K. Busch, *J. Lightwave Technol.*, 17 (1999) 1931.

7. Y. Zijlstra, E. W. J. M. van der Drift, M. J. A. de Dood, E. Snoeks, and A. Polman, *J. Vac. Sci. Technol. B*, 17 (1999) 721.

8. S. Y. Lin, E. Chow, V. Hietala, P. R. Villeneuve, and J. D. Joannopoulos, *Science*, 282 (1998) 274.

9. A. Talneau, L. Le Gouezigou, N. Bouadma, M. Kafesaki, C. M. Soukoulis, and M. Agio, *Appl. Phys. Lett.*, 80 (2002) 547.

10. M. M. Sigalas, R. Biswas, K. M. Ho, C. M. Soukoulis, and D. D. Crouch, *Phys. Rev. B*, 60 (1999) 4426.

11. R. J. Tonucci, B. L. Justus, A. J. Campillo, and C. E. Ford, *Science*, 258 (1992) 783.

12. R. F. Cregan, B. J. Mangan, J. C. Knight, T. A. Birks, P. St. J. Russell, P. J. Roberts, and D. C. Allan, *Science*, 285 (1999) 1537.

13. H. Y. Ryu, H. G. Park, and Y. H. Lee, *IEEE J Sel. Top. Quantum Electron.*, 8 (2002) 891.

14. H. Hirayama, Y. Hamano, and Y. Aoyagi, *Appl. Phys. Lett.*, 69 (1996) 791.

15. S. Noda, N. Yamamoto, M. Imada, H. Kobayashi, and M. Okano, *J. Lightwave Technol.*, 17 (1999) 1948.

16. Y. Tanaka, Y. Sugimoto, N. Ikeda, H. Nakamura, K. Kanamoto, K. Asakawab, and K. Inoue, *Appl. Phys. Lett.*, 86 (2005) 141104.

17. N. Kawai, K. Inoue, N. Carlsson, N. Ikeda, Y. Sugimoto, K. Asakawa, and T. Takemori, *Phys. Rev. Lett.*, 86 (2001) 2289.

18. S. J. McNab, N. Moll, and Y. A. Vlasov, *Opt. Express*, 11 (2003) 2927.

19. M. Loncar, T. Doll, J. Vuckovic, and A. Scherer, *J. Lightwave Technol.*, 18 (2000) 1402.

20. X. D. Wang, C. J. Summers, and Z. L. Wang, *Nano Lett.*, 4 (2004) 423.

21. T. Baba, K. Inoshita, H. Tanaka, J. Yonekura, M. Ariga, A. Matsutani, T. Miyamoto, F. Koyama, and K. Iga, *J. Lightwave Technol.*, 18 (2000) 1402.

22. T. Martensson, P. Carlberg, M. Borgstrom, L. Montelius, W. Seifert, and L. Samuelson, *Nano Lett.*, 4 (2004) 699.

23. S. W. Leonard, H. M. van Driel, K. Busch, S. John, A. Birner, A. P. Li, F. Muller, U. Gosele, and V. Lehmann, *Appl. Phys. Lett.*, 75 (1999) 3063.

24. S. W. Leonard, J. P. Mondia, H. M. van Driel, O. Toader, S. John, K. Busch, A. Birner, U. Gosele, and V. Lehmann, *Phys. Rev. B*, 64 (2000) R2389.

25. H. T. Miyazaki, H. Miyazaki, K. Ohtaka, and T. Sato, *J Appl. Phys.*, 87 (2000) 7152.

26. J. Rybczynski, U. Ebels, and M. Giersig, *Colloids Surf. A*, 219 (2003) 1.

27. S. H. Park and Y. Xia, *Langmuir*, 15 (1999) 266.

28. M. Trau, D. A. Saville, and I. A. Aksay, *Science*, 272 (1996) 706.

29. A. Dimitrov and K. Nagayama, *Langmuir*, 12 (1996) 1303.

30. R. Biswas, M. M. Sigalas, G. Subramania, and K. M. Ho, *Phys. Rev. B*, 57 (1998) 3701.

31. A. A. Zakhidov, R. H. Baughman, Z. Iqbal, C. Cui, I. Khayrullin, S. O. Dantas, J. Marti, and V. G. Ralchenko, *Science*, 282 (1998) 897.

32. A. Blanco, E. Chomski, S. Grabtchak, M. Ibisate, S. John, S. W. Leonard, C. Lopez, F. Meseguer, H. Miguez, J. P. Mondia, G. A. Ozin, O. Toader, and H. M. van Driel, *Nature*, 405 (2000) 437.

33. Y. A. Vlasov, X. Z. Bo, J. C. Sturm, and D. J. Norris, *Nature*, 414 (2001) 289.

34. S. Y. Lin, J. G. Fleming, D. L. Hetherington, B. K. Smith, R. Biswas, K. M. Ho, M. M. Sigalas, W. Zubrzycki, S. R. Kurtz, and J. Bur, *Nature*, 394 (1998) 251.

35. S. Noda, K. Tomoda, N. Yamamoto, and A. Chutinan, *Science*, 289 (2000) 604.

36. J. G. Fleming, S. Y. Lin, I. El-Kady, R. Biswas, and K. M. Ho, *Nature*, 417 (2002) 52.
37. M. Campbell, D. N. Sharp, M. T. Harrison, R. G. Denning, and A. J. Turberfield, *Nature*, 404 (2000) 53.
38. C. Cuisin, A. Chelnokov, J. M. Lourtioz, D. Decanini, and Y. Chen, *Appl. Phys. Lett.*, 77 (2000) 770.
39. G. Feiertag, W. Ehrfeld, H. Freimuth, H. Kolle, H. Lehr, M. Schmidt, M. M. Sigalas, C. M. Soukoulis, G. Kiriakidis, T. Pedersen, J. Kuhl, and W. Koenig, *Appl. Phys. Lett.*, 71 (1997) 1441.
40. O. Toader and S. John, *Science*, 292 (2001) 1133.
41. S. R. Kennedy, M. J. Brett, O. Toader, and S. John, *Nano Lett.*, 2 (2002) 59.
42. H. Morishita and Y. Hatamura, *Proceedings of the IEEE/RSJ International Conference on Intelligent Robots and Systems* (1993) 1717.
43. K. Koyano and T. Sato, *Proceedings of the IEEE Interernational Conference on Robotics and Automation* (1996) 2541.
44. H. Miyazaki and T. Sato, *Adv. Robotics*, 11 (1997) 169.
45. F. Garcia-Santamaria, H. T. Miyazaki, A. Urquia, M. Iisate, M. Belmonte, N. Shinya, F. Meseguer, and C. Lopez, *Adv. Mater.*, 14 (2002) 1144.
46. H. W. P. Koops, J. Kretz, M. Rudolph, M. Weber, G. Dahm, and K. L. Lee, *Jpn. J. Appl. Phys.*, 33 (1994) 7099.
47. K. Aoki, H. T. Miyazaki, H. Hirayama, K. Inoshita, T. Baba, N. Shinya, and Y. Aoyagi, *Appl. Phys. Lett.*, 81 (2002) 3122.

11
Nanoparticles and Colloidal Self-assembly

Gabriel Caruntu, Daniela Caruntu, and Charles J. O'Connor

1. Introduction

Nanotechnology is a burgeoning area of scientific research which comprises the production and manipulation of materials at the nanometer scale. Nanomaterials are commonly defined as solids whose constituent dimensions range between 1 and 100 nm, though the scale is usually extended to several hundreds of nanometers. As an example, the average width of a human hair is on the order of 10^5 nm whereas a single red blood cell is ~5,000 nm and a DNA molecule has a diameter of 2–12 nm. Nanotechnology is a synergism of physical, chemical, biological, and engineering concepts having as a common characteristic the nanometer size. Considered of a pure scientific interest a few years ago nanoparticles are nowadays commonplace for the development of new cutting-edge applications in communications, data storage, optics, energy storage and transmission, environment protection, biology, and medicine due to their relevant optical, electrical, and magnetic properties. Their original properties, not encountered in the case of their bulk counterparts, can lead to a potentially tremendous scientific and technological progress. Moreover, all these potential applications are expected to impact profoundly every aspect of modern living. Unlike their bulk counterparts, nanomaterials present a reduced dimensionality associated with a very high surface/volume ratio that increases with decreasing particle size.

Consequently, a large fraction of the constituting atoms will be found at the surface of the particles, rendering them highly reactive and inducing specific properties. Because the intrinsic properties of the nanoparticles, such as composition, crystallinity, size, and surface topography are crucial for their physical properties, the control of the structural characteristics through the chemical synthesis is highly desired. As the nanoparticle size/shape plays a crucial role on the unique material properties, studies of the particle size and shape evolution could be useful in providing a mechanistic insight into the manipulation and control of their properties and functionalities. Thus, for the design of novel advanced functional nanomaterials the uniformity of the shape and size of the nanoparticles is also a key issue. Nanoparticles are considered monodisperse when their size distribution is <5%. In the past decade, a particular interest has been paid to the development of new synthetic routes enabling the rigorous control of the morphology and size of the nanoparticles. To be used in advanced technologies, magnetic nanoparticles are particularly required to be highly dispersible in various media since they have the tendency to cluster and precipitate, which drastically reduce their efficiency.

The preparation of nanoparticles can be achieved through different approaches, either chemical or physical including gaseous, liquid, and solid media. While physical methods generally tend to approach the synthesis of nanostructures by decreasing the size of the constituents of the bulk material (top-down approach), chemical methods tend to attempt to control the clustering of atoms/molecules at the nanoscale range (bottom-up approach). However, the chemical methods are the most popular, because they present several major advantages over the other conventional methods, i.e., they are highly reliable and cost effective, they allow a much more rigorous control of the shape and size of the nanoparticles and the agglomeration of the resulting particles can be alleviated by functionalization with different capping ligands. Since the literature pertaining to the synthesis of nanoparticles is vast and covers almost all types of materials, including some authoritative reviews on the preparation of metals, semiconductors and magnetic nanomaterials [1–5] an attempt to cover both the synthesis and characterization of the nanostructured materials is pointless. We will therefore give a brief survey on the use of the scanning electron microscopy on the characterization of nanostructured materials as well as providing some relevant examples covering different classes of nanostructured materials.

It is noteworthy to mention that because of its lower magnification, the scanning electron microscopy gives information at a large scale about the size and morphology of the nanoparticles. However, for a complete characterization of the structural characteristics of the nanocrystalline materials, the scanning electron microscopy for nanoparticles is usually performed in conjunction with other experimental techniques, including atomic force microscopy (AFM), transmission electron microscopy (TEM), x-ray diffraction (XRD), as well as adsorption-desorption (BET) isotherms.

2. Metallic Nanoparticles

Metallic nanoparticles are important for both theoretical and practical viewpoints. They represent one of the most promising classes of nanomaterials by virtue of their interesting optoelectronic, thermal, magnetic, and superior catalytic properties, which often rival those of the bulk material. Unlike the bulk metals, in the case of metal nanoparticles the conduction band is absent being replaced by discrete states at the band edge because of the quantum-confinement of the electrons. Consequently, metallic nanoparticles are of great theoretical importance because they can serve as model systems for the development of new theories. As the spectrum of the potential applications of metallic nanoparticles is diversifying every day, there is a tremendous interest in preparation of metal nanoparticles with uniforms sizes and shapes by cost-effective and easily scalable synthetic procedures.

Rakhimov et al. prepared metallic nanoparticles (Fe, Co, and Ni) dispersed into a polymer matrix by exposing reaction mixtures containing metal salt precursors, a polymer and a reducing agent ($NaBH_4$), to the action of elastic wave pulses [6]. As early as 1994, Thadhani introduced the idea of solid-state chemical reactions in powder mixtures induced by compressive shocks [7]. According to this concept, the exposure of a powder to a shock induces major structural rearrangements which, in turn, would lead to an increase in the reactivity of the powder and finally to a high-rate chemical reaction between the powdered precursors. Therefore, two main processes can be distinguished, i.e., shock-assisted and shock-induced reactions, respectively. While the shock-assisted processes occur via solid-state defect-enhanced diffusion after unloading to ambient pressure, the shock-induced reactions occur during shock compression upon mechanical equilibration and before unloading to ambient pressure. The mechanisms of shock-assisted reactions include solid-state defect-enhanced diffusional processes whereas for the shock-induced reactions, mechanisms different from conventional solid-state nucleation and growth processes are dominant. Rheological explosion creates elastic waves which propagate throughout the polymer matrix in which the precursors have been uniformly distributed via extrusion with formation of the desired metallic nanostructures. Figure 11.1 shows typical scanning electron microscope (SEM) micrographs of polymer/metallic nanoparticles composites exposed to the action of an elastic wave with and without adding the reduction agent. While in absence of the reduction agent the mechano-chemical reaction does not occur, upon addition to the reducing agent, particles with a variable size, ranging between 10 and 100 nm are observed (some of them indicated by arrows in Fig. 11.1b).

As inferred from the SEM micrographs, one of the major drawbacks of this method is that the size of the nanoparticles is not controllable through variation of the pressure of the elastic wave and usually the resulted metallic nanoparticles

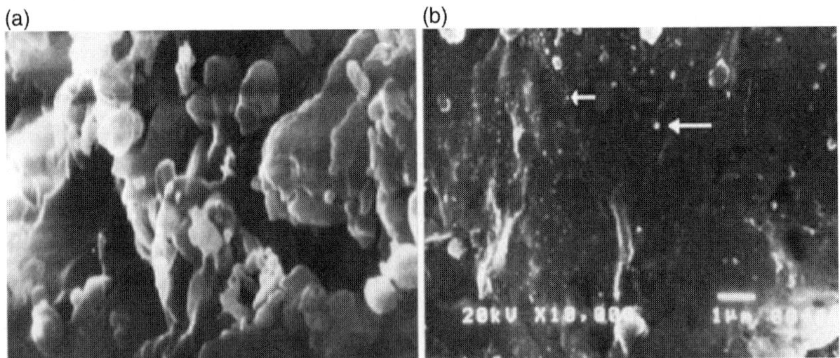

FIGURE 11.1. SEM micrographs of two samples containing a mixture of polyethylene (PE) and $CoCl_2\,6H_2O$ (a) and a mixture of polyethylene, $CoCl_2\,6H_2O$ and $NaBH_4$, respectively. Both samples have been subjected to the action of elastic waves at $P = 250$ kbar. Reproduced from reference [6]. © 2004 American Institute of Physics.

present a wide size distribution although experimental evidence has shown that the pressure influences the yield of the metallic nanoparticles. In addition to nanocrystalline metals, transition metal alloys engineered as thin films with tunable size, shape, and interparticle spacing are very promising materials for the development of the new generation ultra-high-density magnetic recording media. Fiévet and coworkers have investigated the morphology of nanocrystalline Fe–Co–Ni and Co–Ni alloys obtained by the polyol technique [8, 9]. Nanocrystalline alloys were prepared by homogeneous/heterogeneous nucleation from the reduction of the corresponding transition metal acetates mixed in stoichiometric ratio in polyols at moderate temperatures of 195°C for several hours.

Then, NaOH is added to the polyol which will increase the monodispersity of the nanoparticles and, also will accelerate the chemical reactions. As illustrated in Fig. 11.2, the resulted nanoparticles appear as white, uniform, nonaggregated, and nearly monodisperse spherically shaped entities with a size of ~200–260 nm and a relatively narrow size distribution ($\sigma < 15\%$).

Because the energy-dispersive x-ray (EDX) microscopy allows the examination of bigger particles or clusters with a much higher accuracy than the TEM, scanning electron microscopy is a powerful tool to investigate the chemical composition of individual particles, as well as their surface topography. EDX analysis can be used in conjunction with XRD and magnetic measurements to determine the compositional uniformity of the nanostructured materials. Accordingly, Fiévet and coworkers found that the resulted CoNi alloys are roughly chemically homogeneous throughout the whole volume of the sample, although a small concentration gradient can be detected across the particle. Such a compositional difference was ascribed by the authors to the different kinetic mechanisms of the reduction and precipitation of the two metals by the polyol. Insights into the mechanism of the self-assembly of FePt nanocrystals prepared by the thermal decomposition/reduction of a palladium precursor mixed into a stoichiometric ratio with $Fe(CO)_5$ in polyalcohols were gained by Murray and coworkers by using a combination of

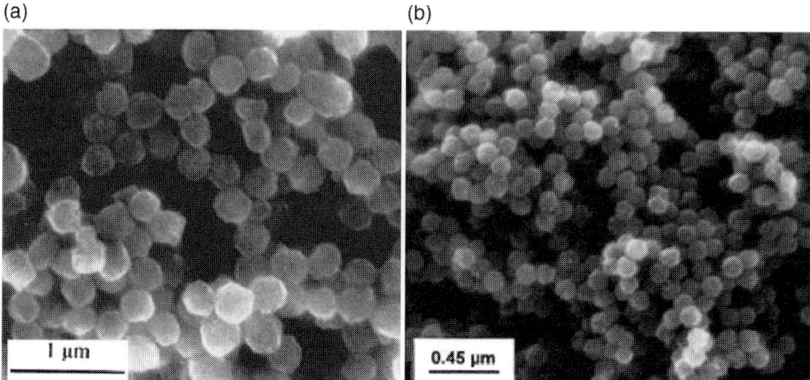

FIGURE 11.2. Typical SEM image of $Co_{65}Ni_{35}$ (a) and $Fe_{14}Co_{43}Ni_{43}$ (b) powders prepared by the polyol process. Reproduced from reference [8]. © 2004 American Institute of Physics.

microscopy techniques, including the transmission and scanning electron microscopy [10]. While the chemical composition of the Fe_xPt_{1-x} nanocrystals $(48 \times <70)$ can be tuned by simply varying the molar ratio of Fe/Pt in the initial reagents, the size of the nanoparticles can be controlled by a seed mediated-growth process. Initial monodisperse ~3 nm FePt nanoparticles were subsequently used as seeds for the preparation of 10 nm nanocrystals. Figure 11.3 illustrates the TEM and SEM micrographs of FePt nanocrystals self-assembled onto a substrate.

Figure 11.3b shows the TEM pattern of a 3D cubic superlattice produced by the deposition of FePt nanocrystals coated with carboxylic acids/amine onto a copper substrate and followed by the slow evaporation of the solvent. The nanocrystals are spherical, equally spaced with an average size of 6 nm, very narrow size distribution

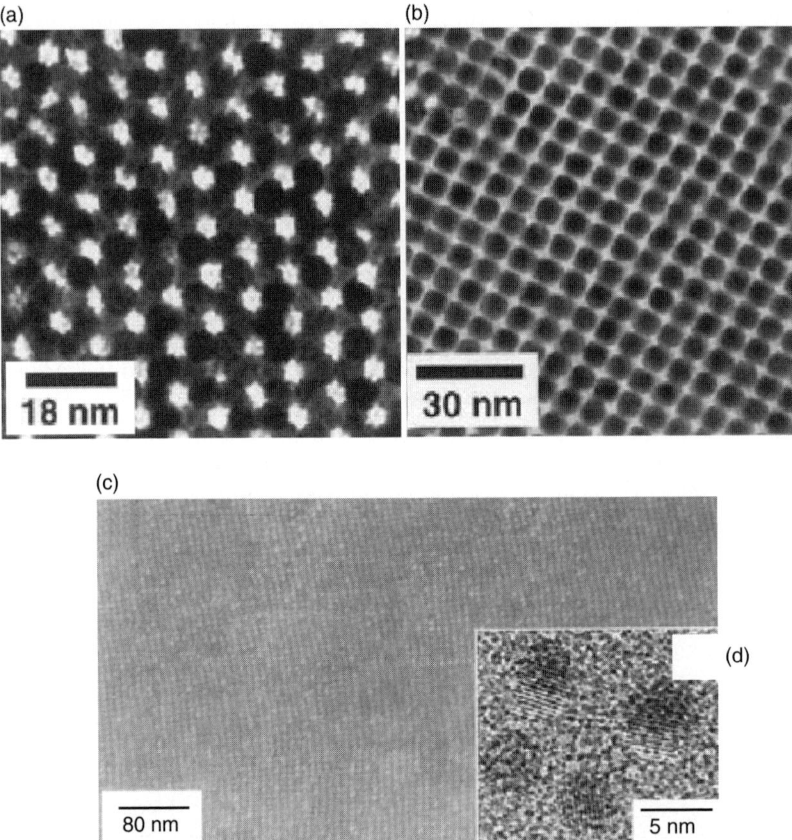

FIGURE 11.3. TEM micrograph of a 3D assembly of 6-nm $Fe_{50}Pt_{50}$ sample after replacing the mixture of oleic acid/oleyl amine with hexanoic acid/hexylamine (a and b). HRSEM image of (a); 180-nm-thick, 4-nm $Fe_{52}Pt_{48}$ nanocrystal assembly annealed at 560°C for 30 min under 1 atm of N_2 gas (c). High-resolution TEM image of 4-nm $Fe_{52}Pt_{48}$ nanocrystals annealed at 560°C for 30 min on a SiO_2-coated copper grid (d). Reproduced from reference [10]. © 2000 American Association for the Advancement of Science.

and an interparticle distance of 1 nm. It is noteworthy to mention that the interparticle distance can be finely tuned from 4 to 1 nm (as shown in Fig. 11.3b) by exchanging the initial capping ligand (oleic acid/oleyl amine) with shorter hydrocarbonated molecules, such as other acids or amines. As revealed by the microscopy studies, in addition to the interparticle distance, the ligand exchange process affects the packing mode of the self-assembled of the nanocrystals; similar to the cobalt nanocrystals when passivated with oleic acid molecules the FePt nanocrystals are hexagonally packed, whereas the replacement of the oleic acid with the hexanoic acid leads to a cubic close packed superlattice of the FePt nanoparticles [11]. Figure 11.3c illustrates a typical high-resolution scanning electron micrograph of a ~180 nm-thickness ensemble of self-assembled FePt nanoparticles subjected to a thermal treatment at $560^{\circ}C$ for 30 min under a protecting atmosphere of N_2 which prevents the oxidation of the sample. While the nanoparticles still remain nonaggregated, the interparticle distance was found to decrease from ~4 to ~2 nm with the onset of a coherent strain of the supperlattice. Such a strain is presumably ascribed to the different thermal coefficients of the FePt nanoparticles and the copper substrate, respectively. Figure 11.3d also represents a high-resolution TEM picture of the individual, heat-treated FePt nanocrystals revealing their high crystallinity and the retention of their spherical shape upon the annealing.

SEM was intensively used as an investigative tool by Penner and coworkers who proposed a general, highly versatile synthetic approach for the size-selective electrodeposition of a wide variety of nanostructured coinage metals (Cd, Cu, Ag, Au, Ni, and Pt) onto conductive surfaces [12]. The growth of the individual nanoparticles is independent on the nearby particles and proceeds slowly upon applying two successive voltage pulses to the corresponding plating solutions. While the first voltage pulse will initiate the nucleation of particles on the substrate, the second pulse will enable them to grow to the desired size. Additionally, when the depletion layer surrounding each particle is reduced or completely eliminated, a much lower nucleation density for the metal nanoparticles will be created. Indeed, the mechanistic pathway which allows such a phenomenon to occur is the lower density of nuclei, which, in turn can be induced by decreasing the deposition time. Spherically shaped metal nanoparticles with different sizes and a narrow size distribution were deposited onto a graphite electrode (Fig. 11.4). While the smallest particles are Cd (d ~100 nm), the biggest nanocrystals observed by scanning electron microscopy are Cu (d ~2,000 nm). By using the oxidative electrodeposition of the metals, the authors were also successful in depositing nanocrystalline metal oxides onto different substrates, as suggested for MoO_2.

The crystalline nature of the metallic nanoparticles can be inferred from the careful examination of the high-resolution SEM micrographs. Penner observed that the Au and Ag nanoparticles deposited electrochemically at low current potentials present a faceted morphology, which provides strong evidence about the formation of single-crystals in the incipient steps of the synthetic process [13]. Furthermore, the existence of the structural defects on the substrate can be exploited by directing the growth process of the metal nanoparticles to occur at the step edges of the graphite. Upon growing, the metal nanoparticles will align at

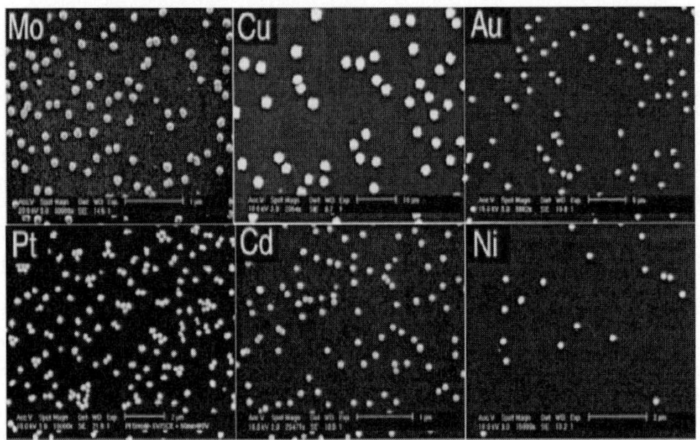

FIGURE 11.4. Typical scanning electron micrographs of metal particles prepared using the slow growth method by using different deposition current densities ranging between 5 and 260 µA cm^{-2}. Reproduced from reference [12]. ©2001 Elsevier Science Ltd.

the step edges due to a higher nucleation density to eventually coalesce and form metallic nanowires. As observed by SEM, the diameter of the nanowires was found to vary between 60 and 750 nm, whereas their length is ~100 µm [14–17]. This method referred to as the "electrochemical step edge decoration" (ESED) consists of applying three successive current pulses to the plating solutions. While a first oxidative "activation" pulse will oxidize the step edges of the substrate before the deposition, the second "nucleation" pulse, with a large overpotential induces the reduction of the corresponding metals from their aqueous salts.

Finally, a small "growth" pulse corresponding to a small overpotential will trigger the growth of the metallic nanoparticles and their self-assemblage into nanowires. The adjacent scheme depicts the synthetic strategy for the fabrication of metallic or metal oxide nanowires through the growth and coalescence of individual particles on a graphite substrate by the "step-edge decoration" process. As shown in Fig. 11.5, the diameter of the copper nanowires varies between 73 and 340 nm with increasing the deposition time from 120 to 2,700 s, whereas their length is ~150 µm. A 2D attachment of metallic nanoparticles deposited by the reduction of the corresponding metallic salts in aqueous solutions can lead to continuous films. These films can be further patterned via lithography for incorporation in various nanoarchitectures suitable for different advanced technological applications. Buriak and coworkers have deposited various noble metal nanoparticles onto germanium substrates via a galvanic displacement process [18]. As shown in Fig. 11.6, continuous, highly uniform and well adherent metal films can be formed via deposition of individual spherical particles with a diameter which can be controlled by the variation of the reaction conditions; that is the concentration of the precursors as well as the deposition time.

Xia and coworkers have extensively used SEM studies to investigate the factors influencing the shape and the size of the metal nanoparticles prepared by the

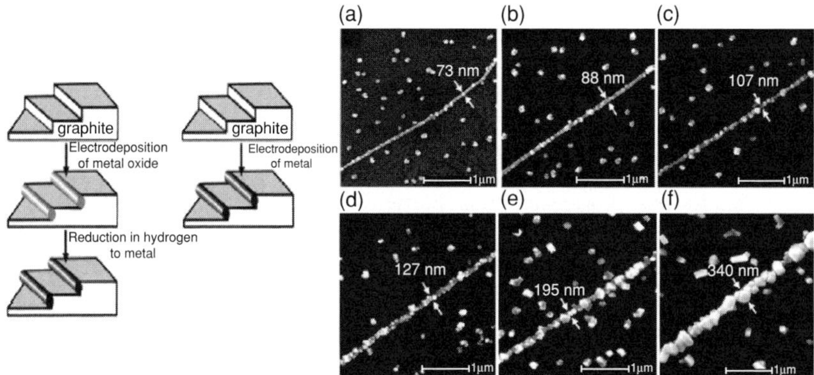

FIGURE 11.5. Typical SEM images of copper nanowires prepared by the "step-edge decoration method." Figures (a–f) show the time evolution of the copper nanowires, where the reaction time was varied from 120 s (a) to 180, 300, 600, 900 and 2,700 s, respectively (Fig. 5a–f). Reproduced from references [14–17]. © 2002 the American Chemical Society and © 2001 the American Association for the Advancement of Science.

FIGURE 11.6. SEM images of nanoparticle films deposited onto a Ge (100) surface: (a) Au; (b) Pd; (c) Pt (deposition time: 90 min); (d) Pt; (e) Zn; and (f) Cu (deposition time: 120 min). Reproduced from reference [18]. © 2004 American Chemical Society.

polyol process [19–22]. They prepared silver nanoparticles by reducing the silver nitrate with ethylene glycol at 148°C both in absence and presence of an ionic salt, such as NaCl or KCl. Experimental evidence has shown a dramatic change in the mechanistic pathways upon addition of an ionic salt to the reaction precursors. The changes occurring during the reaction were monitored by observation of the color changes in conjunction with TEM/SEM investigation of the reaction intermediates. In the absence of the ionic salt and when the reaction was performed in air, after 5 min the reduction of the silver precursor by the polyol generates twinned grains with an irregular shape and a size of ~20 nm, whereas a prolonged heating up to 20 h results into the dissolution of the small initial particles with formation of spherically shaped, nearly monodisperse single crystalline particles.

As the reaction time is extended up to 45 h, the size of the nanoparticles increases to ~ 60 nm, process which is accompanied by a shape evolution from spherical to cubes and tetrahedrons with truncated edges and corners (Fig. 11.7b). However, the poly-(vinylpyrrolidone)-capped nanoparticles preserve their monodispersity, as well as their single-crystalline nature, as inferred from the corresponding selected-area electron micrographs shown in the insets of Fig. 11.7b. After 46 h of reaction, the size of the silver nanoparticles increases to 80 nm and they appear as truncated polyhedrons. The size evolution of the nanoparticles was confirmed from the interpretation of the corresponding UV-Vis spectra, which indicated a broadening of the absorption peak appearing at $\lambda \approx 40$ nm, which has been presumably ascribed to the enhanced scattering of the bigger nanoparticles.

The role of the oxidative environment as well as the presence of the ionic salt in the reaction solution on the morphology of the products is emphasized: whereas when NaCl is dissolved in the polyol under inert atmosphere, the initial twinned nanocrystals subsequently form long, uniform nanowires (Fig. 11.8a), when NaCl is replaced by KCl the resulting Ag nanocrystals preserve the morphology adopted in the case when they are prepared in absence of an ionic salt. The different mechanistic pathways observed in the two cases suggest that the Cl^- ions are involved in the process, rather than the alkali cations. The formation of the nanowires proceeds rapidly (<90 min) and is believed to occur through the alignment and coalescence of the individual truncated particles, whereas in the case of KCl-mediated synthesis the shape evolution of the resulting Ag nanoparticles is similar to that when no ionic salt was added to the reaction solution. Moreover, the conversion of the twinned crystals into the single crystalline nanoparticles is presumably ascribed to their higher reactivity which engenders two concurrent processes: a preferential oxidative dissolution of the twinned crystals and the reduction/nucleation of new particles (Fig. 11.8). The etching process of the twinned Ag nanocrystals can therefore be accelerated upon performing the reaction under oxidative environment, as well as upon adding Cl^- to the reaction solution.

Though the role of the anion is not completely elucidated, the SEM/TEM observations strongly suggest that the Cl^- ions will coordinate the Ag nanoparticles,

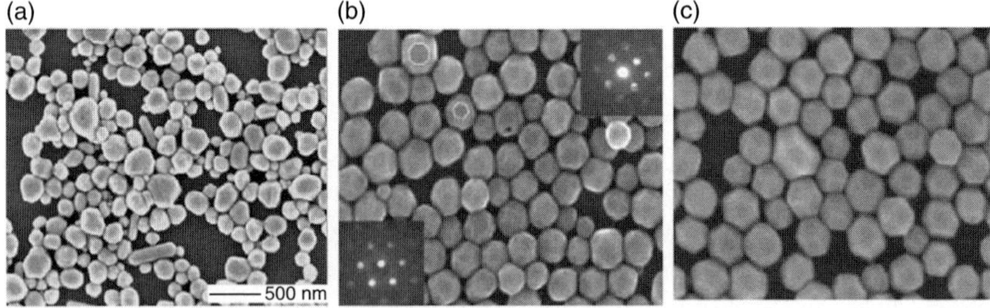

FIGURE 11.7. SEM images of Ag samples collected at 60 min (a), 45 h (b), and 46 h (c). *Insets* show SAED patterns corresponding to different orientations of electron beam: perpendicular to the (100) facet of a truncated cube (*upper right*) and the (111) facet of a truncated tetrahedron (*lower left*). Reproduced from reference [20]. © 2004 American Chemical Society.

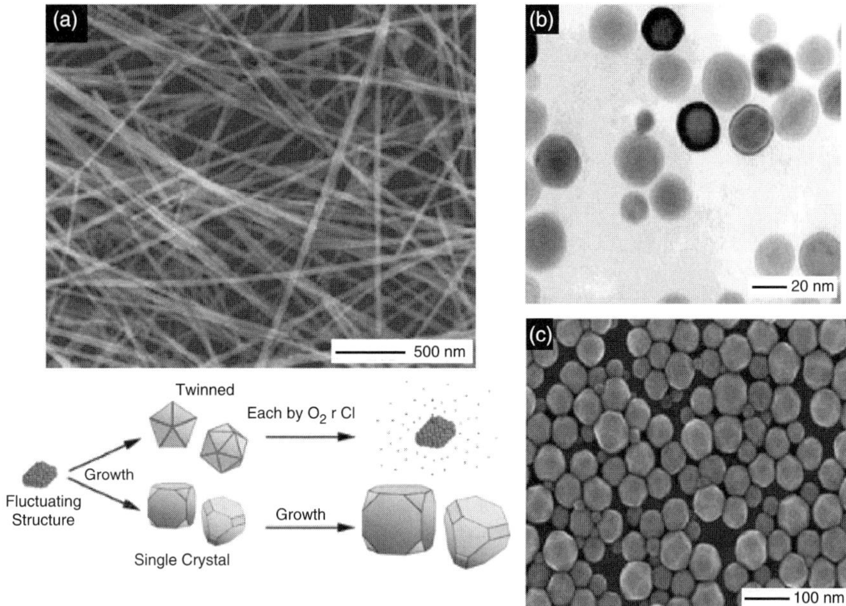

FIGURE 11.8. SEM image of a sample taken from the same reaction mixture at 90 min. In the absence of oxygen, the twinned decahedrons in this system grew into long, uniform nanowires (a) SEM (b), and TEM (c) images of nanoparticles formed at 52 h and 20 min, and 54 h, respectively. Reproduced from reference [20]. © 2004 American Chemical Society.

thereby accelerating the oxidative etching of the multiple twinned nanocrystals and promoting the formation of single-crystalline particles [23–25]. Xia also demonstrated that mastering of the reaction conditions in the polyol method can result in the control of the shape of the resulted metallic nanocrystals. They showed that the shape of silver nanocrystals can be changed from spherical to cubic when the ratio between the reaction precursor/polymer is varied [26, 27]. When the concentration of the $AgNO_3$ was increased threefold in the reaction solution and the molar ration $AgNO_3$/poly(vinyl pyrrolidone) (PVP) was kept at 3:2 nanocubes are obtained instead of wires or truncated polyhedrons. However, when the molar ratio between $AgNO_3$ and PVP exceeded 3:2 or the reaction temperature was increased above 120°C, the reaction resulted in the formation of irregularly-shapped twinned Ag nanocrystals. Consequently, control over the size of the nanocubes can be achieved by varying the reaction time. Figure 11.9 shows low-magnification (Fig. 11.9a) and high magnification (Fig. 11.9b) SEM micrographs of the reaction product which demonstrate the formation regularly faceted PVP-coated Ag nanocubes with an average edge of ~175 nm and a standard deviation $\sigma = 13\%$ which to self-assemble on the substrate to form 2D arrays.

However, the in-depth investigation of the morphology of the Ag nanocubes shows that they present slightly truncated edges (Fig. 11.9b) and can be described

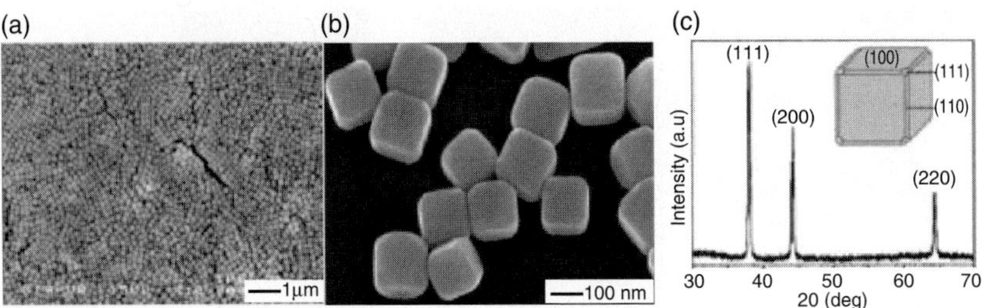

FIGURE 11.9. SEM and TEM images of silver nanocube (a, b). X-ray diffraction pattern of silver nanocubes (c). The *inset* in (c) is a scheme of the nanocube morphology. Reproduced from reference [19]. © 2002 the American Association for the Advancement of Science.

as in the inset of the Fig. 11.9c. Interestingly, the cubic shape of the Ag nanocrystals is presumably ascribed to the preferential adsorption of the poly(vinyl pyrrolidone) on particular crystallographic planes, thus enhancing the growth rate along the <111> directions, which results in the formation of these cubes [28, 29]. Such a morphology agrees well with the well-known tendency of the nanocrystals to grow preferentially along crystallographic directions occurs which minimize the surface energy. Single-crystalline spherical particles with sizes <20 nm are generally considered by being formed by polyhedrons with faces corresponding to high-index crystallographic planes. The higher crystallographic index of the surface crystallographic planes, the higher is the surface energy of the corresponding particle. To reach a thermodynamic equilibrium during the growth process, the system will have the tendency to follow that reaction pathway which minimizes its free energy and, consequently, the initial nuclei will grow preferentially in particular crystallographic directions whose surface energies are minimal. Speculation of the allure of the XRD pattern indicated that the peaks match with those of the fcc bulk Ag with a cell parameter of 4.088 Å, close to the value reported in literature. Although most nanocrystalline metals tend to crystallize into twinned crystals with faces formed by the lowest energy, calculation of the ratio between the (200), (111) and (220), (111) planes evidenced that the faces of the Ag nanocubes are planes of the {100} family. Thus, the obtained values were systematically found to be higher that these observed for the bulk material (0.67 vs. 044 for the first pair of peaks and 0.33 vs. 0.25 for the second pair of XRD peaks). Xia also proposed a new general synthetic approach for the preparation of hollow metallic nanoarchitectures with controllable size and shape by the oxidation of the more electropositive nanocrystyalline metals used as sacrificial templates [26, 27]. As for the shape of the resulting particles, they are a replica of the parent crystals with a size which in general is 20% bigger than that of the templates. This synthetic pathway is summarized in Scheme 11.1.

According to the Scheme 11.1, the galvanic replacement of the more electropositive metal is pictured as a four-step process, including the mixing of the colloidal solution of the more electropositive metal nanoparticles (Ag) with the noble metal (Au), followed by the initiation of the galvanic reaction which will

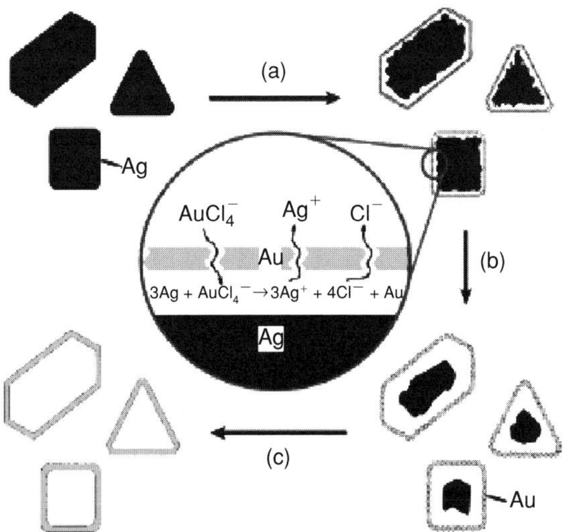

SCHEME 11.1. Synthetic strategy used by Xia for the preparation of metallic hollow nano-structures through galvanic dissolution of more electropositive metals. Reproduced from reference [27]. © 2002 the American Chemical Society.

progressively etch the more active metal with formation of a thin shell of the more noble metal at the surface of the individual particles which serve as a template. As the thickness of these shells increases, the core of such clusters will be consumed during the galvanic reaction with formation of hollow structures of the more noble metal. This general approach can be virtually extended to any couple of metals for which the oxidative dissolution of the more electropositive metal is accompanied by the reductive deposition of shell-like nanostructures of the more noble metal. Xia used this "template-engaged replacement reaction" for the preparation of different nanostructured metallic hollow structures, including Au, Pd, Pt whose aqueous solutions were refluxed in presence of a colloidal solution of Ag nanoparticles, according to the standard reduction potentials presented in Table 11.1. Experimental investigation have evidenced that the metallic shells are permeable to the active chemical species in the solution which promote the oxidation–reduction reactions and which can diffuse through the shells from/to the reaction solution generating the consumption of the more active metal. Figure 11.10 shows typical TEM and SEM micrographs of the starting Ag nanoparticles which appear as having irregular shapes and an average size of 50 nm (Fig. 11.10a) and the corresponding Au hollow nanostructures obtained by the galvanic reaction, respectively.

The resulting Au nanocrystals appear as retaining the shape of the sacrificial templates with surfaces free of defects and having a hollow structure as inferred from the different degrees of darkness observed for their core and edges, respectively.

Because the growth of the hollow nanostructures takes place in the vicinity of each individual sacrificial template, the shape of the resulting metallic shell-like nanocrystals can be varied from spherical to cubic or tube-like. Figure 11.11

TABLE 11.1. Standard reduction potentials for the metals used in the template-engaged replacement reaction to prepare metallic hollow nanostructures

Metal	Reduction reaction	Standard reduction potential (V)
Ag	$Ag^+ + e^- = Ag$	0.80
Pd	$Pd^{2+} + 2e^- = Pd$	0.83
Au	$Au^{3+} + 3e^- = Au$	0.99
Pt	$Pt^{2+} + 2e^- = Pt$	1.2

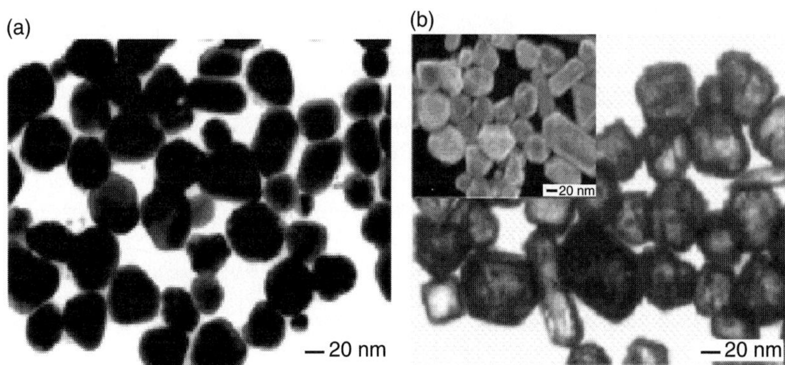

FIGURE 11.10. TEM pattern of silver nanoparticle (a). Representative SEM micrograph of Au nanoparticles with hollow interiors (b). Reproduced from reference [27]. © 2002 American Chemical Society.

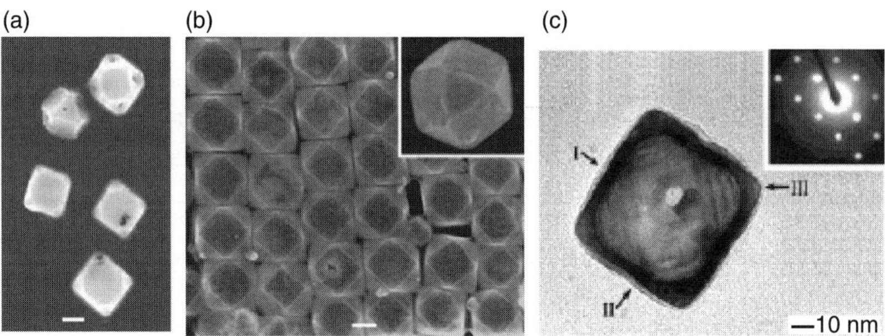

FIGURE 11.11. SEM and TEM images of cubic silver nanoparticles synthesized using a solution-phase approach. The insert shows the SAED pattern recorded by directing the electron beam perpendicular to the top surface of an individual silver nanocube (a, b). Representative TEM image of a cubic gold nanoshell (c). Reproduced from reference [19]. © 2002 American Association for the Advancement of Science.

displays the corresponding micrographs of the Ag nanocubes (Fig. 11.11a) and the resulting Au nanoboxes (Fig. 11.11b).

Figure 11.11a shows the morphology of the original Ag nanocubes after reacting with 0.3 mL of $HAuCl_4$ solution whereas Fig. 11.11b shows the same particles after reacting with 1.5 mL of HAuCl$_4$ 10^{-3} M solution. The SEM micrograph shown in Fig. 11.11a reveals the formation of a thin Au shell on the surface of the Ag nanocubes together with the existence of some black spots which have been identified as pinholes. These pinholes are presumably permeable to the mutual diffusion of the chemical species from the solution, thereby promoting the galvanic reaction. Moreover, the careful investigation of the SEM micrograph shows that the position of these pinholes allowed for the identification of the crystalline planes involved in the galvanic reaction. Thus, the reaction was found to occur in a sequential fashion, involving the crystalline planes {110}, {100} and {111}, which is in good agreement with the decreasing order of their free energies [29]. As seen in the corresponding SEM micrograph, the Au nanoboxes self-assemble spontaneously to form a 2D array on the microscope grid and appear as cubical with truncated edges. This truncated morphology shown in the inset of the Fig. 11.11b suggests the coexistence of two different sets of faces: eight triangular faces, each corresponding to an edge of the original nanocube and six square faces, corresponding to its faces, respectively. The diffraction spots in the selected-area electron-diffraction (SAED) pattern (inset of Fig. 11.11c) indicate that each individual gold nanobox is single-crystalline in nature and presents well-defined faces corresponding to the {100} crystallographic planes. The next step in implementing these highly ordered arrays of nanometer-sized metallic particles consists of their use as various platforms for the design of different nanodevices.

The Murray group at IBM have explored intensively synthetic pathways for the preparation of high-quality metallic nanocrystals and studied their physical properties, including devices based on self-assembled metallic nanoparticles with high potential in high-density magnetic data storage applications. Spherically shaped ε-cobalt nanocrystals with a narrow size distribution were prepared by the reduction of a cobalt precursor with lithium-triethylborohydride at 200°C in presence of oleic acid and trioctylphosphine as capping ligands [30]. Then, the resulting nanoparticles self-assemble spontaneously into hexagonally packed superlattices when the colloidal solution is deposited onto a substrate followed by the slow evaporation of the solvent. Electric measurements were performed by constructing to electrodes on the highly periodically nanocrystalline cobalt arrays by electron beam lithography and lift-off. As illustrated in Fig. 11.12a, the SEM micrograph of the device the ~10 nm clearly shows that the device is constructed by self-assembled cobalt nanoparticles forming one to three monolayers-thick 2D superlattices on top of which the two electrodes are eventually attached. Although the distance between the two electrodes consists of few cobalt nanocrystals it can be varied in a narrow range and thus influencing with a great extent the tunneling properties of the nanodevice.

Upon annealing the sample at 500°C under a reducing atmosphere, the interparticle distance of the initial self-assembled nanocrystals (Fig. 11.12b) decreases from 4 to 2 nm, whereas the size of the nanoparticles is preserved (Fig. 11.12c). Since colloidal self-assembled metallic nanocrystals are of high interest in electronic devices, the shape control, in addition to the tunability of

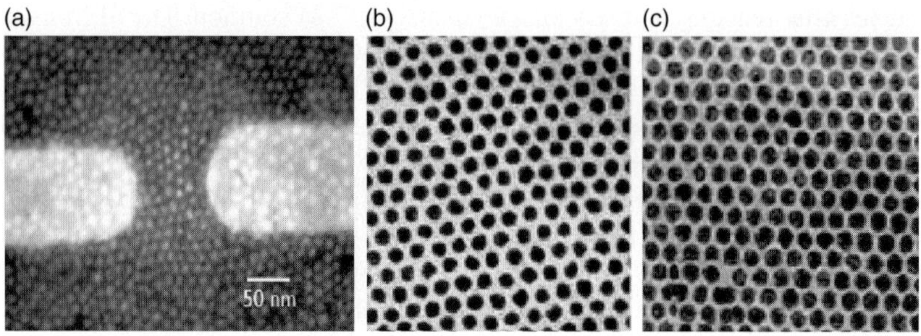

FIGURE 11.12. SEM micrograph of the metallic electrodes fixed onto a cobalt superlattice (a); TEM images of cobalt superlattice before annealing (b); and after annealing, respectively(c). Reproduced from reference [30]. © 2000 American Association for the Advancement of Science.

the interparticle distance has proven critical for the tailoring of the physical properties of such devices. Highly ordered 3D-superlattices formed by ~7 nm cubic iron nanocrystals were studied by Chaudret and coworkers [31]. They proposed a new synthetic route for the preparation of the nanocrystalline iron metal-organic iron precursor, $Fe[N(SiMe_3)_2]_2$, based on the ability of the amido-complexes to form 3D superlattices [32]. The iron precursor was reduced with hydrogen at 150°C in presence of hexadecylamine and oleic acid as capping agents. Then the iron nanocubes self-assemble onto a substrate to form extended 3D superlattices, which can be visualized by both transmission (Fig. 11.13a) and scanning electron microscopy (Fig. 11.13b).

By analogy with the formation of Co nanorods via the thermal decomposition of a cobalt-organic precursor [33], the authors suggested that the growth of the iron nanocubes is based on the clustering and coalescence of spherical nanoparticles initially formed in the reaction solution. The subsequent growth of these nanoparticles results in the formation of iron nanocubes which are primarily organized in two dimensions, but eventually self-assemble spontaneously into 3D-superlattices. The interparticle separation distance was found to vary from 1.6 to 2 nm, which agrees well with the presence of the organic molecules at the surface of the nanocubes. However, unlike the cobalt nanocrystals, where the size of the particles can be controlled by varying the length of the capping ligands (amines or acids), the size of the iron nanocubes the size is the same regardless of the nature of the organic molecules used in the reaction. This suggests that the capping ligand is the hexametyldisilazane resulting from the chemical reaction which eventually remains attached to the surface of the resulting nanocrystals. The 3D-superlattices can be further redissolved in suitable solvents and deposited as monolayers. Active metals are often reactive against the surrounding atmosphere leading to the formation of a thin metal oxide layer on the metal surface, and degrading its physical properties. Johnson and coworkers prepared Fe and Fe_3C nanoparticles encapsulated into carbon cages upon pyrolysis of iron stearate at high temperature (900°C) under inert atmosphere [34].

(a)   (b)

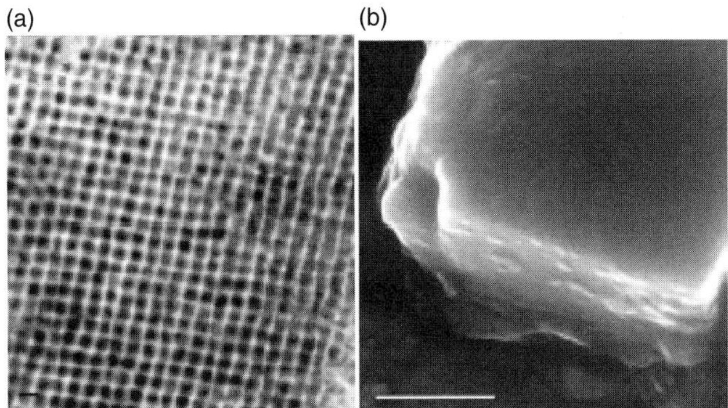

FIGURE 11.13. TEM micrograph of a 3D superlattice as observed from thin slices of the product embedded in an epoxy resin (a). Bar is 10 nm in length; SEM micrograph of a 3D superlattice of nanocubes (b) The length bar is 500 nm. Reproduced from reference [32]. © 2003 American Association for the Advancement of Science.

Speculation of the reaction mechanism involves the presence of the carbon resulted from the chemical reaction thus creating a thin carbon matrix which surrounds the resulting iron nanoparticles. Figure 11.14 shows a typical SEM image of the iron nanoparticles with a size varying between 20 and 200 nm embedded into a graphite matrix.

Upon pyrolysis of the iron organic precursor, the iron nanoparticles form and growth, process which is accompanied by the diffusion and agglomeration of the graphite nanoclusters onto the metallic surfaces. The HRTEM micrograph shows no grain boundaries for the metallic cores, indicating that they are single-crystalline in nature. The carbon protective layer is also single-crystalline, as revealed by TEM investigations when fringes corresponding to the $(10\,l)$ planes are clearly identified (Fig. 11.14b).

(a)   (b)

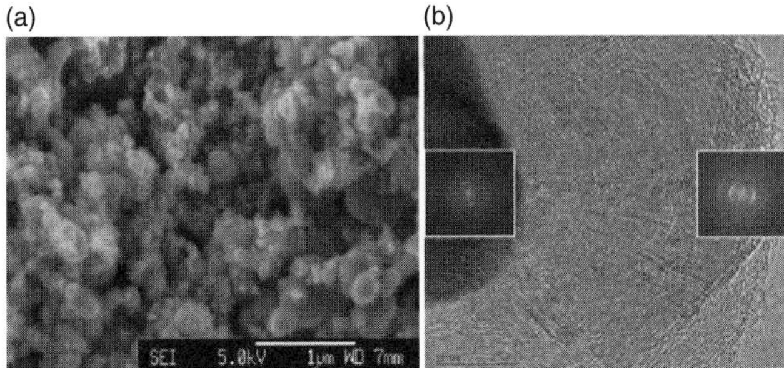

FIGURE 11.14. SEM micrograph of the carbon-encapsulated iron nanoparticles (a); HRTEM image of an iron nanoparticle embedded into the graphite matrix (b). The *insets* are the corresponding SAED patterns of the metallic core and the carbon shell, respectively. Reproduced from reference [34]. © 2004 American Chemical Society.

3. Mesoporous and Nanoporous Metal Nanostructures

A special class of materials in the nanotechnology landscape is represented by the metallic architectures with ordered arrays or uniform nanopores. These materials present the remarkable ability to adsorb high amounts of ions, atoms or molecules and, consequently, can be used as host templates for infiltrating different materials such as metals or semiconductors. The species filling the voids of the host material will retain its symmetry layout via self-organization, thereby facilitating the formation of highly ordered 3D nanoarchitectures. Nanoporous materials find a widespread utility in advanced technological applications by replacing the existing porous materials fabricated via conventional methods such as plasma sputtering or metallurgy. Among their potential applications, they are suitable for the design of new thermoelectric, optical and photonic devices, catalysts, fuel cells, gas sensing and storage media, microfluidic flow control, gas sensing and selective separation. However, a key issue in the fabrication of such nanophase materials is the design of materials with a controllable nanopore size, since particular technological application require materials with a well-defined and uniform pore size. Several effective synthetic routes to design nanoporous metallic materials have been proposed in the last decade, either chemical or electrochemical. Sotiropoulos and coworkers deposited mesoporous nickel into the voids of a microporous polymer matrix template by electroplating followed by the removal of the polymer matrix upon annealing [35]. Although these mesoporous structures present high surface areas, the voids generated during the electrodeposition are not uniform, they possess different cross sections which, in turn, will hamper the free movement of gaseous and liquid species through the pore network and will drastically alter their catalytic properties. Consequently, the design of metallic materials with uniform nanoporous morphology is highly desired for their further integration in the catalytic processes. Colvin and coworkers reported a general synthetic route based on Au-catalyzed electroless deposition of metals to prepare macroporous Ni, Cu, Ag, and Au films with high internal surface area. The method uses colloidal SiO_2 as colloidal templates constructed by a close packed arrangement of ~200 nm spherical particles which are subsequently removed by etching with HF [36]. According to this procedure, colloidal silica is functionalized with a silane 3-mercaptopropyltrimethoxysilane (3-MPTMS) which leaves thiol-terminated surfaces of the colloidal silica (Scheme 11.2).

The colloidal silica nanocrystals are then self-organized onto a substrate (glass) via a convective self-assembly process and ~5 nm gold nanoparticles are subsequently immobilized onto the surface of the SiO_2 nanospheres by the attachment of the thiol moiety when the silane functionalized silica is treated with a diluted gold solution. The resulting Au/SiO_2 nanocomposite material is therefore dried and immersed into an electroless plating solution when the metal infiltrates directly into the interstitial spaces of the Au/SiO_2 nanocomposites [37, 38]. The

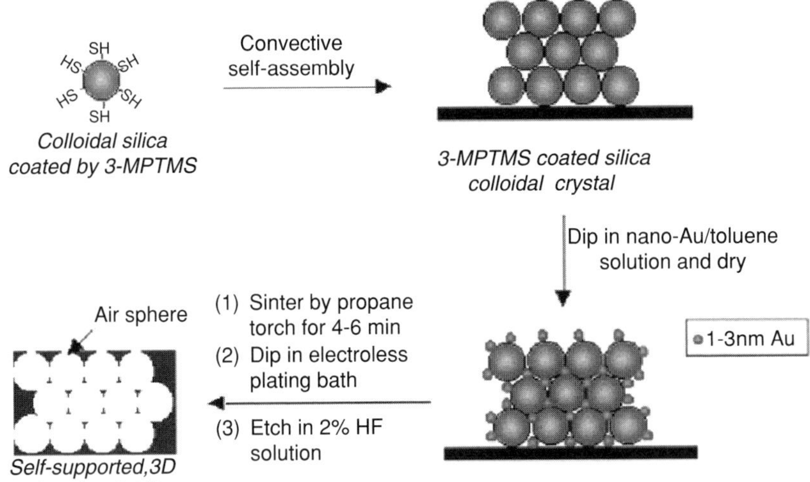

SCHEME 11.2. Synthetic pathway to the preparation of metal nanosphere arrays by using a non-conductive matrix. Reproduced from reference [36]. © 1999 the American Chemical Society.

Au nanoparticles attached to the colloidal SiO_2 spheres serve as a catalyst for the electroless deposition of the metal. The SEM micrographs of the nanoporous metallic films of Pt, Cu, and Ni (Fig. 11.15) demonstrate that they are inverse replica of the parent silica nanospheres with ~318 nm voids for Ni, ~325 nm for Cu, and ~353 nm voids for Ni, respectively.

As shown in the insets of Fig. 11.15a and b, each cavity is connected to its nearest 12 neighbors through pores with a diameter of ~60 nm, indicating that the connecting pores form in the regions where the original silica spheres were in contact [39–41]. However, one of the drawbacks of this method is represented by the fragility against sonication and thermal treatment of the nanoporous metal 3D networks. The metal films can be handled only when they are sufficiently thick and in most cases a post-synthesis thermal treatment results in the fracture of the films and/or destruction of their porosity. However, the porosity of these nanoparticulate films can be greatly decreased upon annealing at relatively high temperatures, thereby increasing the grain size. Wiley and coworkers have adapted this procedure to prepare highly ordered 3D-periodical nanoscale metal (Ni, Pd, and Au) and semiconductor (Pb, Bi, Sb, and Te) arrays. The two-step template-assisted strategy is analog to that used for the fabrication of 1D porous membranes [42–44]. According to the first approach, a close-packed opal structure serves as nanomold for the infiltration of metallic nanospheres by electrodeposition through a multistep synthetic protocol (Scheme 11.3). In the first step, a porous metallic mesh is obtained by electrochemical deposition of the metal into the void spaces of a SiO_2 opal composed by ~290 nm spheres, followed by the dissolution of the opal. The metallic mesh is then converted into the corresponding oxide, by subsequent oxidation in air at 550°C.

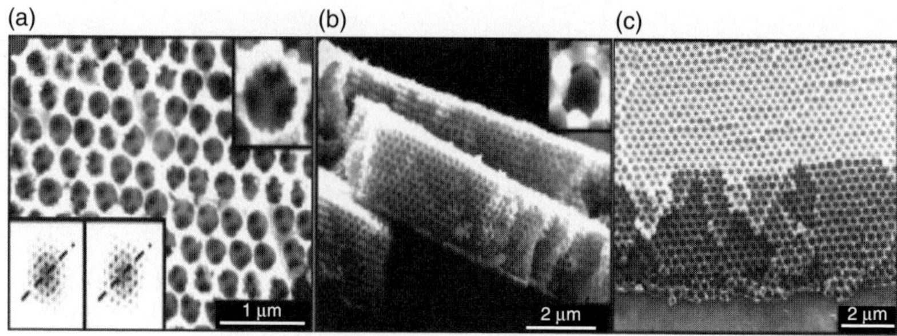

FIGURE 11.15. SEM top-down view of nanosphere metal arrays fabricated by electro-chemical deposition using the void spaced of inverse opal spheres as "nanomolds." Reproduced from reference [36]. © 2004 Royal Society of Chemistry.

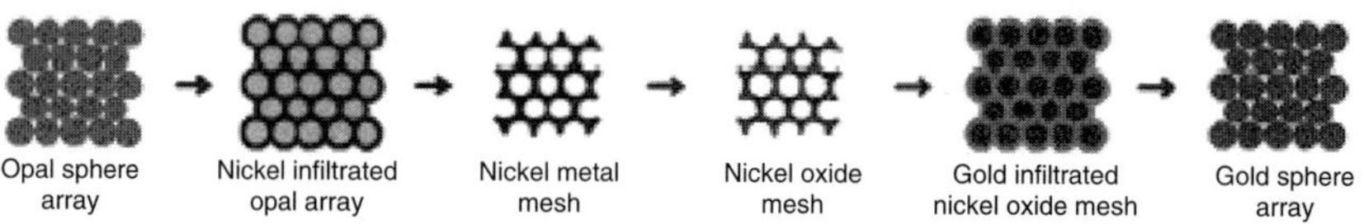

SCHEME 11.3. Synthetic pathway to the preparation of metal nanosphere arrays by using a nonconductive matrix. Reproduced from reference [44]. © 2004 the American Chemical Society.

The oxide is a poor electrical conductor and can be further used as a template for the electrochemical deposition of a second metallic nanosphere array. After the electrodeposition of the second metal, the metal oxide mesh is dissolved upon treatment with a dilute acidic solution. Figure 11.16 depicts the top-down SEM micrographs of highly ordered nanoporous arrays of different metals obtained by this multistep template strategy.

The second method consists of drawing a molten metal directly into the inter-stitial space of a carbon inverse opal obtained by a phenolic process, which acts as a solid matrix during the process (Scheme 11.4). The SEM micrographs of Te and Bi nanosized spheres obtained by melt infiltration are shown in Fig. 11.17. As seen in the images, the metallic crystals consist of a close-packed arrangement of ~200 nm spheres which reproduce with a high fidelity the interstices of the conductive carbon template.

Interestingly, when carbon infiltrates the opal matrix material, in some cases it can coat the internal surfaces, instead of filling the void spaces of the SiO_2 crys-tal. This will result in the formation of two different types of channels in the car-bon mesh that is a percolated spherical array of voids as well as a percolated network of octahedral and tetrahedral interstices [44]. Subsequent melt infiltra-tion of a metal into these two different channels results in the formation of a nanocrystalline metal array which reproduces a cubic close-packed structure sim-ilar to that observed for NaCl. Figure 11.17c shows a SEM micrograph of a Bi sphere array where the bigger spheres are placed in the corners of the unit cell

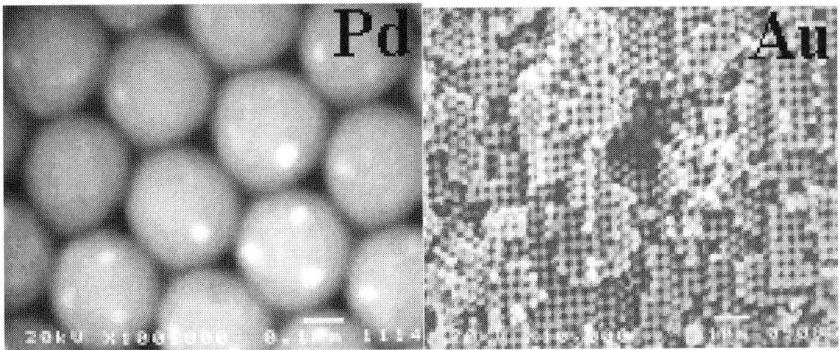

FIGURE 11.16. SEM top-down view of nanosphere metal arrays fabricated by electrochemical deposition using the void spaced of inverse opal spheres as "nanomolds." Reproduced from reference [42]. © 2004 Materials Research Society.

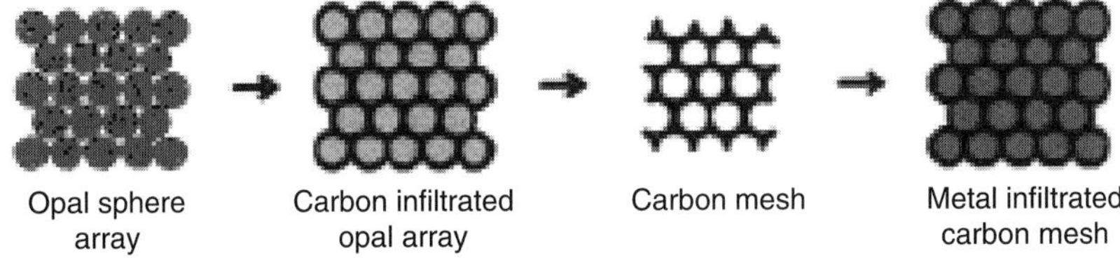

SCHEME 11.4. Synthetic pathway to the preparation of metal nanosphere arrays by using a nonconductive matrix. Reproduced from reference [44]. © 2001 the American Chemical society.

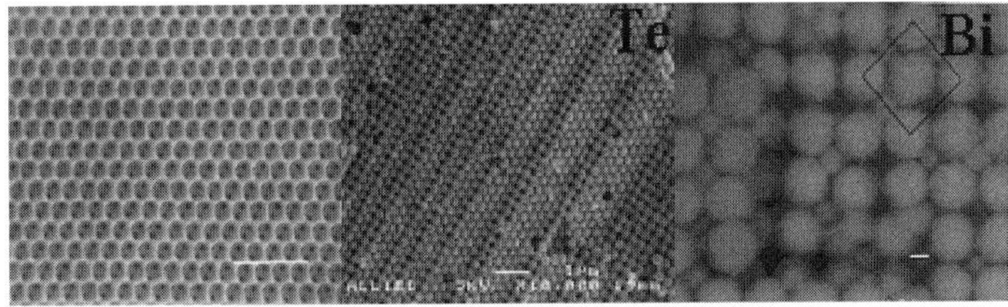

FIGURE 11.17. Fracture surface SEM images of the inverse opal used as a template and Te and Bi nanosphere metal arrays fabricated by melt infiltration. Reproduced from reference [44]. © 2001 the American Chemical Society.

face, whereas the smaller ones are found in the interstitial positions. Control of the size of the metal nanospheres can be achieved through variation of the pressure during the metal infiltration, because the radius of the infiltrated metal decreases upon increasing the pressure. Erlebacher have exploited the selective dissolution (dealloying process) to create nanopores into noble metals [45].

Such nanoporous materials are ideal candidates for the design of high-performance electrodes and devices with potential applications in catalysis and gas sensing [46]. When an alloy is treated with an acid, the more electropositive metal will be dissolved, leaving the inert metal unaltered. The selective deposition of an initial layer of the more electrochemically active metal will create terrace vacancies on the surface, thus leaving the atoms of the less electrochemically active metal with no lateral coordination (adatoms). The adatoms of the noble metal have the tendency to aggregate and cluster thereby forming islands on the surface of the alloy. These islands render the alloy more exposed to the acid solution and will contribute to the acceleration of the dissolution process of the more electrochemically active metal. As the dissolution progresses, an increasing number of adatoms will coalesce and attach to the existing clusters of the noble metal, thus leaving the surface exposed to the electrolyte solution. Figure 11.18a shows a cross-section SEM image of a nanoporous gold sample obtained from the selective dissolution of Ag from a Au/Ag alloy upon treatment with nitric acid, whereas Fig. 11.18b shows the top-down SEM image of the gold film where several 3D pores with a diameter of 20–30 nm and different lengths can be distinguished.

The Monte Carlo simulation of the nanoporous gold agrees well with the experimental SEM observations (Fig. 11.18c). It is worth mentioning that the authors used a complex kinetic model which describes the selective dissolution of Ag by taking into account several physical processes, including the diffusion on the surface and the mass transport of different species in both the solution and throughout the volume of the alloy. A more subtle synthetic approach based on the selective dissolution of the more electropositive metals has been adopted by Erlebacher and allows for the design of nanoporous metals possessing a variable pore size distribution [47]. This method consists of a sequential selective dissolution of the silver from an Au/Ag alloy. Between the two dealloying steps, the resulting nanoporous gold sample is heat-treated to increase the pore size, then silver is subsequently deposited onto the 3D-porous architecture and the resulting alloy is annealed again. While the first heat treatment will produce the larger pores, the second annealing is performed to reconstruct and homogenize the

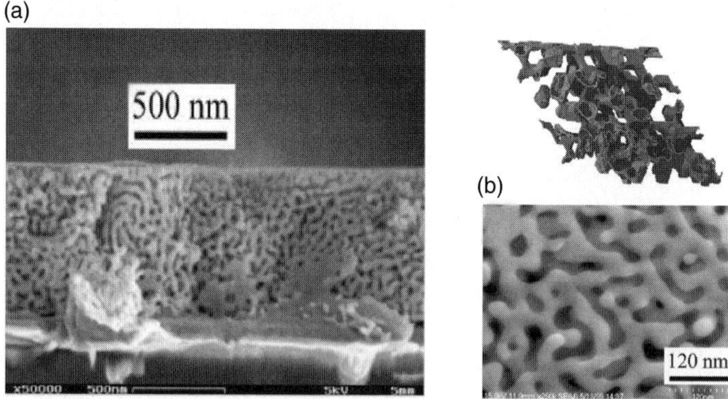

FIGURE 11.18. Representative cross-section SEM image of nanoporous Au obtained by dealloying of Au silver from an Ag–Au alloy (a). Top-down SEM image of the of the nanoporous gold film (b). Simulated gold nanonetworks obtained by kinetic Monte Carlo simulations (c). Reproduced from reference [45]. © 2001 Nature Publishing.

gold–silver interface. In Fig. 11.19 are illustrated plan view SEM images of the nanoporous gold obtained from a commercial 1 μm thick 12-carat gold leaf through this "controlled multimodal pore size distribution method."

As shown in Fig. 11.19a, the dealloying of the gold leaf subjected to the action of the nitric acid for 1 h produces a gold nanoporous structure with a uniform porous network having a diameter of 15 nm. The sample was subsequently annealed at 400°C for 8 h to increase the pore size to ~150 nm and, then the pores were filled with silver reduced from a plating solution under a dilute N_2H_4 atmosphere (Fig. 11.19b). The authors have chosen an unconventional plating process in which the reducing agent is in a gaseous state (N_2H_4) and the nanoporous gold membrane is floating over the plating solution. When the silver is reduced from the solution, it will uniformly fill the porous network of the gold sample leading to the formation of two intertwining networks of gold and silver, respectively. The separation of the two metallic nanonetworks was secured by performing the plating process at room temperature, where any chemical reaction between the two noble metals is precluded. Upon annealing again the sample at 400°C for 4 h and the removal of the plated silver a hierarchical nanoporous architecture possessing

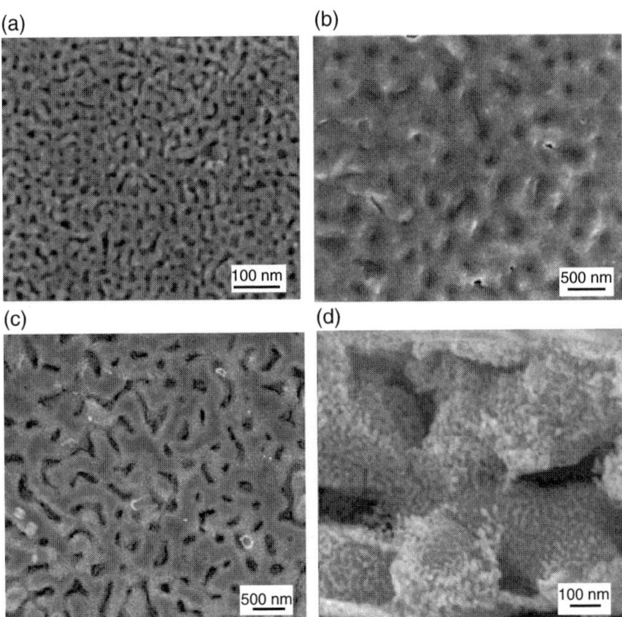

FIGURE 11.19. Typical SEM micrograph of a gold nanoporous network obtained by selective dissolution of Ag from a white gold leaf (a). Large pore NPG membrane made by annealing sample (a) at 400°C for 8 h and filled with silver (b). Hierarchical porous membrane (plan view) made by annealing the sample shown in (b) at 400°C for 4 h, and performing a second dealloying etch in nitric acid for 5 min (c). The image (d) represents the cross-section micrograph of sample (c). Reproduced with permission of the American Chemical Society from Ref. [46].

two different pore-sized nanonetworks will form (Fig. 11.19c). Successive annealing and plating processes could potentially produce nanoporous metallic architectures with multimodal pore size distributions. By using different plating metals, like platinum, Erlebacher also designed highly stable nanoporous membranes through both crystallographically and thermodynamically controlled growth processes. These membranes can be further used for the design of electrodes in proton exchange membrane fuel cells which exhibit performances rivaling those of the commercially available catalytic cells [46].

An alternative two-step approach to produce nanoporous metals has been proposed by Kaneko and coworkers based on the thermal decomposition of transition metal hydroxides under a protective atmosphere [48]. The metal hydroxide was dispersed into poly(vinyl alcohol) (PVA) film by soaking the PVA films into an aqueous solution of the transition metal, followed by the treatment of the resulting samples with a concentrated NaOH solution.

The nanoporous metal/PVA films were obtained by decomposing the metal hydroxides at 700°C under flowing nitrogen. Subsequent annealing of the composite materials results in the complete removal of the poly(vinyl)alcohol with formation of a nanoporous metallic structure. The morphology of the resulting nanoporous metals was studied by SEM and revealed that the films are formed by nanoparticles possessing irregular shapes (Fig. 11.20a). In-depth analysis of the morphology of the nanoparticles was performed by TEM measurements which show that the nanoparticles are aggregated and posses an average size of 10 nm (Fig. 11.20b, c). Furthermore, the morphology of the nickel nanoparticles can also be inferred by combining XRD data with nitrogen BET isotherms. While the crystallite size calculation by using the Scherrer formula [49] indicated a crystallite size of 8.3 nm, the pore distribution curves calculated by using the Dollomore and Heal formalism [50] lead to a particle size of 5.6 nm. These results strongly indicated that each individual ~10 nm nickel nanoparticle posses a porous structure.

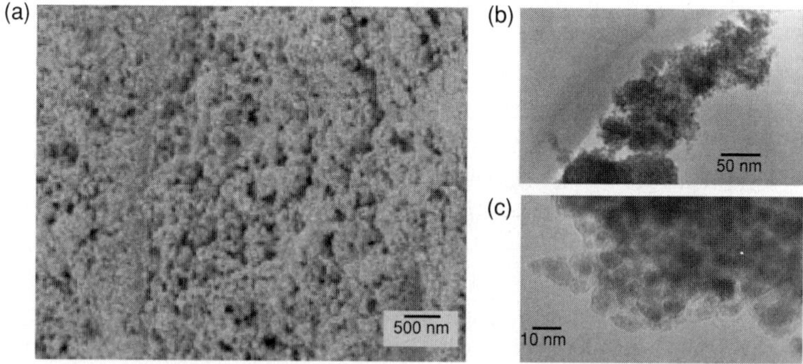

FIGURE 11.20. (a) SEM micrograph of nanoporous nickel obtained from the thermal decomposition of Ni(OH)2 at 700°C; (b) low-magnification TEM micrograph of the nanoporous nickel; and (c) high-magnification TEM micrograph of the sample (b). Reproduced with permission from reference [48]. © 2003 Wiley-VCH Verlag GmbH & Co.

4. Nanocrystalline Oxides

Nanocrystalline oxides are materials of interest for a wide range of technological applications, such as magnetic data storage, catalysis, photonics and communications. Since most of their properties are dependent on the size and shape of the constituting nanoparticles, the study of their morphology by electron microscope techniques serves as an additional tool for development of new synthetic strategies which will allow a better control over their structural characteristics. Thus, a significant progress has been made in the past years on the determination of the morphology of metal oxide nanoparticles by using scanning electron microscopy techniques.

4.1. Nanocrystalline Oxides for Optical Applications

The Colvin group at Rice University has designed photonic crystals with tunable optical properties from close-packed spherical hollow TiO_2 nanocrystals assembled into 3D-architectures possessing controllable size and thicknesses [51]. The hollow ceramic nano and mesoporous crystals assembled into highly ordered arrays are promising materials for photonic applications because they usually exhibit optical properties different to those of the material consisting of the filled spheres. The photonic crystals were obtained through an elegant double-template method which involves the preparation of SiO_2 nanospheres free of defects or grain boundaries whose voids are therefore filled with a macroporous polymer. The silica spheres are then removed and the polymer is further used as a scaffold for the preparation of hollow TiO_2 nanospheres by hydrolysis of a titanium alkoxide. As the polymer is etched from the composite material, the hollow TiO_2 nanospheres will appear as a replica of the polymeric matrix with controllable thickness and degree of overlap. Figure 11.21 illustrates electron microscope images of TiO_2 nanospheres assembled into a 3D architecture.

The TEM image (Fig. 11.21a) shows that the hollow nanoparticles have a diameter of ~150 nm and are spherically shaped, thus possessing uniform surfaces and being free of defects. As shown in Fig. 11.21b, their thickness and degree of overlap can also be controlled since the hydrolysis of the Ti alkoxide

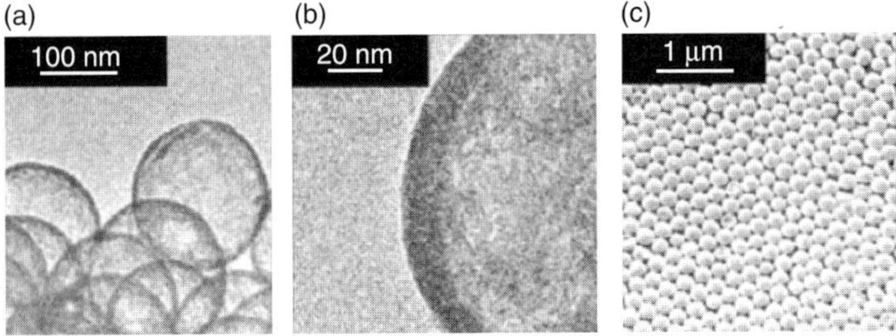

FIGURE 11.21. TEM (a, b) and SEM (c) micrographs of hollow TiO_2 nanospheres fabricated by a double-template technique. Reproduced from reference [51]. © 2001 American Institute of Physics.

occurs into the air voids in the macroporous polymer matrix. Also, the SEM image presented in Fig. 11.21c shows that the TiO_2 nanospheres replicate with high fidelity the shape and the size of the SiO_2 nanoparticles and form a periodic arrangement in the 3D space. Xia and coworkers demonstrated that inverse opal structures can be used as templates for the fabrication of highly ordered porous mesoscale ceramic nanoparticles by using the sol-gel technique [52]. These template-assisted methods arrays exploit the tendency of colloidal particles to self-organize spontaneously into 3D architectures; in general the size control can be achieved by varying the size of the colloidal spheres used as scaffolds. Moreover, the control of the diameter and the wall thickness of the nanoparticles, as well as the quality of the 3D-arrays is the result of the interplay between several factors, including the filling of the interstices of the template, the wetting of the template surface and the strength of de Van der Waals interactions [52]. Similar to the porous metallic nanostructures, highly ordered metal-oxide mesoporous arrays are formed upon the removal of the polystyrene, As seen in Fig. 11.22a, the TiO_2 nanoparticles appear as uniformly spherical, dense and monodisperse spheres with a diameter of several hundreds of nanometers. The TEM image shown in the inset of Fig. 11.22a reveals that each individual nanoparticle has a hollow interior with a wall thickness of ~50 nm. In Fig. 11.22b is illustrated a representative TEM image of the ~200 nm polystyrene beads which are used as templates in the sol-gel process developed by Xia group. The method was first used for the preparation of TiO_2 and SnO_2 arrays of nanoparticles with an average diameter of 380 and 190 nm, respectively, with a wall thickness which can be varied between 30 and 100 nm. The choice of appropriate metal precursors for the sol-gel process allowed the preparation of other nanoporous metal oxides, including SiO_2, SnO_2, ZrO_2, WO_3, Sb_4O_6, WO_3, Al_2O_3, NiO, and Fe_2O_3 [53–59]. Figure 11.23 represents typical SEM micrographs of nonporous SiO_2 and SnO_2 nanoporous obtained by the sol-gel process. They reveal the formation of well-defined structures which consist of the uniform packing of spherical particles with regular interparticle distances which retain the long-range order of the polymeric scaffold.

Scanning electron microscopy is a powerful tool for the investigation of anisotropic nanostructured materials. As an example, ZnO is a hexagonal wide gap semiconductor material whose growth habits are determined by its high crystalline anisotropy, that is its high c/a ratio. Vayssières and coworkers proposed a new synthetic concept for the deposition of nanostructured oxides in the form of nanorods or nanowires by strictly controlling both the kinetics and thermodynamics of the interfacial growth of oxides from aqueous solutions at moderate temperatures [60, 61]. In the case of ZnO, a high potential material for optical, catalytic and electrical applications, they demonstrated that the mastery of its structural characteristics can lead to the formation of nanorods arrayed on the substrate when particles are precipitated from an aqueous solution. Accordingly, they synthesized ZnO nanorods deposited onto various substrates by the thermal decomposition of a Zn–amino complex at mild temperatures. Figure 11.24 shows the SEM micrographs of the resulting ZnO nanorods deposited onto a glass substrate (Fig. 11.24a) and a nanocrystalline ZnO film (Fig. 11.24b). Large arrays of

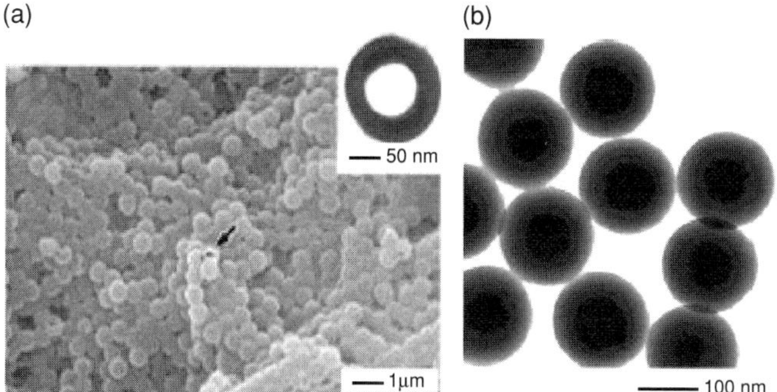

FIGURE 11.22. SEM image of monodispersed TiO_2 hollow spheres produced by templating a sol-gel precursor against a 3D crystalline array of polystyrene beads followed by selective etching in toluene (a). The *inset* shows the TEM image of such a hollow sphere, with a wall thickness of ~50 nm. SEM image of the polystyrene nanospheres used as templates (b). Reproduced from reference [52]. © 2000 Wiley-VCH Verlag GmbH & Co.

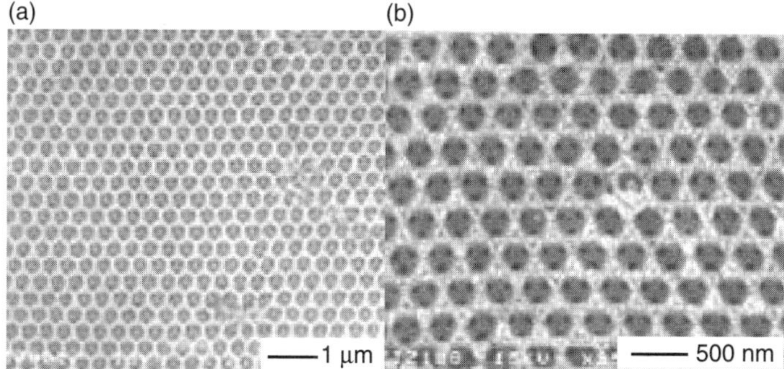

FIGURE 11.23. SEM images of 3D porous materials (a) SiO_2 and SnO_2 (b) prepared by Xia. Reproduced from reference [54]. © 1999 American Chemical Society.

FIGURE 11.24. Typical SEM images of ZnO nanorods deposited by Vayssières onto different substrates—(a) glass and (b) ZnO by chemical growth in aqueous solution. Reproduced from reference [60]. © 2003 Wiley-VCH Verlag GmbH & Co. and reference [63]. © 2001 American Chemical Society.

well-aligned hexagonal ZnO nanorods (inset of Fig. 11.24a) with a length of several microns and a diameter of 100–200 nm uniformly deposited onto the substrate parallel to the (001) crystallographic direction can be distinguished in the picture. Also, the TEM investigation has shown that these nanorods are of single-crystalline in nature. Similar results were reported by Kwon and coworkers [62], who prepared large arrays of pencil-like ZnO hexagonal nanorods via a template-less route based on the oxidation of a Zn powder with H_2O_2 under hydrothermal conditions [62].

By the appropriate control of several reaction parameters, such as the pH and the ionic strength of the reaction solution this method can be extended to the preparation of nanostructured Fe_2O_3 and Cr_2O_3 [63, 64]. Furthermore, the treatment of the β-FeOOH nanorods with a reductive atmosphere will result in the formation of highly oriented iron nanorods [65]. Such a result is absolutely remarkable since nanostructured magnetic materials, exhibit a "magnetic" disorder due to the magnetic interactions and of nanorods or nanowires. Thus, to obtain "magnetically" ordered nanoarchitectures, it often necessitates the presence of an external magnetic field during the synthesis. According to this synthetic approach, β-FeOOH nanorods are first deposited onto various substrates through a hydrolysis-condensation process of an iron precursor dissolved in an aqueous solution. Then, the β-FeOOH nanorods are reduced with hydrogen at 300°C. Figure 11.25 shows the electron microscope images of an array of $\alpha-Fe$ nanotubes oriented parallel to the substrate's surface. Their diameter was estimated to vary in the range of 30–40 nm, whereas their length varies between 800 and 1,000 nm (Fig. 11.25a). Interestingly, the close examination of the individual nanorods performed at a much higher resolution revealed some interesting morphological features of the iron nanorods. Accordingly, each nanorod was found to contain an assembly of rod-like nanoparticles with a length of 15–30 nm and a diameter of 5–10 nm. The particles are stacked together into bigger aggregates possessing a columnar morphology (Fig. 11.25b).

FIGURE 11.25. Typical SEM (a) and TEM (b) images of α-Fe nanorods obtained by reduction of β-FeOOH nanorods deposited on a substrate by a solution chemical route. Reproduced from reference [65]. © 2002 American Chemical Society.

Since it is well known that vapor transport reactions lead to nanocrystalline products with very interesting morphology, the scanning electron microscopy is a powerful tool for the investigation of their structural characteristics. A comprehensive investigation of the morphological features of nanocrystalline ZnO by using SEM was recently given by Ren and coworkers, who demonstrated that a mixture of bulk ZnO and In_2O_3 annealed at high temperature (~800°C) under a temperature gradient will result into beautifully shaped, complex and highly ordered hierarchical nanostructures [66]. The SEM examination of the nanopowders evidenced that the general aspect of the nanopowders consist on smaller ZnO nanocrystals grown on In_2O_3 bigger cores and the reaction products present in some areas of high concentrations of nanowires with a particular morphology. A close investigation of the morphology of each type of particular morphological feature indicated that the symmetry of the individual In_2O_3 primary branches is responsible for the general allure of the hierarchical structure. Three complex morphologies have been identified, that is In_2O_3 cores with attached small ZnO branches possessing sixfold, fourfold, and twofold symmetry. While the diameter of the In_2O_3 skeletons varies between 50 and 500 nm, the diameter of the small ZnO branches ranges from 20 to 200 nm. Figure 11.26 shows some representative SEM micrographs of the areas of the sample where hierarchical nanoarchitectures with a sixfold symmetry predominate. While Fig. 11.26a and b gives us a general view of the complex morphology of the resulted nanopowders, showing the coexistence of multiple branches with different symmetries, Fig. 11.26c illustrates a high-magnification micrograph which is a close-up of a branch presenting a sixfold symmetry.

The presence of branches with a sixfold symmetry agrees well with the tendency of the primary In_2O_3 branches to crystallize into a hexagonal cell, thereby promoting the growth of the secondary ZnO branches in directions perpendicular to each individual face. The hexagonal symmetry of the In_2O_3 primary branches observed by the SEM microscopy is therefore confirmed by the SAED image

FIGURE 11.26. SEM, TEM, and SAED images at different magnifications of the sixfold nanostructures grown by vapor-phase transport. Reproduced from reference [66]. © 2002 American Chemical Society.

(Fig. 11.26j), where single spots ascribable to both hexagonal In_2O_3 and ZnO phase are identified. Interestingly, a close investigation of each individual branch revealed that, in turn, the secondary ZnO branches grow in the same direction and adopt a sixfold symmetry. Furthermore, from the SAED images one can infer the crystallographic direction along which the two kinds of branches will grow. Thus, while the In_2O_3 cores grow along the [110] and [001] zone axes, the secondary ZnO branches prefer to grow along the [0001] crystallographic direction. The main branches were identified by the EDX spectroscopy as pure In_2O_3, whereas the secondary branches correspond to pure ZnO. The SEM images of the branches presenting a fourfold symmetry are predominant allowing the identification of at least five different fourfold symmetry configurations.

Similar to the sixfold symmetrical nanocrystalline branches, the fourfold architectures present smaller ZnO branches growing in the same direction, but in this case, as revealed by the cross section-bright field TEM image presented in Fig. 11.27h, the symmetry of the In_2O_3 branches is lowered to a fourfold symmetrical architecture. Moreover, the SEM micrographs (Fig. 11.27a, b) suggest that the primary In_2O_3 branches present a rectangular symmetry and the number of the secondary ZnO branches attached to the In_2O_3 will increase with the length of the primary branches. Another interesting feature revealed by the SEM investigation is that the secondary branches are not always perpendicular to the primary In_2O_3 core. Finally, some hierarchical twofold nanostructures have been observed, as shown in Fig. 11.28. As illustrated by Fig. 11.28b, the ZnO branches are perfectly parallel to each other and are perpendicularly grown to the In_2O_3 nanowires. Although the mechanism of growth is not yet elucidated, it seems that the concurrent growth of In_2O_3 and ZnO nanorods under vapor transport in a temperature gradient is mainly dominated by the shape of the In_2O_3 nanowires which will dictate the general symmetry of the hierarchical nanostructure. Speculation of the growth mechanism presupposes that the In_2O_3 will evaporate and condense

FIGURE 11.27. SEM, TEM, and SAED images at different magnifications of the fourfold nanostructures grown by vapor-phase transport. Reproduced from reference [66]. © 2002 American Chemical Society.

FIGURE 11.28. SEM images at different magnifications of the twofold nanostructures grown by vapor-phase transport. Reproduced from reference [66]. © 2002 American Chemical Society.

first to form the main nanowires, whereas the subsequent evaporation and condensation of ZnO nanorods will occur onto the pre-formed In_2O_3 nanowires.

This fact would preclude the simultaneous growth of ZnO and In_2O_3 upon condensation on the cooler edges of the furnace, because in this case a simple speculation of the mechanism of growth will lead to an isotropic growth of the nanorods. However, such a situation will correspond to the case when these hierarchical topologies will not be observed experimentally. Following a similar synthetic pathway, but using a graphite foil as a collector, Ren prepared ZnO nanobridges and nanonails having a complex morphology [67]. The investigation of the ZnO nanopowders by scanning electron microscopy revealed that nanometer-sized nanorods grow epitaxialy on micron-sized ZnO bridges (Fig. 11.29a, b). A view-side of the nanobridges reveals that they are uniform and present hexagonally shaped nanorods with an average diameter of 150 nm perpendicularly oriented to the nanobridges (Fig. 11.29b). The SAED patterns of both the bridges and the nanorods have shown well-defined spots which can be indexed into a cubic lattice, which indicates that the ZnO nanobridges are single crystalline in nature. The identification of the crystallographic planes from the SAED diffraction image also suggested that each hexagonal nanorod is grown epitaxialy parallel to the [0001] direction on the (0001) plane and/or along the [0001 h] direction on the (0001 h) plane of the nanobridge. As shown in Fig. 11.29a–d, some other interesting variations of the basic nanobelt-type morphology were observed.

When the nanobelt is tangled and the nanorods grow perpendicular to it, a "roller-coaster" type of nanobridge is obtained (Fig. 11.29a), whereas when two individual belts are joined perpendicular to each other, an interesting architecture pictured in Fig. 11.29b is observed. Although the growth of the nanorods or nanobelts is in general regular, some discrepancies from the general trend are also observed; when the nanobelts are too thick the nanorods are found to grow on all the four edges of the nanobelts (Fig. 11.29c) and sometimes the density and the orientation of the nanorods with respect to the parent nanobelts is found to vary with a great extent (Fig. 11.29d). Furthermore, the variation of the growth conditions allowed for the synthesis of highly regular ZnO nanonails with a variable

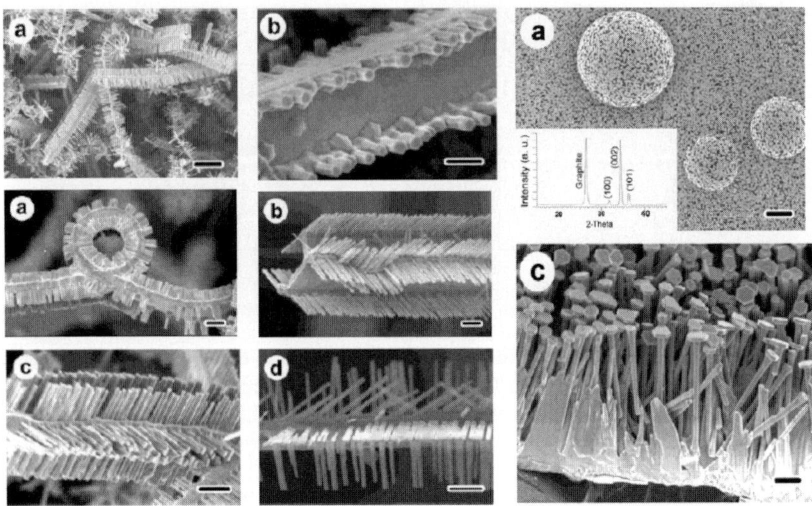

FIGURE 11.29. SEM images at different magnifications of the ZnO nanostructures grown by vapor-phase transport/condensation method proposed by Ren et al [67]. Reproduced from reference [67]. © 2003 American Chemical Society.

size which are assembled in nanonail-flower-like architectures (Fig. 11.29a, c). SEM observation has shown that the nanonails are oriented perpendicular to the surface with a diameter of ~150–200 nm and have a length which does not exceed several microns. The vapor phase transport process was also used by Yang and coworkers to prepare highly ordered ZnO nanowires; however, in this case they used an Au film as a catalyst [68]. The catalytic role of the Au film is inferred from the SEM images of the resulted ZnO film, which show that ZnO nanowires form only on the Au-plated portions of the sapphire's substrate (Fig. 11.30a–c).

ZnO nanowires are found to have a variable length of 2–10 mm, which is adjustable via the variation of the deposition time, are parallel to each other and perpendicularly oriented to the substrate's surface. The vertical orientation of the wires experimentally observed by the SEM was tentatively attributed to the interplay between the strong preference of ZnO to grow in the [0001] direction and the direct proportionality (by a factor of 4) between the a-axis of ZnO and the c-axis of sapphire. This latter feature will eventually result into a unique vertical growth configuration. However, the diameters of the prepared ZnO nanowires were found to vary in a wide range; while more than 95% present a diameter between 70 and 100 nm, for a small fraction the diameter was found to range between 20 and 150 nm. The large dispersity of the nanowire's diameter has been ascribed to the fact that the gold particles deposited onto the substrate will change their dimensions when the system is brought to a high temperature thereby resulting in nanowires with a larger dispersity.

The propensity of ZnO to crystallize into a hexagonal structure is given by its internal crystallographic anisotropy. These characteristics could be exploited for

FIGURE 11.30. SEM images at different magnifications of the twofold nanostructures grown by vapor-phase transport. Reproduced from reference [68]. © 2001 American Association for the Advancement of Science.

the synthesis of nanocrystalline ZnO tretrapods, nanobelts and nanonails via a modified vapor transport method [69]. As shown in Fig. 11.31, the nanocrystalline ZnO appears under SEM investigation as uniform, highly crystalline tretrapod-like clusters of 600–1,000 nm hexagonal nanorods with an angular separation of 110° and a length and a diameter of 80–100 nm (Fig. 11.31b). TEM investigation coupled with XRD have shown that the ZnO tetrapods are single crystals adopting a wurtzite-type structure and they are grown parallel to the (002) planes of the hexagonal ZnO. Another n-type semiconductor with a wide band gap and interesting a wide potential in optical applications such as the design of solar cells, flat panel displays, optoelectronic devices and gas sensors is In_2O_3. Qian and coworkers prepared single crystal In_2O_3 nanocubes by the thermal decomposition of $In(OH)_3$, which was obtained by reacting hydrothermally metallic indium with H_2O_2 in alkaline medium at 200°C. Figure 11.32a and b shows two representative SEM images of the $In(OH)_3$ and In_2O_3 nanopowders, respectively. Both nanocrystalline products are found to be composed by agglomerated cubic-shaped nanoparticles with well-defined faces belonging to the {001} family. The edge is found to vary between 80 and 120 nm for $In(OH)_3$, which is slightly shortened in the case of In_2O_3 nanocrystals. The SAED micrographs have shown that each particle is single crystalline in nature, whereas a closer examination of the individual particles revealed that their edges are truncated. Speculation of the growth mechanism supposedly assumed that two factors are responsible for the formation of the cubically shaped $In(OH)_3$ nanoparticle. Furthermore, the nanoparticles preserve their morphology upon heat treatment at 400°C, when the nanocrystalline $In(OH)_3$ is converted into In_2O_3. One factor would be the intrinsic isotropy of $In(OH)_3$, which crystallizes into a cubic lattice, whereas the second factor is related to the existence of an oxygen-rich environment, for the nanocrystals during the growth process, which will result into a low plane-index faceted final morphology. The FESEM investigation evidenced the influence of the temperature over the particle size (Fig. 11.33). Accordingly, while at a low reaction temperature of 150°C, the size of the resulted $In(OH)_3$

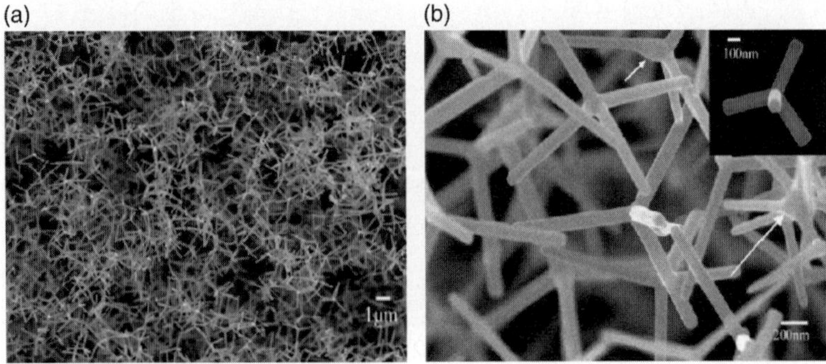

FIGURE 11.31. SEM images at different magnifications of the twofold nanostructures grown by vapor-phase transport. Reproduced from reference [69]. © 2005 American Chemical Society.

FIGURE 11.32. Field-emission SEM images of nanocrystalline $In(OH)_3$ (*left columns*) and In_2O_3 (*right column*) prepared hydrothermally in the presence of H_2O_2 in an alkaline medium. Reproduced from reference [69b]. © 2005 American Chemical Society.

cubic nanocrystals is 50–80 nm, an increase of the temperature to 250°C will lead to a fivefold increase of the particle size. However, despite this dramatic change in the particle size, during the heat treatment, the morphology and crystallinity of the nanoparticles are preserved. These results clearly suggest that the scanning electron microscopy is a powerful investigative tool allowing, not only for exploring the morphology of the nanocrystalline products, but also for gaining an insight into the reaction mechanism and, eventually, suggesting synthetic strategies to be followed for a better control of the size and shape of the nanoparticles.

Another important semiconductor material with several technological applications in catalysis and energy storage/conversion is Cu_2O. Murphy and coworkers has prepared highly uniform Cu_2O nanocubes by reducing $CuSO_4$ in air with

FIGURE 11.33. Field-emission SEM images of nanocrystalline In_2O_3 (a) and $In(OH)_3$ (b and c) hydrothermally in the presence of H_2O_2 in an alkaline medium. Reproduced from reference [69b]. © 2005 American Chemical Society.

sodium ascorbate at 55°C in the presence of cetyltrimethylammonium (CTAB) as a surfactant and an alkaline medium [71].

Again, the shape of the nanocrystals was found to be the effective result of the intrinsic cubic crystalline structure of Cu_2O. As shown in Fig. 11.34, the resulted Cu_2O nanocrystals appear under the electron microscope investigation as cubically shaped with an average edge length of 450 nm and a very high monodispersity. Thus, a closer inspection of the nanocubes by high magnification TEM, suggested that the particles have partially hollow interiors, present rough surfaces and are constructed by smaller clusters. Furthermore, complementary TEM observations evidenced the role played by the surfactant upon the size and the shape of the Cu_2O nanoparticles: the nanoparticles can be easily changed from irregular aggregates to well-defined nanocubes by simply increasing the concentration of the surfactant in the reaction solution. Another interesting class of nanostructured materials for optical applications is represented by the core-shell phosphors, where the core is silica and the shell is a phosphor material. Such complex nanoarchitectures are expected to be cheaper than conventional phosphor materials and to have a tunable size of both the silica core and the phosphor shell. Fang investigated the structure and morphology of YVO_4/Eu coated silica spheres obtained by a sol-gel process [72]. Figure 11.35 shows some typical SEM images of the SiO_2 nanoparticles (Fig. 11.34a) and the pure Eu-doped YVO_4 powder, respectively. As expected, the ~500 nm SiO_2 nanoparticles are nonaggregated, spherically shaped, nearly monodisperse with well-defined contours, whereas the phosphor powder consists of irregularly shaped, aggregated clusters with a wide size distribution (150–500 nm). Investigation of the SEM images of the phosphor coated-SiO_2 nanoparticles revealed the absence of any irregularly shaped particles, which is indicative that the phosphors have been adsorbed onto the surface of SiO_2 nanoparticles (Fig. 11.36b–e).

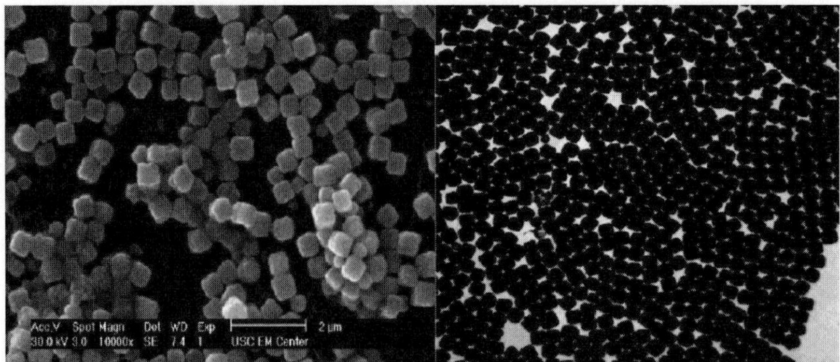

FIGURE 11.34. SEM and TEM images of monodisperse Cu_2O nanocrystals prepared by reductive growth in alkali conditions. Reproduced from reference [71]. © 2003 American Chemical Society.

(a)   (b)

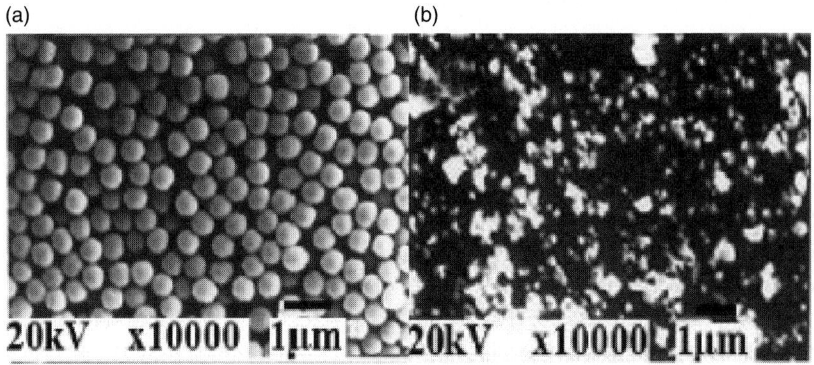

FIGURE 11.35. Representative SEM images of SiO_2 nanoparticles and Eu-doped YVO nanopowders. Reproduced from reference [72]. © 2005 American Chemical Society.

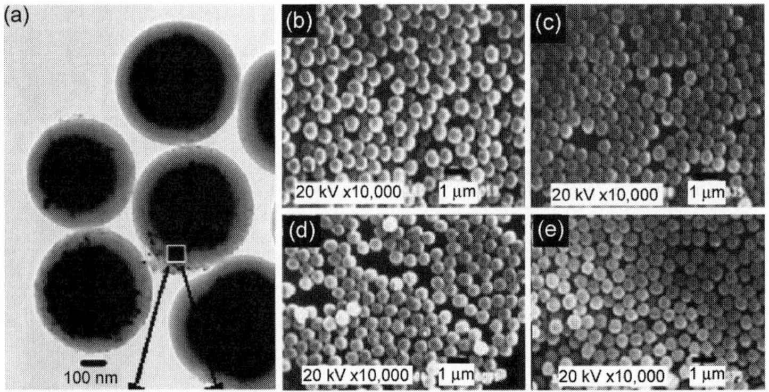

FIGURE 11.36. TEM (a) and SEM images (b–e) of monodisperse SiO_2/Eu:YVO_4 nanophosphors. Reproduced from reference [72]. © 2005 American Chemical Society.

The existence of the core-shell structure of the resulting nanoparticles has been furthermore confirmed by TEM microscopy. Figure 11.36a shows a typical TEM micrograph, where a well-defined, dark SiO_2 core with an average diameter of 500 nm can be easily distinguished, whereas the phosphor shell appears as a gray contour around the core and have an average thickness of 60 nm. HRTEM investigation evidenced that such a nanocomposite contains an amorphous SiO_2 core and a well-crystallized phosphor shell.

4.2. Nanocrystalline Magnetic Oxides

Magnetic ceramics are another important class of technological materials, due to their remarkable properties which render them suitable for many applications in the next generation of electronics, catalysis and magnetic information storage. Although ferrites MFe_2O_4 are traditionally prepared in bulk, the miniaturization of magnetic and electronic devices has demanded advanced materials with new forms and shapes, such as nanoparticles or thin films. As the magnetic properties at nanometer scale are more than ever influenced by the microstructure of the material, in terms of their particle size, morphology and aggregation level, electron microscopy represents a very important tool in investigating these structural characteristics.

The production of nanocrystalline ferrites can be achieved through a wide variety of chemical methods. Kima and coworkers have studied the morphology and magnetic properties of Co–Zn ferrites prepared by a sol-gel method [73]. According to their procedure, the transition metal salts are mixed in ethylene glycol and the resulting solutions are refluxed to obtain a gel. The gel is dried and thermally treated at high temperature with conversion of the reaction intermediates into the final spinel-type nanostructured oxides. Investigation of the morphology of the obtained nanopowders performed by scanning electron microscopy revealed that the samples consist of agglomerated nanoparticles with a spherical shape and a size which increases progressively with the Zn content from ~55 to 85 nm (Fig. 11.37 a, b).

FIGURE 11.37. Representative SEM micrographs of $Co_{1-x}Zn_xFe_2O_4$ nanocrystalline ferrites prepared by a sol-gel process. Reproduced from reference [73]. © 2002 American Institute of Physics.

Our group intensively used the scanning electron microscopy to perform a mechanistic investigation of the morphology of single-phase nanocrystalline ferrite films prepared by a soft solution processing method. The method, called, the liquid phase deposition is based on the controlled hydrolysis of the transition metal fluorides in aqueous solutions at moderate temperatures and followed by the elimination of the fluoride ions in presence of a fluoride scavenger [74]. The investigation of the morphology of the ferrite films with similar thicknesses revealed important changes in their microstructure, depending on both the nature and the concentration of the transition metal ion. While cobalt ferrite films are constructed by spherical nanoparticles with a mean diameter of 200 nm (Fig. 11.38a), the morphology changes when cobalt is replaced by copper or zinc. Accordingly, copper and zinc ferrite films are constructed from rod-like particles with variable lengths and a diameter of few hundreds of nanometers and with a different orientation with respect to the substrate's plane. As seen in Fig. 11.38b, the copper ferrite films are formed by rod-like particles which are aligned parallel to the plane of the substrate, whereas in the case of zinc ferrite films the elongated particles are aligned perpendicular to the plane of the substrate (Fig. 11.38c).

These experimental observations are presumably related to the intrinsic crystallographic characteristics of the binary transition metal oxides MO (M = Co, Cu, and Zn), which form in the first stage of the annealing of the as-prepared films and then react with the iron precursor to give the spinel-type oxides. While CoO crystallizes into a cubic lattice, CuO adopts a tetragonal symmetry and ZnO crystallizes into a hexagonal lattice, the nonunitary c/a ratio in these latter case could be considered as a key factor influencing the morphology of the nanocrystalline transition metal ferrites. As the films form through the concurrent diffusion of the iron and the transition metal ions during this solution process, the concentration of the cationic species in the treatment solution is expected to

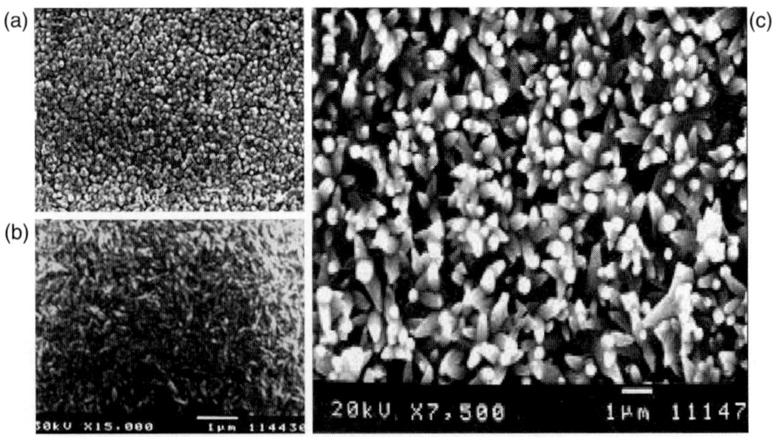

FIGURE 11.38. Representative SEM micrographs of MFe_2O_4 nanocrystalline ferrites. M = Co (a), Cu (b), and Zn (c) prepared by the liquid phase deposition method. Reproduced from reference [74]. © 2004 Royal Society of Chemistry.

influence the morphology of the deposited ferrite films. Figure 11.39 presents the representative SEM micrographs of the $Zn_xFe_{3-x}O_4$ films for different concentrations of the Zn^{2+} ions in the treatment solution. The concentration of the zinc ions in the treatment solution was varied between 0 and 0.5 M, whereas the zinc composition of the resulted ferrite films, as determined by inductive-coupled plasma spectroscopy varies between 0.31 and 0.88. The SEM micrographs clearly show that the ferrite films are highly homogeneous and are formed by large arrays of nanoparticles whose morphology changes upon increasing of the zinc content in the resulted ferrite films. Although there is not clear evidence about the role played by Zn^{2+} ions on the morphology of the resulting films, the SEM data show that the increase in the Zn^{2+} concentration is accompanied by a change of the morphology of the deposited films from a spherical to a rod-like type.

As commonly reported for the liquid phase deposition-based processing of thin films, it seems that the deposition proceeds through the heterogeneous nucleation of primary particles in solution followed by a surface-directed growth. While in absence of the zinc ions the resulted α-Fe_2O_3 film is constructed by spherical nanoparticles with a diameter of ~150 nm, with increasing of the zinc content the morphology changes to columnar, the corresponding films being constructed by perpendicularly oriented rod-like nanoparticles with an average diameter of 200 nm. Moreover, in some cases, the SEM investigation of the films revealed the presence of cracks which result from the thermal treatment of the ferrite films as

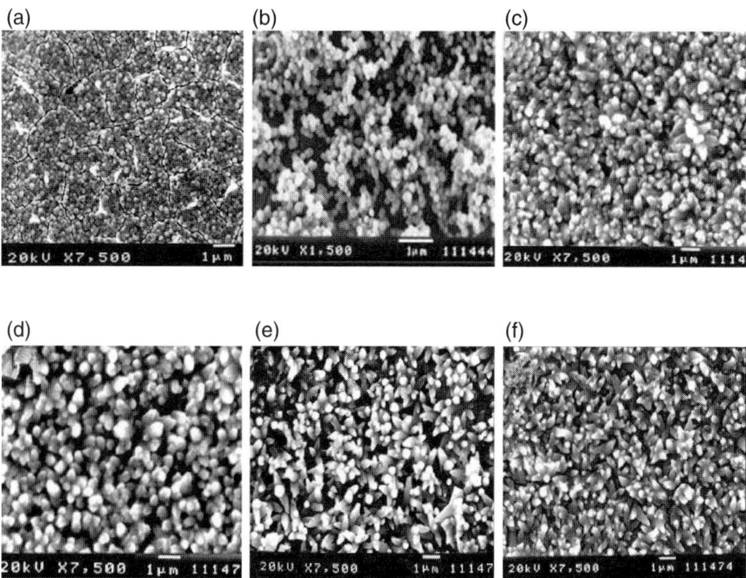

FIGURE 11.39. SEM images of the $Zn_xFe_{3-x}O_4$ nanocrystalline films having different zinc contents. Reproduced from reference [74]. © 2004 Royal Society of Chemistry.

a result of the different values of the thermal coefficients of the film and of the substrate, respectively. An interesting study on the morphology of nanocomposites of SiO_2 particles with a magnetic core of $ZnFe_2O_4$ was performed by Etourneau and coworkers [75]. While the zinc ferrite nanoparticles were prepared by direct precipitation of the transition metal salts in aqueous solution, the silica nanoshells were obtained by a low-temperature water-in-oil microemulsion method. The morphology of the resulting nanocomposites was imaged by electron microscope techniques, including SEM and TEM. As shown in Fig. 11.40, the SEM photographs reveal that the shape of the resulted particles strongly depends on the nature of the surfactant used in the microemulsion process.

Specifically, the SEM pictures show that, excepting for the bis[2-ethylhexyl] sulfosuccinate sodium salt (AOT) which lead to shapeless particles for the other surfactants, that is polyoxyethylene (4) lauryl ether (Brij30), dodecyl sulfate (SDS) in propanol or the mixture of AOT and Brij 30, the nanocrystalline $ZnFe_2O_4/SiO_2$ nanocomposites are well-defined, spherically shaped with narrow size distribution. The irregular shape of the nanocomposites obtained when AOT was used as a surfactant is presumably ascribed to the destabilization of the surfactant molecules attached to the nanoparticle's surface. This is a direct result of their hydrolytic decomposition eventually promoted by the NaOH added to precipitate the initial ferrite nanoparticles. The diameter of the particles was found to vary between 40 and 60 nm with a magnetic core of 4–6 nm, which reproduces with high fidelity both the size and the spherical shape of the reverse micelles. Although that in most cases nanocomposites are spherically shaped, a close analysis of the SEM pictures reveal that when a mixture of surfactants was used

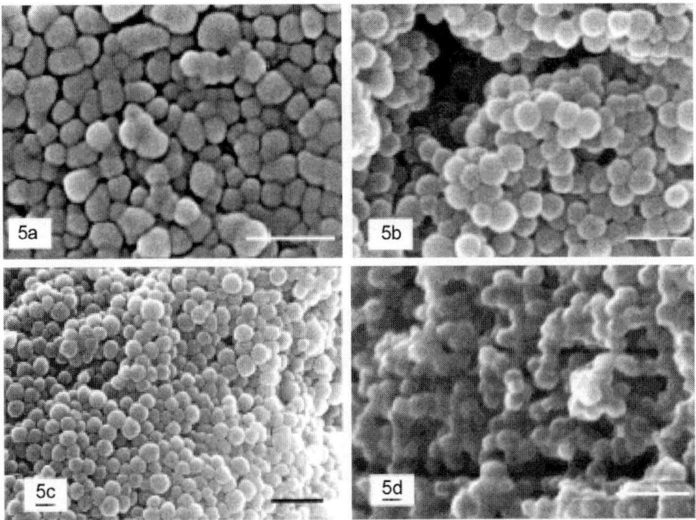

FIGURE 11.40. SEM images of the $SiO_2/Zn_xFe_{3-x}O_4$ nanocrystalline composites obtained by a microemulsion method using different surfactants. Reproduced from reference [75]. © 2002 American Chemical Society.

instead of pure surfactants, the nanoparticles present the highest shape uniformity (Fig. 11.40c). Nevertheless, such a behavior has been tentatively ascribed to the fact that a mixture of surfactants will render the interface water–oil of the inverse micelles more flexible, thereby lowering the interfacial tension. In turn, a lower interfacial tension will result in a more facile attachment of the silica to the surface of magnetic nanoparticles. Another interesting feature revealed by the experimental observation by SEM is related to the sample obtained by using dodecyl sulfate in propanol (Fig. 11.40d). In this case, in addition to the agglomerated $SiO_2/ZnFe_2O_4$ nanocomposites, 30–50 nm spherical SiO_2 nanoparticles can be observed in the corresponding SEM micrograph. While the agglomeration of the resulted nanocomposites could be associated to the interparticle interactions, the formation of the bare SiO_2 nanoparticles can be ascribed to the increased stability of the inverse micelles with increasing the water and oil ratio. In-depth analysis of the morphology of the nanocomposites was performed by TEM.

The TEM observations confirmed the SEM investigations, that is, unlike the case when a mixture of surfactants was used, the nanocomposites contain more than a magnetic core or empty silica shells. In the case of the nanocomposites prepared by using a mixture of AOT and Brij 30, the resulted nanoparticles are nearly spherical and contain a single magnetic core located at the center of the silica shell (Fig. 11.41). Wong and coworkers have proposed a new, rapid, molten-salt synthetic approach for the large-scale preparation of submicron-sized

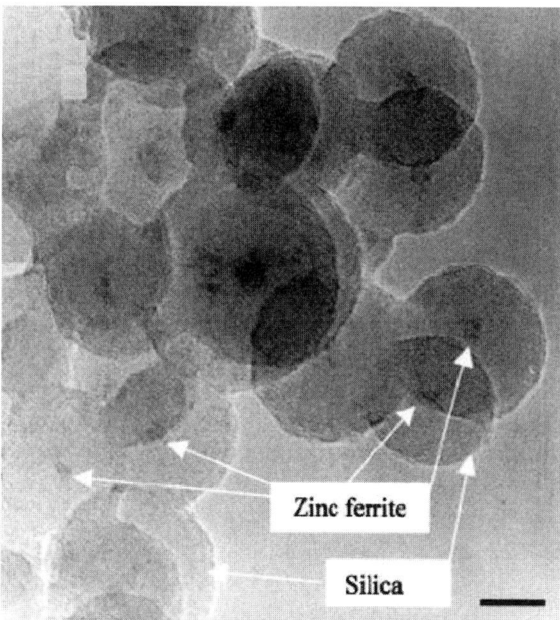

FIGURE 11.41. TEM images of the $SiO_2/Zn_xFe_{3-x}O_4$ nanocrystalline composites obtained by a microemulsion method using different surfactants. Reproduced from reference [75]. © 2002 American Chemical Society.

$Bi_2Fe_4O_9$ particles with different morphologies [76]. This method consists in the annealing of appropriate amounts of the precursors placed into a quartz tube and heated at high temperature for several hours. The investigation of the morphology of the reaction products by scanning electron microscopy revealed that two major factors influence the size and the shape of the resulted crystals. These include the precursor ratio in the reaction mixture, and the nature of the surfactant. As shown in Fig. 11.42a the resulted powders consist of well-defined cubical nanoparticles possessing a variable size and smooth faces. The edge length of the cubic particles is found to range from 166 to 833 nm with an average of 386 nm and a standard deviation of 146 nm. The SEM image of a single cubic particle revealed that the crystals present both the edges and the corners truncated and they can be described by the morphology drawn in the inset of Fig. 11.42b.

The chemical composition of the barium ferrite powders, as obtained from the elemental analysis performed by the energy dispersive x-ray spectroscopy on different areas of the samples have confirmed that the resulting ferrite powders are free of impurities and present a stoichiometric composition. Furthermore, the TEM investigation revealed that the nanocrystals are single crystalline in nature and confirmed the morphology of the faces, edges and corners, as inferred from the SEM analysis (Fig. 11.42d).

FIGURE 11.42. SEM and TEM images of the $Bi_2Fe_4O_9$ particles prepared by a molten salt. Reproduced from reference [76]. © 2005 Royal Society of Chemistry.

The symmetry of the particles was obtained from the high-resolution transmission electron micrograph displayed in Fig. 11.42e which indicates again that each individual nanocube is single crystalline and adopts an orthorhombic symmetry with a lattice parameter close to that reported in the literature for the bulk crystalline $Bi_2Fe_4O_9$.

5. Nanostructured Semiconductor and Thermoelectric Materials

In the past years, a significant progress has been made in the synthesis of nanocrystalline transition metal chalcogenides with predictable architectures and controllable morphology. This special class of inorganic compounds is renowned as excellent semiconductor and thermoelectric materials and can potentially serve as platform for various applications in the semiconductor industry and the storage and conversion of the energy. Thus, upon bringing their dimensions to the submicrometric scale, the microscopic techniques have proven again not only to be a powerful investigative tool in determining the morphology of the nanoparticles but also they can offer unique opportunities to gain insights into the mechanism of the formation of these nanoarchitectures., Odom and coworkers reported the preparation of conducting $NbSe_2$ nanocrystals with different dimensionalities (nanowires (1D) and nanoplates (2D)) through a solution-based one-pot reaction. According to the experimental procedure, the $NbSe_2$ nanocrystals form as a result of the chemical reaction between $NbCl_5$ and Se dissolved in appropriate ratios in high boiling point amines (oleylamine or dodecylamine) at a moderate temperature (280°C), maintained for several hours under inert atmosphere [77]. The morphological investigation of the nanocrystals by using the scanning electron microscopy revealed that regardless of the organic solvent used in the reaction, the shape of the nanoparticles is strongly dependent on the thermal history of the sample. This can be ascribed to the variation of the ratio between the capping ligand and the intermediate; specifically, when the reaction solution was left to cool slowly to the room temperature, 2D $NbSe_2$ nanoplates are formed. Conversely, when the reaction mixture is quenched nanowires, instead of nanoplates are observed to form in the solution. Figure 11.43 illustrates the SEM micrographs of the nanocrystalline $NbSe_2$ with two different morphologies obtained from the thermal decomposition of the molecular precursors. The SEM observations in the case of $NbSe_2$ corroborate the results obtained from XRD indicating that the structure of the nanoplates is a replica of the bulk material. While $NbSe_2$ nanoplates are found to present a regular morphology with lateral dimensions ranging between 500 and 1,000 nm and a thickness between 10 and 70 nm (Fig. 11.43a), by quenching the reaction mixture polycrystalline $NbSe_2$ nanowires with a diameter of 2–25 nm and a length of tens of microns are formed (Fig. 11.43b). Fang and coworkers have extensively investigated the morphology of hexagonal Bi_2Te_3 nanoplates prepared by the thermal decomposition of appropriate precursors in high boiling point organic solvents [78]. Dark single

FIGURE 11.43. Typical SEM images of nanocrystalline NbSe$_2$ platelets synthesized in oleylamine (*top inset*). NbSe$_2$ nanoplates grown in dodecylamine (*bottom inset*). HRTEM of an individual nanoplate on its edge. Reproduced from reference [77]. © 2005 American Chemical Society.

crystalline hexagonal nanoplatelets were prepared by reacting bismuth-2-ethyl-hexanoate with metallic tellurium in hot phenyl ether at 150°C under inert atmosphere. The microstructure of the Bi$_2$Te$_3$ nanoplatelets observed by both the scanning electron microscopy and transmission electron microscopy revealed that the nanocrystals present uniform hexagonal shapes with an edge length of 200–300 nm, whereas their thickness is close to 15 nm (Fig. 11.44a, b). Since the bulk Bi$_2$Te$_3$ is a hexagonal layered material consisting of successive atomic layers stacked along the [001] direction, the SEM investigation shows that the shape of the nanoplatelets originates principally from a specific growth mechanism that is controlled with a great extent by the intrinsic crystal properties of the material.

Since the growth of the nanoparticles, in general, occurs along crystallographic directions so that the surface energy of the nanocrystals is minimized, nano-hexagonal Bi$_2$Te$_3$ platelets will be favored to form because they are associated with the low diffraction index faces which develop during the growth process. In addition to the insight into the morphology of the nanocrystals, the SEM observation provides informations about the role of different reaction parameters, such as the concentration of the reactants, the concentration of the capping agents, and the reaction temperature on the morphology of the nanoplatelets. Thus, the temperature has been found to play a crucial role on the tunability of both the size and the shape of the nanoplates; whereas the increase of the temperature results in much more uniformly shaped nanoplatelets, their thickness also increases with increasing reaction temperature. The hexagonal shape of the nanocrystals was further confirmed by transmission electronmicroscopy experiments (Fig. 11.45a). A closer inspection of the TEM pattern shows that the nanocrystals present undulations on their surfaces, which is ascribed to a strain originating from a slight bending of the nanoplatelets on their surface. The mechanism of growth of these nanoplatelets is also confirmed by selected area electron diffraction (SAED), which shows that the nanoplatelets are free of defects single crystals which are terminated by {11–20} faces, whereas the slowest growth direction was identified as <0001>.

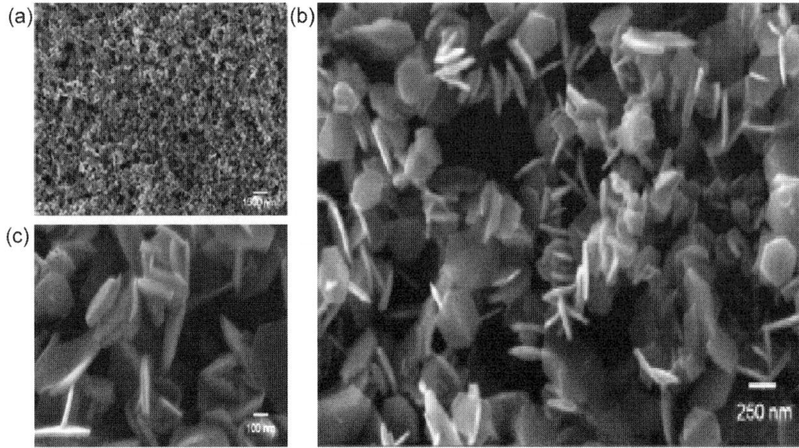

FIGURE 11.44. SEM images with different magnifications of nanocrystalline Bi_2Te_3 nanoplatelets obtained through an organic-solution route. Reproduced from reference [78]. © 2005 American Chemical Society.

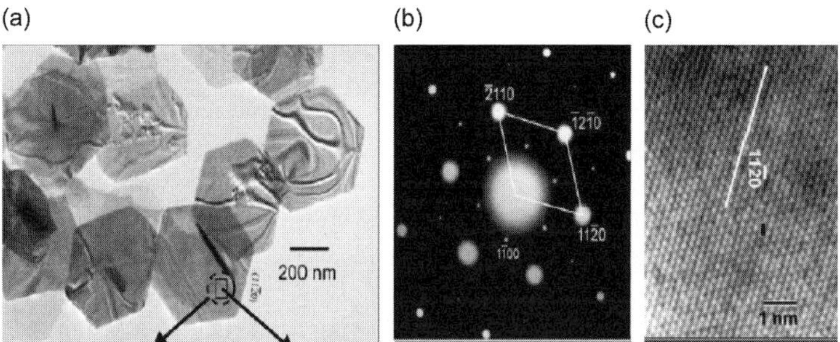

FIGURE 11.45. TEM, SAED, and HRTEM patterns of nanocrystalline Bi_2Te_3 nanoplatelets obtained through a two-step epitaxial growth process. Reproduced from reference [78]. © 2005 American Chemical Society.

Interestingly, when the same reaction route is followed in the presence of a small amount of selenium, the mechanistic pathway leading to Bi_2Te_3 nanoplatelets becomes more complicated. In addition, the investigation of the morphology of the resulted nanocrystals by electron microscopy allows elucidating the mechanism of growth. As shown in Fig. 11.46a, the resulted nanocrystals present a complex morphology, being formed by small nanoplatelets that are attached perpendicularly to longer nanorods. The EDS analysis performed on different areas of the samples revealed that selenium is absent in these nanocrystals, whereas the long rods contain only a very small amount of bismuth and the attached smaller nanoplatelets consist of Bi_2Te_3. Speculation on such a major change in the reaction mechanism would be presumably attributed to the different stabilities of the complexes formed in the reaction solution by metallic bismuth and selenium with trioctylphosphine, respectively. Since selenium

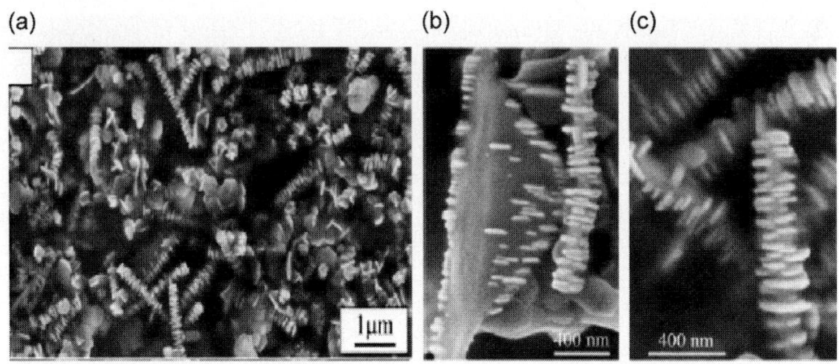

FIGURE 11.46. SEM patterns of nanocrystalline Bi_2Te_3 nanoplatelets obtained in presence of a small amount of selenium. Reproduced from reference [78]. © 2005 American Chemical Society.

possesses a smaller size, the covalent bonds formed with electron-donor ligands will be much more stable than those of its bismuth counterpart. This observation corroborates well with the more pronounced metallic character shown by tellurium which is much easily precipitated from the solution containing the tellurium and the selenium complexes with trioctylphosphine. Scanning electron microscopy was systematically used by Cheng and coworkers to investigate the morphology of star-shaped PbS nanocrystals with a hierarchical architecture, as well as the reaction conditions that favor a particular morphology [79]. Star-shaped PbS nanocrystals were prepared by using a simple solution-based synthetic route based on the thermal decomposition of the thioacetamide (TAA) in aqueous solutions of $Pb(CH_3COO)_2$ and acetic acid. As shown in Fig. 11.47a–d, the PbS nanocrystals present a hierarchical morphology, being constructed by star-shaped units formed by eight arms with a dendritic architecture, which reproduce the eight small diagonals of a cubic lattice with a separation distance between two adjacent corners close to ~2 μm. Each individual branch is composed by an ensemble of three leaflets perpendicularly oriented to the arm. Scanning electron microscopy performed for different orientations of the sample indicated that each leaflet possesses an average length of 400 nm and a thickness of 80–10 nm. A closer investigation of the morphology of these PbS nanostructures was performed by TEM. Figure 11.47e shows a representative bright field TEM image of these nanoarchitectures, which confirms the result obtained by SEM investigation related to the star-shaped architecture. Moreover, the nanocrystals are single crystalline and the orientation of the each individual symmetric arms is parallel to the <111> crystallographic directions of the cubic lattice, whereas the three units which form the arm are found to be parallel to the $(1\bar{1}0)$, $(01\bar{1})$, and $(10\bar{1})$ planes, respectively. In-depth investigation of the morphological features of the PbS nanopowders by electron microscopy suggested that three main parameters influence the morphology of the PbS nanocrystals namely the concentration of the pH of the reaction solution, the concentration of the acetate ions and as the reaction temperature.

FIGURE 11.47. Representative SEM (a–d) and TEM (e) images of star-shaped PbS crystals obtained from thioacetamide and lead acetate in aqueous solution at moderate temperatures. Reproduced from reference [79]. © 2005 American Chemical Society.

Since the PbS nanocrystals form as a result of the reaction between $Pb(CH_3COO)_2$ and S^{2-} ions released in solution by the tioacetamide, the reaction equilibrium and, hence, the morphology of the nanoparticles will be affected with a great extend by the pH of the solution. As seen in Fig. 11.48, the ideal pH value for the formation of the star-shaped PbS nanocrystals is 3.8, whereas more acidic medium results in the formation of PbS nanocubes instead of star-shaped particles (Fig. 11.48c, d) and more basic conditions will lead to the formation of sheet-like PbS nanocrystals (Fig. 11.48a). A key factor influencing the shape of the PbS nanocrystals was found to be the acetate ions present in the solution. The acetate ions presumably influence the reaction mechanism as a result of the low degree of dissociation of $Pb(CH_3COO)_2$, as well as the several possible complexes that the acetate ions can form with the Pb^{2+} ions in the reaction solution, including $Pb(CH_3COO)^+$, $Pb(CH_3COO)_2$, and $Pb(CH_3COO)_3^-$, respectively. As shown in Fig. 11.49b, in the absence of the acetate ions in the reaction solution, the highly symmetric morphology is not longer preserved and the nanocrystalline PbS formed from the reaction between $Pb(NO_3)_2$ and TAA present irregular shapes and a size ranging between 100 and 200 nm.

Lastly, the reaction temperature also was found to play an important role on the morphology of the nanocrystals. SEM investigation of the morphology of the nanocrystals obtained at different temperatures revealed that the highly symmetric eight-arm shaped PbS nanocrystals form at 80°C, whereas the decreasing of the temperature will lead to the formation of irregularly shaped nanoparticles. The lowest temperature at which the nanocrystals still preserve their hierarchical architecture is 40°C. At this temperature and below, the SEM micrographs indicated the formation of 60–150 nm nanostrips. The reaction temperature was found to have a twofold influence: first it dramatically influence the complexation of the Pb^{2+} ions with the H_3CCOO^- and second, the decrease of the temperature results into a decrease of the S^{2-} released in solution by the decomposition of the TAA. Moreover, at a given reaction temperature, the molar ratio of the

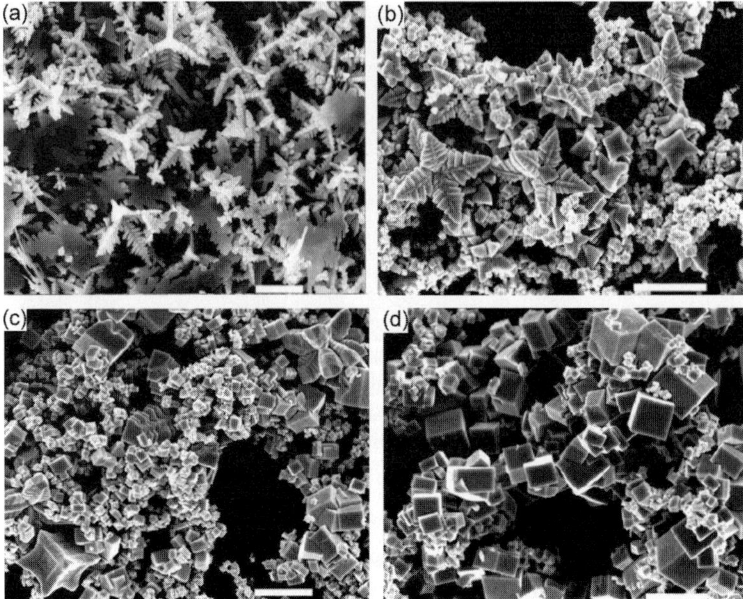

FIGURE 11.48. SEM images of PbS crystals obtained at 80°C in 0.1 M [H$_3$CCOO$^-$] solution at different pH values: (a) pH) 4.5, (b) pH) 2.4, (c) pH) 1.9, and (d) pH) 1.7, respectively. The scale bar is 2 μm (a) and 5 μm (b–d). Reproduced from reference [79]. © 2005 American Chemical Society.

FIGURE 11.49. SEM images of PbS crystals in the presence (a) and in the absence (b) of H$_3$CCOO$^-$ ions in the reaction solution. The scale bar is 1 μm. Reproduced from reference [79]. © 2005 American Chemical Society.

reaction precursors was also found to have a critical role in the morphology of the nanopowders.

Figure 11.50 shows the SEM patterns of the PbS nanopowders obtained at 40°C with different molar ratios of the Pb^{2+} ions and TAA. While for a [Pb^{2+}]/[TAA] ratio of 1:2 the resulted product has a rod-like morphology with nanorods with an average diameter of 60 nm (Fig. 11.49a), the increase of the molar ratio to 2:1, results into a change of their morphology to lamellar nanosheets with a thickness of 34 nm (Fig. 11.49b). Interestingly, when nearly

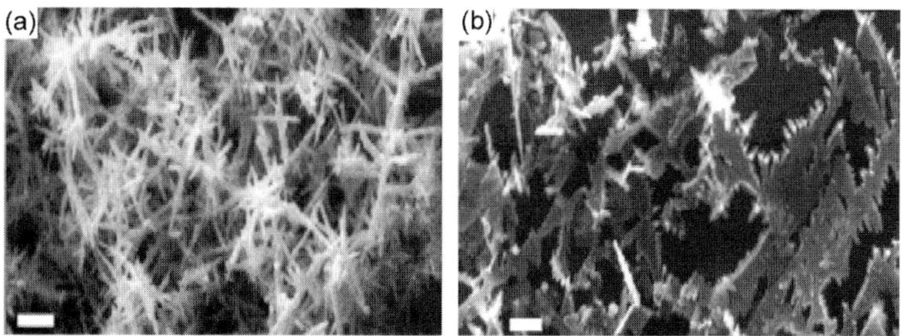

FIGURE 11.50. SEM images of PbS crystals prepared at 40°C with different concentrations of the Pb^{2+} ions and thioacetamide, respectively: (a) $[Pb^{2+}] = 2.7 \times 10^{-2}$ M, [TAA] = 5.4×10^{-2} M; and (b) $[Pb^{2+}] = 5.4 \times 10^{-2}$ M, [TAA] = 2.7×10^{-2} M. The scale bar is 1 μm. Reproduced from reference [79]. © 2005 American Chemical Society.

monodisperse PbS nanoparticles are prepared by precipitation from $Pb(CH_3COO)_2$ and NaS in neutral aqueous solutions containing water-soluble polymers, such as poly(ethylene oxide) (PEO), PVA or sodium dodecyl sulfate (SDS) and surfactants, the resulting nanoparticles self-assemble spontaneously into layered superstructures which finally lead to the formation of nanotubes [80]. The formation of such nanoarchitectures from individual PbS nanoparticles occurs in solutions where the concentration of Pb^{2+} is slightly higher than that of the S^{2-} ions. The assemblage of individual PbS nanoparticles into tubular nanoarchitectures was explained by the interpenetration of the surfactant molecules adsorbed onto the surface of adjacent PbS nanoparticles ultimately leading to their alignment and the formation of nanotubes. Consequently, as a result of the strong interaction exerted between the Pb^{2+} existing in the solution and the dodecyl sulfate ions, the initially formed layers have the tendency to bend during the formation of the nanotubes, the fact which was confirmed experimentally by TEM investigations. The representative SEM micrograph pictured in Fig. 11.51a shows tube-like PbS nanoarchitectures obtained from nanoparticles left in the reaction solution for several weeks. As revealed by transmission electron microscopy, individual nanoparticles with a size of 2–4 nm and presenting an interparticle distance of 2–3 nm will self-assemble over the time at room temperature to form tubes with a diameter of 100 nm and a length of 20–400 nm. Again, a more detailed investigation was performed by TEM and revealed that the PbS nanotubes present smooth surfaces and are constructed from layers of individual PbS nanocrystals (Fig. 11.51b).

6. Conclusions

Since the physical properties of nanoparticles differ markedly from those of their bulk counterparts and are size and shape dependent, scanning electron microscopy is a modern, fundamental and well-suited nondestructive investigative tool in the characterization of nanoparticles. In addition to its main ability to

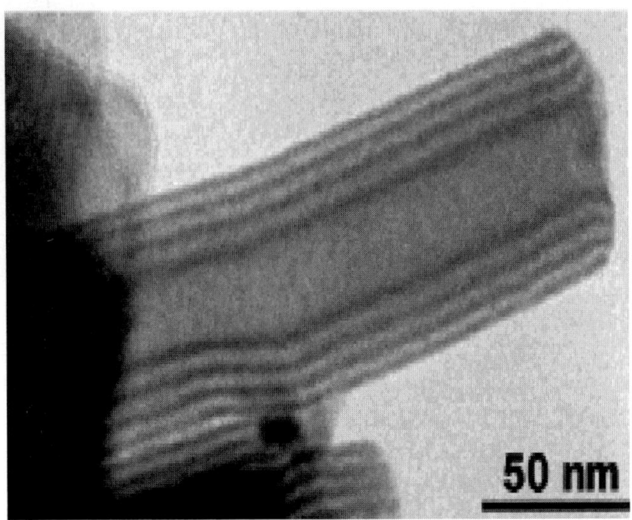

FIGURE 11.51. SEM micrograph of PbS nanotubes formed by a self-assembly process of individual PbS nanoparticles in the poly(ethylene oxide) (PEO) and sodium dodecyl sulfate (SDS)-containing reaction solution TEM image of an individual nanotube. Reproduced from reference [80]. © 2003 American Chemical Society.

offer a 2D visualization of large areas of the sample, scanning electron microscopy can provide diversified qualitative and quantitative information on many physical properties including the size, morphology, surface texture, roughness, and chemical composition of the nanocrystals. Moreover, advanced manipulation of the samples during the scanning electron microscopy experiments, coupled with high-resolution microscopy techniques, like the transmission electron microscopy can provide key information about the morphology of the nanocrystals at the nanometer scale, as well as insights into the different reaction pathways.

References

1. M. A. Willard, L. K. Kurihara, E. E. Carpenter, S. Calvin, and V. G. Harris, *Intern. Mater. Rev.*, 49 (2004) 3.
2. B. L. Cushing, V. L. Kolesnichenko, and C. J. O'Connor, *Chem. Rev.*, 104 (2004) 3893.
3. O. Masala and R. Seshandri, *Ann. Rev. Mater. Res.*, 34 (2004) 41.
4. K. M. Kulinovski, P. Jiang, H. Vaswani, and V. L. Colvin, *Adv. Mater.*, 12(11) (2000) 833.
5. T. Hyeon, *Chem. Comm.*, (2003) 927.
6. R. R. Rakhimov, E. M. Jackson, J. S. Hwang. A. I. Prokofiev, I. A. Alexandrov, and A. Y. Karmilov, *J. Appl. Phys.*, 95(11) (2004) 7133.
7. N. N. Thadani, *J. Appl. Phys.*, 76 (1994) 2129.
8. G. Viau, F. Ravel, O. Acher, F. Fievet-Vincent, and F. Fièvet, *J. Appl. Phys.*, 76(10) (1994) 6570.

9. P. Toneguzzo, O. Acher, G. Viau, F. Fievet-Vincent, and F. Fievet, *J. Appl. Phys.*, 81(8) (1997) 5546.

10. S. Sun, C. B. Murray, D. Weller, L. Folks, and A. Moser, *Science*, 287 (2000) 1989.

11. H. Doyle, *Mat. Res. Soc. Symp. Proc.*, 577 (1999) 385.

12. H. Liu, F. Favier, K. Ng, M. P. Zach, and R. M. Penner, *Electrochim. Acta*, 47(5) (2001) 671.

13. H. Liu and R. M. Penner, *J. Phys. Chem. B*, 104 (2000) 9131.

14. E. C. Walter, B. J. Murray, F. Favier, G. Kaltenpoth, M. Grunze, and R. M. Penner, *J. Phys. Chem B*, 106 (2002) 11407.

15. E. C. Walter, F. Favier, and R. M. Penner, *Anal. Chem.*, 74 (2002) 146.

16. F. Favier, E. C. Walter, M. P. Zach, T. Benter, and R. M. Penner, *Science*, 293 (2001) 2227.

17. M. P. Zach, K. H. Ng, and R. M. Penner, *Science*, 290 (2000) 2120.

18. L. A. Porter, J. H. C. Choi, J. M. Schmeltzer, A. E. Ribbe, L. L. C. Elliott, and J. M. Buriak, *Nano Lett.*, 2(12) (2002) 1369.

19. Y. Sun and Y. Xia, *Science*, 298 (2002) 2176.

20. Y. Sun, B. Gates, B. Mayers, and Y. Xia, *Nano Lett.*, 2 (2002) 165.

21. Y. Sun and Y. Xia, *Adv. Mater.*, 14 (2002) 833.

22. F. Kim, S. Connor, H. Song, T. Kuykendall, and P. Yang, *Angew. Chem. Int. Ed. Engl.*, 116 (2004) 3759.

23. D. L. Van Hyning and C F. Zukoski, *Langmuir*, 14 (1998) 7034.

24. H. S. Shin, H. J. Yang, S. B. Kim, and M. S. Lee, *J. Colloid Interface Sci.*, 274 (2004) 89.

25. A. Henglein, T. Linner, and P. Mulvaney, *Ber. Bunsen-Ges. Phys. Chem.*, 94 (1990) 1449.

26. Y. Sun and Y. Xia, *Adv. Mater.*, 15 (2003) 641.

27. Y. Sun, B. T. Mayers, and Y. Xia, *Nano Lett.*, 2(5) (2002) 481.

28. C. Ducamp-Sanguesa, R. Herrera-Urbina, and M. Figlarz, *J. Solid State Chem.*, 100 (1992) 272.

29. Z. L. Wang, *J. Phys. Chem. B*, 104 (2000) 1153.

30. C. T. Black, C. B. Murray, R. L. Sandstrom, and S. Sun, *Science*, 90 (2000) 1131.

31. F. Dumestre, B. Chaudret, C. Amiens, P. Renaud, and P. Fejes, *Science*, 303(6) (2004) 821.

32. K. Soulantica, A. Maissonat, C. M. Fromen, M. J. Cassanove, and M. Chaudret, *Angew. Chem. Int. Ed. Engl.*, 42 (2003) 1945.

33. F. Dumestre, *Angew. Chem. Int. Ed. Engl.*, 41 (2002) 4286.

34. J. Geng, D. A. Jefferson, and G. Johnson, *Chem. Commun.*, (2004) 2442.

35. I. J. Brown and S. Sotiropoulos, *J. Appl. Electrochem.*, 30 (2000) 107.

36. P. Jiang, J. Cizeron, J. F. Bertone, and V. L. Colvin, *J. Am. Chem. Soc.*, 121 (1999) 7957.

37. E. Kim, Y. Xia, and G. M. Whitesides, *J. Am. Chem. Soc.*, 118 (1996) 5722.

38. G. O. Mallory and J. B. Hajdu, *Electroless Plating: Fundamentals and Applications*; American Electroplaters and Surface Finishers Society: Orlando, 1990.

39. J. E. G. Wijnhoven and W. L. Vos, *Science*, 281 (1998) 802.

40. S. A. Johnson, P. J. Olivier, and T. T. Mallouk, *Science*, 283 (1999) 963.

41. S. H. Park and Y. Xia, *Adv. Mater.*, 10 (1998) 1045.

42. L. Xu, A. Zhakidov, R. H. Baughman and J. B. Wiley, *Mat. Res. Soc. Symp. Proc.* Vol. EXS-2, (2004), M5.17.1.

43. L. B. Xu, W. L. L. Zhou, C. Frommen, R. H. Baughman, A. A. Zakhidov, L. Malkinski, and J. Q. Wang, *Chem. Commun.*, (2000) 997.

44. L. Xu, W. L. Zhou, M. E. Kozlov, I. Ilyas, I. Khayrullin, A. Udod, A. A. Zakhidov, R. H. Baughman, and J. B. Wiley, *J. Am. Chem. Soc.*, 123 (2001) 763.

45. J. Erlebacher, M. J. Aziz, A. Karma, N. Dimitrov, and K. Sieradzki, *Nature*, 410 (2001) 450.

46. Y. Ding, M. Chen, and J. Erlebacher, *J. Am. Chem. Soc.*, 126 (2004) 6876.

47. Y. Ding and J. Erlebacher, *J. Am. Chem. Soc.*, 125(6) (2003) 7772.

48. Y. Hattori, T. Konishi, H. Kanoh, S. Kawasaki, and K. Kaneko, *Adv. Mater.*, 15(60) (2003) 529.

49. B. D. Cullity and S. R. Sock, *Elements of X-ray Diffraction*, Prentice-Hall (2001).

50. D. Dollomore and G. R. Heal, *J. Appl. Chem.*, 14 (1964) 109.

51. R. Rajesh, J. Peng, V. L. Colvin, and D. Mittlemanb, *Appl. Phys. Lett.*, 77(22) (2000) 3517.

52. Y. Xia, B. Gates, Y. Yin, and Y. Lu, *Adv. Mater.*, 12(10) (2000) 693.

53. S. H. Park and Y. Xia, *Chem. Mater.*, 10 (1998) 1745.

54. B. Gates, Y. Yin, and Y. Xia, *Chem. Mater.*, 99 (1999) 2827.

55. S. H. Park and Y. Xia, *Adv. Mater.*, 10 (1998) 1045.

56. B. T. Holland, C. F. Blanford, and A. Stein, *Science*, 281 (1998) 538.

57. J. E. G. J. Wijnhoven and W. L. Vos, *Science*, 281 (1998) 802.

58. J. S. Yin and Z. L. Wang, *Adv. Mater.*, 11 (1999) 469.

59. H. Yan, C. F. Blanford, B. T. Holland, M. Parent, W. H. Smyrl, and A. Stein, *Adv. Mater.*, 11 (1999) 1003.

60. L. Vayssières, *Adv. Mater.*, 15(5) (2003) 464.

61. L. Vayssières, K. Keis, A. Hagfeldt, and S. E. Lindquist, *Chem. Mater.*, 13(12) (2001) 4395.

62. Y. Zhao and Y. U. Kwon, *Chem. Lett.*, 133(12) (2004) 1578.

63. L. Vayssières, N. Beermann, S. E. Lindquist, and A. Hagfeldt, *Chem. Mater.*, 13(2) (2001) 234.

64. L. Vayssières and A. Manthiram, *J. Phys. Chem. B.*, 107 (2003) 2623.

65. L. Vayssières, L. Rabenberg, and A. Manthiram, *Nano Lett.*, 2(12) (2002) 1393.

66. J. Y. Lao, J. G. Wen, and Z. F. Ren, *Nano Lett.*, 2(11) (2002) 1287.

67. J. Y. Lao; J. Y. Huang, D. Z. Wang, and Z. F. Ren, *Nano Lett.*, 3(2) (2003) 235.

68. M. H. Huang, S. Mao, H. Feick, H. Yan, Y. Wu, H. Kind, E. Weber, R. Russo, and P. Yang, *Science*, 292 (2001) 1897.

69. W. Yu., X. Li, and X. Gao, *J. Cryst. Growth Des.*, 5(1) (2005) 151.

70. Q. Tang, W. Zhou, W. Zhang, S. Qu, K. Jiang, W. Yu, and Y. Qian. 2005, 5(1), 147.

71. L. Gou and C. J. Murphy, *Nano Lett.*, 3(2) (2003) 231.

72. M. Yu, J. Lin, and J. Fang, *Chem Mater.*, 17 (2005) 1783.

73. S. W. Lee, Y. G. Ryu, K. J. Y. Kwang-Deog Jung, S. Y. An, and C. S. Kima, *J. Appl. Phys.*, 91(10) (2002) 7610.

74. G. Caruntu, G. G. Bush, and C. J. O'Connor, *J. Mater. Chem.*, 14 (2004) 2753.

75. F. Grasset, N. Labhsetwar, D. Li, D. C. Park, N. Saito, H. Haneda, O. Cador, T. Roisnel, S. Mornet, E. Duguet, J. Portier, and J. Etourneau, *Langmuir*, 18 (2002) 8209.

76. T. J. Park, G. C. Papaefthymiou, A. R. Moodenbaugh, Y. Maoa, and S. S. Wong, *J. Mater. Chem.*, 15 (2005) 2099.

77. P. Sekar, E. C. Greyson, J. E. Barton, and T. W. Odom, *J. Am. Chem. Soc.*, 127 (2005) 2054.

78. W. Lu, Y. Ding, Y. Chen, Z. L. Wang, and J. Fang, *J. Am. Chem. Soc.*, 127 (2005) 10112.

79. Y. Ma, L. Qi, J. Ma, and H. Cheng, *J. Cryst. Growth Des.*, 4(2) (2004) 351.

80. E. Leontidis, M. Orphanou, T. Kyprianidou-Leodidou, F. Krumeich, and W. Caseri, *Nano Lett.*, 3(4) (2003) 569.

12
Nano-building Blocks Fabricated through Templates

Feng Li and John B. Wiley

1. Introduction

Nanoscience involves studying matter in a scale range of 1–100 nm. Scientists have tried to achieve dramatic, innovative enhancements in the properties and performance of materials by tuning their nanoscale structural features [1]. This area of research is growing at an ever increasing rate. The ability of nanotechnology to affordably fabricate structures at the nanometer scale will enable new approaches and processes for developing novel, more reliable, lower cost, higher performance, and more flexible electronic, magnetic, optical, and mechanical devices [1–5]. Nanoscience and nanotechnology extend across a wide range of science and engineering [6–8]. Chemists, physicists, biologists, and engineers all have a role to play in their development and implementation. The recognized applications of nanotechnology include:

- Ultrasmall and highly parallel computers
- Image information processors
- Novel communication and computation devices
- New data storage devices with enhanced ability
- Development of lasers and detectors
- Integrated optical, chemical, and biosensors for improved surveillance and targeting
- Catalysts for enhancing and controlling energetic reactions
- Designer materials with combinations of properties that do not exist currently

Toward this end, a variety of nanosized building blocks have been developed for use in the construction of functional microdevices. Especially exciting are the advances in the fabrication of new nanoscale objects—particles, wires, tubules, spheres, rings, ribbons, meshes, etc., which hold the promise of serving as components in new functional architectures [9–16]. This is not simply the idea of miniaturizing known devices, but the realization of completely unique, technologically significant constructs. To craft such systems, new concepts in device design need to be developed and efficient approaches to their fabrication need to be realized.

Due to the advantage of oriented assemblies, researchers over the past several years have strived to develop 1D nanostructures for a variety of applications [17–26]. The reported methods include arc growth, vapor phase, solution methods, electrodeposition, etc. Notably, porous membranes such as anodic alumina membranes (AAMs), porous polycarbonate membranes (PPMs) and porous silicon membranes (PSMs) can be used as hard templates to direct the growth of 1D nanostructures including simple solid nanowires, superlattice nanowires, and core-shell nanowires; nanowires of elements or compounds are readily prepared [27–30]. Especially intriguing is the prospect of microdevices extensively based on 1D nanoscale components and the fabrication of devices, such as transistors, sensors, field emitters, motors, and bar-code devices, through self-assembly [31–36]. Using a combination of chemical and electrochemical methods, we have focused on developing strategies to control nanostructures from 1D to 3D [37–44]. The ability to tune the topologies of materials will allow one to affect dimensionally dependent properties such as electronic, magnetic, optical, etc. Such materials can then be used as nanoscale building blocks in constructing microdevices.

2. Materials and Methods

2.1. Fabrication of Porous Membranes

AAM, which possess self-organized fine structures with a nanopore array of columnar hexagonal cells, has attracted extensive interest as host materials for the production of magnetic recording media, optical devices, and electroluminescence display devices. AAM can be synthesized by anodic oxidation of aluminum in an acidic electrolyte [45–47]. Two-step anodization process has been proved to be effective in fabricating the pore structures with a well-defined hexagonal array. Figure 12.1 shows field emission scanning electron microscope (FESEM) images of AAM with pores of about 40 nm at different magnifications. The pore size of

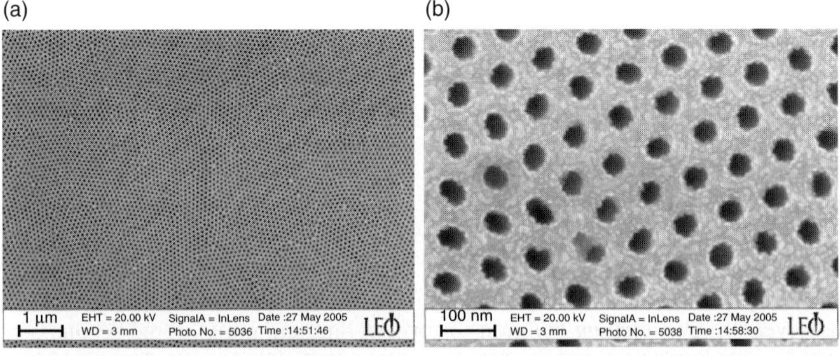

FIGURE 12.1. Anodized alumina membrane (AAM) with pores of about 40 nm prepared by anodic oxidation.

AAM is dependent on the applied potential and the acid used during anodization. Thus AAM with pores of a few nanometers to hundreds of nanometers have already been fabricated. In order to obtain clear images of the insulating AAM, a conductive layer must first be coated to the surface before SEM observation. Choosing an "in-lens" detector during the SEM observation, one can obtain images with much higher resolution compared to those taken with maximum likelihood parameter estimation (MLPE) detector.

Using the porous materials as templates, one can synthesize nanowires and nanotubes by physical, chemical, and electrochemical methods. Microdevices including sensors, transistors, and motors have been fabricated based on the 1D nanosized building blocks synthesized with this method. However, AAM is also limited by its working mediums and temperature. The amorphous Al_2O_3 is very easily dissolved both in acidic and in basic solutions, and its pore structure deforms after annealing at temperatures higher than 500°C. Additionally, it cannot readily dissolve in acidic and basic solutions after annealing at temperatures higher than 800°C, due to the formation of crystalline γ-Al_2O_3 phase. Thus other porous materials may be needed in synthesis if the template has to work at such extreme experimental parameters—one possible choice is PSM.

A photo-assisted electrochemical etching (ECE) procedure can be used to fabricate PSMs [48,49]. Initially, a photolithographic step defines openings in thermally grown oxide on the surface of a silicon membrane. These openings are hexagonally distributed at the surface with a definite center-to-center spacing. Then, anisotropic etching of the silicon in potassium hydroxide (KOH) forms pits (inverted pyramids), which serve to initiate the pore growth during ECE. The ECE experiments were performed at room temperature in a home-designed cell using a 20-W halogen lamp for carrier generation, a bias of 2 V, and a 5.45-wt % HF solution [49].

The pores in Si membranes range in size from several hundreds of nanometers to tens of microns. The preparation of porous Si membranes with smaller, well-organized pores is still a challenge. Because the HF used in the fabrication of Si membrane is corrosive and toxic, the whole operation must be protected carefully in a system constructed with polymer materials such as teflon or polyvinyl chloride (PVC). The silicon materials have two main advantages. They can work at high temperatures up to 1,100°C, and can be removed conveniently in hot KOH solution. Also, they can allow access to large pores with controlled pore structure.

Polymer membranes with self-assembled pore structures can also be employed as a template to direct the growth of high-density 1D nanostructures [50]. An ion track etching technique can also be used to fabricate polymer membranes with cylinder pores in a two-step process [51]. In the first step, a thin polymer film is exposed to charged particles from a nuclear pile. As these particles pass through the polymer materials, they leave sensitized tracks. The polymer tracks are then dissolved with an etching solution to form cylindrical pores in the second step. The pore sizes of polymer membranes, which can be from tens of nanometers to microns, can be controlled precisely through varying the temperature and strength of the etching solution and the exposure time to them. The produced polymer membranes are a thin, translucent, and microporous film with a smooth, flat surface. They can work in acid or base solutions, but are limited in working temperature and sensitive to organic solvents.

2.2. *Synthesis of 3D Colloid Crystals*

3D nanosized building blocks can also be fabricated by template-directed growth method. One of such templates is colloid crystal, which can be synthesized by published methods such as sedimentation and perpendicular growth. For example, silica spheres can form close-packed lattices through a sedimentation process over several months. After sintering at 120°C for 2 days and then 750°C for 4 h, robust opalescent colloid crystals can be fabricated and can be readily cut into smaller sections for applications. However, the drawbacks of the technique are also very apparent. There are many defects in the synthesized colloid crystals and thus only polycrystals can be produced even when grown slowly for months. A perpendicular surface growth procedure [52] has thus been developed to fabricate single-crystal opals of controlled thickness as shown in Fig. 12.2. The single crystals of centimeter square size can be synthesized conveniently in weeks. Recently, many other techniques such as spin-coating technique have been developed for fast fabrication of colloid crystals [53].

2.3. *Electrochemical Deposition*

A variety of chemical and electrochemical procedures have been developed to fabricate 1D and 2D nanostructures from templates. We mainly use electrochemical deposition to synthesize 1D nanowires and nanotubes, as well as 3D nanomeshes [37–43]. First, the templates must be converted to an electrode for deposition as shown in Fig. 12.3. For this end, a thin layer of metal such as Au, Ag, or Pd–Au of about 200-nm thick is deposited onto one side of the samples by magnetron sputtering. Then a silver wire is attached to the metal surface of the samples with silver paste (PELCO Colloidal Silver Paste, Ted Pella Inc., USA) and the metal side of the sample is sealed off with adhesive glue (Scotch Super Strength, 3M, USA). Electrochemical growth can be carried out within the pores of the porous templates at room temperature by a constant current method at $0.1\ \mathrm{mA\cdot cm^{-2}}$ or pulse method (rest time: 1 min; run time: 1 s; and cycle to control the length of nanowires) at $5\ \mathrm{mA\cdot cm^{-2}}$ on a Princeton Applied Research VMP2. The cell used in electrodeposition consists of Pd–Au/template cathode, a platinum anode, and a platinum counter electrode. Commercial Ni, Au, Ag, and Zn plating solution from Technics Inc., USA, are employed in the plating process.

2.4. *SEM and TEM Observation*

The structure characterization of the related materials was carried out on LEO 1530 VP FESEM and JEOL JSM 5410 SEM. TEM was performed on a JEOL EM 8291 electron microscope with a 200-kV accelerating voltage.

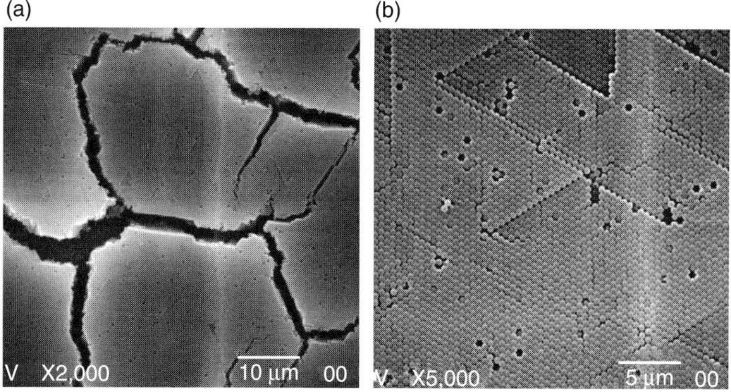

FIGURE 12.2. Scanned electron microscopy (SEM) image of colloid crystal prepared by a perpendicular growth process.

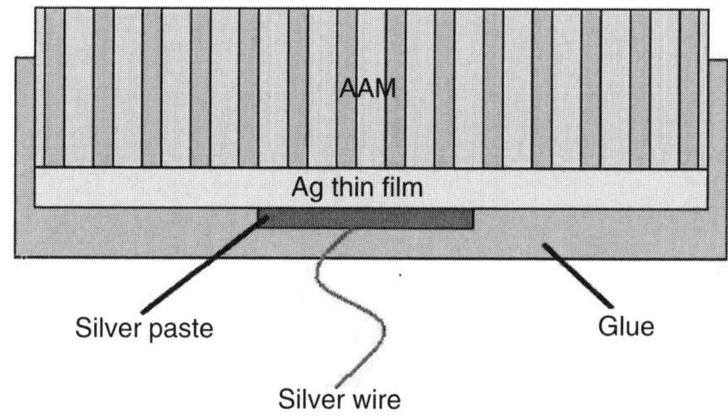

FIGURE 12.3. The structure of microelectrode (see color insert).

3. Nano-Building Blocks

3.1. Nanowires from Porous Templates

A variety of nanowires and nanotubes have been synthesized by template-directed growth. Using electrochemical deposition and AAM as template, we can tune the structures of nanowires by adjusting the deposition time and switching plating solutions as shown in Fig. 12.4. 1D superlattices can be synthesized conveniently. Due to the contrast difference corresponding to different materials such as Au and Ni, the superlattice can be observed conveniently through SEM observation (Fig. 12.4c). The synthesized 1D superlattice nanowires have already been turned into transistors, biosensors, bar codes, and nanomotors [31–36].

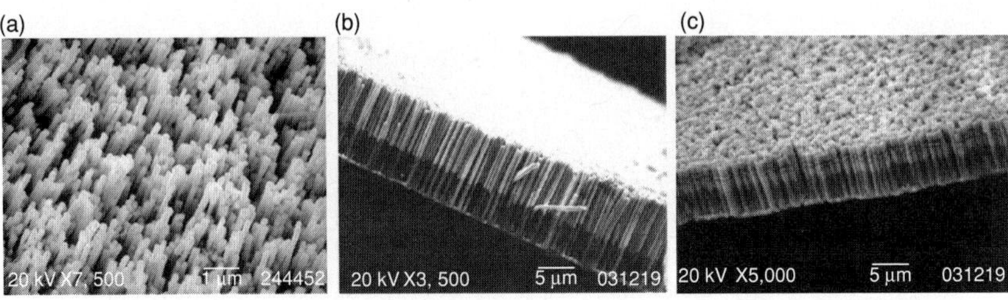

FIGURE 12.4. Nanowire array from template-directed growth: (a) Top view of Au nanowire array; (b) Cross-sectional view of Ni–Au nanowire array; and (c) Ni–Au superlattice nanowire array.

3.2. Nanotubes from Glue Wire–Modified Templates

In order to further control the structure of nanowires, a combination of chemical and electrochemical methods has been developed to direct the growth of nanowire arrays. After pore structure modifications, AAM can be used to direct the growth of 1D nanostructures such as nanotubes. The general procedure used for fabricating nanotubes with glue wires–modified template is outlined in Fig. 12.5. AAMs are initially sputtered on one side with a thin coating of metal. The sputtering process is controlled such that the 1D pores of the membrane are only partially closed. The membrane is then converted to an electrode assembly, which in the process introduces glue into the pores. Controlled electrodeposition into these pores allows the growth of nanoscale metal tubes.

It is a critical step in this fabrication to keep the membrane pores open, after AAM is coated with a Pd–Au thin film. The commercial alumina membranes used in the process have an average pore size of 300 nm, which can be partially filled by magnetron sputtering of Pd–Au film. A Pd–Au thin film of about 200 nm can reduce the pore size of AAM to approximately 150 nm. A wire lead is attached to the metal coating with silver paste, before sealing this side with glue adhesive. Meanwhile, the glue resin is found to diffuse well along the length of the pores before setting and the nanowires of glue can be produced inside the pores. The glue wires do not impede the deposition of metals. Although normally this glue product becomes quite rigid on drying, exposure of the adhesive to water makes the material more pliable, allowing metal to deposit between the pore wall and the glue wire.

Figure 12.6a and b shows top and side views of Ni nanotube array. The tubes made in this method are about 3 µm in length and 300 nm in diameter with a wall thickness of 25 nm. Glue nanowires have already been removed from the system with dichloride methane. Figure 12.6c shows the back view of such fabricated Ni–Au nanotube array with glue nanowires still kept inside the nanotubes. A piece of porous mesh due to the Pd–Au coating on the surface of AAM can be observed in this SEM image, which highlights the opened pores of AAM after magnetron sputtering.

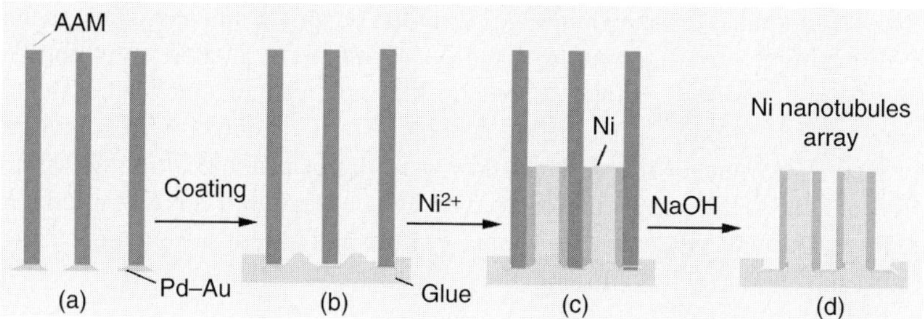

FIGURE 12.5. Processing scheme for fabrication of Ni nanotubes from glue nanowires– modified template. After partially coating the anodic alumina membrane (AAM) with Pd/Au (a) and attaching a metal wire contact (not shown), the metal surface is coated with glue that diffuses into the open pores of the membrane to produce glue wires (b). The glue wires help to direct the electrodeposition of nickel along the wall of the pores producing nickel nanotubes (c). Ni nanotube array after removal of AAM and glue wires (d). (see color insert).

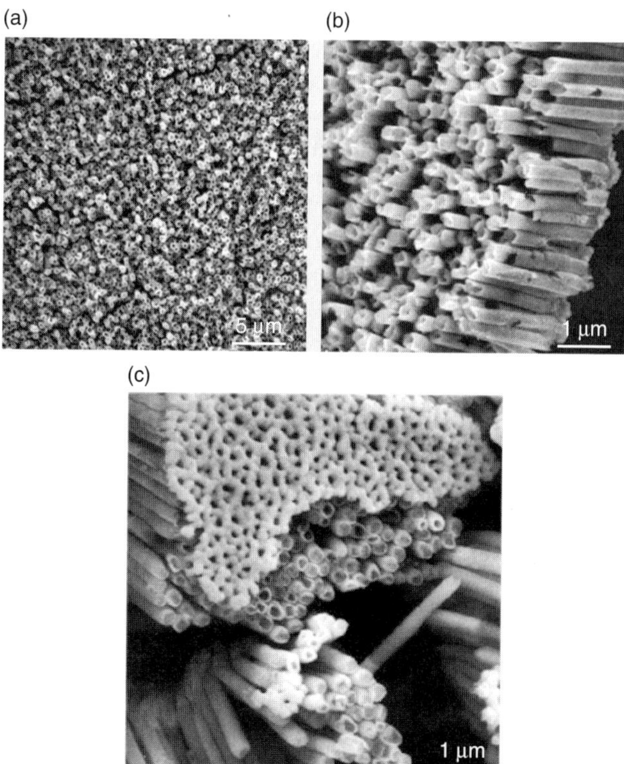

FIGURE 12.6. Scanned electron microscopy (SEM) images of (a) top view at low magnification, and (b) tilted view at higher magnification of Ni nanotube array. (c) SEM image of back view of Ni–Au nanotubes with glue wires inside. A piece of Pd–Au mesh is attached to the nanotubes.

Ni–Au nanotubes, as shown in Fig. 12.6c, can be turned into very short segment of Au nanotubes by dissolution of the Ni segments in an HCl solution. TEM images of the produced Au nanotubes (Fig. 12.7) indicate that the wall of the nanotubes consists of polycrystalline Au nanoparticles. Compared to SEM images in Fig. 12.6, TEM images reveal the hollow structure of nanotubes through contrast difference due to the sample thickness. The TEM observation can give microstructure information of the materials, while SEM method has the advantage in characterizing the surface and 3D morphology of nanosized building blocks.

3.3. Nanowires with Structured Tips from Nanotube–Modified Templates

The composite of glue nanowires, Ni nanotubes, and AAMs can be further used as a template to synthesize nanowires with structured tips. The fabrication process produces interesting nanostructures as shown in Fig. 12.8. After soaking in water for 3 days, Ni nanotubes can be oxidized into nonconductive $Ni(OH)_2$ tubes (Fig. 12.8b). The oxidized nickel component serves to modify the shape of the alumina pore and, consequently, affect the shape of the resulting wires (Fig. 12.8c). Removal of the template and the nickel component results in gold wires with structured tips (Fig. 12.8d) [41]. Figure 12.9a shows an array of wires still attached to the Pd–Au backing. Here the wires are about 7-μm long with the main part of the wire body about 5-μm long and 300 nm in diameter; the structured wire tips are about 2 μm with diameters that average ~200 nm. Sonication can be used to release the wires from the Pd–Au backing. A TEM image of dispersed gold wires with relatively shorter bodies (bodies about 1 μm and the tips over 1 μm) is shown in Fig. 12.9b. The selective area electron diffraction (SAED)

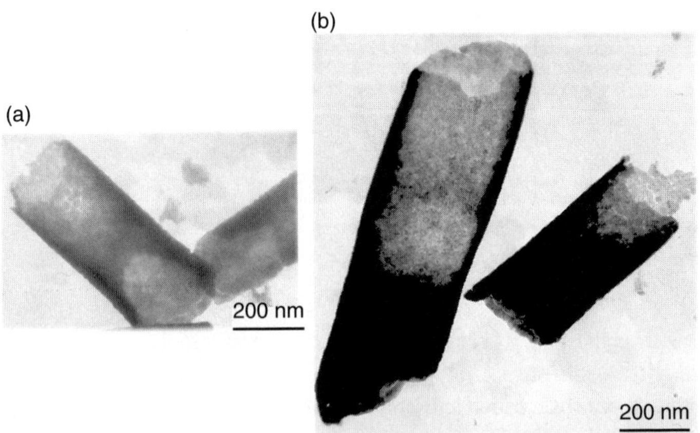

FIGURE 12.7. Transmission electron microscopy (TEM) images of Au nanotubes with different morphologies.

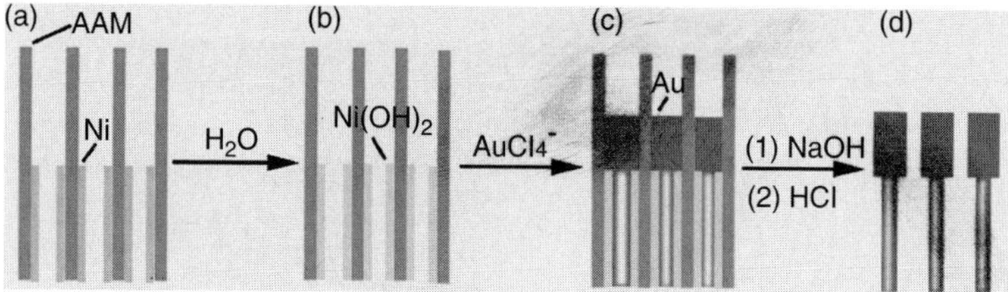

FIGURE 12.8. Scheme of directed growth of nanowires with structured tips (see color insert).

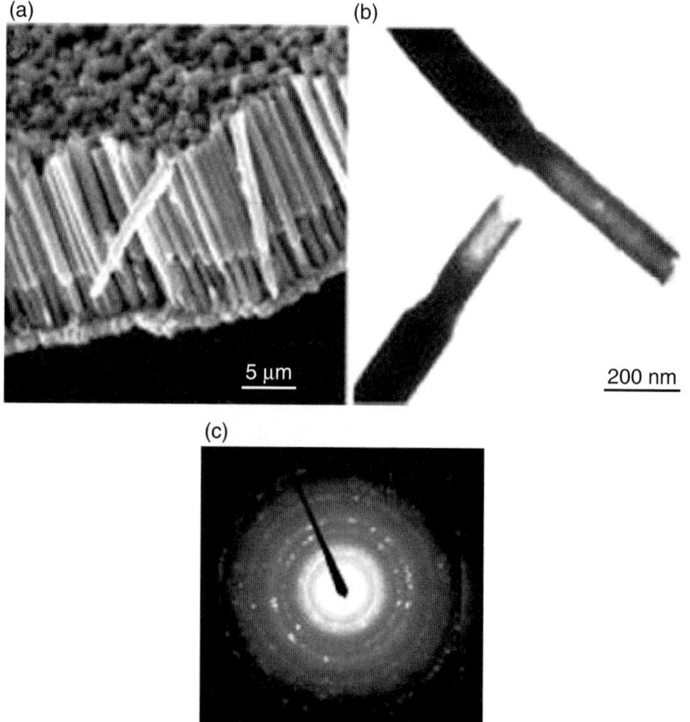

FIGURE 12.9. (a) Scanned electron microscopy (SEM) image of Au nanowires with structured tips; (b) transmission electron microscopy (TEM) image of Au nanowires with hollow tips; (c) selective area electron diffraction (SAED) pattern of the tips.

pattern (Fig. 12.9c) of the structured tips can further reveal that the tips are still hollow and consist of polycrystalline Au nanoparticles. The TEM image of one nanowire with whole structured tip is shown in Fig. 12.10a. The hollow structure of the tip can be observed clearly. If we repeat electrodeposition to get Ni–Au superlattice and then oxidize Ni segments into nonconductive NiO, the composite can be used as template to fabricate nanowires with sculptured grooves and tips as shown in Fig. 12.10b.

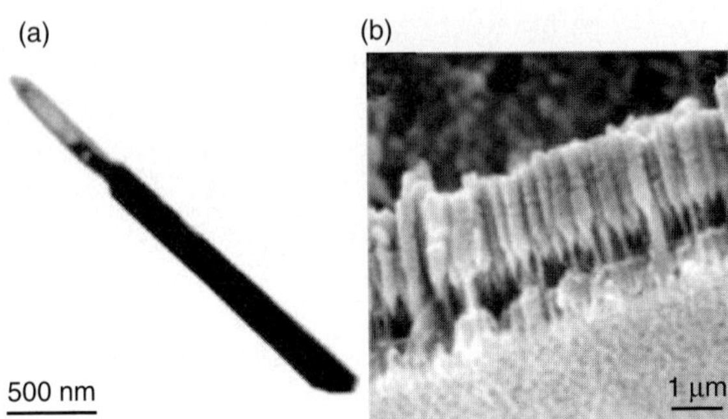

FIGURE 12.10. (a) Transmission electron microscopy (TEM) image of Au nanowire with structured tips and (b) nanowires with sculptured grooves and tips.

It is found that glue wires have played an important role in the production of gold wires with structured tips. First, in the formation of nickel tubes, these wires appear to direct the growth of nickel along the pore wall. If deposition methods are exercised without the wires present, solid nickel wires, are formed, not tubes. At first it seems unusual that the glue resin would allow the growth of nickel tubes along the pore wall. One might expect that the resin would effectively fill the pore and, on drying, be impermeable to the plating solutions. On exposure to water, however, we have found that the glue component becomes pliable. This may allow for displacement of the wire in the nickel-plating process. Also, the extended exposure (3 days in the nickel oxidation step) could have a similar effect, allowing for the deposition of gold between the oxidized nickel component and the glue wires. Many of the fabricated wires had hollow tips due to the presence of the glue wires. The efficient structure control of nanowires and nanotubes will make it possible to tune their properties for possible incorporation into novel microdevices.

3.4. Colloid Crystal Wires and Porous Wires from Directed Assemblies

Colloid crystals have received considerable attention in the past few years due to their periodical dielectric structures, which can result in these systems offering the possibility of controlling photons in a manner analogous to electrons in semiconductors. Investigations in this area could lead to a variety of technologically significant materials including new types of optical filters, optical switches, and mirrors [54–59]. Further interest arises from the importance of such structures to biological systems, including biomaterials and phyllotaxis, the study of geometric features in plants. Much of the current synthetic effort in this area have focused on the fabrication of colloid crystals in one, two, and three dimensions, as well as inverse colloid crystals created by infiltrating different materials into the void space of colloid crystals. However, compared with the synthesis of 2D and 3D colloid crystals and inverse colloid crystals, the fabrication of 1D colloid crystals can be much more difficult due to the required geometric control. Based on the directed assembly technique developed in our laboratory, we have successfully modified the structure of porous membranes with colloid spheres and directed them into assemblies with a variety of geometries. Freestanding colloidal crystal wires with tubular-like packing and porous wires have already been synthesized conveniently by this method [40]. Figure 12.11 outlines the procedure used in the fabrication of colloidal crystal wires and porous wires. For the fabrication of photonic crystal wires, two pieces of porous membranes (Fig. 12.11a), porous silicon and AAM, are initially placed flush to each other. Then silica spheres of about 1.4 μm are infiltrated into the pores of silicon membrane by vacuum filtration (Fig. 12.11b). The pores of the AAM are smaller than those of Si and smaller than the diameter of colloidal spheres. The spheres can thus be blocked in the pores of top Si membrane after filtration. Then the composite of silicon membrane and colloid sphere was treated with silane solution

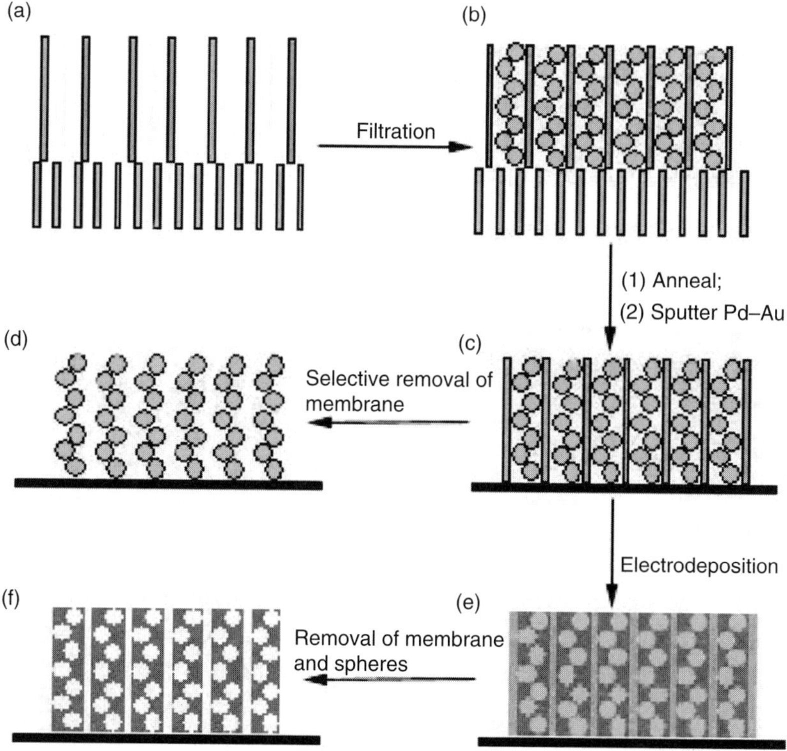

FIGURE 12.11. Fabrication procedure of colloidal crystal wires and porous wires: (a) two porous membranes are placed flush with each other; (b) colloid spheres are infiltrated into top membrane; (c) metal deposition to convert the composite into electrode and obtain metal base; (d) remove template to get colloidal crystal wires on metal base; (e) electrodeposition in colloid spheres-modified template; and (f) removal of templates to release porous wires (see color insert).

(3-methacryloxypropyl-trimethoxysilane, 1% in ethanol), dried at room temperature overnight and annealed at 900°C for 8 h under Ar to fuse the colloid spheres together. In order to provide a support for the colloidal wires after the silicon membrane was removed, the composite was then coated on one side by metal by sputtering, followed by electrodeposition (Fig. 12.11c). After selective removal of Si membrane in KOH solution (6 M) at 85°C, freestanding colloidal crystal wires were released from the template (Fig. 12.11d). If the silane and annealing treatment is repeated several more times, a silica sheath could be found to form on the outside of the wires and thus got peapod-like structure.

Figure 12.12a and b shows a porous Si membrane before and after filling with SiO$_2$ spheres of about 1.4 μm. The porous Si membrane of 200-μm thick with

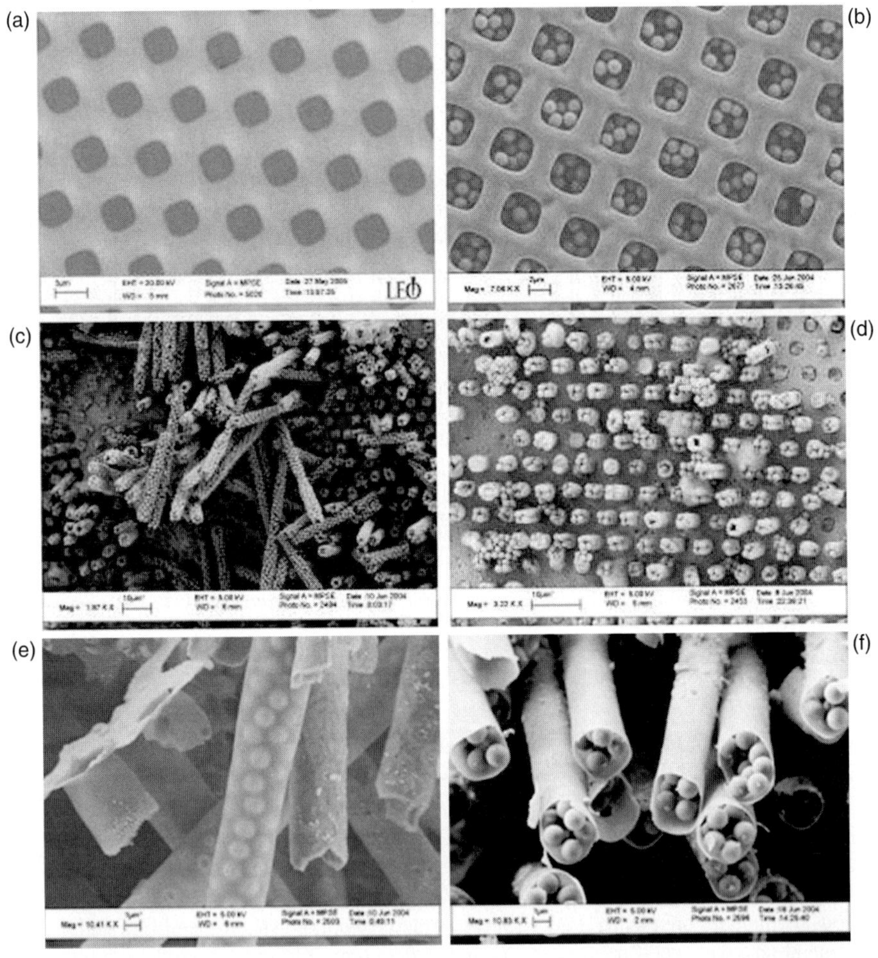

FIGURE 12.12. SEM images of (a) porous silicon membrane; (b) silicon membrane with assembled silica spheres; (c) colloidal crystal wires after release from template; (d) a pattern of short colloid wires.

pores of 3.4 µm is fabricated with a photo-assisted ECE process. Figure 12.12c shows the photonic crystal wires after released from Si membrane with KOH solution. The freestanding colloid wires are typically 60-µm long. Due to the fact that the pores in the silicon membranes used in this study are not perfectly uniform, this results in different sphere-packing schemes within the colloidal wires (Fig. 12.12c). Typically, the spheres pack to produce wires with six helical strands, though wires can also be observed with four to seven strands. Because the wires are somewhat *fragile*, often breaking off at the base of the wire, most of the colloid wires have been cut and washed away from the substrate during fabrication process. Figure 12.12d shows an array of short colloid wires connected with the metal base. The colloidal crystal wires can be encased in a silica sheath by cycling the silane treatment and annealing process several times. The sheaths can appear transparent, though these were coated with metal to minimize charging during the imaging process. Figure 12.12e and f shows a series of silica sheaths. The spheres packed into the sheath can be seen from the top view of the peapod-like structure (Fig. 12.12f). Figure 12.13 shows representative colloid wires having various packing schemes. Figure 12.13a contains four parallel strands while Fig. 12.13b shows a wire with predominately six helical strands. Intermediate to these two packing schemes, one can find chiral arrangements. Figure 12.13c shows a top view of some freestanding colloidal wires; the wire in the lower right contains six helical strands where the packing clearly shows a right-handed helical twist. A series of colloidal crystals with tubular packing have been fabricated through directed assembly. The structure of colloid wires can differ as the diameter of the membrane channels varies relative to the sphere size. While in this study we have observed several packing geometries, there are clearly a large number of both chiral and achiral packings that can be targeted in the future.

By filling porous membranes with nanospheres, we can thus modify the pore structure of templates. A nanosphere-membrane composite allows the production of wires with extended pore structures [42]. The general procedure used to fabricate porous wires is also shown in Fig. 12.11. Initially, two porous membranes with different channel diameters are placed flush to each other (Fig. 12.11a).

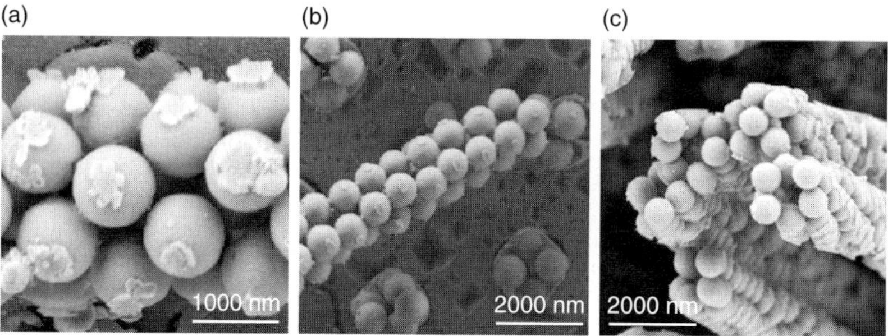

FIGURE 12.13. Examples of different packing geometries. (a) Wire contains six helical strands ($x = 3.0$); (b) wire with four parallel strands ($x = 2.22$); and (c) wire exhibiting chirality with a set of right-handed helices ($x = 2.73$).

Colloid spheres are then infiltrated into the upper membrane by vacuum filtration as shown in Fig. 12.11b. They are locked in the pores of top membrane with the help of bottom membrane that has smaller pores (~10 x). These two steps are similar to the fabrication of colloid crystal wires. The membrane-sphere composites are then converted to an electrode for deposition of gold or nickel metal (Fig. 12.11e). Finally, the membrane and spheres are dissolved to produce porous wires as shown in Fig. 12.11f. Porous wires can be grown from AAMs or polycarbonate membranes (PCMs) containing either silica or polystyrene (PS) spheres.

Figure 12.14 shows porous wires produced from PCM (1-μm channels) and silica spheres. Spheres of about 500 nm in diameter, in and out of the channels, (Fig. 12.14a) can be observed prior to the growth of the porous wires. Electrodeposition within the modified membranes readily produces porous gold wires. Varying the size of the spheres allows one to control the pore size within the wires. Figure 12.14b and c present micrographs of porous wire arrays made from 500- and 300-nm silica spheres, respectively. Evidence for extended pore structures can be seen along the length and tips of the wires as well as from cross-section images (insets of Fig. 12.14b and c). The arrangement of the spheres in the membrane channels dictates the pore structure in the wires. Contacts between the spheres and the channel wall result in pores on the surface of the wires, and contacts between the spheres themselves produce openings between adjacent pores.

Nickel and gold wires with smaller pores, prepared from 300-nm channel alumina templates and 140-nm silica spheres, are presented in Fig. 12.15. Figure 12.15a shows cross-sectional view of AAM together with SiO_2 nanospheres of 140 nm in its channels. Arrays of these wires are obtained after removal of the alumina and silica components (Fig. 12.15b). Although most of the wire surfaces show extensive pore structures, smooth surfaces are occasionally observed indicating that the wire is either solid in these regions or that the pores do not penetrate to the surface. TEM images (Fig. 12.15c and d) of porous gold wires show interconnects between adjacent pores across the width of the wires.

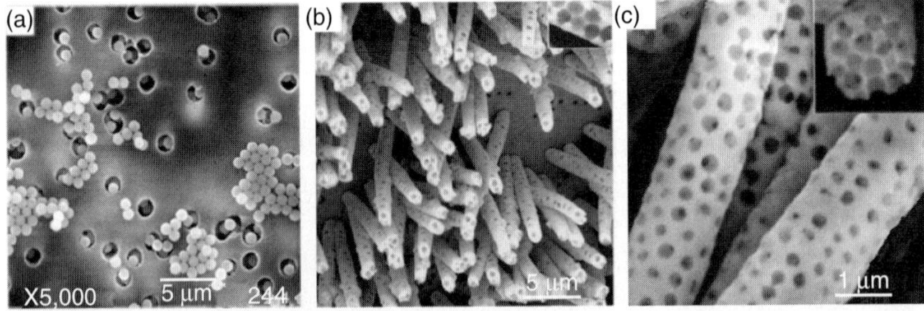

FIGURE 12.14. Scanned electron microscopy (SEM) images of (a) polycarbonate membrane with 500-nm SiO_2 beads and (b) top view of porous Au wires arrays with 500-nm pores. *Inset* in (b): cross section along the length of a porous wire. Side view (c) of porous Au wires array with 300-nm pores. *Inset* in (c): cross section along width of 300-nm pore wire.

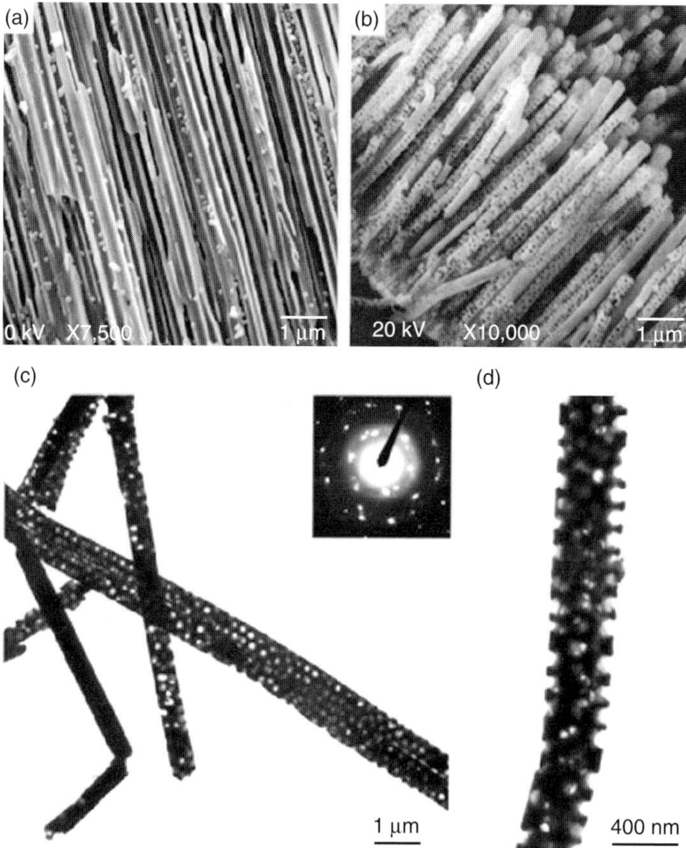

FIGURE 12.15. (a) Cross-section image of anodic alumina membranes (AAM) with SiO_2 nanospheres of 140 nm; (b) scanned electron microscopy (SEM) image of an array of 300-nm porous Ni wires with 140-nm pores and (c,d) transmission electron microscopy (TEM) images of porous Au wires with 140-nm pores. The *inset* in (c) shows selective area electron diffraction (SAED) pattern of Au porous nanowires.

The packing of colloid spheres in the channels of porous membrane can be described as consisting of various numbers of helices. Using formalism similar to that of Pickett *et al.*, we define the relative channel size, *x*, as the ratio d_c/d_s [60]. A plot can then be created with the number of helices vs. relative channel size (Fig. 12.16). It should be realized that these packings could be designated by more than one helix; here we use the set of helices with the greatest pitch to define the number of helices in a particular packing. Both chiral and achiral packings are observed; while the symmetric achiral structures have only one structure associated with them, the chiral structures exhibit both right- and left-handed packing arrangements. One set of structures contains parallel sets of helical strands—in this case the pitch is actually infinite. These structures occur at

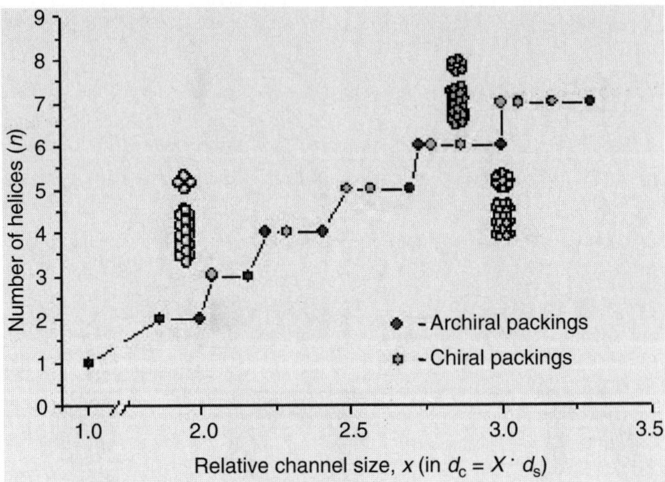

FIGURE 12.16. Number of helices vs. relative channel size for the packing of spheres within channels. Both chiral and achiral packings are indicated (see color insert).

$x = 1.0$, 1.866, 2.22, and 2.73. For structures with helices $n > 1$ and the largest relative channel diameter, symmetric (achiral) structures also occur; this is at $x = 2.0$, 2.16, 2.41, 2.7, 3.0, and 3.3; alternatively, these structures can be described as being based on sets of rings with n-members (for $n \geq 3$) where the angle between the ring members is $360/n$. All the indicated structures intermediate to the achiral ones show chirality. The colloid crystal wires shown in Fig. 12.13 are also highlighted in Fig. 12.16. The experimental results can be modeled exactly with the suggested packing behaviors of colloid spheres. A series of interesting chiral and achiral packings can be achieved by tuning the diameter ratio of colloid sphere and pore precisely.

The porous wires also exhibit a variety of packing geometries duplicated from the colloid wires in the porous membranes. As the channels of the template used in the directed assemblies are not even, the pores along the body of the wires deviate from the ideal, long-range packing in the most instances. In the system prepared from PCM with 1-μm channels and 500-nm spheres (i.e., $x = 2$), one might expect simple packing like hexagonal closest packing (hcp). The PCM channel diameters, however, vary in size (Fig. 12.14a) and, consequently, the packings are varied. Further, packing faults appear quite common, likely due to the rapid infiltration of spheres. Some ordered regions are still observed either as pores along the surface of a wire or within cross sections. Zigzag packing, for example, is seen (inset, Fig. 12.14b); this packing structure is consistent with a channel diameter smaller than twice the sphere diameter ($x \approx 1.87$). As the relative channel size increases ($1.87 < x < 2$; $2 < x < 2.15$), chiral packings could be expected; the arrangement of pores along the surface of several wires gives some indication of this, though packing defects or simply random arrangements of spheres are apparent (Fig. 12.14b). On further increase of the relative channel size

($x > 3$; Fig. 12.14c), short-range-order packing predominates. In wires prepared from 140-nm spheres and 300-nm AAM ($x \approx 2.14$; Fig. 12.15), similar levels of order occur though evidence for the less-efficient, achiral packing ($x = 2.22$) is observed as a series of adjacent pores, especially clear along the right edge of the wire as shown in Fig. 12.15d. The structure of porous nanowires and colloid wires may be improved through close packing of the colloid spheres in membranes with channels that are smooth and uniform.

The extensive pore structures of the wires influence their physical properties. If we assume nearly ideal packing in, for example, the 1-µm wires with 300-nm spheres, the porosity (% void volume) of the porous wires will be about 70% and their relative increase in surface area vs. a solid wire will be about 4×. The short-range order predominating in these systems would, however, work to lower these values slightly. The high porosity of the nanowires has also resulted in a decrease in the mechanical strength of the wires. It was found that the smaller wires (Fig. 12.15b) are especially fragile.

By varying the components used in the fabrication of colloidal crystal wires, diverse materials with interesting properties could be obtained. Further, surface modification of this system will be potential in the design of microdevices including biosensors and selective reactors. The porous wires were prepared by electrodeposition of Au and Ni; a variety of other elements and compounds formed either chemically or electrochemically could also be fabricated as porous wires. These porous materials could be used for applications in catalysis, electrolysis, and sensor arrays. The higher surface areas of the porous materials should serve to increase reactivity/sensitivity, and also control of the pore morphologies might influence selectivity in chemical conversions. A further application for such materials might come in the form of inverse photonic wires. Although the rapid infiltration of spheres into the pores impeded the formation of the highly ordered structures desired in photonics, sphere precipitation into membrane channels over extended periods should lead to more order. Such wires could offer interesting properties and prospect of photonic wires, especially those based on chiral packing, is particularly intriguing. While there are still many challenges that have to be met in the fabrication of various colloidal crystal and porous wire structures, success in these systems will effectively lead to important new technologies.

3.5. 1D, 2D, and 3D Inverse Colloid Crystals from 3D Colloid Crystals

An electrochemical deposition method has also been developed to synthesize 3D inverse colloid crystals from 3D colloid crystals as templates in our laboratory [38]. The procedure for this method is shown in Fig. 12.17. Colloid crystal slabs of about 1 cm^2 were first coated with a Pd–Au thin film of about 200-nm thickness to make electrode. An Ag wire was then connected to the metal-coated surface and protected with glue. A low current density (0.50 mA cm^{-2}) was used in an effort to achieve even deposition within the pores of photonic crystals. Inverse

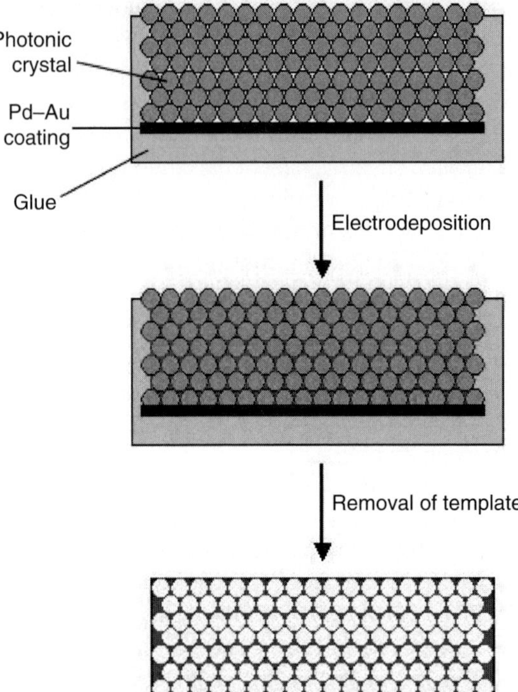

FIGURE 12.17. Procedure scheme of 3D metal inverse photonic crystals (see color insert).

colloid crystals with well-ordered pore structures duplicated from the photonic crystals were produced after removal of templates. To remove the silica matrix, the metal–opal pieces were soaked in 2% HF solution for 24 h.

The structures of 3D colloid crystals can be duplicated conveniently with this method. Figure 12.18 shows a section of a nickel reverse colloid crystal (metal mesh) corresponding to (100) of colloid crystal, after dissolution of the silica

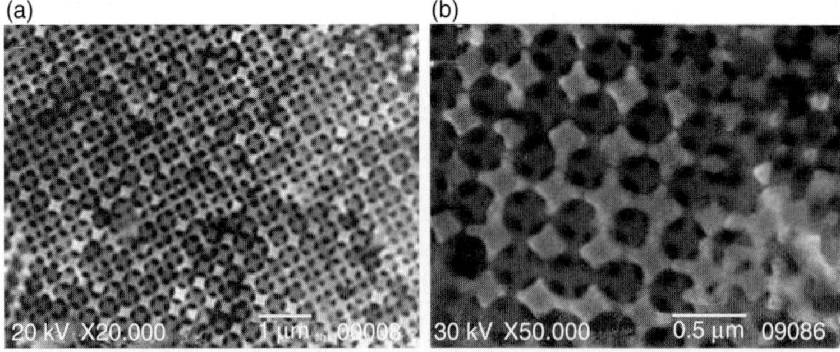

FIGURE 12.18. Scanned electron microscopy (SEM) images of Ni mesh at (a) low and (b) high magnifications.

spheres. The higher-magnification SEM image of the sample (Fig. 12.18b) shows the square features, which are essentially cubes with concave sides that arise from filling the octahedral sites in the close-packed structure. Each cube is connected to eight other cubes through its vertices via tetrahedra. The structure is akin to the fluorite structure (CaF_2) where the calcium ions, representing the cubes, are eight coordinate and the fluoride ions, representing the tetrahedra, are four coordinate. Based on the diameter of the close-packed spheres of ~300 nm, one would predict minimum diameters of ~50, 70, and 120 nm for the interconnects, tetrahedra, and cubes, respectively, and this is what is observed. Since the metal mesh crystals are interpenetrating air- and metal-phase networks, they are metallo-dielectric photonic crystals, which are predicted to have an unusual metallicity band gap. This technique has the major advantage of readily producing well-defined metal meshes from materials with high melting points where melt infiltration would not be possible due to the instability of templates at higher temperatures.

Interestingly, the composite of colloid crystal and metal mesh can cleave in a similar fashion to single crystals. The phenomenon allows for an alternate, mechanical route to the fabrication of reverse photonic crystals in different dimensions. A polishing method for the controlled cleavage of colloid crystal and metal-mesh composites was developed based on this idea in our group [43]. A piece of SiO_2 photonic crystalline slab, which had been partially infiltrated with nickel metal (~100 μm) by an electrochemical method, was mounted onto a polishing wand and protected on four sides by pieces of single-crystal silicon. A piece of ~15-μm thick black composite was produced by polishing both sides with silicon carbide sand paper for 20–30 min. Figure 12.19a shows an image of a typical silica surface after polishing. The polishing process readily allows access to large area arrays with even topologies. Figure 12.19b shows a top view of a smooth extended section of metal mesh. Figure 12.19c and d shows cross-sectional images of mesh pieces before and after dissolution of the silica with HF solution, respectively. Inverse structures of face-centered closest (fcc) SiO_2 colloid crystals are readily apparent (Fig. 12.19d, inset of Fig. 12.19b).

Thin 2D nickel-mesh nanostructures, typically one- to three-layers thick, can also be fabricated when the polishing procedure is continued for longer times. Figure 12.20 presents a series of thin mesh pieces after treatment with HF. The low-magnification SEM image (Fig. 12.20a) indicates that large sections of 2D inverse colloid crystals are accessible by this approach. A view along the edge of this piece at higher magnification (Fig. 12.20b) shows that it is ~300-nm thick. Higher-magnification images clearly show sections of one- and two-layers thick (Fig. 12.20c and d). The (111) and (100) orientations from the colloid crystal template are replicated in these 2D nanonetworks as shown in Fig. 12.20c and d as well as the inset in Fig. 12.20a.

A further significant structural feature observed in these systems is the formation of 1D components. Figure 12.21 shows examples of corrugated wires and ribbons. Unlike the extended metal-mesh arrays, these features were far less common. It is expected that they form from the shearing of a metal-mesh layer

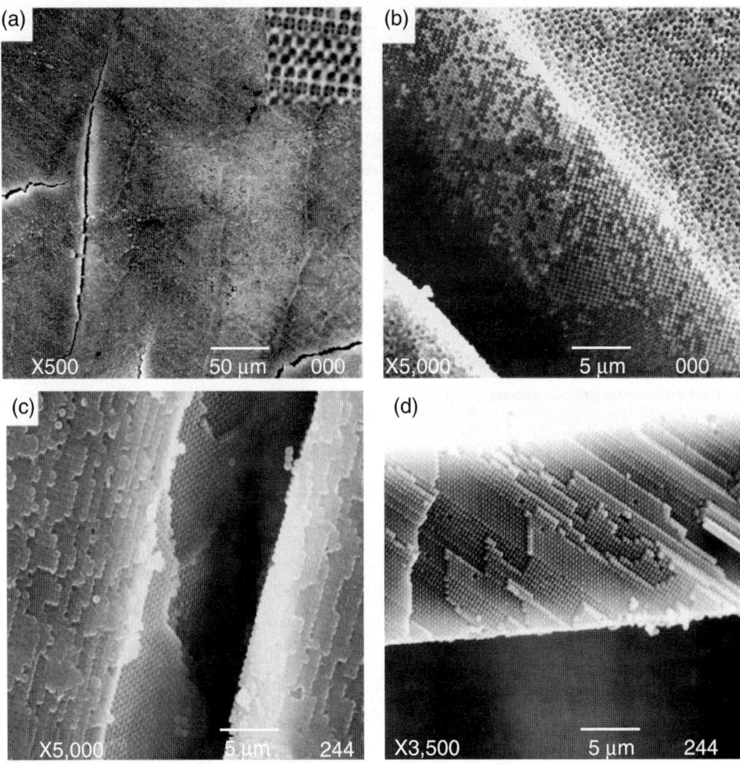

FIGURE 12.19. (a) Top view of colloid-crystal metal-mesh composite after polishing, (b) low-magnification image of inverse nickel mesh after removal of silica, (c) cross-sectional view of composite after polishing, (d) cross-sectional image of thick nickel mesh piece. *Inset* of (b) shows high-magnification image of the surface of nickel mesh.

through an additional opposing movement of a colloid crystal plane, possibly adjacent to a packing fault.

If the polishing process is done with a precision-ion polishing system, a thin photonic-crystal metal-mesh composite piece with an extremely smooth surface can be fabricated. Approximately 10-μm thick sections can be produced when samples are polished on both sides for about 3.5 h. The resulting pieces readily refract light of varying wavelengths depending on the orientation of the piece (Fig. 12.22). This phenomenon indicates, as observed in the SEM images, mechanical processing of colloid-crystal metal-mesh composites by this simple polishing method, which allows for the controlled fabrication of thin sections of inverse metal-mesh colloid crystals with smooth surfaces. Interest in such capabilities may come from the need to process photonic materials for applications where freestanding, thin photonic crystals are desirable. The ability to prepare corrugated nanowires by this approach is also of interest. The periodic shape offered by these 1D and 2D materials is expected to lead to unusual transport, magneto transport, and mechanical properties.

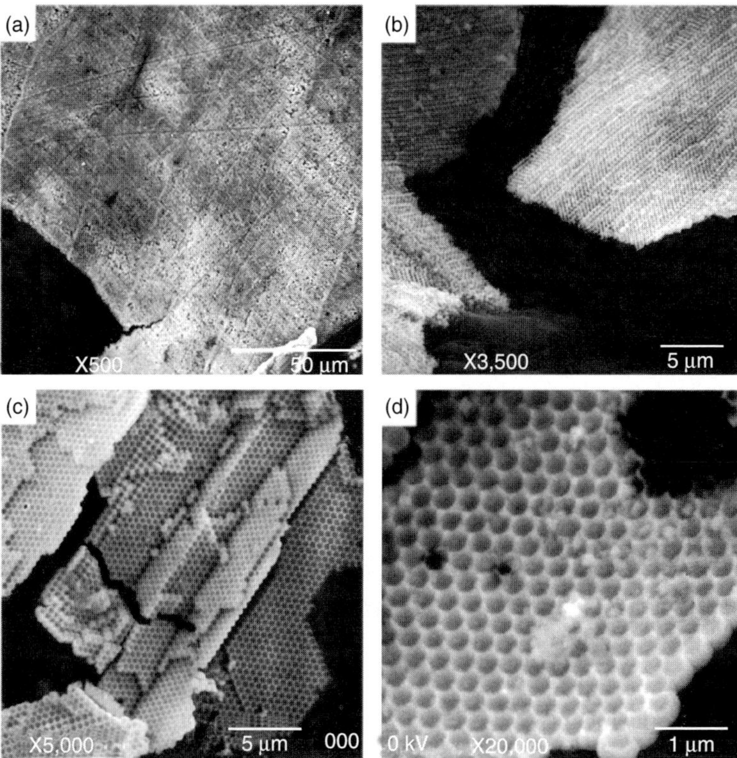

FIGURE 12.20. (a) Top view of photonic-crystal metal-mesh composite after polishing, (b) low-magnification image of inverse nickel mesh after removal of silica, (c) cross-sectional view of metal mesh, (d) cross-sectional view of thick nickel mesh piece.

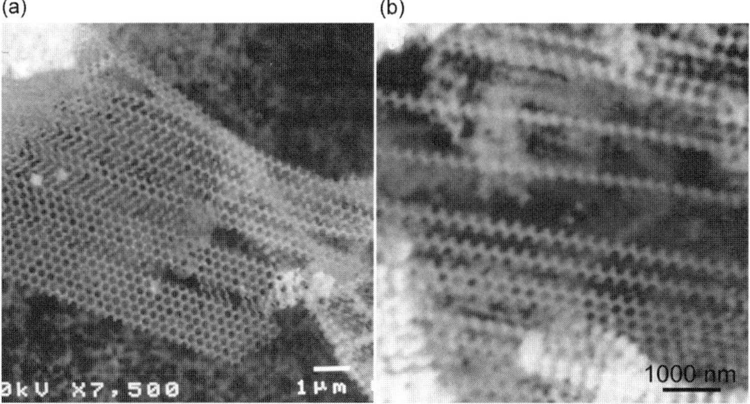

FIGURE 12.21. Scanned electron microscopy (SEM) images of 1D nanostructures produced from mechanical polishing. (a) A set of 1D corrugated nickel nanowires and (b) ribbonlike nanonetworks with 1D nanowires.

FIGURE 12.22. Color changes of photonic-crystal metal-mesh composites as a function of angle. The thin composite piece is displayed on green paper with a line of text to help indicate changes in orientation. The observed colors are green, brown, yellow, and blue and black, for the relative orientation angle of 30°, 60°, 90°, 120°, and 150°, respectively. Red, purple, and other colors (not shown) can also be observed at other angles (see color insert).

3.6. Fabrication of 3D Metal Sphere Colloid Crystals from Inverse Colloid Crystals

The inverse structures duplicated from colloid crystals can be further employed as templates to fabricate metal sphere colloid crystals, which have potential applications including photonics, thermoelectrics, and magnetics [37,39]. Since these inverse colloid crystals replicate the void space of ordinary SiO_2 colloid crystals, they can be used as nanomolds for "casting" crystals of metal nanospheres as shown in Fig. 12.23. A sequence of electrochemical and chemical steps was used to prepare metal sphere arrays within a nonconductive matrix. Initially, a fcc SiO_2 photonic crystal slab containing 290-nm diameter spheres (Fig. 12.23a) was electrochemically infiltrated [11] with nickel (Fig. 12.23b). After removal of the colloid crystal template with a 2% HF solution (Fig. 12.23c), the nickel mesh was slowly oxidized in air at 550°C for 8 h to nonconductive NiO (Fig. 12.23d). The resulting NiO mesh was then used as a nanomold for the electrochemical growth of a gold nanosphere array (Fig. 12.23e). Finally, the NiO template was removed in dilute sulfuric acid to produce an ~30-μm thick array of gold nanospheres (Fig. 12.23f).

Figure 12.24 shows SEM images of NiO mesh (Fig. 12.24a) and Au nanosphere array (Fig. 12.24b) fabricated with this method. However, one of the problems is that the shape of inverse colloid crystals is deformed after oxidization of the NiO as shown in Fig. 12.24a. Compared to the shape of original SiO_2 spheres in colloid crystals, the shape of the fabricated metal spheres is also changed. This difficulty can be overcome by using polymer inverse colloid crystal (Fig. 12.25) as template, which can be synthesized conveniently by immersing SiO_2 colloid crystal slab in methyl methacrylate (MMA) monomer with 1 wt % benzoyl peroxide (BPO) as an initiator. Polymerization was initially carried out at 40°C for 10 h and then at 60°C for 12 h. The fabricated polymethyl methacrylate (PMMA)

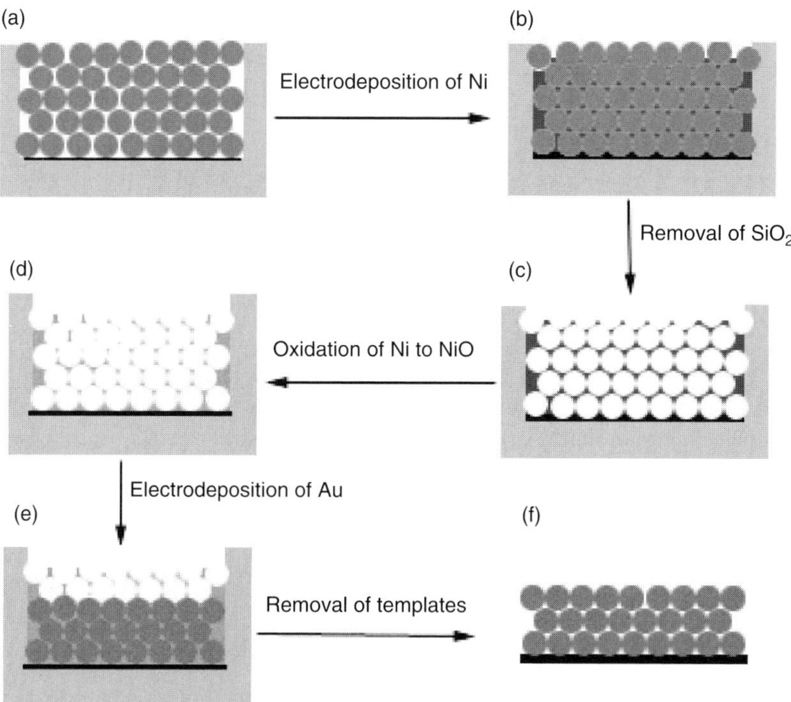

FIGURE 12.23. Procedure for the preparation of Au nanosphere arrays in a nonconductive matrix (see color insert).

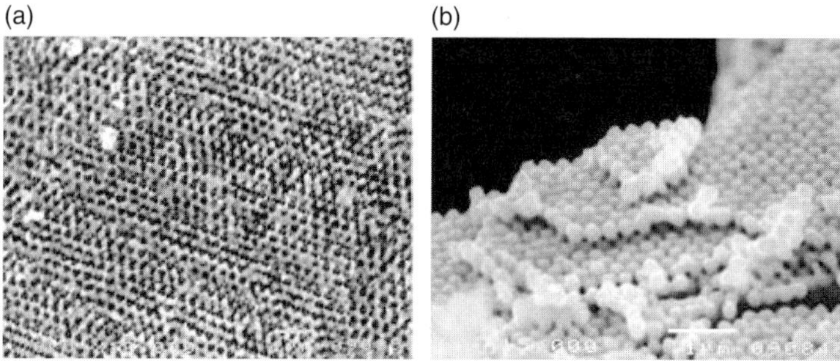

FIGURE 12.24. (a) Scanned electron microscopy (SEM) of NiO mesh used in electrochemical nanomolding and (b) Au nanospheres array from NiO mesh.

inverse colloid crystal possesses well-ordered structure duplicated from SiO$_2$ colloid crystals (Fig. 12.25a). Electrochemical deposition can duplicate the pore structure of polymer reverse colloid crystals and thus produce Ni sphere photonic crystals with well-controlled structures as shown in Fig. 12.25b. Since these metal sphere arrays are metallodielectric photonic crystals, new properties should arise, which are absent for dielectric photonic crystals such as the plasmon gap

(a)   (b)

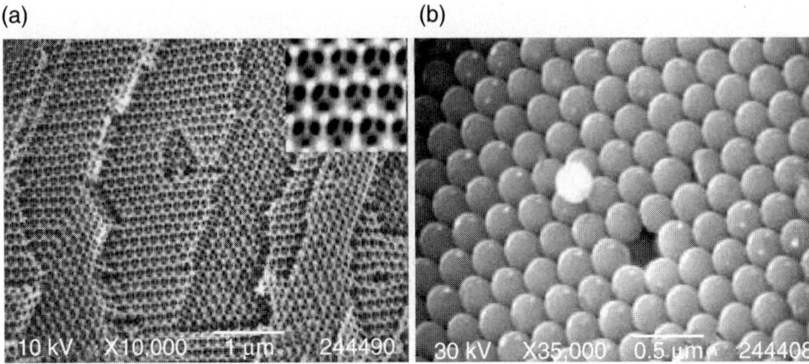

FIGURE 12.25. Scanned electron microscopy (SEM) image of (a) polymethyl methacrylate (PMMA) inverse colloid crystal and (b) Ni nanosphere array with (111) orientation. *Inset* in (a): higher-magnification image of the PMMA inverse colloid crystal.

observed in the microwave or infrared for metal meshes. These fabrication methods are applicable to various metals, semiconductors, and insulators, and can be used to make nanocomposites. By applying similar methods to optimally selected thermoelectrics and to combinations of hard and soft ferromagnets, it may be possible to engineer high-efficiency thermoelectrics and magnets with high-energy storage capabilities.

4. Conclusions

Template methods have proved to be an efficient technique in controlling nanostructures. A variety of 1D, 2D, and 3D nanomaterials have been already prepared with this method through combining chemical, electrochemical, and physical processes in our laboratory. Some unique nanostructures such as nanowires with structured tips, porous nanowires, and colloid crystal wires have been fabricated with by these novel methods. However, there are still some challenges in the idea of modifying template pore structures and thus further tuning the nanostructures for potential applications in microdevices. Further, characterization of these materials may lead to interesting photonic properties and composite materials fabricated with these structures, especially in the case of the sheathed colloidal crystals where interesting magnetic or electronic materials could result.

Acknowledgements

This material is based upon work supported by the National Science Foundation under Grant No. 0403673.

References

1. M. S. Gudiksen, L. J. Lauhon, J. Wang, D. C. Smith, and C. M. Lieber, *Nature*, 415 (2002) 617.
2. L. J. Lauhon, M. S. Gudikson, D. Wang, and C. M. Lieber, *Nature*, 420 (2002) 57.
3. J. Hu, Y. Bando, Z. Liu, T. Sekiguchi, D. Golberg, and J. Zhan, *J. Am. Chem. Soc.*, 125 (2003) 11306.
4. D. Whang, S. Jin, Y. Wu, and C. M. Lieber, *Nano Lett.*, 3 (2003) 1255.
5. W. F. Paxton, K. C. Kistler, C. C. Olmeda, A. Sen, S. K. St Angelo, Y. Cao, T. E. Mallouk, P. E. Lammert, and V. H. Crespi, *J. Am. Chem. Soc.*, 126 (2004) 13424.
6. (a) U. Srinivasan, M. A. Helmbrecht, R. S. Muller, and R. T. Howe, *Optics Photonics News*, 13 (2002) 20; (b) Y. Murakami, K. Idegami, H. Nagai, T. Kikuchi, Y. Morita, A. Yamamura, K. Yokoyama, and E. Tamiya, *Mater. Sci. Eng. C: Biomimetic Supramolecular Syst.*, C12 (2000) 67.
7. K. F. Böhringer, *J. Micromech. Microeng.*, 13 (2003) S1–S10.
8. E. E. Kneller and R. Hawig, *IEEE Trans. Magn.*, 27 (1991) 3588.
9. Y. Sun, B. Mayers, and Y. Xia, *Adv. Mater.*, 15 (2003) 641.
10. M. Steinhart, J. H. Wendorff, A. Greiner, R. B. Wehrspohn, K. Neilsch, J. Schilling, J. Choi, and U. Gösele, *Science*, 296 (2002) 1997.
11. S. Sharma and M. K. Sunkara, *J. Am. Chem. Soc.*, 124 (2002) 12288.
12. M. Brorson, T. W. Hansen, and C. J. H. Jacobsen, *J. Am. Chem. Soc.*, 124 (2002) 11582.
13. Z. W. Pan, Z. R. Dai, and Z. L. Wang, *Science*, 291 (2001) 1947.
14. X. Y. Kong, Y. Ding, R. S. Yang, and Z. L. Wang, *Science*, 303 (2004) 1348.
15. F. Li, Y. Ding, P. Gao, X. Xin, and Z. L. Wang, *Angew. Chem. Inter. Ed.*, 116 (2004) 5350.
16. (a) F. Li, H. Zheng, D. Jia, X. Xin, and Z. Xue, *Mater. Lett.*, 53(4–5) (2002) 282; (b) F. Li, J. Xu, X. Yu, Z. Yang, and X. Xin, *Sensors Actuators, B: Chem.*, 81 (2002) 165; (c) F. Li, L. Chen, Z. Chen, J. Xu, and X. Xin, *Mater. Chem. Phys.*, 73(2–3) (2002) 335; (d) F. Li, X. Yu, L. Chen, H. Pan, and X. Xi, *J. Am. Chem. Soc.*, 85 (2002) 2177; (e) F. Li, Z. Hu, J. Su, H. Zheng, and X. Xin, *J. Inorg. Chem. (China)*, 17 (2001) 315; (f) F. Li, X. Yu, H. Pan, M. Wang, and X. Xin, *Solid State Sci.*, 2 (2000) 767.
17. S. G. Roa, L. Huang, W. Setyawan, and S. Hong, *Nature*, 425 (2003) 36.
18. Y. Huang, X. Duan, Q. Wei, and C. M. Lieber, *Science*, 291 (2001) 630.
19. M. Tanase, L. A. Bauer, A. Hultgren, D. M. Silevitch, L. Sun, D. H. Reich, P. C. Searson, and G. J. Meyer, *Nano Lett.*, 1 (2001) 155.
20. P. A. Smith, C. D. Nordquist, T. N. Jackson, T. S. Mayer, B. R. Martin, J. Mbindyo, and T. E. Mallouk, *Appl. Phys. Lett.*, 77 (2000) 1399.
21. D. Whang, S. Jin, Y. Wu, and C. M. Lieber, *Nano Lett.*, 3 (2003) 1255.
22. A. Tao, F. Kim, C. Hess, J. Goldberger, R. He, Y. Sun, Y. Xia, and P. Yang, *Nano Lett.*, 3 (2003) 1229.
23. B. H. Hong, S. C. Bae, C.-W. Lee, S. Jeong, and K. S. Kim, *Science*, 294 (2001)348.
24. Y. Xia, P. Yang, Y. Sun, Y. Wu, B. Mayers, B. Gates, Y. Yin, F. Kim, and H. Yan, *Adv. Mater.*, 15 (2003) 353.
25. S. Vaddiraju, H. Chandrasekaran, and M. K. Sunkara, *J. Am. Chem. Soc.*, 125 (2003) 10792.
26. C. Ji and P. C. Searson, *J. Phys. Chem. B*, 107 (2003) 4494.
27. K. Nielsch, F. Müller, A.-P. Li, and U. Gösele, *Adv. Mater.*, 12 (2000) 582.
28. H. Cao, Z. Xu, H. Sang, D. Sheng, and C. Tie, *Adv. Mater.*, 13 (2001) 121.

29. M. Martin-Gonzalez, G. J. Snyder, A. L. Prieto, R. Gronsky, T. Sands, and A. M. Stacy, *Nano Lett.*, 3 (2003) 973.

30. D. Xu, X. Shi, G. Guo, L. Gui, and Y. Tang, *J. Phys. Chem. B*, 104 (2000) 5061.

31. N. I. Kovtyukhova and T. E. Mallouk, *Chem. Eur. J.*, 8 (2002) 4354.

32. G. Goglio, S. Pignard, A. Radulescu, L. Piraux, I. Huynen, D. Vanhoenacker, and A. Vander Vorst, *Appl. Phys. Lett.*, 75 (1999) 1769.

33. T. R. Kline, W. F. Paxton, T. E. Mallouk, and A. Sen, *Angew. Chem. Int. Ed.*, 44 (2005) 744.

34. S. R. Nicewarner-Peña, R. G. Freeman, B. D. Reiss, L. He, D. J. Peña, I. D. Walton, R. Cromer, C. D. Keeting, and M. J. Natan, *Science*, 294 (2001) 137.

35. R. Gasparac, B. J. Taft, M. A. Lapierre-Devlin, A. D. Lazareck, J. M. Xu, and S. O. Kelley, *J. Am. Chem. Soc.*, 126 (2004) 12270.

36. N. I. Kovtyukhova, B. K. Kelley, and T. E. Mallouk, *J. Am. Chem. Soc.*, 126 (2004) 12738.

37. L. Xu, W. Zhou, M. E. Kozlov, I. I. Khayrullin, I. Udod, A. A. Zakhidov, R. H. Baughman, and J. B. Wiley, *J. Am. Chem. Soc.*, 123 (2001) 763.

38. L. Xu, W. L. Zhou, C. Frommen, R. H. Baughman, A. A. Zakhidov, L. Malkinski, J.-Q. Wang, and J. B. Wiley, *Chem. Commun.*, 12 (2000) 997.

39. L. Xu, L. D. Tung, L. Spinu, A. A. Zakhidov, R. H. Baughman, and J. B. Wiley, *Adv. Mater.*, 15 (2003) 1562.

40. F. Li, X. Badel, J. Linnros, and J. B. Wiley, *J. Am. Chem. Soc.*, 127 (2005) 3268.

41. F. Li and J. B. Wiley, *J. Mater. Chem.*, 14 (2004) 1387.

42. F. Li, J. He, W. Zhou, and J. B. Wiley, *J. Am. Chem. Soc.*, 125 (2003) 16166.

43. F. Li, L. Xu, W. Zhou, J. He, R. H. Baughman, A. A. Zakhidov, and J. B. Wiley, *Adv. Mater.*, 14 (2002) 1528.

44. Y. Du, W. L. Cai, C. M. Mo, J. Chen, L. D. Zhang, and X. G. Zhu, *Appl. Phys. Lett.*, 74 (1999) 2951.

45. C. S. Cojocaru, J. M. Padovani, T. Wade, C. Mandoli, G. Jaskierowicz, J. E. Wegrowe, A. F. Morral, and D. Pribat, *Nano Lett.*, 5 (2005) 675.

46. A.-P. Li, F. Müller, A. Birner, K. Nielsch, and U. Gösele, *Adv. Mater.*, 11 (1999) 483.

47. R. C. Furneaux, W. R. Rigby, and A. P. Davidson, *Nature*, 337 (1989) 147.

48. V. Lehmann, *J. Electrochem. Soc.*, 140 (1993) 2836.

49. X. Badel, Ph.D. dissertation, KTH Royal Institute of Technology, Kista, Sweden (2005).

50. T. Thurn-Albrecht, J. Schotter, G. A. Kästle, N. Emley, T. Shibauchi, L. Krusin-Elbaum, K. Guarini, C. T. Black, M. T. Tuominen, and T. P. Russell, *Science*, 290 (2000) 2126.

51. M. E. T. Molares, V. Buschmann, D. Dobrev, R. Neumann, R. Scholz, I. U. Schuchert, and J. Vetter, *Adv. Mater.*, 13 (2001) 62.

52. P. Jiang, J. F. Bertone, K. S. Hwang, and V. L. Colvin, *Chem. Mater.*, 11 (1999) 2132.

53. P. Jiang and M. J. McFarland, *J. Am. Chem. Soc.*, 126 (2004) 13778.

54. A. van Blaaderen, *MRS Bull.*, 29 (2004) 85; (b) V. N. Manoharan and D. J. Pine, *MRS Bull.*, 29 (2004) 91; (c) V. C. Colvin, *MRS Bull.*, 26 (2001) 637.

55. (a) P. Jiang, J. F. Bertone, and V. L. Colvin, *Science*, 291 (2001) 453; (b) K. P. Velikov, C. G. Christova, R. P. A. Dullens, and A. van Blaaderen, *Science*, 296 (2002) 106; (c) L. Ramos, T. C. Lubensky, N. Dan, P. Nelson, and D. A. Weitz, *Science,* 286 (1999) 2325; (d) O. D. Velev, A. M. Lenhoff, and E. W. Kaler, *Science*, 287 (2000) 2240; (e) S. Wong, V. Kitaev, and G. A. Ozin, *J. Am. Chem. Soc.*, 125 (2003) 15589; (f) F. Caruso, R. A. Caruso, and H. Möhwald, *Science*, 282 (1998) 1111; (g) V. N.

Manoharan, M. T. Elsesser, and D. J. Pine, *Science*, 301 (2003) 483; (h) S. H. Im, Y. T. Lim, D. J. Suh, and O. O. Park, *Adv. Mater.*, 14 (2002) 1367; (i) V. Kitaev and G. A. Ozin, *Adv. Mater.*, 15 (2003) 75; (j) Y. Lu, Y. Yin, B. Gates, and Y. Xia, *Langmuir*, 17 (2001) 6344; (k) Y. Xia, B. Gates, and Z. Y. Li, *Adv. Mater.*, 13 (2001) 409.

56. (a) Y. Yin, Y. Lu. B. Gates, and Y. Xia, *J. Am. Chem. Soc.*, 1230 (2001) 8718; (b) Y. Yin, Y. Lu, and Y. Xia, *J. Am. Chem. Soc.*, 123 (2001) 771; (c) E. Kumacheva, R. K. Golding, M. Allard, and E. H. Sergeant, *Adv. Mater.*, 14 (2002) 221; (d) M. Allard, E. H. Sargent, P. C. Lewis, and E. Kumacheva, *Adv. Mater.* 16 (2004) 1360. (e) J. P. Hoogenboom, C. Rétif, E. de Bres, M. van de Boer, A. K. van Langen-Suurling, J. Romijn, and A. van Blaaderen, *Nano Lett.*, 4 (2004) 205.

57. (a) P. Yang, A. H. Rizvi, B. Messer, B. F. Chmelka, G. M. Whitesides, and G. D. Stucky, *Adv. Mater.*, 13 (2001) 427; (b) H. Míguez, S. M. Yang, N. Tétreault, and G. A. Ozin, *Adv. Mater.*, 14 (2002) 1805.

58. Y. Yin and Y. Xia, *J. Am. Chem. Soc.*, 125 (2003) 2048.

59. J. H. Moon, S. Kim, G.-R. Yi, Y.-H. Lee, and S. M. Yang, *Langmuir*, 20 (2004) 2033.

60. (a) R. O. Erickson, *Science*, 181 (1973) 705; (b) G. Pickett, M. Gross, and H. Okuyama, *Phys. Rev. Lett.*, 85 (2000) 3652.

13
One-dimensional Wurtzite Semiconducting Nanostructures

Pu Xian Gao and Zhong Lin Wang

1. Introduction

As for the definition of nanostructures, a widely accepted one is that "nanostructure" represents a system or object with at least one dimension in the order of 100 nm or less. Typical oxide nanostructures are of three types based on dimensional categories: 0D, 1D, and 2D. 0D nanostructures such as nanoparticles and 2D thin films have been extensively explored and utilized in many applications. A recent emerging of wire-like nanostructures is the 1D oxide nanostructures.

1D nanostructure is a material with two physical dimensions in the nanorange (1–100 nm), while the third one can be very large. Typically four types of 1D nanostructure configurations are reported in literature, including nanotubes [1], nanowires [2–5], nanorods [6–8], and nanobelts (nanoribbon) [9–13]. These nanostructures have potential applications in nanoelectronics, nanooptoelectronics, and nanoelectromechanical systems [3,4,14–23]. Figure 13.1 shows SEM images resenting the four different typical morphologies made by ZnO. Among them, nanobelt (nanoribbon) is the latest one being recognized and investigated extensively since its discovery in 2001 [10]. By utilizing a dramatically increased surface-to-volume ratio and the novel physical properties brought by the nanoscale structure, nanoscale sensors and transducers with superior performance can be achieved [24].

In this chapter, a review will be provided focusing on the emerging growth techniques and morphological control of 1D oxide nanostructures. ZnO will be taken as a main example for demonstrating the novelty of oxide nanostructures.

2. Synthesis and Fabrication of 1D Nanostructures

During the past decade, 1D metal oxide nanostructures have been extensively investigated. Most of the studies have been focused on the synthesis and fabrication of the nanostructures. Two categories of synthesis and fabrication techniques are generally used. One is the "bottom-up" techniques using vapor phase deposition,

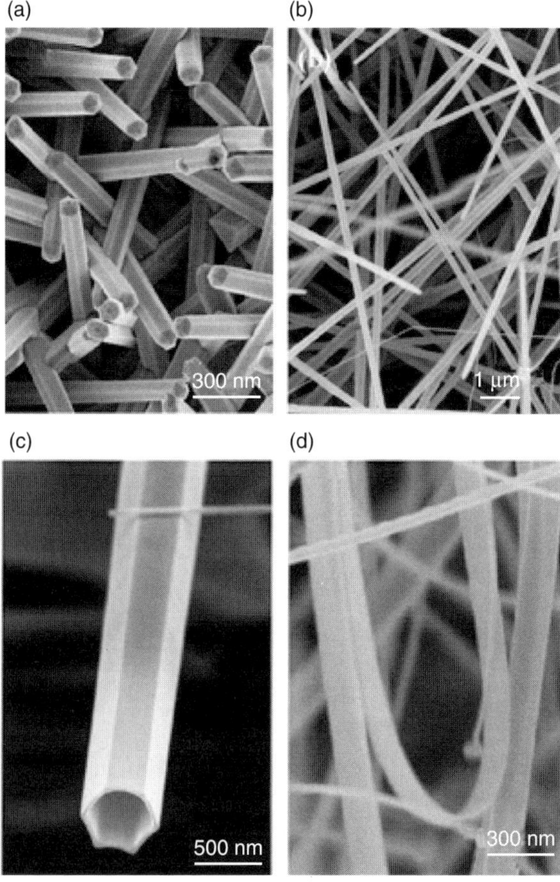

FIGURE 13.1. Four types of 1D nanostructures of ZnO.

chemical synthesis, self-assembly, and location manipulations. The other is the "top-down" approach utilizing the lithography and precision-engineered tools like cutting, etching, and grinding to fabricate nanoscale objects out of bulk materials. In the "bottom-up" category, several approaches have been well established, which include an extensively explored vapor phase deposition method, including chemical vapor deposition (CVD) and physical vapor deposition (PVD), and liquid phase deposition (solution synthesis approach).

2.1. Vapor Phase Deposition Method

Vapor phase deposition method is the most versatile method for synthesis and fabrication of 1D nanostructures in the past few years. CVD and PVD are two typical vapor phase deposition methods. Both methods can be subclassified based on the source of energy used, as summarized in Table 13.1 [25].

TABLE 13.1. Vapor phase deposition techniques for nanostructure fabrication

Vapor phase deposition techniques	Thin film and coatings	1D nanostructures
Physical vapor deposition processes (PVD)	All fit	1. The most popular method, low cost, pure products, but operation temperature could be very high
1. Thermal evaporation		2. Very few use
2. Electron beam PVD		3. Nanorods synthesis
3. Sputtering		4. Carbon nanofibers synthesis
Magnetron sputtering		5. Used for complex and high melting point compound
Direct current diode sputtering		
Radio frequency sputtering		
4. Cathodic arc discharge		
5. Pulse laser deposition		
Chemical vapor deposition processes (CVD)	All fit	(4–7) are very suitable for superlattice and core-shell nanostructures fabrications
1. Thermal CVD		
2. Low pressure CVD (LPCVD)		
3. Plasma-enhanced CVD (PECVD)		
4. Photochemical and laser CVD (LCVD)		
5. Metal–organic CVD (MOCVD)		
6. Molecule beam epitaxy (MBE)		
7. Atomic layer deposition (ALD)		

2.1.1. Physical Vapor Deposition Method

The traditional PVD method is dedicated to thin film and coating technology, which generally rely on the physical vaporization and condensation process of materials. Based on the source of energy, five typical methods have been developed (Table 13.1) [25]. The first one is "thermal evaporation" process, which generally uses resistor heater in a furnace to heat and facilitate the vaporization of source materials and nanostructure deposition process. This method has been fully developed as one of the most versatile methods for 1D nanostructure fabrication [4,9–13]. Examples of 1D nanostructures grown by thermal evaporations are rather diverse, including the elemental group such as Si and Ge, the II–VI group (ZnS, CdS, ZnSe, and CdSe), III–V group (GaAs, InAs, InP, GaP, and GaN), and semiconducting oxides (ZnO, CdO, PbO, In_2O_3, Ga_2O_3, SnO_2, and SnO). The second type of PVD is the electron-beam PVD, which has been extensively used for fabrications of ceramic thin film and coatings; but so far few reports has been found about its application in 1D nanostructure synthesis.

The third type of method is the "sputtering process." Sputtering is the removal of surface atoms due to bombardment from a flux of impinging energetic particles. The consequences of the interaction between an incident particle and a solid surface are mostly dictated by the kinetic energy of the projectile particle, although its internal energy may also play a significant role. Surface diffusion is usually used for explanation of the nanoscale islands or rods growth during the sputtering process. Sputter methods have been utilized for some nanorods synthesis such as W [26], Si [27], B [28,29], CNx [30], B_2O_3, and TiO_2.

The cathodic arc discharge is another important PVD method. It is famous for synthesis of the first batch of carbon nanotubes back to 1993 [1,31]. It is a very popular method for fabrication of carbon-related nanostructures, but there are very few applications in growth of oxide nanostructures.

Pulse laser deposition (PLD) [32] is a developed method for creation of nanostructures back to the seminal work on $YBa_2Cu_3O_{7-x}$ thin film growth by Venkatesan and coworkers [33] in 1987. The process is that the source material is rapidly evaporated from a bulk target in a vacuum chamber by focusing a high-power laser pulse on its surface. A large number of material types have been deposited as thin films using PLD. For some complex and high-melting point (T_m) compounds, there is no doubt that using PLD is the simpler and easier method to receive better quality of nanostructures. PLD has been used for synthesis of 1D nanostructures since the early 1990s. It has been successfully used in carbon nanotube synthesis [34] with high yield instead of using cathode arc discharge despite of a large amount of impurity. In the last decade, using PLD method, a variety of 1D complex compounds and core-shell structured nanostructures have been synthesized [35–37]. Outside the auspices of high-performance semiconductor applications, PLD is a promising research tool for investigating new growth phenomena and systems as any other technique, and can offer some unique features.

2.1.2. Chemical Vapor Deposition

CVD can be classified as six or more categories based on the difference of the vacuum control level, heating source, and reactant gas types, as shown in Table 13.1. Thermal CVD, LPCVD, LCVD, and MOCVD have been used to synthesize several types of nanowires. Similar to the thermal evaporation method in PVD approach, thermal CVD involves chemical reaction during the deposition process rather than a physical evaporation and deposition process. It has produced high-quality carbon nanotube [38] and nanowires including GaN [39,40] and GaAs. Like thermal evaporation in PVD, a solid source could be placed upstream in thermal CVD also. The solid is vaporized through heating, which is transported by a carrier gas that feeds the catalyst and source material for nanowire growth.

LPCVD is a technique for fabrication of thermal oxide layer on semiconductor wafers in electronic industry. This approach uses commercial oven or furnace as a main reaction chamber, which is generally composed of the reactant gas control and supply system and temperature/pressure control system. The pressure control is normally under 0.1 Torr. Despite the old technique for oxide film growth in electronic industry, recently the LPCVD is a successful method for synthesis of various nanostructures such as carbon nanotube [41,42] and Si nanowire [43–45].

In Laser CVD, the CVD setup has been modified for laser ablation with controlled doses, leading to a control over nanowire lengths and diameters. Such a precision control made it possible to form superlattices [46] or core-shell [47] structures with doping. MBE has been used to grow 1D nanostructure of III–V group semiconductor compounds such as GaN [48], GaAs [49], InGaAs [42], and AlGaAs [50]. ALD seemly has been used for fabrication of templates and ordered core-shell nanostructures [51–54] due to its precise control on atomic layer-by-layer deposition.

2.2. Solution-based Chemical Synthesis Methods

As a generic method for synthesis of nanostructures, chemical synthesis has been a powerful method for producing nanomaterials. Since the emergence of 1D nanomaterials, solution-based chemical synthesis methods have been extensively explored for fabrication of 1D nanostructures. Many materials such as elemental noble metals, e.g., Au, Pt, and Ag [4] and tri- or multielemental compounds like $Ba(Sr)TiO_3$ [55,56], $LaPO_4$ [57], $CePO_4$ [58], and BaH_2SiO_4 [59], in nanowire-like shape have been successfully synthesized by solution-based chemical synthesis methods, which cannot be easily fabricated by using vapor phase methods. The 1D nanostructures synthesized by chemical synthesis can be easily achieved in a large yield, but the large amount of impurities usually contained in the products always dramatically hinder its application, although complex postfiltering and purification is useful for increasing its purity. From the cost point of view, a low-cost chemical synthesis with high yield undoubtedly is ideal for future commercialization of 1D nanostructures.

2.3. Conjunctional Methods Involving Lithography Patterning

As described above, almost all of the techniques associated with the traditional thin film and coating process have somehow been successfully found applications in fabrication of 1D nanostructures. Lithography is a key technology in semiconductor industry for processing devices and circuits at wafer level in a designed sequence. It involves the feature patterning on a surface through exposure to a variety of sources such as light, ions, or electrons, and then a subsequent etching and/or deposition of materials to produce the desired devices. The ability to pattern features in the nanometer range is a key to the future nanoelectronics.

Figure 13.2 shows the convergence of the top-down and bottom-up production approaches. It is worth noting that there is an overlapping area between the top-down and bottom-up approaches since the late 1990s in the extrapolated area. Then, hybrid methods combining both approaches should be the desired techniques in the future. To date, more and more efforts have been devoted in using the conjunctional synthesis methods for fabricating hierarchical or ordered building blocks of nanomaterials. An example is the development of soft lithography technique, which has taken the advantage of both the improved knowledge of self-assembly and nanolithography process.

Currently most of the conjunctional synthesis methods utilize the lithographically patterned substrate for catalysts patterning. It has been reported that by taking advantage of self-assembled gold or Ni particles, patterned growth of carbon nanotubes [60] and ZnO nanowire arrays [61] for 2D photonic crystal applications have been successfully achieved. By utilizing the Brownian motion of nanowires in fluids and also the lithographical microfluidic channels on the

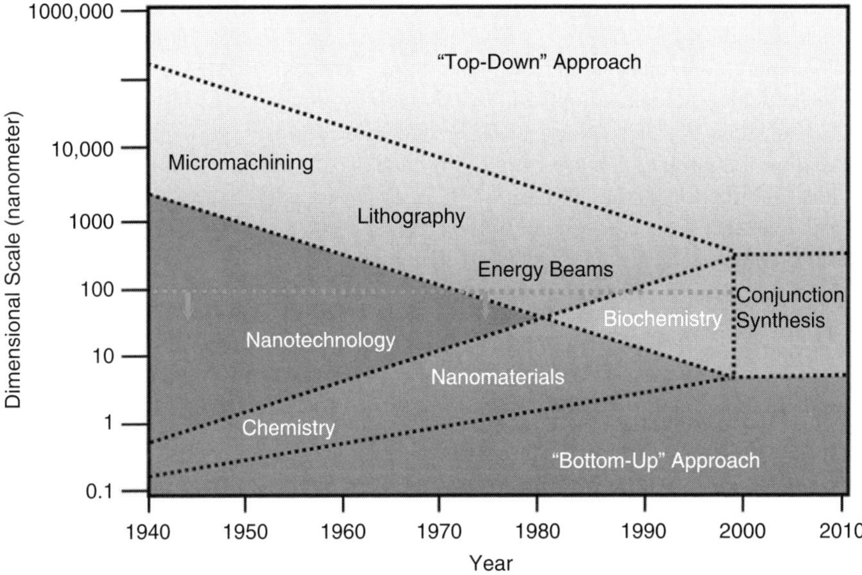

FIGURE 13.2. A schematic diagram showing the convergence of the top-down and bottom-up approaches.

deposition substrate, the positional-aligned nanowire arrays along the microfluidic channels can be obtained at a large scale. Using the same strategy, a large-scale crossbar arrays [62] of nanowires with a certain robustness and repeatability for logic circuits have been successfully achieved.

In the mean time, some researchers have been working on the conjunction of vapor phase deposition, chemical synthesis, and lithography techniques for synthesis of hierarchical building blocks of functional materials for nanodevices.

3. 1D Metal Oxide Nanostructures

Metal oxide is one of the most important functional material candidates for chemical and biological sensing and transduction. Its unique and tunable physical properties have made itself an excellent candidate for electronic and optoelectronic applications. As a result, nanostructures of metal oxides have emerged as a frontier field in advanced science and technology since the microelectronics revolution in the 1970s; as the thin film technology has been comprehensively developed and pushed to commercialization for ultra-large-scale integrated circuits (ULSI).

Now in the early 21st century, 1D metal oxide nanostructures have become a major research topic in the current nanoscience and nanotechnology. As described in the introduction, three typical morphological configurations have been clearly defined: wire-like, tube-like, and belt (ribbon)-like nanostructures. Metal oxide nanowires, nanotubes, and nanobelts are fascinating groups of materials that have enormous potential for novel applications. Electronic applications include transistors [63–66], chemical sensors [67–69], biological sensors [70], integrated circuits [71,72], and logic devices. NEMS, or nanoelectromechanical systems, and optical devices such as photonics, LEDs, and lasers are also possible. For these purposes, 1D metal oxide nanostructures will function as the most important category of building blocks. The rational control and hierarchical fabrication of these building blocks so far are still challenging. In this section, a brief review is given about the rational control and hierarchical synthesis of 1D metal oxides nanostructures involving the bottom-up self-assembly and top-down physical assembly processes.

3.1. Oxide Nanowires

For synthesis of nanowires, the bottom-up synthesis is the most cost-effective means for producing nanowires in large quantities. Nevertheless, randomly oriented assemblies of nanowires have been a hindrance for massively parallel device applications. To meet the requirement of massive production and integration, hierarchical and ordered nanostructure consisting of a large number of individual nanowires is an optimum choice. To date, the invention and development of techniques such as microfluidics, field-driven self-assembly, and template-confined growth have led to successful creations of alignment and architecture assembly of various nanowires.

Quite a lot of reports on the metal oxide nanowires such as ZnO, In_2O_3 [73], SnO_2 [74], and SiO_2 [75] have been seeking the rational and controllable synthesis of these functional nanoscale materials. Generally, there are mainly two strategies for achieving the alignment of 1D nanostructures. One approach is to use patterned catalysts to grow oriented nanowires epitaxially on substrates with matched lattices. The other is to use arrays of nanochannels on substrates such as porous anode alumina membranes (PAAMs) as templates for growing metal oxide nanowire arrays [76]. The other strategy is to physically manipulate or assemble the randomly distributed nanowires into ordered and patterned forms using external fields such as electric, magnetic, and hydro force.

A typical example of aligned growth is for ZnO, one of the most important metal oxides for electronic and optoelectronic applications. In 2001, aligned growth of ZnO nanowires on single crystal substrate has been achieved by using thermal CVD method [5]. In the report, the aligned ZnO nanowire arrays were grown on (11–20) sapphire substrates. A 1–3.5-nm thick layer of Au was coated on the substrate to function as catalysts. For photonics applications, some new approaches have been implemented for growing patterned nanowire arrays with surface coating. As mentioned before, using self-assembled gold or Ni patterning, patterned growth of carbon nanotubes [77] and ZnO nanowire arrays [78] for 2D photonic crystal applications have been successfully achieved.

It is worth noting that most of the nanowires synthesis and fabrication research have been on elemental and binary compound system, there are only a few reports on success of 1D nanostructures of trielemental or more complex compounds, even though ternary and complex metal oxides are an important class of oxides for structural and functional applications. For example, the perovskite oxides such as $BaTiO_3$ and $SrTiO_3$ [57, 58] are important ferroelectric materials. So far, it is a challenge to synthesize 1D nanostructures of these materials although $BaTiO_3$ nanowires have been synthesized using a solution-based method. Recently, 1D nanostructures of Zn_2SnO_4 has been prepared by a reaction among elements Sn, ZnS, and $Fe(NO)_3$ at $1,350°C$ in a high-temperature tube furnace [79]. $ZnGa_2O_4$ has been synthesized into nanowire shape using thermal CVD method [80].

3.2. Oxide Nanotubes

Tubular structure is a special and important morphological configuration in nature. Synthetic nanoscale tubular structures have caught great research attention since the discovery of carbon nanotube. In addition to famous carbon nanotube, some important engineering and functional compounds such as MoS_2 [81], MoS_3 [82], and BN [83] have been synthesized into tubular structures. Using tungsten-comprised precursors, Al-rich mullite as an important engineering material has been fabricated in tubular shape [84]. GaN, an important blue light emission material, has been fabricated into nanotube arrays using a conjunctional of MOCVD and thermal CVD, in which ZnO nanowire arrays were taken as the template for growing GaN nanotube arrays [85]. By combustion-assisted CVD method, nanotube arrays of SnO_2 have also been achieved [86], which could be of importance in nanosensing.

Using nanotube, nanobelt, and nanowires as templates, tubular structures can be synthesized for a wide range of materials. Most of these materials are inert to the environment, which can be removed later, or can be completely convertible by a chemical reaction. In this section, ZnO will be taken as a typical example for illustrating the self-assembly process of tubular oxide nanostructures during a simple solid–vapor deposition process.

To date, ZnO has been one of the most extensively explored inorganic materials besides carbon in the nanoscale science and technology. Recently, there have been quite a few reports about the successful synthesis of ZnO microscale tubes [87,88], and also some nanotubes [89–92] with polycrystalline or textured crystal structures.

Chemically, solid ZnO is an ionic-bonded crystal. The bonding between cation Zn^{2+} and O^{2-} is so strong that the melting point of ZnO is as high as $1,975°C$. The melting point of Zn metal is as low as $411°C$ and can be easily oxidized in air [93]. Both Zn and ZnO have a similar hexagonal crystal structure. The low melting point of Zn and the high melting point of ZnO, and their hexagonal crystal structures have given a great advantage for designing and fabricating various metal–semiconductor core-shell nanostructures and tubular structures [94].

3.2.1. ZnO Microtubes and Nanotubes [94]

Using a thermal evaporation approach, and taking advantage of selective surface oxidation on the as-formed Zn nanoobject, tubular ZnO nanostructures have been synthesized. We now examine how the oxidation rate of Zn depends on the surface energy. During a systematic growth study of ZnO nanodiskettes, it is interesting to find that short microtubes were formed with clean hexagonal-faceted surface. Figure 13.3a shows an inclined view of a typical hexagonal tube of ZnO about 3–5 µm in length, 2–3 µm in width, and ~200 nm in wall thickness. An enlarged open end of the tube is illustrated in Fig. 13.3b, where the inner surface of the tube is stepped, rough, and without facets, revealing a possible sublimation process of the material from the center. Figure 13.3c displays the front view of a ZnO tube of ~2 µm width and ~4 µm length with a rather sharp inner surface, and the hexagonal facets are clean but not perfect. In Fig. 13.3d, a tube with ~4 µm height and ~4 µm diameter is shown. The hollow core of the tube can be clearly seen, where the inner diameter at the entrance is larger than that in the middle of the tube, suggesting that the original material filling the core was not completely sublimated. It is evident from Fig. 13.3 that the template for forming the tube was a hexagonal Zn rod; the oxidation of the surface formed the wall of the tube, and the resublimation of the Zn core created the hollow tube.

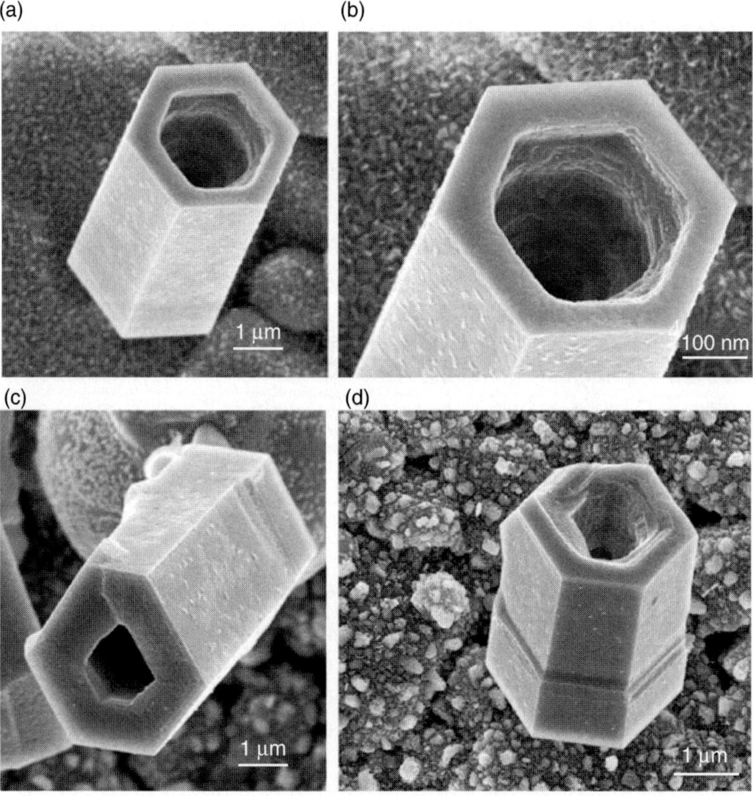

FIGURE 13.3. A series of SEM images displaying individual short microtubes with rough outer and inner surfaces. (a, b) Low magnification SEM images showing the outer hexagonal faceted surfaces and the inner rough and stepped surfaces of the ZnO tubes. (c, d) Two short micron-sized tubes with hexagonal facets.

Adjusting the pressure in the synthesis has led the growth of high-quality tubular structures. Figure 13.4a and b shows two microtubes of ZnO with irregular shaped cross sections, respectively from the side and front views. The two microtubes have possible circular cross section, and rough side surfaces. The stepped inner surface of the microtube in Fig. 13.4b revealed that, despite a possible single crystal structure, the formation of microtube still follows a possible sublimation process of core material. Figure 13.4c gives a clear description of an intact, smooth, and long freestanding microtube of ZnO, which should be of single crystal. Two typical TEM images showing a Zn–ZnO core-shell nanocable and a ZnO nanotube with remaining Zn core inside are displayed in Fig. 13.4d and e, respectively. The diameters of the two Zn–ZnO nanocable and nanotube are 50–100 nm. The partially left or residual Zn inside the tube and inner wall is apparent in Fig. 13.4e. The inset select area nanodiffraction patterns indicate that they are single crystals with the same orientation and growth along c-axis.

3.2.2. ZnO Nanotube Arrays on ZnO Cages [94]

As illustrated above, the low melting point of Zn and the high melting point of ZnO have given hints to design and fabricate various functional nanostructured

FIGURE 13.4. (a, b) SEM images displaying individual microtubes of ZnO with irregular shaped cross sections. (c) Hexagonal faceted tubes of ZnO. (d, e) TEM images showing a Zn–ZnO core-shell nanocable and a ZnO nanotube with residual Zn core, the *inset* electron diffraction patterns indicate their single crystal structures.

ZnO nanoobjects. Controlling the growth kinetics would allow us to control the formation process of the Zn–ZnO system. Lower-temperature synthesis of 1D ZnO nanostructures is very attractive because it facilitates to integrate the process with other synthesis and fabrication processes. Here, we show a case of synthesis of nanotube arrays on large shells.

In the deposition region at growth temperature around 450°C, ZnO nanotube arrays on ZnO cages were observed, as shown in Fig. 13.5a. Microspheres of ZnO with sizes ~30 μm grew freestanding on alumina substrates. Radical branched tubes surround the central microsphere (Fig. 13.5b). A broken microsphere in Fig. 13.5c reveals that the microsphere in fact is a hollow shell-like structure, inside which, radially distributed nanotube arrays with a uniform diameter of ~1.5 μm grew directly out of the inner cage surface. The outer surface of ZnO cage is covered by radially extruding nanotube arrays of ZnO (Fig. 13.5d). Further, close examination at the inner surface of ZnO cage found that, besides the hexagonal nanotubes, some wavy ZnO nanowalls are present. A typical TEM image of a broken nanotube from hollow ZnO cages is shown in Fig. 13.5f, where the hollow cage inside is in a visible bright contrast. Its corresponding electron diffraction pattern shown in the inset (bottom) proves its single-crystal growth along [0001] direction.

It is interesting to find out in Fig. 13.6 that the outer surfaces of some ZnO cages are enclosed by regularly arranged hexagonal pyramid-like domains; the six intersection lines of the pyramid are visible from the front (Fig. 13.6b–d). Each pyramid is about 2 μm in width and 1–1.5 μm in height. On the tip of each ZnO pyramid, there is a single standing out nanorod of ~300 nm width and ~600 nm height (Fig. 13.6b and d). A single nanotube with diameter of ~200 nm has been observed sometimes at the tip of the pyramid (Fig. 13.6a and c).

3.2.3. Growth Mechanism of ZnO Microtubes and Nanotubes [94]

There are three important factors on the growth of ZnO tubular structures: the kinetics in the Zn and ZnO system, the lower energy facets, and the oxidation rates of different surfaces. The oxidation rates on different crystal surfaces and at different temperature regions are distinct. The Zn surface with lowest energy such as $\{0001\}$ planes tend to be most stable and may resist to be oxidized, while the higher energy surfaces such as $\{10\bar{1}0\}$ are likely to have a higher oxidation rate. It is evident from the above experimental results that the selective and epitaxial oxidation growth of ZnO on Zn rod is an effective way to design and synthesize tubular nanostructures of ZnO in a dimension ranging from nanometers to micrometers.

Similar to the growth mechanism proposed on ZnO nanoiskettes [94] and nanocages [95], a growth mechanism on ZnO nanotubes is illustrated in Fig. 13.7. The first step (Fig. 13.7a) is the necessary vapor transfer of reduced Zn droplets by Ar gas to the deposition area of temperatures 400–500°C. A faster growth along c-axis direction leads to the formation of hexagonal-based Zn rods (Fig. 13.7b). Since the Zn $\{0001\}$ surfaces have lower energy than the $\{10\bar{1}0\}/\{2\bar{1}\bar{1}0\}$ surfaces,

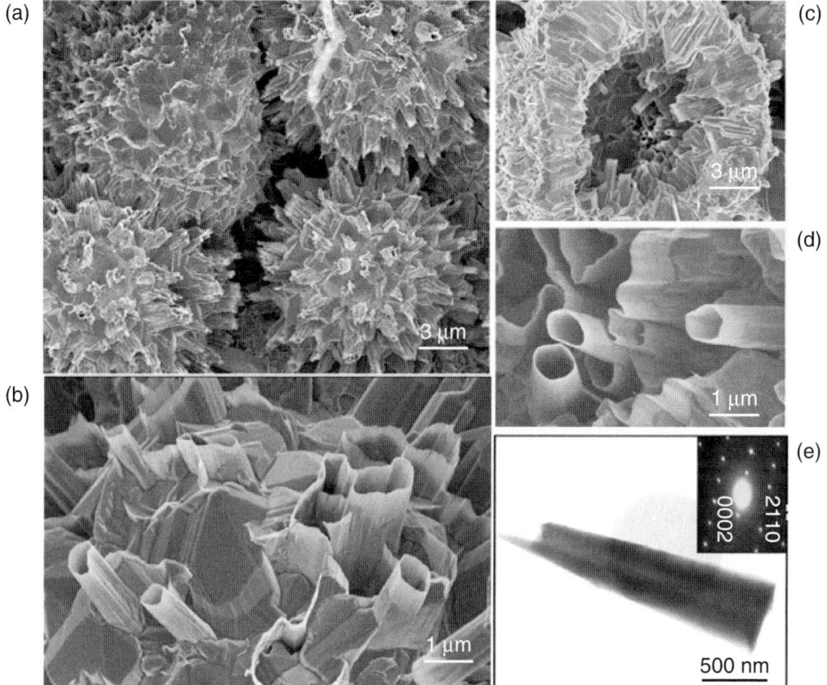

FIGURE 13.5. (a–d) A series of SEM images showing ZnO nanotube arrays distributed on the inner and outer surfaces of a large-size hollow shell-like structure. (e) A typical TEM image of a broken ZnO nanotube from hollow shell-like structure, the *right* side corresponding electron diffraction pattern indicates its single-crystal growth direction along [0001].

they are more stable and resistive to oxidation. As a result, the side surfaces of the hexagonal nanorod are oxidized first, forming a ZnO shell. At the same time, the low melting point Zn has a lower vaporization temperature, the remaining Zn inside the rod can be resublimated through the open ends at the {0001} surfaces, leaving behind a ZnO tube that has relatively smooth facets as defined by the Zn rod (Fig. 13.7c). The sublimation may not be complete, thus, the inner surface of the tube is rough possibly with some residual Zn. If the sublimation is terminated at the early stage, the half-tubular structures will be formed.

The growth of the structures in Fig. 13.6 is more complex, but they are still governed by the process illustrated in Fig. 13.7. It is likely that a large Zn liquid droplet was deposited on the substrate, which was too large to be a single crystal, thus, a polycrystalline Zn ball was formed, but with many hexagonal pyramids forming on the surface, each of which was initiated from a Zn nuclei. Then, the small Zn vapor could deposit on the surface at the tips of the pyramids to reduce the local surface energy, forming a hexagonal rod at the tip. Then, an oxidation and subsequent sublimation process as illustrated in Fig. 13.7 is possible, forming the tubular structure at the tip (Fig. 13.6c).

FIGURE 13.6. (a) An SEM image displaying a collection of hexagonal nanotubes and (b) nanorods of Zn–ZnO; on each island there is only one nanotube or grown nanorod on its top, closer views in (c) and (d), respectively.

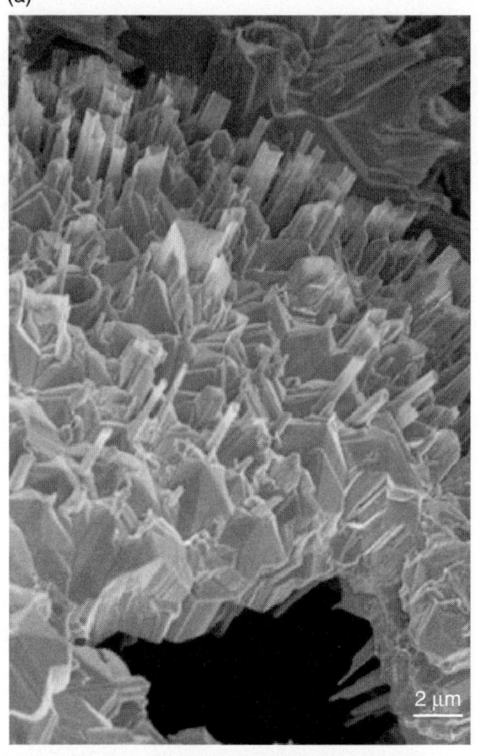

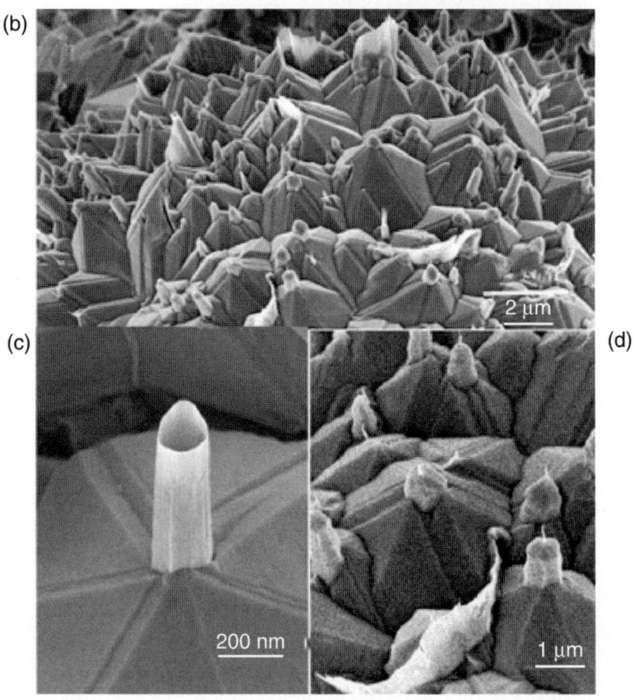

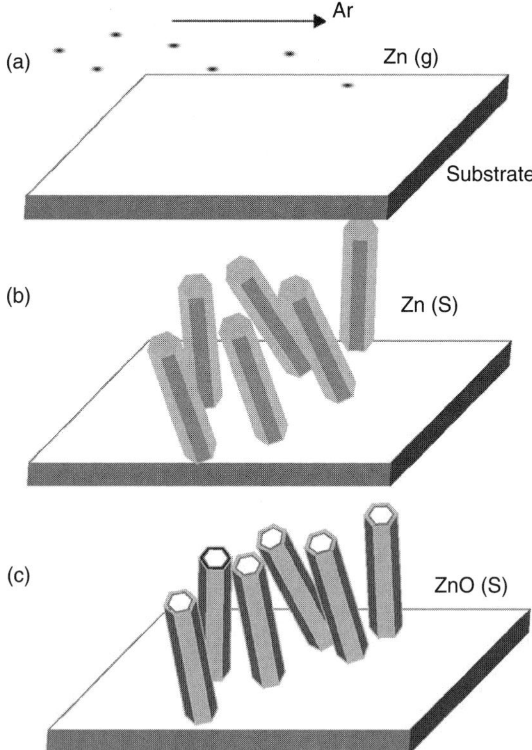

FIGURE 13.7. A proposed formation model of freestanding or rooted ZnO nanotubes.

3.3. Oxide Nanobelts

As inspired by carbon nanotubes, research in 1D nanomaterials is the forefront in nanotechnology due to their unique electronic, optical, mechanical, and chemical properties. For noncarbon-based 1D nanostructures, devices have been fabricated utilizing semiconductor nanowires, such as Si and InP. Piezoelectric effect, an important phenomenon that couples electromechanical behaviors, appears to be forgotten in nanomaterials research. None of the well-known nanostructures, such as quantum dots and carbon nanotubes, exhibit an intrinsic piezoelectricity. This is an area that remains to be explored because piezoelectric effect at nanoscale is critical in electromechanical sensors, actuators, and resonators.

In 2001, semiconducting oxide nanobelts (nanoribbons) have been discovered, giving rise to a brand new configuration with respect to the nanowire configuration. The nanobelt is an 1D nanostructure that has a rectangular cross section with well-defined crystal surfaces. As inspired by this discovery, research in functional oxide-based 1D nanostructures is rapidly expanding and is becoming a forefront research in nanotechnology. In the last 4 years, synthesis of belt-like oxides, sulfides, and selenide-based 1D nanostructures has drawn a lot of attention due to their unique geometrical structure and functional properties and their potential applications as building blocks for nanoscale electronic and optoelectronic devices. It is worth mentioning that a variety of functional nanodevices such as field effect transistors and ultrasensitive nanosize gas sensors, nanoresonators, and nanocantilevers have been fabricated based on individual nanobelts. In this section, the main focus will be on the belt-like ZnO nanostructures.

3.3.1. Polar Surface-dominated ZnO Nanobelts [96,97]

Wurtzite ZnO has a hexagonal structure (space group $C6mc$) with lattice parameters $a = 0.3296$ and $c = 0.52065$ nm, which has a number of alternating planes composed of tetrahedrally coordinated O^{2-} and Zn^{2+} ions, stacked alternatively along the c-axis (Fig. 13.8a). The tetrahedral coordination in ZnO results in non-central symmetric structure, and consequently the piezoelectricity and pyroelectricity. Another important characteristic of ZnO is the polar surfaces. The most common polar surface is the basal plane. The oppositely charged ions produce positively charged Zn-(0001) and negatively charged O-(000$\bar{1}$) surfaces, resulting in a normal dipole moment and spontaneous polarization along the c-axis.

For the (0001) polar surface-dominated (PSD) nanobelt, its structure can be approximated to be a capacitor with two parallel charged plates (Fig. 13.8b). The polar nanobelt tends to roll over into an enclosed ring to reduce the electrostatic energy. A spiral shape is also possible for reducing the electrostatic energy. The formation of the nanorings and nanohelices can be understood from the nature of the polar surfaces. If the surface charges are uncompensated during the growth, the spontaneous polarization induces electrostatic energy due to the dipole moment, but rolling up to form a circular ring would minimize or neutralize the overall dipole moment, reducing the electrostatic energy. On the other hand, bending of the nanobelt produces elastic energy. The stable shape of the nanobelt is determined by the minimization of the total energy contributed by spontaneous polarization and elasticity. If the nanobelt is rolled uniradically loop-by-loop, the repulsive force between the charged surfaces stretches the nanohelix, while the elastic deformation force pulls the loops together; the balance between the two forms the nanohelix/nanospring. In all of the cases illustrated here, the polar axis c is pointing towards the center.

A new structure configuration is introduced if the direction of the spontaneous polarization is rotated for 90°. The nanobelt has polar charges on its top and bottom surfaces (Fig. 13.8c). If the surface charges are uncompensated during growth, the nanobelt may tend to fold itself as its length gets longer to minimize the area of the polar surface. One possible way is to interface the positively charged Zn-(0001) plane (top surface) with the negatively charged O-(000$\bar{1}$) plane (bottom surface), resulting in neutralization of the local polar charges and the reduced surface area, thus, forming a loop with an overlapped end. The long-range electrostatic interaction is likely to be the initial driving force for folding the nanobelt to form the first loop for the subsequent growth. As the growth continues, the nanobelt may be naturally attracted onto the rim of the nanoring due

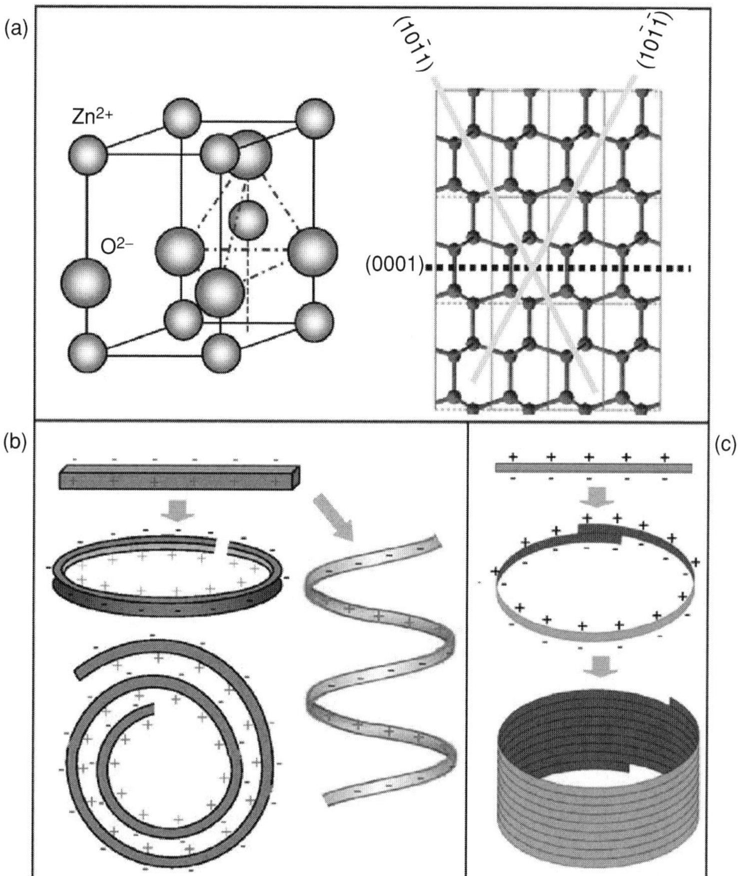

FIGURE 13.8. (a) 3D structure model of the unit cell of ZnO, and the corresponding projection along a-axis, showing the polar surfaces terminated with Zn and O, respectively. (b) Schematic model of a polar nanobelt, and the possible nanostructures created by folding the polar nanobelt with its polar direction pointing towards the center, forming nanoring, nanospiral, and nanohelices. (c) Model of a polar nanobelt and its self-coiling process for forming a multilooped ring by folding the nanobelt with its polar direction pointing to the axial direction of the ring. The nanoring is initiated by folding a nanobelt into a loop with overlapped ends due to long-range electrostatic interaction among the polar charges; the short-range chemical bonding stabilizes the coiled ring structure; and the spontaneous self-coiling of the nanobelt is driven by minimizing the energy contributed by polar charges, surface area, and elastic deformation.

to electrostatic interaction and extends parallel to the rim of the nanoring to neutralize the local polar charge and reduce the surface area, resulting in the formation of a self-coiled, coaxial, uniradius, multilooped nanoring structure.

Structurally, ZnO has three fast growth directions:$<2\bar{1}\bar{1}0>$ ($\pm[2\bar{1}\bar{1}0]$, $\pm[\bar{1}2\bar{1}0]$, $\pm[\bar{1}\bar{1}20]$); $<01\bar{1}0>$ ($\pm[01\bar{1}0]$, $\pm[10\bar{1}0]$, $\pm[1\bar{1}00]$); and $\pm[0001]$. Together with the polar surfaces due to atomic terminations, ZnO exhibits a wide range of novel structures that can be grown by tuning the growth rates along these directions. These are the fundamental principles for understanding the formation of numerous ZnO nanostructures.

The PSD ZnO nanostructures were grown by a high temperature solid–vapor deposition process. The raw material could be pure ZnO or a mixture of ZnO (melting point 1,975°C) with a certain amount of doping materials, such as indium oxide and lithium carbonate powders, which was placed at the highest temperature zone of a horizontal tube furnace. Before heating to a desired temperature, the tube furnace was evacuated to $\sim 10^{-3}$ Torr to remove the residual oxygen. ZnO decomposes into Zn^{2+} and O^{2-} at high temperature (1,350–1,400°C) and low pressure ($\sim 10^{-3}$ Torr), and this decomposition process is the key step for controlling the anisotropic growth of the nanobelts. The Ar carrier gas was introduced at a flux of 25–50 sccm (standard cubic centimeters per minute). The condensation products were deposited onto an alumina/silicon substrate placed in a lower temperature zone under Ar pressure of 200–500 Torr.

3.3.2. Nanospiral and Nanosprings [97, 98]

Helix or spiral is a unique shape from the nature. A good example is the DNA double helices [99]. In the artificial nanoregime, a number of inorganic materials have been achieved in helical/spiral shape. For example, carbon nanotube coils [100] are created when paired pentagon–heptagon atomic rings arrange themselves periodically within the hexagonal carbon network [101]. Formation of nanosprings of amorphous silica has been thought to occur through a perturbation during the growth of a straight nanowire [102]. Helical structures of SiC [103] are proposed to be a screw-dislocation-driven growth process. Using the difference in surface stress on the two surfaces, rings and tubes of strained bilayer thin films, e.g., Si/SiGe, have been made [104].

ZnO is a good example to show a unique capability to display a spiral shape. It is known that, due to differences in surface energies among (0001), $\{01\bar{1}0\}$, and $\{2\bar{1}\bar{1}0\}$, freestanding nanobelts and nanowires of ZnO are usually dominated by the lower energy, nonpolar surfaces such as $\{01\bar{1}0\}$ and $\{2\bar{1}\bar{1}0\}$. There are two types of synthesis that can produce the single crystal nanospirals. One is through introducing doping, such as In and/or Li, ZnO nanobelts dominated by the (0001) polar surfaces can be grown [97]. The other is using pure ZnO as the only source, but without introducing carrier gas in the growth [99]. In either cases (0001) planar defects appear to be present for stabilizing the polar surfaces.

In the doping case, the nanobelt grows along $[2\bar{1}\bar{1}0]$ (the a-axis), with its top/bottom large surface $\pm(0001)$ and the side surfaces $\pm(01\bar{1}0)$. The typical thickness is of 5–20 nm and the belt has a large aspect ratio of $\sim 1:4$, allowing a great flexibility and toughness for the nanobelts. A PSD nanobelt can be approximated to be a capacitor with two parallel-charged plates (Fig. 13.9a). The polar nanobelt tends to roll over into an enclosed ring to reduce the electrostatic energy (Fig. 13.9b). A spiral shape is also possible for reducing the electrostatic energy

(Fig. 13.9c). The formation of the nanorings and nanohelices can be understood from the nature of the polar surfaces. If the surface charges are uncompensated during the growth, the spontaneous polarization induces electrostatic energy due to the dipole moment, but rolling up to form a circular ring would minimize or neutralize the overall dipole moment, reducing the electrostatic energy. On the other hand, bending of the nanobelt produces elastic energy. The stable shape of the nanobelt is determined by the minimization of the total energy contributed by spontaneous polarization and elasticity.

If the nanobelt is rolled loop-by-loop, the repulsive force between the charged surfaces stretches the nanospring, while the elastic deformation force pulls the loops together; the balance between the two forms the nanospring that has elasticity (Fig. 13.9d). The nanospring has a uniform shape with radius of ~500–800 nm and evenly distributed pitches. Each is made of a uniformly deformed single-crystal ZnO nanobelt.

For the nondoping synthesis case [99], by controlling the growth kinetics through refining parameters such as temperature, pressure, and duration of time, especially the pregrowth pressure level, single crystal ZnO nanosprings could be synthesized at high yield (>50%). The result revealed that doping is not necessary for forming the PSD nanosprings and nanoloops, but a pregrowth low pressure as low as ~10^{-3} Torr seems to be the key. In this case, the nanobelts that form the nanosprings grow along [$2\bar{1}\bar{1}0$], and the ones that form the nanoloops/nanospirals grow along [$10\bar{1}0$]. The comprised PSD ZnO nanobelts typically have a much larger thickness of ~50 nm with respect to those of doping case. Using dark field

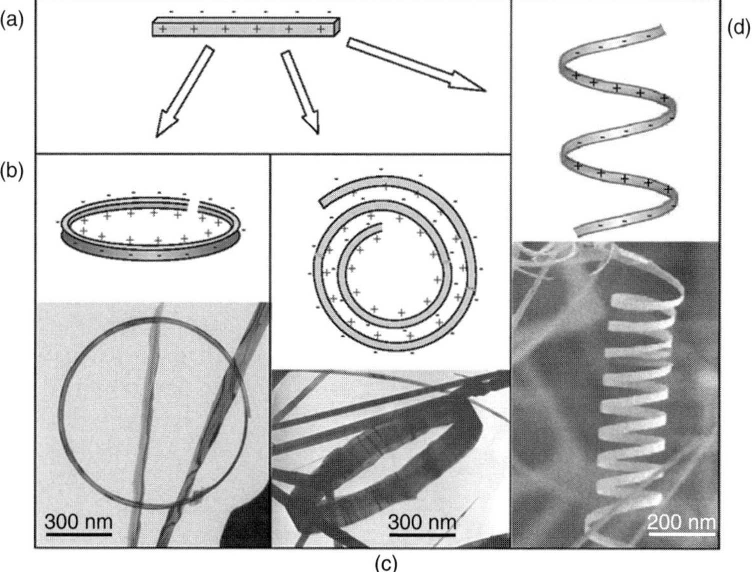

FIGURE 13.9. (a) Model of a polar nanobelt. Polar surface-induced formation of (b) nanoring (c) nanospiral, and (d) nanohelices of ZnO and their formation processes.

TEM imaging, it is found that planar defects are usually present in the PSD nanobelts, which could have played a key factor for the growth process. The study has made it possible to synthesize high purity, high yield nanosprings, opening the door for systematically understanding the properties and exploring the applications of semiconducting and piezoelectric nanosprings of ZnO.

3.3.3. Nanobows [105,106]

Nanobows are novel nanostructures found recently in growth of nanorings. Continuous and uniform bending of nanobelts into semirings is a characteristic of all nanobows. Two types of nanobows have been found. One is the freestanding nanobows extended from nanowires or nanorods. The other is the nanobows directly rooted on the deposition substrate. Figure 13.10a shows a hexagonal ZnO rod with a ZnO nanobelt grown from one of its six primary crystallographic facets, ended on the other side facet of the hexagonal nanorod. The rod grows along [0001] with side surfaces of $\{2\bar{1}\bar{1}0\}$ or $\{01\bar{1}0\}$. Based on the characteristics of ZnO nanostructures, the inner arc of the nanobow will be Zn-terminated or O-terminated. In Fig. 13.10b, two other types of nanobows of ~3 μm width are directly rooted at polycrystalline alumina substrate. Figure. 13.10c shows a circular platelet dominated by $\{0001\}$ polar surfaces. Figure 13.10d displays a conjunctional structure of a nanobow and a bow-like platelet.

A mechanism about the formation of nanorings has been proposed as following [97]. For a thin, straight PSD nanobelt, the spontaneous polarization-induced electrostatic energy decreases upon rolling into a circular ring due to the neutralization of the dipole moment. However, the elastic energy introduced during deformation increases. If the nanobelt is sufficiently thin (~10 nm), the former can overcome the latter, so that the total energy reduces by forming a ring. The stable shape of the ring is dictated by the minimization of the total energy contributed by spontaneous polarization and elasticity. This is the electrostatic polar charge model.

3.3.4. Seamless Nanorings [107,108]

By adjusting the raw materials with the introduction of impurities, such as indium, a seamless nanoring structure of ZnO has been synthesized (Fig. 13.11). TEM imaging and diffraction analysis have proved that the nanoring is a single-crystal entity with circular shape. The single-crystal structure referred here means a complete nanoring that is made of a single crystalline ribbon bent evenly at the curvature of the nanoring. Although the radius of the ring is large, its thickness could range from <10 nm to >50 nm. The nanoring is made of a loop-by-loop coaxial, uniradius, epitaxial coiling of a nanobelt, which is shown in the dark-field TEM image in Fig. 13.11b, the inset enlarged image corresponding to the white squared area shows the contrast due to the periodical looping circles.

The growth of the nanoring structures can be understood from the polar surfaces of the ZnO nanobelt. The polar nanobelt shown in Fig. 13.11b grows along [10$\bar{1}$0], with side surfaces ±(1$\bar{2}$10) and top/bottom surfaces ±(0001), and has a

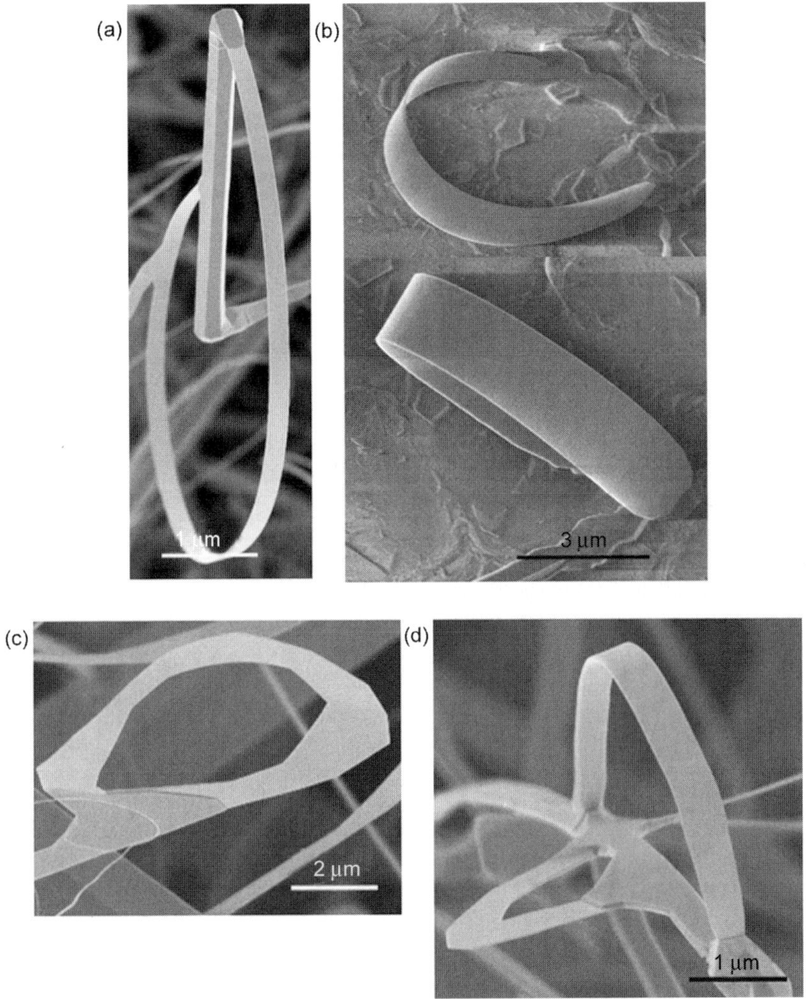

FIGURE 13.10. (a) ZnO nanobows made from individual polar surface-dominated single-crystal nanobelt in conjunction with a hexagonal ZnO nanowire. (b) Two ZnO nanobow directly rooted on polycrystalline alumina substrate. (c) Bow-like circular platelet of ZnO. (d). A combination nanoarchitecture of a nanobow and a bow-like circular platelet of ZnO.

typical width of ~15 nm and thickness of ~10 nm. The nanobelt has polar charges on its top and bottom surfaces (Fig. 13.12a). If the surface charges are uncompensated during growth, the nanobelt may tend to fold itself as its length getting longer to minimize the area of the polar surface. One possible way is to interface the positively charged Zn-(0001) plane (top surface) with the negatively charged O-(0001) plane (bottom surface), resulting in neutralization of the local polar charges and the reduced surface area, thus, forming a loop with an overlapped end

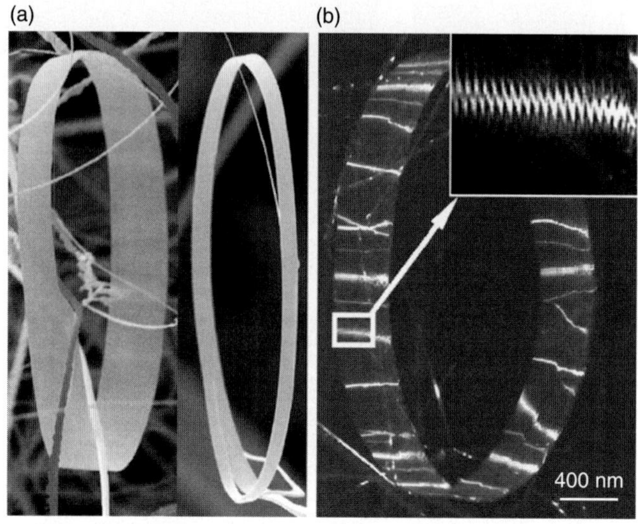

FIGURE 13.11. Seamless single-crystal nanorings of ZnO. (a) SEM images of two nanorings. (b) A dark field TEM image of a nanoring, the *inset* in (b) is an enlargement of a local region, displaying the contrast produced by the uniradius, loop-by-loop self-coiling nanobelt.

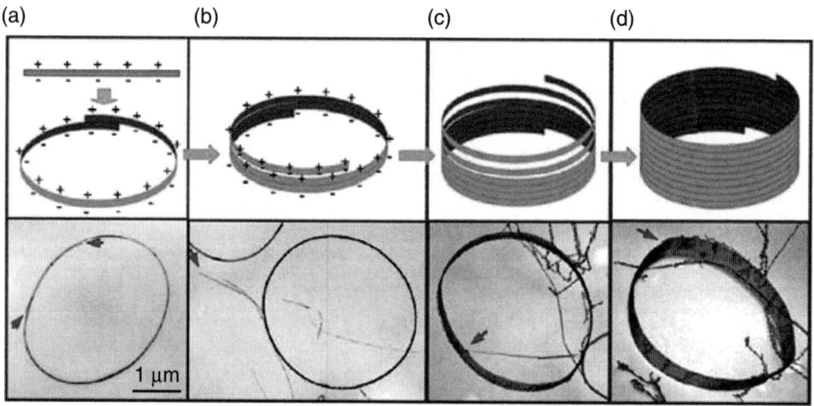

FIGURE 13.12. Growth process of self-coiled seamless single crystal nanorings of ZnO.

(Fig. 13.12b). This type of folding is 90° with respect to the folding direction for forming the nanospring or nanospiral, possibly due to the difference in aspect ratio of the nanobelts and relative size of the polar surfaces. The radius of the loop may be determined by the initial folding of the nanobelt at the initial growth, but the size of the loop cannot be too small to reduce the elastic deformation energy. The total energy involved in the process coming from polar charges, surface area and elastic deformation. The long-range electrostatic interaction is likely to be the initial driving force for folding the nanobelt to form the first loop for the subsequent growth. This is the nucleation of the nanoring.

As the growth continues, the nanobelt may be naturally attracted onto the rim of the nanoring due to electrostatic interaction and extends parallel to the rim of the nanoring to neutralize the local polar charge and reduce the surface area, resulting in the formation of a self-coiled, coaxial, uniradius, multilooped nanoring structure (Fig. 13.12c). The self-assembly is spontaneous, which means that the self-coiling along the rim proceeds as the nanobelt grows. The reduced surface area and the formation of chemical bonds (short-range force) between the loops stabilize the coiled structure. The width of the nanoring increases as more loops winding along the nanoring axis, and all of them remain in the same crystal orientation (Fig. 13.12d). Since the growth was carried out in a temperature region of 200–400°C, "epitaxial sintering" of the adjacent loops forms a single-crystal cylindrical nanoring structure, and the loops of the nanobelt are joined by chemical bonds as a single entity. A uniradius and perfectly aligned coiling is energetically favorable because of the complete neutralization of the local polar charges inside the nanoring and the reduced surface area. This is the "slinky" growth model of the nanoring. The charge model of the nanoring is analogous to that of a single alpha helix.

3.3.5. Deformation-free Nanohelix [109]

As mentioned in Fig. 13.9, the nanosprings usually have a radius of 0.5–1.5 μm, which cannot be too small as constrained by the elastic deformation energy. Recently a nanohelical structure was found that has radius as small as ~50 nm (Fig. 13.13a), much smaller than the nanospring as presented in Fig. 13.9. In order to understand this structure, the intrinsic crystal structure of the nanohelices has been investigated. TEM image presents the uniform shape and contrast of the nanohelix (Fig. 13.13b). High-resolution transmission electron microscopy (HRTEM) imaging reveals that the nanohelix has an axial direction of [0001], although the growth direction of the nanowire changes periodically along the length. Detailed HRTEM images from the regions labeled c and d in Fig. 13.13b are displayed in Fig. 13.13c and d, respectively, which shows that the nanowire that constructs the nanohelix grows along [01$\bar{1}$1]. Because the incident electron beam is parallel to [2$\bar{1}$$\bar{1}$0], the two side surfaces of the nanowire are ±(0$\bar{1}$11). No dislocations were found in the nanohelices. It is important to note that the image recorded from the "twist" point of the nanohelix shows no change in crystal lattice (Fig. 13.13c). The traces of the two sides are visible, indicating the non-twisted or non-deformation single-crystal structure of the entire nanohelix.

The nature of the ±{0$\bar{1}$11} planes can be understood from the atomic model of ZnO. By projecting the structure along [2$\bar{1}$$\bar{1}$0] (the a-axis), beside the most typical ±(0001) polar surfaces that are terminated with Zn and oxygen, respectively, ±(01$\bar{1}$$\bar{1}$) and ±(0$\bar{1}$11) are also polar surfaces (Fig. 13.8b). From the structure information provided by Fig. 13.12, the structure of the nanowire that self-assemble to form the nanohelix can be constructed. The nanowire grows along [01$\bar{1}$1], the two end surfaces being $\pm(0001)$, side surfaces being $\pm(01\bar{1}\bar{1})$, $\pm(1\bar{1}01)$, and $\pm(10\bar{1}\bar{1})$, as shown in Fig. 13.13e. This segment of nanowire has a pair of polar surfaces $[\pm(10\bar{1}\bar{1})]$ across its thickness.

The structure model presented in Fig. 13.13e is the basic building block/ segment for constructing the nanohelix by a self-coiling process during the growth, but without introducing any deformation. Because there are six crystallographic equivalent $<0\bar{1}11>$ directions: $[01\bar{1}1]$, $[\bar{1}101]$, $[\bar{1}011]$, $[0\bar{1}11]$, $[1\bar{1}01]$, and $[10\bar{1}1]$, and there is a $60°$ rotation between the two adjacent directions, there are six equivalent orientations to stack the building block along the [0001] axial direction without introducing deformation or twist. A realistic 3D model of the nanohelix is presented in Fig. 13.13f, which is a stacking of the building blocks around the [0001] axis following the sequences of the six directions described above. The interface between the two building blocks is perfectly coherent and the same piece of crystal, without mismatch, translation, or twist. If viewing along $[2\bar{1}\bar{1}0]$ from the side of the nanohelix, its projected structure is given in Fig. 13.13g, where the red and blue lines indicate the surfaces terminated with Zn^{2+} and O^{2-}, respectively.

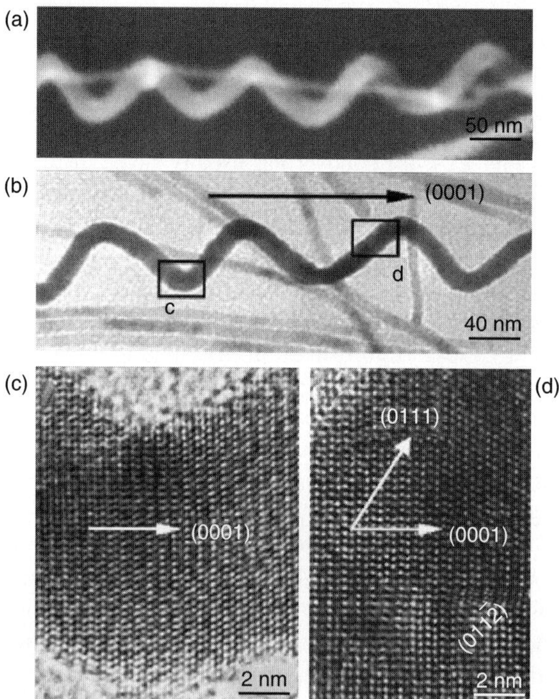

FIGURE 13.13. (a) SEM image of a left-handed nanohelix. (b) A bright-field TEM image of a nanohelix. No significant strain contrast is found (apart from the overlap effect between the nanohelix and nanowires). (c, d) HRTEM images recorded from the c and d areas labeled in (a), respectively, showing the growth direction, side surfaces, and dislocation-free volume. (e–g) structure of the deformation-free nanohelix.

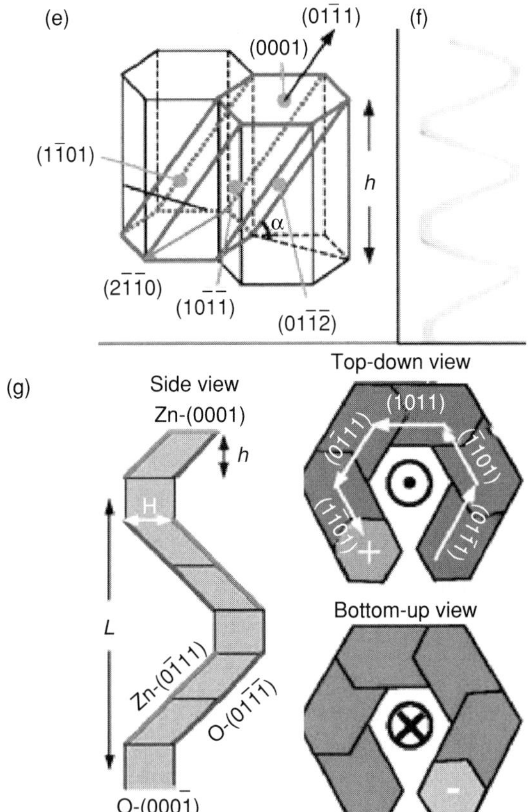

FIGURE 13.13. (*Continued*) (e) The fundamental building block of the nanowire (in purple) for constructing the nanohelix, and its growth direction and surfaces. (f) A model of the nanohelical structure. (g) The side-, top-down and bottom-up views of the nanohelical model. The distribution of charges on the surfaces of the nanohelical is analogous to the charge model of an alpha helix.

The distribution of the polar charges on the surfaces of the polar nanowire is analogous to the charge model of a single alpha helix model, and it is best seen through the top and bottom views of the model (Fig. 13.13g). If viewing the nanohelix from top-down, the Zn-(0001) at the front and the O-(011̄1̄) at the sides are directly seen. The Zn-terminated (0001) leads to the nanowire growth due to self-catalysis. The six growth directions of the building blocks are indicated. If viewing the nanohelix from bottom-up, the O-(0001̄) at the end and the Zn-(01̄1̄) at the sides are seen. It is important to point out that there is no deformation introduced in the hexagonal screw-coiling stacking process, thus, no dislocations are needed to accommodate deformation.

The structure model presented in Fig. 13.13 is supported by the energy calculation for the surface charge. By folding a polar nanowire into a helix without introducing deformation, the change in total energy drops more than 15%.

It is very interesting that a similar structure model has been proposed in a latest report on $ZnGa_2O_4$ helical structures [83]. Bae et al. synthesized the helical $ZnGa_2O_4$ single crystal nanowires by using a two-step thermal evaporation method. Two types of helical zinc gallate ($ZnGa_2O_4$) nanostructures were achieved: helical $ZnGa_2O_4$ nanowire rolls either on a straight ZnSe nanowire support, which makes them look like "vines," or without any support, in which they have the appearance of "springs." The two step methods are listed as following: (1) High-purity single-crystalline ZnSe nanowires were synthesized using CoSe/ZnO on an Au nanoparticles-deposited Si substrate at 800°C. (2) The pre-grown ZnSe nanowires were placed near the ZnO/Ga, and a temperature of 600 or 900°C was maintained for 10–60 min, thereby producing the $ZnGa_2O_4$ nanowires. For the explanation of the growth process, instead of six equivalent < 01–11 > growth direction switching, due to the cubic structure of the spinel zinc gallate, four equivalent growth directions < 011 > switching led to a square-staired elevation instead of a hexagonal stair elevation.

3.4. Hierarchical Oxide Nanostructures

Generally, crystallographic characteristics will determine the growth habit of the specific nanocrystal growth. For example, structurally ZnO has three fast growth directions: $<2\bar{1}\bar{1}0>$ ($\pm[2\bar{1}\bar{1}0]$, $\pm[\bar{1}2\bar{1}0]$, $\pm[\bar{1}\bar{1}20]$); $<01\ \bar{1}0>$ ($\pm[01\bar{1}0]$, $\pm[10\bar{1}0]$, $\pm[1\bar{1}00]$); and $\pm[0001]$. Together with the polar surfaces due to atomic terminations, ZnO could exhibit a wide range of novel structures that can be grown by tuning the growth rates along these directions. These are the fundamental principles for understanding the formation of numerous ZnO nanostructures. For hexagonal nanorods, the fast growth direction is along [0001], one of the major easy axis of wurtzite structured ZnO. While for ZnO nanobelts, three fast growth directions become choices for directional growth. There is a large amount of space for playing the crystal growth of nanobelts.

3.4.1. Aligned Nanopropellers and Junction Arrays [110,111]

Modifying the composition of the source materials can drastically change the morphology of the grown oxide nanostructure. A mixture of ZnO and SnO_2 powders in a weight ratio of 1:1 could be used as the source material to grow a complex ZnO nanostructure. Figure 13.14a shows an SEM image of the as-synthesized products rooted on polycrystalline alumina substrate with a uniform feature consisting of sets of central axial nanowires, surrounded by radial-oriented nanoblade nanostructures. The morphology of the branch appears like a "propeller blade," and the axial nanowire is the "propeller axis," which has a uniform cross section with dimension in the range of a few tens of nanometer. The nanoblades generally have spherical Sn catalyst balls at the tips (Fig. 13.14a), and the branches display a ribbon shape. The ribbon branches have a fairly uniform thickness. There is another type of nanopropellers that could be free-standing. Figure 13.14b shows a string of freestanding nanopropellers, the

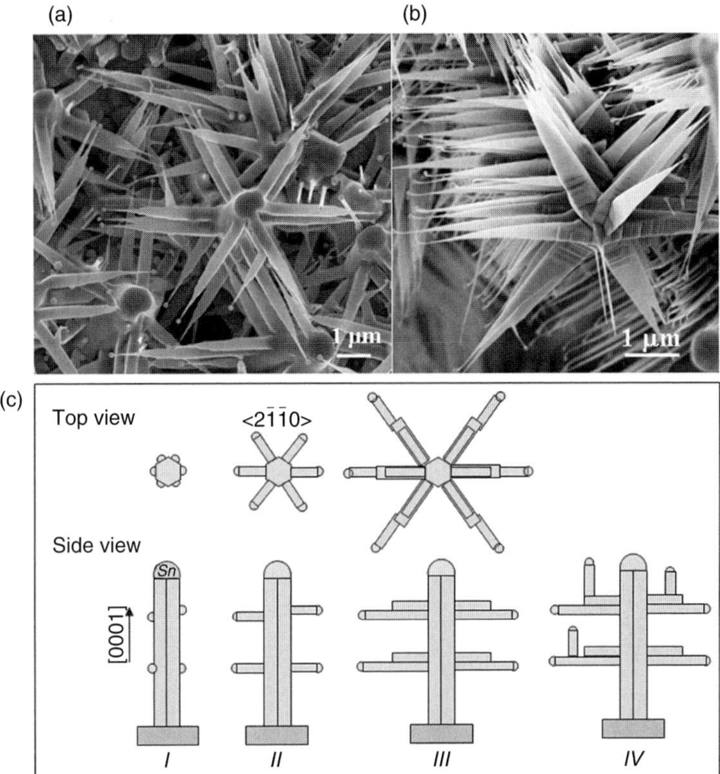

FIGURE 13.14. Nanopropeller arrays of ZnO. (a) SEM image of Sn catalyzed growth of aligned nanopropellers based on the six crystallographic equivalent directions. (b) A freestanding nanopropellers with a secondary grown nanoneedle. (c) Stages I–IV represent the growth processes of the nanopropellers.

secondary growth on the propeller surface leads to another nanowires on top of the nanoblade surface in addition to the central nanowire (Fig. 13.14b).

It is known that SnO_2 can decompose into Sn and O_2 at high temperature, thus, the growth of the nanowire–nanoribbon junction arrays is the result of vapor–liquid–solid (VLS) growth process, in which the Sn catalyst particles are responsible for initiating and leading the growth of ZnO nanowires and nanobelts. The growth of the novel structure presented here can be separated into two stages. The first stage is a fast growth of the ZnO axial nanowire along [0001] (Fig. 13.14c I). The growth rate is so high that a slow increase in the size of the Sn droplet has little influence on the diameter of the nanowire, thus the axial nanowire has a fairly uniform shape along the growth direction. The second stage of the growth is the nucleation and epitaxial growth of the nanoribbons due to the arrival of the tiny Sn droplets onto the ZnO nanowire surface (Fig. 13.14c II).

This stage is much slower than the first stage because the lengths of the nanoribbons are uniform and much shorter than that of the nanowire. Since Sn is in liquid state at the growth temperature, it tends to adsorb the newly arriving Sn species and grows into a larger size particle (i.e., coalescing). Therefore, the width of the nanoribbon increases as the size of the Sn particle at the tip becomes larger, resulting in the formation of the tadpole-like structure observed in TEM (Fig. 13.14c III). The Sn liquid droplets deposited onto the ZnO nanowire lead to the simultaneous growth of the ZnO nanoribbons along the six equivalent growth directions: $\pm[10\bar{1}0]$, $\pm[0\bar{1}10]$, and $\pm[\bar{1}100]$. Secondary growth along [0001] results in the growth of the aligned nanowires on the surfaces of the propellers (Fig. 13.14c IV).

3.4.2. Nanoarchitectures of ZnO

As discussed in Section 3.4.1, ZnO has two important structure characteristics: the multiple and switchable growth directions: $<01\bar{1}0>$, $<2\bar{1}0>$, and $<0001>$; and the $\{0001\}$ polar surfaces. With the three types of stable facets of $\{01\bar{1}0\}$, $\{2\bar{1}\bar{1}0\}$, and $\{0001\}$ and the 13 fastest growth directions, a diversity of morphologies have been created. Nanobelts of different crystal facets have been synthesized. Another key factor is the polar surfaces. The Zn-terminated (0001) surface is positively charged, and the O-terminated $(000\bar{1})$ surface is negatively charged, thus, there is a spontaneous dipole moment along the c-axis. If the crystal thickness is large, the effect of the dipole is small, but once the nanobelt is as thin as 10–50 nm, the spontaneous polarization can produce a few unusual growth features, such as nanorings, nanobows, nanohelices, and nanospirals, which are the results of minimizing the electrostatic energy due to the polarization. Recent report on a combined ZnO nanoarchitectures has shown that these fundamental growth features coexist during the growth, and their recombination produces a diversity group of nanoarchitectures including several types of nanorings, nanobows, platelet circular structures, Y-shape split ribbons, and crossed ribbons. In Section 3.4.2.1, the switch growth phenomena of ZnO nanostructures will be elaborated.

3.4.2.1. Switched Growth in Circular Architectures

Figure 13.15 shows a series of SEM images indicating the switching growth directions of ZnO circular architectures. The morphology shown in Fig. 13.15a gives a good example of a nanobelt that switched its growth direction during the syntheses. The belt grew along a specific direction, then it switched for ~30° into another direction, and finally rotated 60° and then self-coiled into a full ring. Figure 13.15b is a case of a switching growth mode within a nanoring, in which seemly the starting end of the nanoring could not find the end point when it crosses the curved nanobelt, so a growth along a switched direction can be a choice for leading to the eventual meeting with the original nanobelt. As shown in the right, the enlarged section shows a three-time switching growth made the loop closed. Figure 13.15c is a case in which three branches grew first along three $<01\bar{1}0>$ directions, and then one of the two side branches crossed the middle

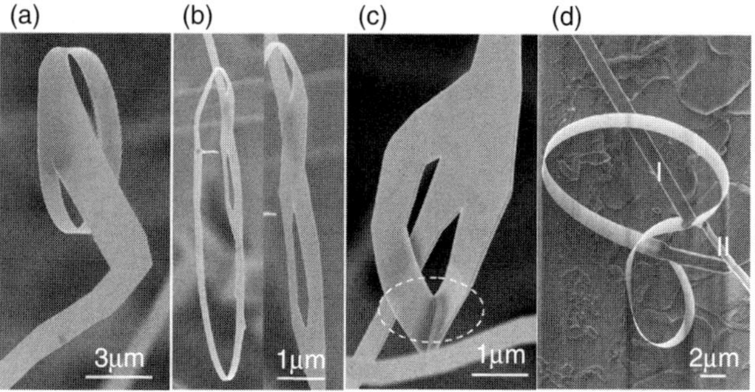

FIGURE 13.15. Switching growth directions in formation of ZnO nanoarchitectures. (for details, see the text.)

branch underneath its surface, and then rejoin the other branch. The joining point is not flat due to the mismatch (as indicated by a dotted ellipse) caused by passing the middle branch. The structure of this kind is a pair of "bow and arrow." It is worth mentioning that such a combined structure is still single crystalline, although at the interface there is apparently a mismatch. Under TEM, it has been found that the two side ribbons could miss each other at the joining point. And the top and bottom surfaces have been revealed to be the {0001} polar surfaces under TEM. It must be pointed out that the planar circular-like structure presented here is formed by changing the growth direction among the six crystallographic equivalent directions, rather than the minimization of electrostatic energy. As seen in Fig. 13.15d, the nanobelt could twist itself into a configuration that matches the local crystallography and surface charge. Secondary growth is responsible for the unusual twisting configuration of the nanobelt.

An architecture can be made by combining the growth configurations presented in Figs. 13.10 and 13.11. Figure 13.16a and b is a combination of a nanorod, a semicircular bow and a perfect ring. The nanorod grows along [0001] and is enclosed by six $\{01\bar{1}0\}$ side facets. A nanobelt grows perpendicularly to the nanorod along $[01\bar{1}0]$ and bends uniformly and finally ends at the nanorod, forming a nanobow. This nanobelt is dominated by the {0001} polar surfaces and the polar charges enforce the bending of the nanobelt. The joint point between the nanobelt and the rod is a neck structure. A complete ring on the top of the nanorod has a solid conjunction with the rod, and its large surfaces are $\pm(01\bar{1}0)$, and its top and bottom rims are the {0001} polar surfaces. The growth direction of the self-coiled belt is $[2\bar{1}\bar{1}0]$, similar to the seamless nanoring reported previously (Fig. 13.11b).

The combined growth structure presented in Fig. 13.16 is the first observation of its kind. This type of combination directly proves the coexistence of the polar belt rolled rings/bow and the polar belt self-coiled seamless nanoring. Figure 13.16c is a schematic model for this integrated growth phenomenon. The first step is the growth of a hexagonal rod along [0001]. The top facet is dominated by

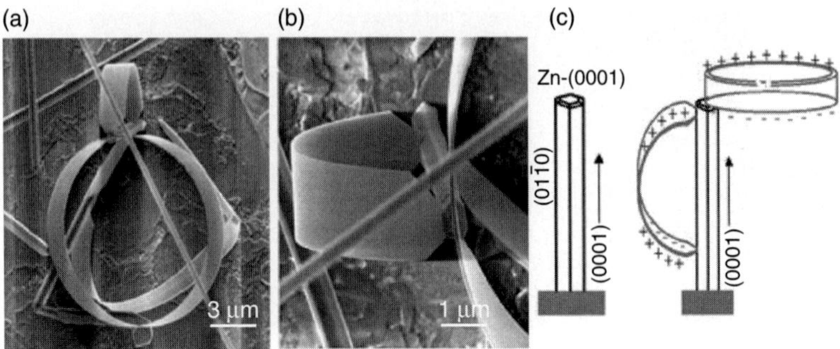

FIGURE 13.16. (a) A nanoarchitecture composed of a rod, a bow, and a ring. (b) Magnified SEM images to capture the orientation relationships among the three components. (c) A structure model of the architecture.

Zn-terminated (0001) plane, which is self-catalytically active, and the side facets are likely to be the six crystallographically equivalent {01$\bar{1}$0} surfaces. As growth continues, the stepped side facets as shown in Fig. 13.16b would function as a good nucleation site for the secondary growth, leading to the formation of a belt along a normal direction of [01$\bar{1}$0]. Due to the spontaneous polarization of the ZnO nanobelt across its thickness, bending into a ring would reduce the electrostatic energy, provided the thickness-to-radius ratio is less than ~4%. But the bending is ended if the nanobelt reaches the rod and rejoins the single crystal rod, forming a semicircular bow. On the other hand, with the growth of the rod, the growth front of Zn-(0001) plane could be the nucleation site for the other fast growth belt along [2$\bar{1}\bar{1}$0] parallel to the side facet of the rod, and it tends to fold back to form a loop, such as the model shown in Fig. 13.8c. A perfect ring is formed by epitaxial self-coiling of a polar belt. The bottom surface of the ring tends to be O-(000$\bar{1}$) surface, while the top surface is Zn-(0001).

3.4.2.2. Y Shape Single-crystal Nanoribbons

Splitting of a single crystal could occur and the side branch growth leads to the formation of Y-shape nanoarchitecture of ZnO. Figure 13.17 gives a typical example for the split growth phenomena. Figure 13.17a shows that a ribbon of large aspect ratio (around 5–10) and smooth polar surfaces splits into two high aspect ratio ribbons of equal width but half in thickness. The ribbon was too thick to be bent by polar charge prior splitting, but bending is possible after splitting (bottom enlarged section). In Fig. 13.17b, TEM imaging and diffraction show that the splitting is due to an initiation of a side branch along [01$\bar{1}$0] away from the original branching direction of [11 2 0], but both branches are still dominated by {0001} polar surfaces. Figure 13.17c shows that a curly ribbon splits out of a straight ribbon, later it rejoins the straight ribbon. The curly ribbon is due to its small thickness, and is due to the surface polar charge. These growth features are rather unique.

Joining two nanostructures together could occur in two ways. One is a natural crystallographic matching so that two of them are integrated as a single crystal. The other is a sintering during growth. Figure 13.18a and b shows two nanobelts that are

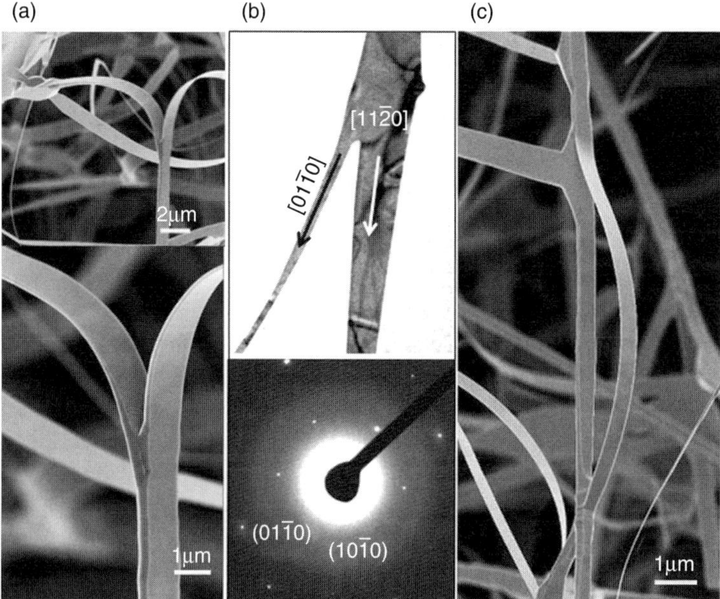

FIGURE 13.17. (a) SEM images of two tangential split ribbons. (b) TEM image and corresponding [0001] diffraction pattern of a Y structure, which is produced by separating the growth along [01$\bar{1}$0] from the original [2$\bar{2}\bar{1}$0] direction. (c). A bow structure formed by the split growth of a belt and its later rejoining to the straight ribbon.

joined together. It is possible that both crystallographic bonding and sintering play a role in holding them together. But a case shown in Fig. 13.18c is likely due to sintering. The nanostructure shown in Fig. 13.18d is a result of self-splitting.

3.4.2.3. Crossed Nanoribbon Architectures

Crossed ribbon structures have been observed with some unusual configurations. One way of creating crossed structures is the switching/splitting in growth directions among <01$\bar{1}$0 > and < 2$\bar{1}\bar{1}$0 >. These type of structures usually have a smooth surface and all of the branches share the same flat {0001} surface (Fig. 13.19a and b). The two-branched nanoribbon in Fig. 13.19c grow along [1$\bar{2}$10] and [$\bar{2}$1$\bar{1}$0]. For the three-branched nanoribbons in Fig. 13.19d, electron diffraction pattern shows that the three nanoribbons are a single crystal and are along [1$\bar{1}$00], [01$\bar{1}$0], and [$\bar{1}$010].

The other group of crossed structure is complex because there is a crystal at the joint point of the nanobelts, and sometime the two nanobelts do not share a common flat surface (Fig. 13.19e–g). This type of structure could involve the formation of twins at the central crystal and/or sintering. The two ribbons could grow perpendicularly from a cross junction area, and they do not share a common flat surface. Figure 13.19h is a more complicated case, where multiple ribbons grew from one junction, forming a 3D complex structure. These nanostructures demonstrate the diversity of ZnO growth.

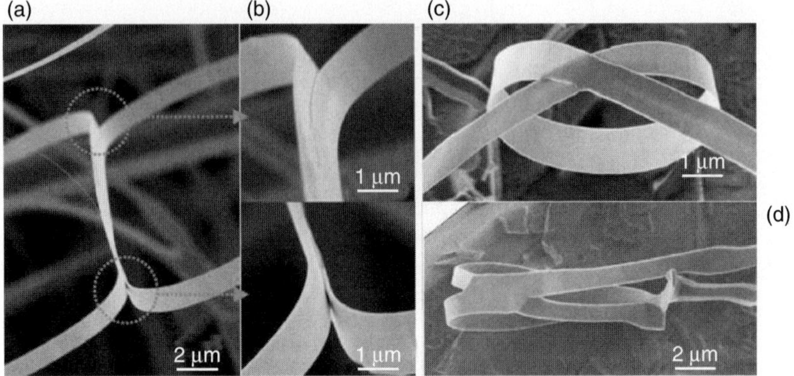

FIGURE 13.18. Joining of nanostructures by high temperature sintering. (a) Two micron-sized ribbons merged in parallel in the contact region and (b) the enlarged views of the top and bottom sides of the joining points, respectively. (c) A self-bent and twisted ribbon into a closed loop. (d) A polar surface-dominated full-ring split out another nanoribbon, which overlapped and merged with the original loop.

3.4.2.4. Zigzag Chain Architectures

Zigzag chain type of nanostructures has also been found for ZnO. Figure 13.20 shows two typical configurations of the zigzag structures. The chain structures are dominated by the large {0001} polar surfaces, but its growth direction changes periodically, as shown in Fig. 13.20c, leading to the formation of chain structures. For the same family of growth directions of <$01\bar{1}0$>, e.g., there are six equivalent directions and there is a $60°$ rotation between the two adjacent directions. Therefore, a periodic change in growth direction following a periodicity of $[1\bar{1}00] \rightarrow [10\bar{1}0] \rightarrow [01\bar{1}0] \rightarrow [10\bar{1}0]$, and a fixed length growth results in the formation of zigzag chain structure, similar to the one reported previously. The junction shown in Fig. 13.20b is due to a split growth of the nanoribbon along one of the directions. Figure 13.20d shows a TEM image of a closed loop with extension, the nanoribbon is in a zigzag shape, the inset TEM diffraction pattern proved the switching growth model in Fig. 13.20c.

4. Growth Mechanisms

Rational growth control of 1D nanostructure is of great importance for achieving high-yield, uniform, and repeatable building blocks of various device applications. It is essential to understand the physical and chemical insight of the growth process of 1D nanostructure. Since 1950s, two growth mechanisms for whiskers and later 1D nanostructures have been generalized based on the presence or absence of the metal catalysts in the growth process. One is VLS process and the other is vapor–solid (VS) mechanism. Recently, polar surface-induced growth phenomena studies on ZnO, ZnS, and CdSe have added more insight on understanding the physical and chemical process of the 1D nanostructures growth.

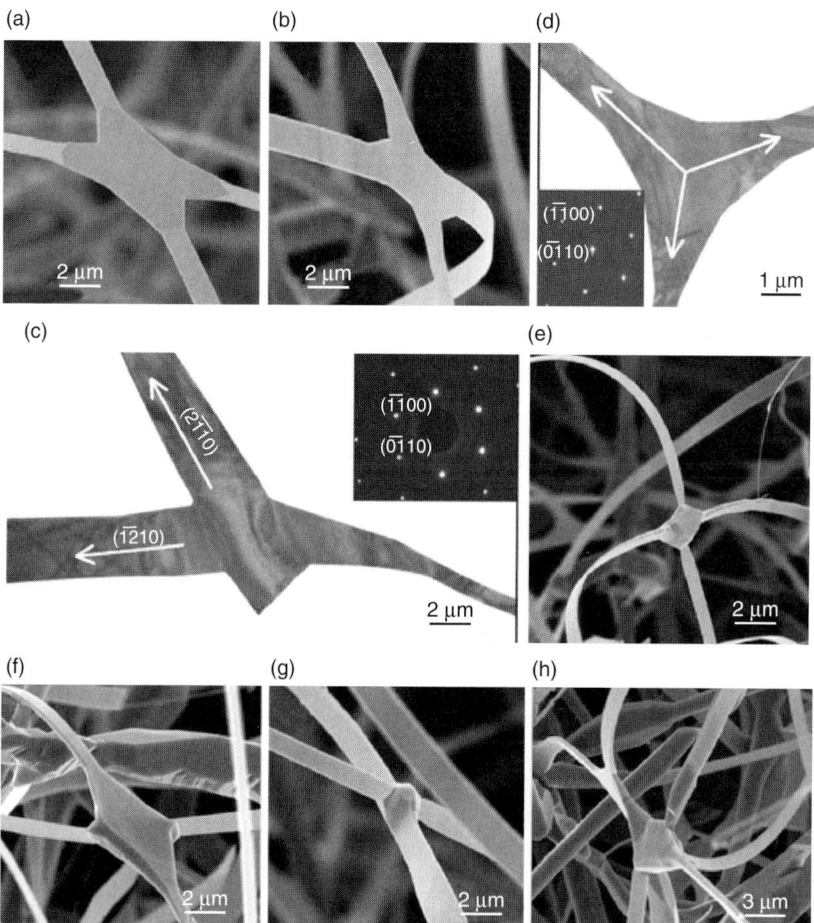

FIGURE 13.19. Crossed junctions of ZnO. (a–d) SEM and TEM images of the crossed nanobelts that share the same polar surface of (0001). The junction branches could be along [011̄0] or [21̄1̄0]. (e–g) Cross junctions of nanobelts with twisted ribbons, so that they do not share a common polar surface. (h) A multijunction structure of over four ribbons.

4.1. Catalyst-involved Vapor–Liquid–Solid Growth Process

VLS process has been proposed for nearly four decades [112], and has been widely used for describing the growth of 1D nanostructures involving catalysts, where the catalysts take the form of nanoparticles at the ends of 1D nanostructures. The process includes three steps: the dissolution of vapor deposition materials in the catalysts liquid droplets at local high temperature, saturation and supersaturation of

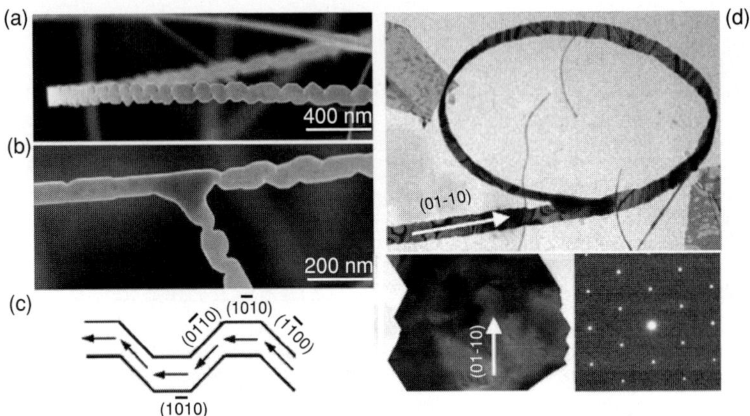

FIGURE 13.20. (a, b) Polar surface-dominated ZnO ribbons with zigzag chain structure, which are formed by a periodic change in growth direction. (c) A model of the chain structure. (d) Nanorings made of zigzag nanoribbon, where the electron diffraction analysis proves the validation of chain model.

deposition materials in liquid droplets, then precipitation of solid deposition materials along specific direction. Figure 13.21 shows the growth process of ZnO whisker along [0001] on Si substrate, where Au functions as the catalyst.

Catalysts-stimulated growth has been widely utilized for taking advantage of the VLS growth process. There are various ways to create nanoscale catalysts particles. For example, using laser ablation, under controlled temperature and pressure environment, the irradiation of a target composed of catalysts and reactant materials will generate size-controlled clusters. Using chemical techniques, monodispersive nanoparticles could be synthesized followed by drying and dispersion on a substrate as catalysts. Catalysts could also be obtained by sputter or CVD coating on substrate, followed by heat treatment, forming islands of nanoscale catalysts. During the decomposition process of some systems, such as ZnO–Sn and SnO_2 (SnO), catalyst nanoparticles are formed during the redox:

$$2ZnO\ (s) + C\ (s) \rightarrow 2Zn\ (g) + CO_2\ (g), \tag{13.1}$$

$$SnO_2\ (s) + C\ (s) \rightarrow SnO\ (g) + CO\ (g), \tag{13.2}$$

$$SnO_2\ (s) \rightarrow Sn\ (l) + SnO\ (g), \tag{13.3}$$

$$2SnO\ (g) \rightarrow Sn\ (l) + SnO_2\ (s), \tag{13.4}$$

It is noted that low solubility of deposition materials in the catalyst liquid droplets is the prerequisite for achieving a high level of supersaturation that will lead to the precipitation and growth of 1D nanostructures along a specific direction.

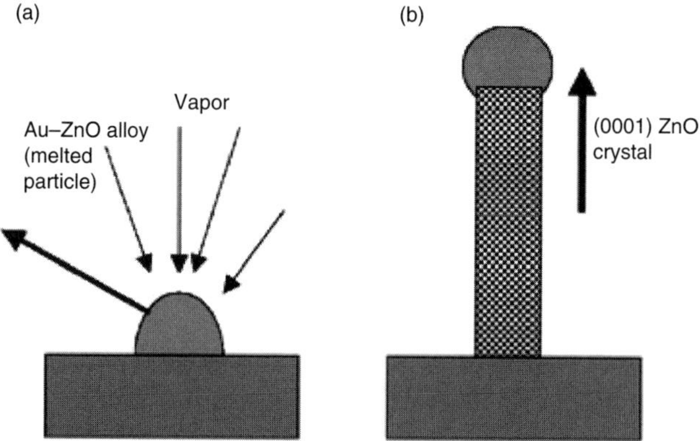

FIGURE 13.21. The growth process of ZnO along [0001] with Au functions as the catalyst.

4.1.1. Crystallographic Orientation Alignment of ZnO Nanorods and Nanobelts [113]

A chemical vapor transport process was used for synthesis of aligned ZnO nanorods and nanobelts. A mixture of commercial ZnO, SnO_2, and graphite powders in certain ratio (Zn:Sn:C = 2:1:1) was placed in an alumina boat as the source material and positioned at the center of the alumina tube. To investigate the duration time dependence on the growth of desired nanostructures, deposition process was conducted at 1,150°C for 5, 15, 30, and 60 min, respectively, under a constant pressure of 200 mbar and Ar flow rate of 20 sccm. The entire length of the tube furnace is 60 cm. The desired nanostructures were deposited onto alumina substrate located 21 cm away from the furnace center in a temperature range of 550–600°C.

Figure 13.22 is a group of typical SEM images of the as-synthesized nanostructures on an alumina substrate at 1,150°C, showing the initial growth and grown characteristics of the desired nanostructure. Microsize ZnO rods (microrods) of irregular side surfaces but uniform flat (0001) surfaces of ~3 μm are formed on the alumina substrate, which sometimes show faceted side surfaces. On each microrod, there are a number of nanorods oriented perpendicular to the (0001) plane, as clearly depicted in Fig. 13.22a and b, which is, respectively, the top and side view of the oriented ZnO nanorods. At the tip of each nanorod and nanobelt there is a comparable sized Sn ball. The oriented nanorods dispersive distribute on the flat c-plane of the microrods with uniform diameters ranging around 20–40 nm and an average height of about 40–80 nm.

The data suggest that the diameters of the nanorods are well confined by the size of the metallic Sn ball on the tip, and the growth time has a little influence on the diameter of the nanorods. Figures 13.22c and d gives a good description of the oriented growth of the nanorods on the c-plane of the ZnO micron rods. Figures show that each nanorod tends to grow along the same crystallographic orientation as the micron rod with six parallel faceted surfaces of $\{2\bar{1}\bar{1}0\}$, which

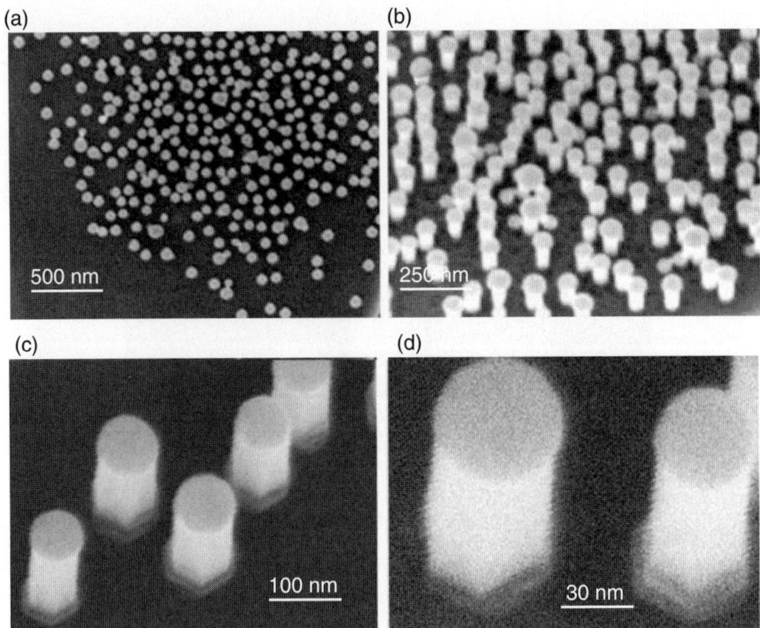

FIGURE 13.22. Growth of oriented nanorods using metallic Sn as the catalyst at 1,150°C. (a, b) The top and side view of nanorods after initial growth of 5 min on a single crystal microrod. (c, d) Show the epitaxial growth of the ZnO nanorods, forming arrays with an identical crystallographic orientation, as represented by the parallel side surfaces.

indicates an orientation alignment of the ZnO nanorods and a coherent epitaxial relationship with the base ZnO micron rod.

4.1.2. Orientation Guidance of Sn-catalyst on Growth of 1D ZnO Nanostructures [114]

To further probe the crystallographic orientation alignment phenomena as described in Section 4.1.1, a detailed investigation on crystallographic relationship between ZnO nanostructures and Sn particles has been carried out using x-ray diffraction (XRD) analyses and HRTEM. Firstly, we use sapphire single crystal substrate with (0001) orientation for the growth of oriented nanorods catalyzed by Sn, which later proven to be an idea sample for XRD analysis. Figure 13.23a is a typical SEM image of the as-grown aligned ZnO nanorods on the sapphire (0001) substrate. It is clearly seen that the ZnO nanorods grow normal to the substrate with spherical Sn ball at the tips (Fig. 13.23b). The corresponding XRD pattern in Fig. 13.23c revealed that only strong peaks with multiple of {0002} are strongly diffracted; the {200} of β-Sn peaks are rather weak; and the alumina peaks did not show up because the entire substrate has been covered by a layer of ZnO nanorods. It is suggested that there must be

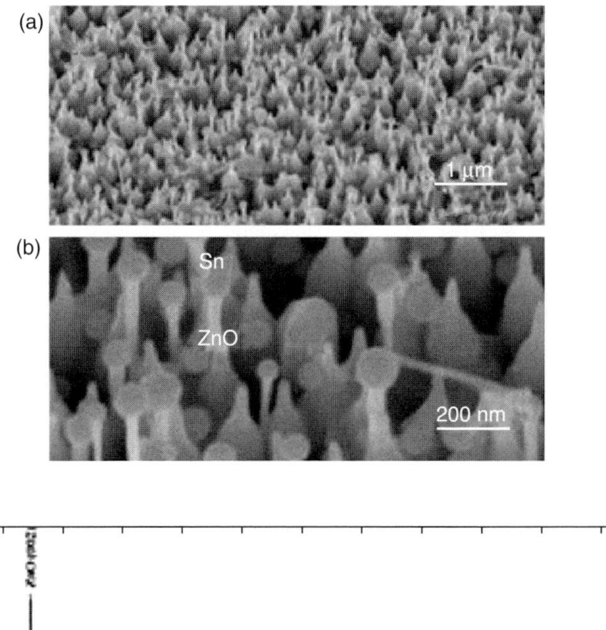

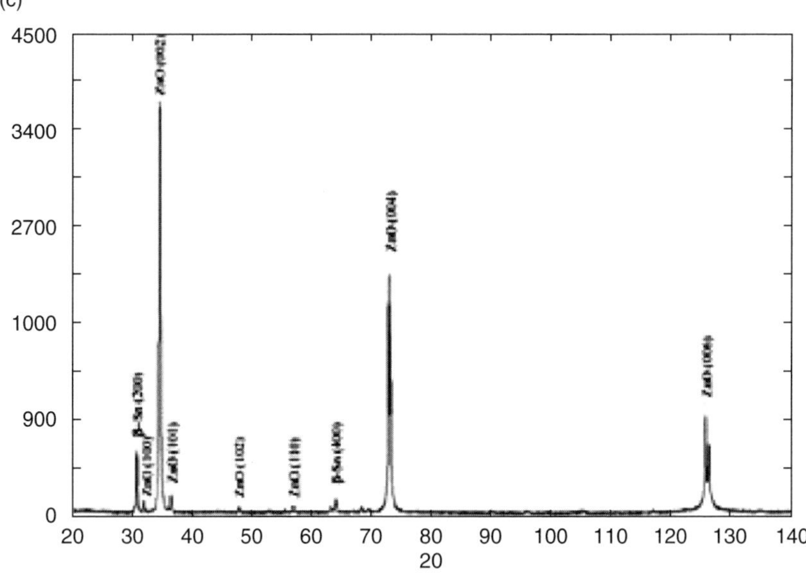

FIGURE 13.23. (a, b) A low magnification and an enlarged SEM image of as-grown aligned ZnO nanorods on sapphire (0001) substrate, respectively and (c) its corresponding x-ray diffraction pattern.

certain orientation relationship between ZnO nanorods and their Sn catalysts. To clarify this question, HRTEM is necessary to be carried out on the crystallographic relationship between ZnO and Sn.

In the VLS process, a metal catalyst is rationally chosen. It is generally believed that the metal particle is a liquid droplet during the growth and its crystal structure in solid may have no influence on the structure of the nanowires/nanobelts to be grown. Using tin particle-guided growth of ZnO 1D nanostructures as a model system, we show that the interface between the tin particle and the ZnO nanowires/nanobelts could be partially crystalline or ordered during the VLS growth although the local growth temperature is much higher than the melting point of the bulk tin, and the crystallographic structure and lattice mismatch at the interface is important in defining the structure characteristics of the grown nanowires/nanobelts. The interface prefers to take the least lattice mismatch, thus, the crystalline orientation of the tin particle may determine the growth direction and the side surfaces of the nanowires/nanobelts. The results may have an important impact in understanding the physical and chemical processes involved in the VLS growth.

Figure 13.24a and b shows bright-field and dark-field images of nanowires with a particle at its tip. The selected area electron diffraction (SAED) patterns recorded from the particle, the wire, and both the wire and the particle are shown in Fig. 13.24c–e, respectively. The interface orientation relationship between the ZnO wire and the Sn particle is $(020)_{Sn} \parallel (0001)_{ZnO}$, $[\bar{1}01]_{Sn} \parallel [2\bar{1}\bar{1}0]_{ZnO}$. Figure 13.24f is an HRTEM image of the white rectangle enclosed area in Fig. 13.24a, which depicts the interface structure between ZnO nanowire and the Sn particle.

The epitaxial orientation relationships between the Sn particle and the ZnO 1D nanostructure can be explained from the *lattice mismatch* at the interface. The interface of [0001] growth nanowire is composed of $(0001)_{ZnO}$ and $(020)_{Sn}$ planes. Two sets of $\{01\bar{1}0\}_{ZnO}$ planes match with the $\{101\}_{Sn}$ planes with a lattice mismatch as small as 0.7% with reference to ZnO, the left $\{01\bar{1}0\}_{ZnO}$ matches to the $\{200\}_{Sn}$ with a lattice mismatch of 3.6%. Thus, hexagon is preferred to be the cross section at the interface, corresponding to a wire/rod morphology.

4.2. Self-catalyzed Growth Mechanism

4.2.1. Nanocombs and Nanosaws [115–119]

"Comb-like" structures of ZnO have been reported for wurtzite structures, but the mechanism that drives the growth was not elaborated until recently. By using thermal evaporation method, the comb structures have been synthesized (Fig. 13.25a and b) with the comb teeth growing along [0001] (which can be directly seen from the inset in Fig. 13.25a), the top/bottom surfaces being $\pm(01\bar{1}0)$, and side surfaces $\pm(2\bar{1}\bar{1}0)$. Using convergent beam electron diffraction (CBED), which relies on dynamic scattering effect and is an effective technique for determining the polarity of wurtzite structure, we have found that the comb structure is an asymmetric growth along Zn-[0001] (Fig. 13.25b). This conclusion is received by comparison with the experimentally observed CBED pattern and the theoretically calculated pattern by matching the fine detailed structure features in the (0002) and (000$\bar{2}$) diffraction disks (Fig. 13.25d and e). The positively charged Zn-(0001) surface is chemically active and the negatively charged O-(000$\bar{1}$) surface is relatively inert, resulting in a growth of long fingers along [0001]. Using HRTEM, we found that the Zn-terminated (0001) surface has tiny Zn clusters, which could lead to self-catalyzed growth without the presence of foreign catalyst. The chemically inactive (000$\bar{1}$) surface typically does not grow nanobelt structure.

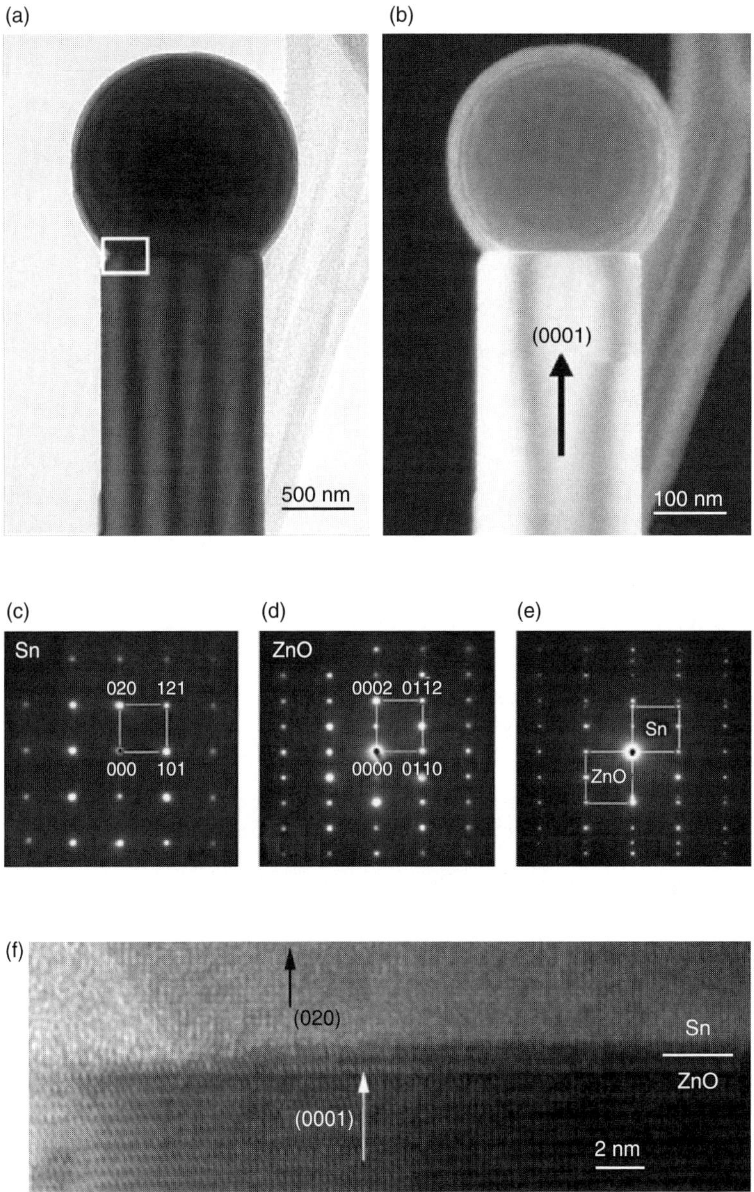

FIGURE 13.24. (a, b) Bright-field and dark-field images of a [0001] growth ZnO nanowire with an Sn particle at the growth front. (c–e) SAED patterns recorded from the particle, the rod, and both the particle and the rod in (a), respectively. (f) An HRTEM image from the white rectangle enclosed area in (a) from the interface region.

FIGURE 13.25. (a) SEM image of "comb-like" cantilever arrays of ZnO, which is the result of surface polarization-induced growth due to the chemically active Zn-(0001), the *inset* shows the pointed hexagonal nanocantilevers on hexagonal base. (b) High-resolution TEM recorded from the tip of the comb teeth, showing possible evidence of Zn segregation at the growth front, which is likely to be effective for driving self-catalyzed growth. (c) Structure model of ZnO projected along [01$\bar{1}$0], showing the termination effect of the crystal. (d, e) Experimentally observed and theoretically simulated convergent beam electron diffraction patterns for determining the polarity of the nanocombs, respectively.

The anisotropic growth appears to be a common characteristic for the wurtzite family. Similar sawteeth growth has been observed for ZnS and CdSe, and it is suggested to be induced by the Zn- and Cd-terminated (0001) surfaces, respectively.

5. Summary

Metal oxides possess a broad range of functional properties that are of great importance in industrial applications. Because of these properties, metal oxides have been widely studied. In the field of nanoscience and technology, metal oxide nanostructures especially 1D nanostructures have drawn a lot of interest.

In this chapter, a review is given about 1D metal oxide nanostructures, about the structure formation, morphology control, and catalytic process. Various nanostructures of ZnO were taken as examples for illustrating the 1D nanostructure. It is suggested that, from growth technique point of view, a conjunction between the bottom-up and top-down techniques is necessary for the future applications of nanostructures.

Acknowledgment. The results reviewed in this paper were partially contributed by our group members and collaborators: Xiangyang Kong, Yong Ding, Rusen Yang, Xudong Wang, William Hughes, Christopher Ma, and Daniel Moore, to whom we are very grateful. Research was supported by NSF, NASA, and DARPA.

References

1. S. Iijima and T. Ichihashi, *Nature*, 363 (1993) 603.
2. Z. L. Wang (Ed.), *Nanowires and Nanobelts—Materials, Properties and Devices; Vol. I: Metal and Semiconductor Nanowires*, Kluwer Academic, Dordrecht (2003).
3. C. R. Martin, *Chem. Mater.*, 8 (1996) 1739.
4. Y. N. Xia, P. D. Yang, Y. G. Sun, Y. Y. Wu, B. Mayers, B. Gates, Y. D. Yin, F. Kim, and Y. Q. Yan, *Adv. Mater.*, 15 (2003) 353.
5. M. H. Huang, S. Mao, H. Feick, H. Yan, Y. Wu, H. Kind, E. Weber, R. Russo, and P. Yang, *Science*, 292 (2001) 1897.
6. E. W. Wong, B. W. Maynor, L. D. Burns, and C. M. Lieber, *Chem. Mater.*, 8 (1996) 2041.
7. P. Yang and C. M. Lieber, *Science*, 273 (1996) 1836.
8. L. Vayssieres, *Adv. Mater.*, 15 (2003) 464.
9. Z. L. Wang (Ed.), *Nanowires and Nanobelts—Materials, Properties and Devices; Vol. II: Nanowires and Nanobelts of Functional Materials*, Kluwer Academic, Dordrecht (2003).
10. Z. W. Pan, Z. R. Dai, and Z. L. Wang, *Science*, 291 (2001) 1947.
11. X. Y. Kong, Y. Ding, R. Yang, and Z. L. Wang, *Science*, 303 (2004) 1348.
12. Z. L. Wang, *Ann. Rev. Phys. Chem.*, 55 (2004) 159.
13. Z. R. Dai, Z. W. Pan, and Z. L. Wang, *Adv. Funct. Mater.*, 13 (2003) 9.
14. P. Avouris, *Acc. Chem. Res.*, 35 (2002) 1026.
15. S. J. Tans, A. R. M. Verschueren, and C. Dekker, *Nature*, 393 (1998) 49.

16. W. A. Deheer, A. Chatlain, and D. Ugarte, *Science*, 270 (1995) 1179.
17. S. Frank, P. Poncharal, Z. L. Wang, and W. H. De Heer, *Science*, 280 (1998) 1744.
18. J. Kong, N. R. Franklin, C. W. Zhou, M. G. Chapline, S. Peng, K. J. Cho, and H. Dai, *Science*, 287 (2000) 622.
19. X. F. Duan, Y. Huang, Y. Cui, J. F. Wang, and C. M. Lieber, *Nature*, 409 (2001) 66.
20. M. Arnold, P. Avouris, and Z. L. Wang, *J. Phys. Chem. B*, 107 (2002) 659.
21. E. Comini, G. Faglia, G. Sberveglieri, Z. W. Pan, and Z. L. Wang, *Appl. Phys. Letts*, 81 (2002) 1869.
22. X. D. Bai, P. X. Gao, Z. L. Wang, and E. G. Wang, *Appl. Phys. Letts*, 82 (2003) 4806.
23. V. Sazonova, Y. Yaish, H. Ustunel, D. Roundy, T. A. Arias, and P. L. McEuen, *Nature*, 431 (2004) 284.
24. Z. L. Wang (Ed.), *Nanowires and Nanobelts–Materials, Properties and Devices; Vol. II: Nanowires and Nanobelts of Functional Materials*, Kluwer Academic, Dordrecht (2003).
25. J. Singh and D. E. Wolfe, *J. Mater. Sci.*, 40 (2005) 1.
26. T. Karabacak, A. Mallikarjunan, J. P. Singh, D. Ye, G.-C. Wang, and T.-M. Lu, *Appl. Phys. Letts*, 83 (2003) 3096.
27. B. Marson and K. Sattler, *Phys. Rev. B*, 60 (1999) 11593.
28. L. M. Cao, Z. Zhang, L. L. Sun, C. X. Gao, M. He, Y. Q. Wang, Y. C. Li, Y. Zhang, G. Li, J. Zhang, and W. K. Wang, *Adv. Mater.*, 13 (2001) 1701.
29. L. M. Cao, K. Hahn, C. Scheu, M. Rühle, Y. Q. Wang, Z. Zhang, C. Gao, Y. C. Li, Y. Zhang, M. He, L. L. Sun, and W. K. Wang, *Appl. Phys. Letts*, 80 (2002) 4226.
30. K. Suenaga, M. P. Johansson, N. Hellgren, E. Broitman, L. R. Wallenberg, C. Colliex, J. E. Sundgren, and L. Hultman, *Chem. Phys. Letts*, 300 (1999) 695.
31. D. S. Bethune, C. H. Klang, M. S. de Vries, G. Gorman, R. Savoy, J. Vazquez, and R. Beyers. *Nature*, 363 (1993) 605.
32. P. R. Willmott, *Prog. Surf. Sci.*, 76 (2004) 163.
33. D. Dijkkamp, T. Venkatesan, D. Wu, S. A. Shaheen, N. Jisrawi, Y. H. Minlee, W. L. Mclean, and M. Croft, *Appl. Phys. Letts*, 51 (1987) 619.
34. A. Thess, R. Lee, P. Nikolaev, H. Dai, P. Petit, J. Robert, C. Xu, Y. H. Lee, S. G. Kim, A. G. Rinzler, D. T. Colbert, G. Scuseria, D. Tomanek, J. E. Fisher, and R. E. Smalley, *Science*, 273 (1996) 483.
35. A. M. Morales and C. M. Lieber, *Science*, 279 (1998) 208.
36. L. J. Lauhon, M. S. Gudiksen, D. Wang, and C. M. Lieber, *Nature*, 420 (2002) 57.
37. S. Han, C. Li, Z. Liu, B. Lei, D. Zhang, W. Jin, X. Liu, T. Tang, and C. W. Zhou, *Nano Lett.*, 4 (2004) 1241.
38. C. J. Lee and J. Park. *Appl. Phys. Letts*, 77 (2000) 3397.
39. J. C. Wang, S. Q. Feng, D. P. Yu, *Appl. Phys. A*, 75 (2002) 691.
40. Z. Zhong, F. Qian, D. Wang, and C. M. Lieber, *Nano Lett.*, 3 (2003) 343.
41. B. Zheng, C. Lu, G. Gu, A. Makarovski, G. Finkelstein, and J. Liu, *Nano Letts*, 2 (2002) 895.
42. C. M. Hsu, C. H. Lin, H. J. Lai, and C. T. Kuo, *Thin Solid Films*, 471 (2005) 140.
43. X. F. Duan and C. M. Lieber, *Adv. Mater.*, 12 (2000) 298.
44. Y. Cui, L. J. Lauhon, M. S. Gudiksen, J. Wang, and C. M. Lieber, *Appl. Phys. Letts*, 78 (2001) 2214.
45. X. F. Duan, C. M. Niu, V. Sahi, J. Chen, J. W. Parce, S. Empedocles, and J. L. Goldman, *Nature*, 425 (2003) 274.
46. M. S. Gudiksen, L. J. Lauhon, J. F. Wang, D. C. Smith, and C. M. Lieber, *Nature*, 415 (2002) 617.

47. L. J. Lauhon, M. S. Gudiksen, D. Wang, and C. M. Lieber, *Nature*, 420 (2002) 57.
48. Y. H. Kim, J. Y. Lee, S. H. Lee, J. E. Oh, and H. S. Lee, *Appl. Phys. A*, 80 (2005) 1635.
49. H. Hasegawa and S. Kasai, *Phys. E*, 11 (2001) 149.
50. Z. H. Wu, M. Sun, Y. Mei, and H. E. Ruda, *Appl. Phys. Letts*, 85 (2004) 657.
51. Y. S. Min, E. J. Bae, K. S. Jeong, Y. J. Cho, J. H. Lee, W. B. Choi, and G. S. Park, *Adv. Mater.*, 15 (2003) 1019.
52. M. S. Sander, M. J. Cote, W. Gu, B. M. Kile, and C. P. Tripp, *Adv. Mater.*, 16 (2004) 2052.
53. H. Shin, D. K. Jeong, J. Lee, M. M. Sung, and J. Kim, *Adv. Mater.*, 16 (2004) 1197.
54. X. D. Wang, E. Graugnard, J. S. King, Z. L. Wang, and C. J. Summers, *Nano Lett.*, 4 (2004) 2223.
55. J. J. Urban, W. S. Yun, Q. Gu, and H. Park, *J. Am. Chem. Soc.*, 124 (2002) 1186.
56. J. J. Urban, J. E. Spanier, O. Y. Lian, W. S. Yun, and H. Park, *Adv. Mater.*, 15 (2003) 423.
57. W. B. Bu, Z. L. Hua, L. Zhang, H. R. Chen, W. M. Huang, and J. L. Shi, *J. Mater. Res.*, 19 (2004) 2807.
58. W. B. Bu, Z. L. Hua, H. R. Chen, L. Zhang, and J. L. Shi, *Chem. Letts*, 33 (2004) 612.
59. Q. W. Chen and T. Zhu, *Chem. Phys. Letts*, 375 (2003) 167.
60. K. Kempa, B. Kimball, J. Rybczynski, Z. P. Huang, P. F. Wu, D. Steeves, M. Sennett, M. Giersig, D. V. G. L. N. Rao, D. L. Carnahan, D. Z. Wang, J. Y. Lao, W. Z. Li, and Z. F. Ren, *Nano Lett.*, 3 (2003) 13.
61. X. D. Wang, C. J. Summers, and Z. L. Wang, *Nano Lett.*, 4 (2004) 423.
62. Z. Zhong, D. Wang, Y. Cui, M. W. Bockrath, and C. M. Lieber, *Science*, 302 (2003) 1377.
63. S. J. Tans, R. M. Verschueren, and C. Dekker, *Nature*, 393 (1998) 49.
64. P. L. McEuen, M. Bockrath, D. H. Cobden, Y.-G. Yoon, and S. G. Louie, *Phys. Rev. Lett.*, 83 (1999) 5098.
65. A. Bachtold, P. Hadley, T. Nakanishi, and C. Dekker, *Science*, 294 (2001) 1317.
66. Y. Cui, Z. Zhong, D. Wang, W. Wang, and C. M. Lieber, *Nano Lett.*, 3 (2003) 149.
67. J. Kong, H. T. Soh, A. Cassell, C. F. Quate, and H. Dai, *Nature*, 395 (1998) 878.
68. J. Kong, N. R. Franklin, C. Zhou, M. G. Chapline, S. Peng, K. Cho, and H. Dai, *Science*, 287 (2000) 622.
69. Y. Cui, Q. Wei, H. Park, and C. M. Lieber, *Science*, 293 (2001) 1289.
70. F. Patolsky, G. F. Zheng, O. Hayden, M. Lakadamyali, W. Zhuang, and C. M. Lieber, *Proc. Natl Acad. Sci. USA*, 101 (2004) 14017.
71. M. C. Mcalpine, R. S. Friedman, and C. M. Lieber, *Proc. IEEE*, 93 (2005) 1357.
72. R. S. Friedman, M. C. McAlpine, D. S. Ricketts, D. Ham, and C. M. Lieber, *Nature*, 434 (2005) 1085.
73. M. J. Zheng, L. D. Zhang, G. H. Li, Y. Zhang, and F. Wang, *Appl. Phys. Letts*, 79 (2001) 839.
74. Z. R. Dai, J. L. Gole, J. D. Stout, and Z. L. Wang, *J. Phys. Chem. B*, 106 (2002) 1274.
75. Z. W. Pan, Z. R. Dai, L. Xu, S. T. Lee, and Z. L. Wang, *J. Phys. Chem. B*, 105 (2001) 2507.
76. C. A. Huber, T. E. Huber, M. Sadoqui, J. A. Lubin, S. Manalis, and C. B. Prater, *Science*, 263 (1994) 800.
77. K. Kempa, B. Kimball, J. Rybczynski, Z. P. Huang, P. F. Wu, D. Steeves, M. Sennett, M. Giersig, D. V. G. L. N. Rao, D. L. Carnahan, D. Z. Wang, J. Y. Lao, W. Z. Li, and Z. F. Ren, *Nano Lett.*, 3 (2003) 13.
78. X. D. Wang, C. J. Summers, and Z. L. Wang, *Nano Lett.*, 4 (2004) 423.
79. Y. Li and X. L. Ma, *Phys. Stat. Sol. A*, 202 (2005) 435.
80. S. Y. Bae, J. Lee, H. Jung, J. Park, and J.-P. Ahn, *J. Am. Chem. Soc.*, 127 (2005) 10802.

81. R. Tenne, L. Margulis, and M. Genut, *Nature*, 360 (1992) 444.

82. Y. Feldman, E. Wasserman, D. J. Srolovitz, and R. Tenne, *Science*, 267 (1995) 222.

83. E. J. M. Hamilton, S. E. Dolan, C. E. Mann, and H. O. Colliex, *Science*, 260 (1993) 649.

84. X. Y. Kong, Z. L. Wang, and J. S. Wu, *Adv. Mater.*, 15 (2003) 1445.

85. J. Goldberger, R. R. He, Y. F. Zhang, S. W. Lee, H. Q. Yan, H. J. Choi, and P. D. Yang, *Nature*, 422 (2003) 599.

86. Y. Liu, J. Dong, and M. Liu, *Adv. Mater.*, 16 (2004) 353.

87. L. Vayssieres, K. Keis, A. Hagfeldt, and S. E. Lindquist, *Chem. Mater.*, 13 (2001) 4386.

88. J. P. Cheng, R. Y. Guo, and Q. M. Wang, *Appl. Phys. Letts*, 85 (2004) 5140.

89. J. J. Wu, S. C. Liu, C. T. Wu, K. H. Chen, and L. C. Chen, *Appl. Phys. Letts*, 81 (2002) 1312.

90. J. Zhang, L. D. Sun, C. S. Liao, and C. H. Yan, *Chem. Commun.*, 3 (2002) 262.

91. Y. J. Xing, Z. H. Xi, Z. Q. Xue, D. Zhang, J. H. Song, R. M. Wang, J. Xu, Y. Song, S. L. Zhang, and D. P. Yu, *Appl. Phys. Letts*, 83 (2004) 1689.

92. B. P. Zhang, N. T. Binh, K. Wakatsuki, Y. Segawa, Y. Yamada, N. Usami, M. Kawasaki, and H. Koinuma, *Appl. Phys. Letts*, 84 (2004) 4098.

93. R. V. Parish, *The metallic Elements*, Longman, New York (1977).

94. P. X. Gao, C. S. Lao, Y. Ding, and Z. L. Wang, *Adv. Funct. Mater*, 16 (2006) 53.

95. P. X. Gao and Z. L. Wang, *J. Am. Chem. Soc.*, 125 (2003) 11299.

96. X. Y. Kong and Z. L. Wang, *Nano Lett.*, 3 (2003) 1625.

97. X. Y. Kong and Z. L. Wang, *Appl. Phys. Letts*, 84 (2004) 975.

98. P. X. Gao and Z. L. Wang, *Small*, 1 (2005) 945.

99. T. Murata, I. Yamato, Y. Kakinuma, A. G. W. Leslie, and J. E. Walker, *Science*, 308 (2005) 654.

100. S. Amelinckx, B. Zhang, D. Bernaerts, F. Zhang, V. Ivanov, and J. B. Nagy, *Science*, 265 (1994) 635.

101. R. P. Gao, Z. L. Wang, and S. S. Fan, *J. Phys. Chem. B*, 104 (2000) 1227.

102. H. F. Zhang, C. M. Wang, E. C. Buck, and L. S. Wang, *Nano Lett.*, 3 (2003) 577.

103. H. F. Zhang, C. M. Wang, and L. S. Wang, *Nano Lett.*, 2 (2002) 941.

104. O. G. Schmidt and K. Eberl, *Nature*, 410 (2001) 168.

105. W. Hughes, Z. L. Wang, *J. Am. Chem. Soc.*, 126 (2004) 6703.

106. P. X. Gao and Z. L. Wang, *J. Appl. Phys.*, 97 (2005) 044304.

107. X. Y. Kong, Y. Ding, R. Yang, and Z. L. Wang, *Science*, 303 (2004) 1348.

108. Y. Ding, X. Y. Kong, and Z. L. Wang, *Phys. Rev. B*, 70 (2004) 235408.

109. R. S. Yang, Y. Ding, and Z. L. Wang, *Nano Letts*, 4 (2004) 1309.

110. P. X. Gao and Z. L. Wang, *J. Phys. Chem. B*, 108 (2002) 12653.

111. P. X. Gao and Z. L. Wang, *Appl. Phys. Letts*, 84 (2004) 2883.

112. R. S. Wagner and W. C. Ellis, *Appl. Phys. Letts*, 4 (1964) 89.

113. P. X. Gao, Y. Ding, and Z. L. Wang, *Nano Lett.*, 3 (2003) 1315.

114. Y. Ding, P. X. Gao, and Z. L. Wang, *J. Am. Chem. Soc.*, 126 (2004) 2066.

115. Z. L. Wang, X. Y. Kong, and J. M. Zuo, *Phys. Rev. Letts*, 91 (2003) 185502.

116. C. Ma, D. F. Moore, Y. Ding, J. Li, and Z. L. Wang, *Int. J. Nanotechnology*, 1 (2004) 431.

117. D. F. Moore, C. Ronning, C. Ma, and Z. L. Wang, *Chem. Phys. Letts*, 385 (2004) 8.

118. C. Ma, D. F. Moore, J. Li, and Z. L. Wang, *Adv. Mater.*, 15 (2003) 228.

119. C. Ma, Y. Ding, D. F. Moore, X. D. Wang, and Z. L. Wang, *J. Am. Chem. Soc.*, 126 (2004) 708.

14
Bioinspired Nanomaterials

Peng Wang, Guobao Wei, Xiaohua Liu, and Peter X. Ma

1. Introduction

Biomaterials are defined as substances other than food or drugs contained in therapeutic or diagnostic systems that are in contact with tissue or biological fluids [1]. Biomaterials are utilized to manufacture various medical devices, diagnostic products, and pharmaceutical preparations. Many medical devices, such as catheters, heart pacemaker, and all kinds of sensors, have already been widely used in the health industry. Body parts [2], such as hip [3] and other joints [4], can be replaced with biomaterial implants. Drug release rate can be controlled by drug delivery system made of biomaterials [5,6]. In light of the broadness of applications, it is not surprising that biomaterials include a big family of materials, such as polymers, metals, ceramics, glasses, composites, and various natural materials including proteins. A comprehensive review of biomaterials or devices is beyond the scope of this chapter. Instead, this chapter focuses on synthetic polymeric biomaterials (SPM) with nanometer-size features. SPM posses a unique set of desirable properties compared to natural materials and other synthetic materials. Unlike natural materials, SPM are free of pathogens. They tend to have good process ability. Many processes have been successfully developed to change the surface chemistry of polymeric materials to improve their biocompatibilities or incorporate specific biological functions. Many polymers are readily degradable under physiological environment and the degradation rate can be tailored by varying molecular structures or processing parameters.

A living body is a complicated and delicate system that puts strict requirements on properties of biomaterials. A lot of research efforts have been focused on controlling materials' chemical, morphological, and mechanical properties so that they can better serve specific purposes in a human body. Recent studies show that materials with controlled properties at nanometer scale may posses more desirable properties such as better protein adsorption and cell adhesion. For example, 3D nanofibrous poly(L-lactic acid) (PLLA) scaffolds can significantly enhance adsorption of proteins, which mediates cell interactions with scaffolds [7]. Interestingly, the enhancement of adsorption on the 3D nanofibrous scaffolds shows certain selectivity in proteins. Another advantage of nanofibers lies in their porosity and surface area, which makes them a good candidate for scaffolds to support growth of cells. Nanostructured polymers were also designed as synthetic bladder constructs that mimic the topography of natural bladder tissue [8]. Results from in vitro experiments show that adhesion of bladder smooth muscle cells is enhanced as feature dimensions of polymer surface decrease into the nanometer scale.

It is generally hypothesized that the more desirable properties observed from nanomaterials are attributed to the capability of mimicking of natural extracellular matrix (ECM) in size. For example, collagen, which is found in every major tissue that requires strength and flexibility, has a unique triple-helix structure. The most abundant type of collagen is type I. Each type I collagen protein molecule forms a triple-helix structure made of three polypeptide chains. The triple-helix structure is stabilized by intermolecular hydrogen bonds. Collagen molecules further assemble into collagen fibrils. Collagen fibers are formed by the assembly of collagen fibrils. The diameters of collagen fibers are within the range of 50–500 nm. The average diameter of the fiber prepared by Ma et al. [9] is ~160 nm, which falls within the range of the size of natural collagen fibers. This type of biomimetic materials holds great potential in tissue engineering applications.

The study on nanomaterials requires a powerful microscope with the magnification higher than optical microscopes. Scanning electron microscope (SEM) and atomic force microscope (AFM) are two major options. Each technique has its own advantages and disadvantages.

Inside the vacuum chamber of SEM a beam of electrons is generated by an electron gun. The beam scans across sample surface. The primary imaging method is based on secondary electrons released from the sample. The secondary electrons are detected by a scintillation material that produces flashes of light from the electrons. The light flashes are then detected and amplified by a photomultiplier tube. An image can be formed by correlating the sample scan position with the resulting signal.

AFM works by scanning a sharp tip positioned at the end of a cantilever across sample surfaces. The cantilever deflects accordingly with the change of forces between the sample surface and the tip. The deflection is captured by a beam of laser, which is reflected from the end of the cantilever to a photodiode. AFM can be operated at contact mode and noncontact mode with the latter having a larger distance between sample surface and tip. Generally, noncontact mode is more favorable for samples with softer surfaces.

Theoretically, both SEM and AFM can achieve a resolution of several nanometers, or an even higher resolution. Compared to AFM, the advantages of SEM are as following. First, SEM is capable of handling samples with complicated 3D structures, while it is difficult to use AFM to characterize those samples. For example, AFM cannot detect undercut features. It is also difficult to use AFM to characterize samples with very rough surfaces. Second, to some extent SEM is less vulnerable to artifacts than AFM. For example, the feature size in the images from AFM tends to vary with the value of "setpoint," which is manually set by users. The setpoint controls the distance between the AFM tip and the sample surface. It can be tricky for a novice to find an appropriate value for the setpoint. AFM users also need to bear in mind that the quality of AFM images is heavily dependent on the quality of the AFM tip. Third, many SEM instruments have an element analysis capability that AFM lacks.

SEM has its own shortcomings. Traditional SEM requires samples to be conductive. Hence nonconductive samples are usually coated with a thin layer of conductive materials, which might cause artifacts on samples. This explains the major difficulty of using SEM to characterize nonconductive polymer samples, especially when high magnifications are needed. Traditional SEM also requires samples to be tested under high vacuum. The aforementioned limitations can be partly overcome using an environmental SEM (ESEM). ESEM can be operated at a higher pressure so that the specimens can be characterized under more natural environmental conditions. On the downside, however, higher operational pressure inevitably causes a drop of resolution. Last but not least, the electron beam scanned across sample surface might cause damages to some sensitive organic materials. Users can easily test the vulnerability of the sample by focusing the electron beam at one location and observing the changes of the image over time.

It is worthwhile to point out that SEM is probably the most often used microscopic technique for biomaterial characterization despite of its limitations. It is largely because that AFM cannot handle complicated 3D structures. This chapter introduces readers to some recent research progress on biomaterials. SEM is used in each example and provides important morphological information. The chapter includes three sections. The first section focuses on the fabrication of fibrous materials for tissue engineering using either phase separation or electro-spinning method. The second section emphasizes the applications of nanoparticles in bone tissue engineering. This section covers recent work on polymer/hydroxyapatite (HAP) nanocomposites, drug-loading nanoparticles capable of controlled drug release. The last section of this chapter presents a new method for surface modification of nanofibrous biomaterials.

2. Nanofibers

2.1. Nanofibers Made from Phase Separation

A homogeneous polymer solution may become thermodynamically unstable under certain conditions such as lowering temperature, and being separated into more than one phase to lower the total free energy of the system. After removal of the solvent, polymer in the polymer-rich phase solidifies. Phase separation techniques have been used for preparation of porous membranes for purification and separation purposes [10]. The phase separation process usually is induced thermally or by a nonsolvent. A novel phase separation process was developed recently to fabricate nanofibers to mimic natural collagen fibers [9]. Scaffolds prepared based on the nanofibers have extremely high porosity and surface area per unit volume.

The nanoscale PLLA fibers were fabricated using a five-step protocol [9]: polymer dissolution, phase separation and gelation, solvent extraction, freezing, and freeze drying under vacuum. The optimum process conditions may vary with the solvent used for polymer dissolution. Here, the process using tetrahydrofuran (THF) as the solvent of PLLA is introduced. Similar processes using other solvents can be found from the same literature [9].

First, PLLA and THF with desired ratio are loaded into a glass vial. The mixture is heated to 55°C and stirred with a magnetic stirrer until the polymer totally dissolves and a homogeneous solution forms. Hot polymer solution is then transferred into a Teflon vial. The vial is quickly moved into a refrigerator set at a chosen temperature for phase separation.

After phase separation and gelation at lower than room temperature, the vial containing the gel is immersed into a large distilled water reservoir to extract THF from the gel. After extraction, the gel is removed from the water and blotted with a piece of filter paper to dry free water at surface of the gel. The gel is then frozen and freeze-dried at about −5°C under a vacuum lower than 0.5 mmHg.

SEM was utilized to characterize the morphologies of the polymeric fibers obtained and was found to be a convenient tool to analyze the highly porous sample. The specimens were first cut into suitable sizes using a razor blade or fractured by liquid nitrogen. Samples were fixed on top of SEM stubs with double-sided tape. Because PLLA is not conductive, samples were coated with gold using a sputter coater (Desk-II; Denton Vacuum Inc., USA). The coating time was 200 s with the gas pressure lower than 50 mTorr and the current of 40 mA. A Philips XL30 SEM was used to characterize polymeric fibers. The Philips XL30 SEM is one of a new generation of SEMs that is completely controlled from a computer workstation. The filament is made of Zirconated Tungsten. The machine's resolution can reach as small as 2 nm.

After loading the sample into the vacuum chamber of SEM, users need to choose a magnification value and correct the focus to make the image as sharp as possible. If the image moves when changing focus, the aperture needs centering. Astigmatism may need to be corrected at the beginning of testing a new sample or after changing kiloVolt, spot size, or working distance.

Figure 14.1 shows SEM images of PLLA fibers prepared from PLLA/THF solutions with different PLLA concentrations at a gelation temperature of 8°C. It can be seen that PLLA forms highly porous (up to 99% porosity) nanoscale fibrous matrices. The matrices have a unique 3D continuous structure that is similar to a natural collagen matrix. The diameter of the PLLA fibers ranges from 50 to 500 nm, the same diameter range of natural collagen fibers. The average fiber diameter of the matrices does not change statistically with the concentration of polymer solution used to fabricate the matrices within the investigated range. However, the average unit length (the fiber length between two conjunction points) decreases. With increase of polymer concentration, the size of pores decreases and becomes more uniform.

Gelation temperature is an important factor controlling the morphology of the matrices. Figure 14.2 shows SEM images of matrices prepared at different gelation temperatures. It is evident that the matrix structure formed by gelation at 23°C or 19°C is different from that formed at lower gelation temperatures. At

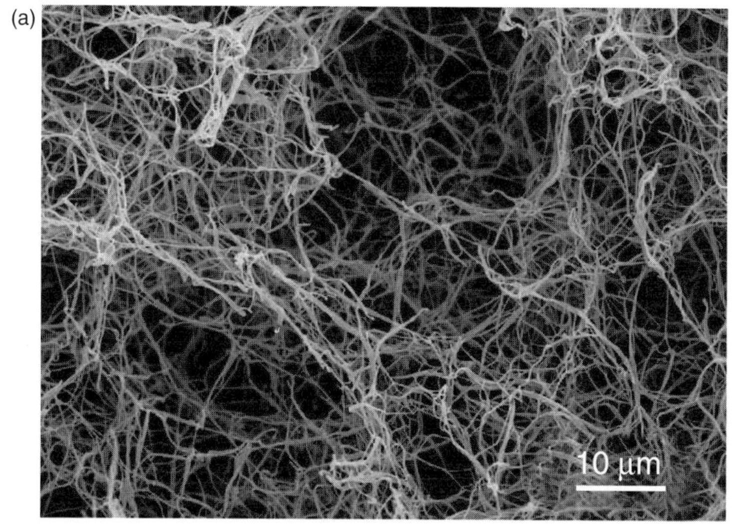

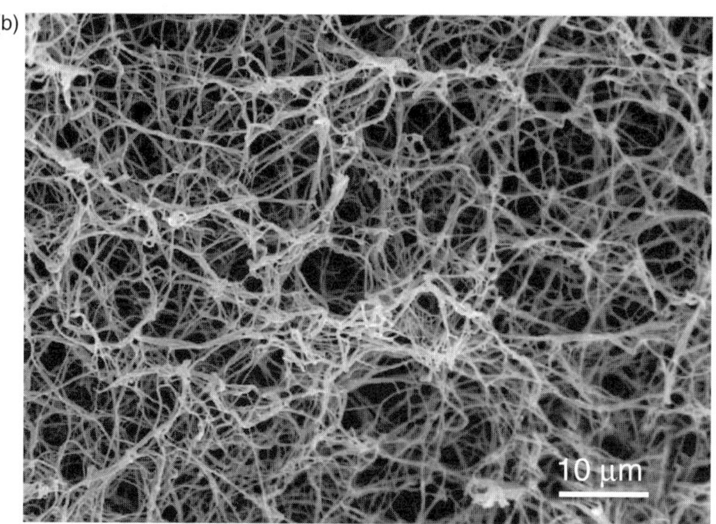

FIGURE 14.1. SEM micrographs of PLLA fibrous matrices prepared from PLLA/THF solution with different PLLA concentrations at a gelation temperature of 8°C: (a) 1.0% (wt/v), ×2.0K; (b) 2.5% (wt/v), ×2.0K;

higher temperature PLLA forms platelet-like structure and the crossover gelation temperature is 17°C. Matrices formed by gelation at 17°C have both platelet-like and nanofibrous structures. For the matrices formed at lower temperatures, a continuous 3D structure is observed. The diameter of fibers does not change statistically with temperature within this temperature range. The interfiber spacing becomes more uniform at lower temperatures.

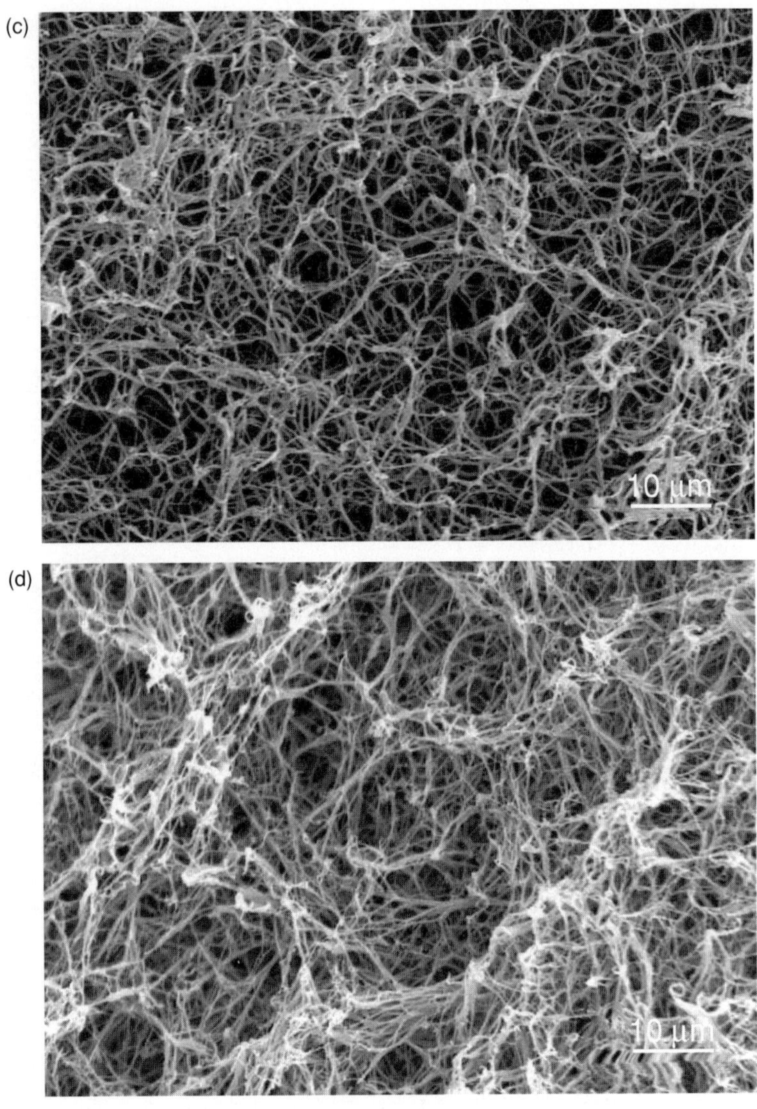

FIGURE 14.1. (*Continued*) (c) 5.0% (wt/v), ×2.0K; (d) 7.5% (wt/v), ×2.0K. (Adapted from [9] with permission. Copyright (1999) John Wiley & Sons.)

Differential scanning calorimeter (DSC) (Perkin–Elmer, Norwalk, CT, USA) was used to measure the enthalpy of melting ΔH. The degree of crystallinity X can be calculated using the following equation:

$$X = \Delta H \,/\, \Delta H^0 \qquad (14.1)$$

where ΔH^0 is the enthalpy of melting of 100% crystalline polymer.

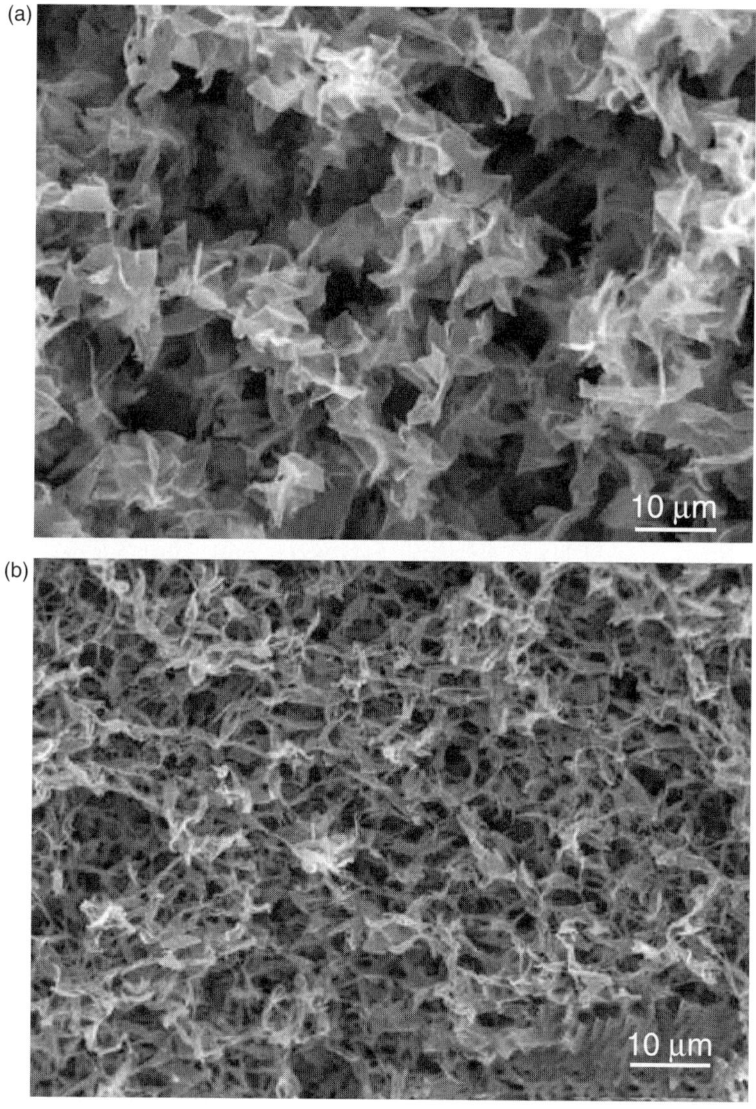

FIGURE 14.2. SEM micrographs of PLLA matrices prepared from 5.0% (wt/v) PLLA/THF solution at different gelation temperatures: (a) 23°C, ×2.0K; (b) 19°C, ×2.0K;

The melting point, enthalpy of melting, and degree of crystallinity of the matrices prepared by phase separation PLLA/THF at different temperatures (gelation temperature) are listed in Table 14.1. The melting point and degree of crystallinity of matrices do not change significantly with polymer concentration at a gelation temperature of −18°C. The degree of crystallinity does not change significantly

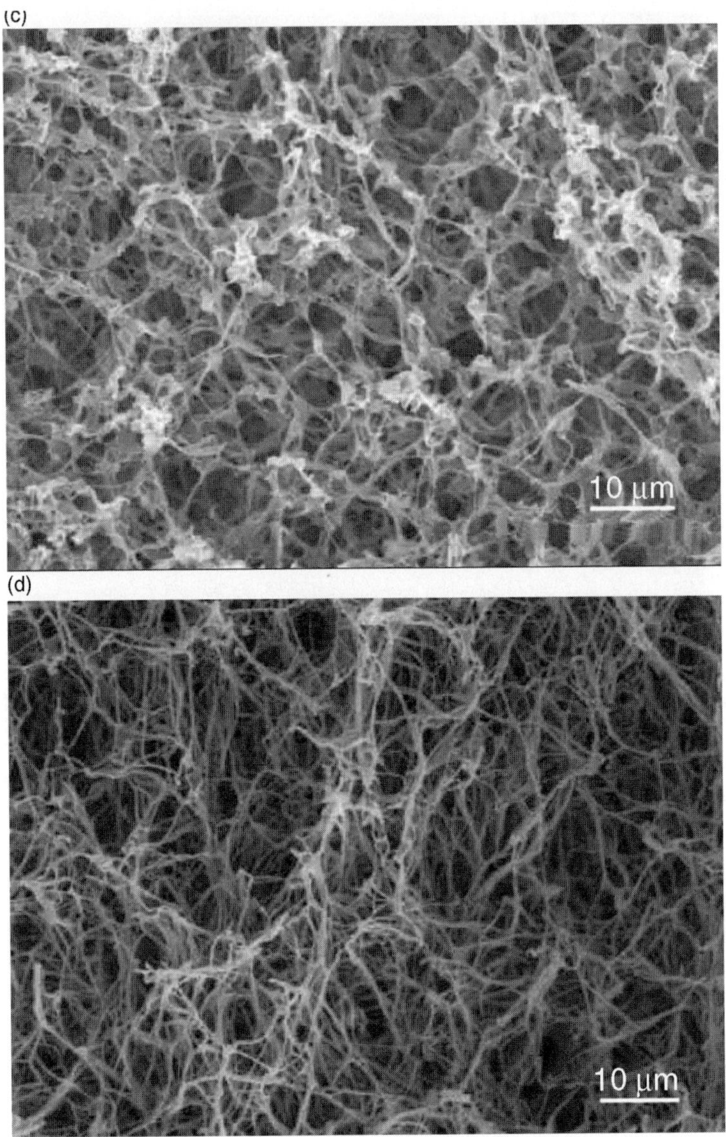

FIGURE 14.2. (*Continued*) (c) 17°C, ×2.0K; (d) 15°C, ×2.0K;

with gelation temperature either in the lower temperature range (<15°C). However, the matrices formed at a higher temperature have a higher degree of crystallinity than the matrices formed at a lower temperature. The change of crystallinity accompanies the change of morphology, which has been mentioned earlier. PLLA films were prepared by casting PLLA/THF solution at room temperature. The degree of crystallinity of cast films is close to that of matrices prepared at the same

(e)

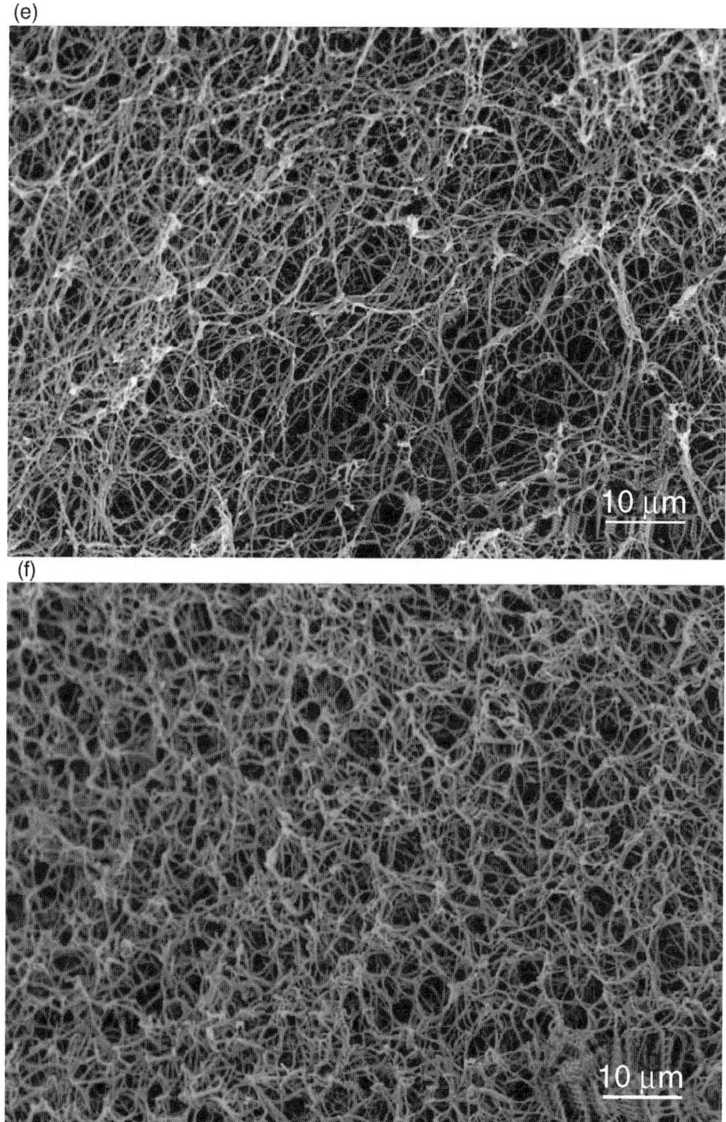

(f)

FIGURE 14.2. (*Continued*) (e) −18°C, ×2.0K; (f) liquid nitrogen, ×2.0K. (Adapted from [9] with permission. Copyright (1999) John Wiley & Sons.)

temperature (23°C). The higher degree of crystallinity could be caused by easier rearrangement of PLLA chains at higher temperature. The difference in structure of matrices prepared at different temperatures suggests that phase separation occurs through a mechanism different from that at a lower temperature. It is hypothesized that phase separation at higher temperature (19°C or higher) is due

TABLE 14.1. Results of DSC measurements of PLLA matrices and cast film

PLLA/THF concentration (%, wt/v)	Gelation temperature (°C)	Tm (°C)	ΔH (J/g)	X
1.0	−18	180.5	49.5	24.4
2.5	−18	181.6	55.3	27.2
5.0	−18	179.1	56.0	27.5
7.5	−18	177.0	53.3	26.2
5.0	Liquid nitrogen	180.7	56.8	27.9
5.0	8	183.4	53.2	26.2
5.0	15	180.2	57.6	28.3
5.0	23	182.5	74.2	36.5
PLLA film	23	179.3	68.2	33.5

Source: Adapted from [11].

to crystal nucleation and growth process. In other words, solid-liquid phase separation occurs at a higher temperature. The platelet-like (aggregates of many tiny crystals) structure appears to support this argument. The nanofibrous structure formed at lower temperatures is hypothesized to be caused by spinodal liquid-liquid phase separation and consequential polymer crystallization. The explanation is further supported by results of a recent light scattering study of the PLLA/THF gelation process.

The structure of matrices is dependent on phase separation procedure. Consequently, one would expect that the structure of matrices might change with the solvent or the polymer since the phase diagrams of different polymer/solvent combinations are unlikely to be the same. THF is replaced with a mixture of dioxane and methanol, or PLLA is replaced with poly(D,L-lactic acid) (PDLLA) or poly(D,L-lactic-co-glycolic acid) (PLGA) to explore the richness of 3D structures generated by phase separation. Figure 14.3 shows SEM images of PLLA matrices prepared from PLLA/dioxane/methanol solution using the same procedure as mentioned earlier. Nanofibers can also be found in the matrices prepared from PLLA/dioxane/methanol solution. However, the fibers tend to aggregate together. While for the matrices prepared from PLGA/dioxane/H_2O or PDLLA/dioxane/H_2O, the structure is evidently different (Fig. 14.4). The structure is neither nanofibrous nor platelet-like. PLGA and PDLLA are not crystallizable. This, again, indicates that polymer crystallization plays an essential role in controlling the structure of the matrices.

Among the several matrix structures mentioned earlier, the nanofibrous structure is of special interest due to its resemblance to the structure of natural ECM. The nanofibrous structure has a much higher surface/volume ratio than other structures. The method applied for fabrication of a nanofibrous matrix is simple compared to the nonwoven fabric processing in the textile technology and does not require any sophisticated equipment. The porosity of the matrix can be easily controlled by varying the concentration of polymer solution.

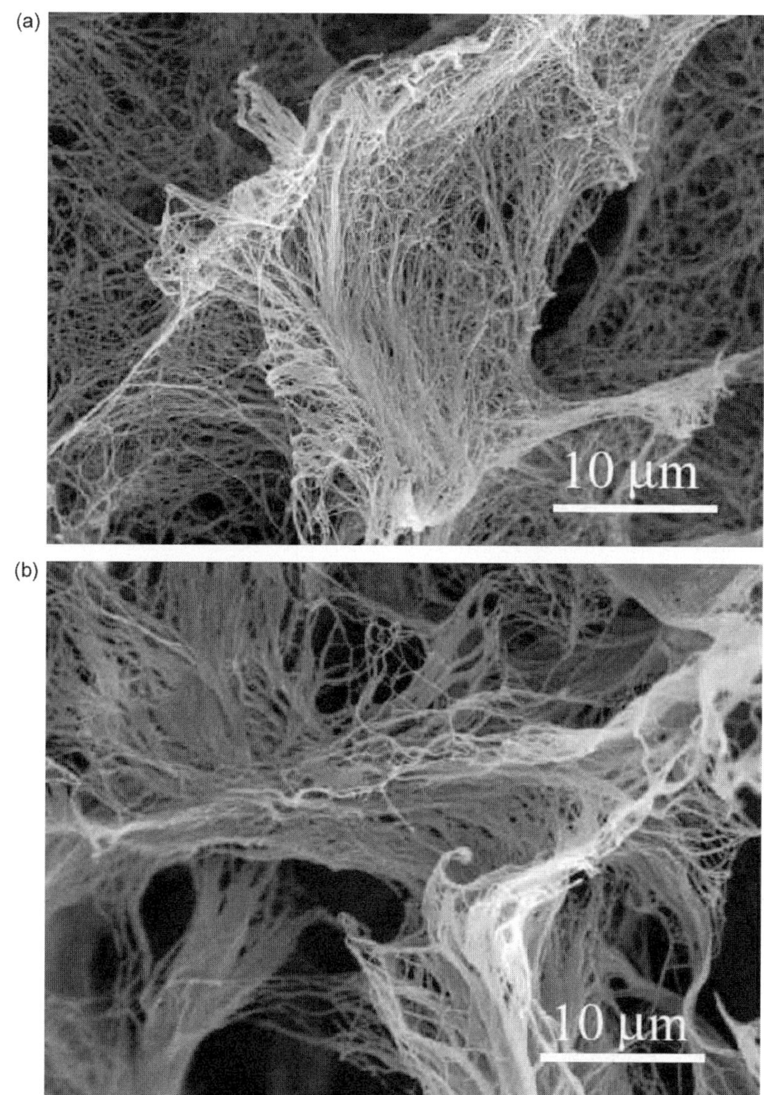

FIGURE 14.3. SEM micrographs of PLLA matrices prepared from 2.5% (wt/v) PLLA/dioxane/methanol (dioxane/methanol = 80/20) solution with a gelation temperature of $-18°C$: (a) with water extraction, ×2.0K; (b) without water extraction, ×2.0K. (Adapted from [9] with permission. Copyright (1999) John Wiley & Sons.)

2.2. Three-Dimensional Nanofibrous Scaffolds with Predesigned Macropores

In the human body, every tissue and organ has its unique 3D architecture. Appropriate spatial organization of cells is essential for tissues and organs to be fully functional. Consequently, scaffolds, which are developed to support growth

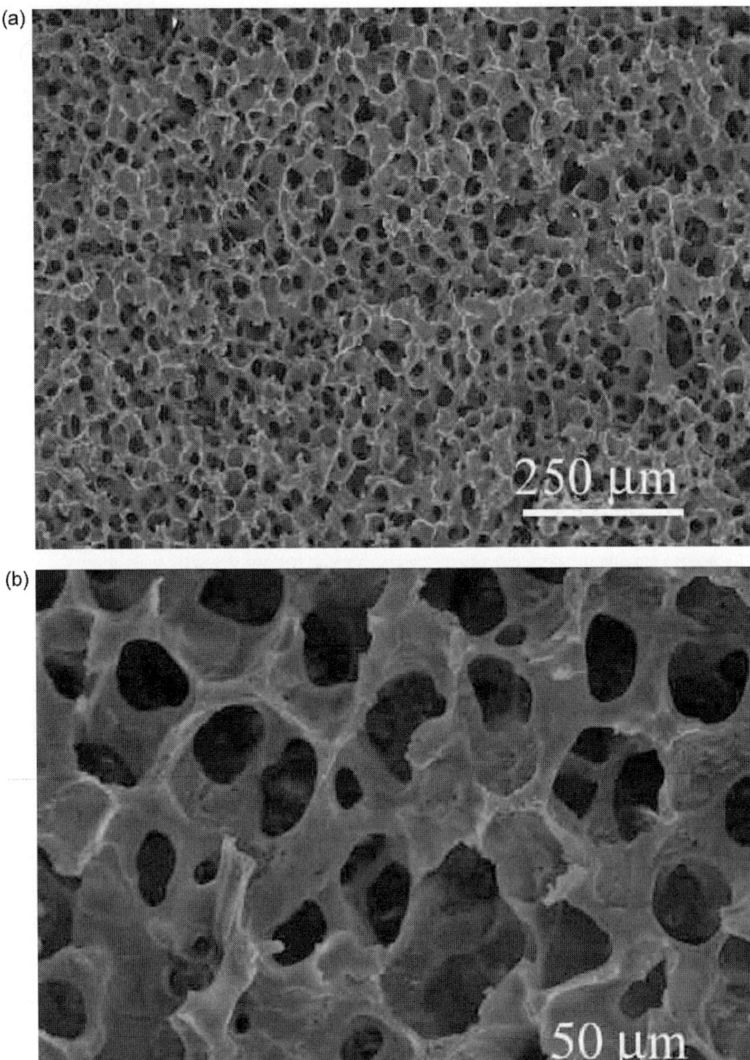

FIGURE 14.4. SEM micrographs of porous matrices prepared from uncrystallizable aliphatic polyester solutions at a gelation temperature of −18°C: (a, b) 10% PLGA (85/15)/dioxane/H_2O (dioxane/H_2O = 80/20);

of cells, need to posses certain 3D structures to allow cells to grow into desired shape and to facilitate mass transport during cell culture. A few methods have been developed to fabricate porous scaffolds for tissue engineering. Among them, the salt-leaching technique is one of the most frequently used methods. The technique uses salt particles as porogen since the salt can be washed away using water

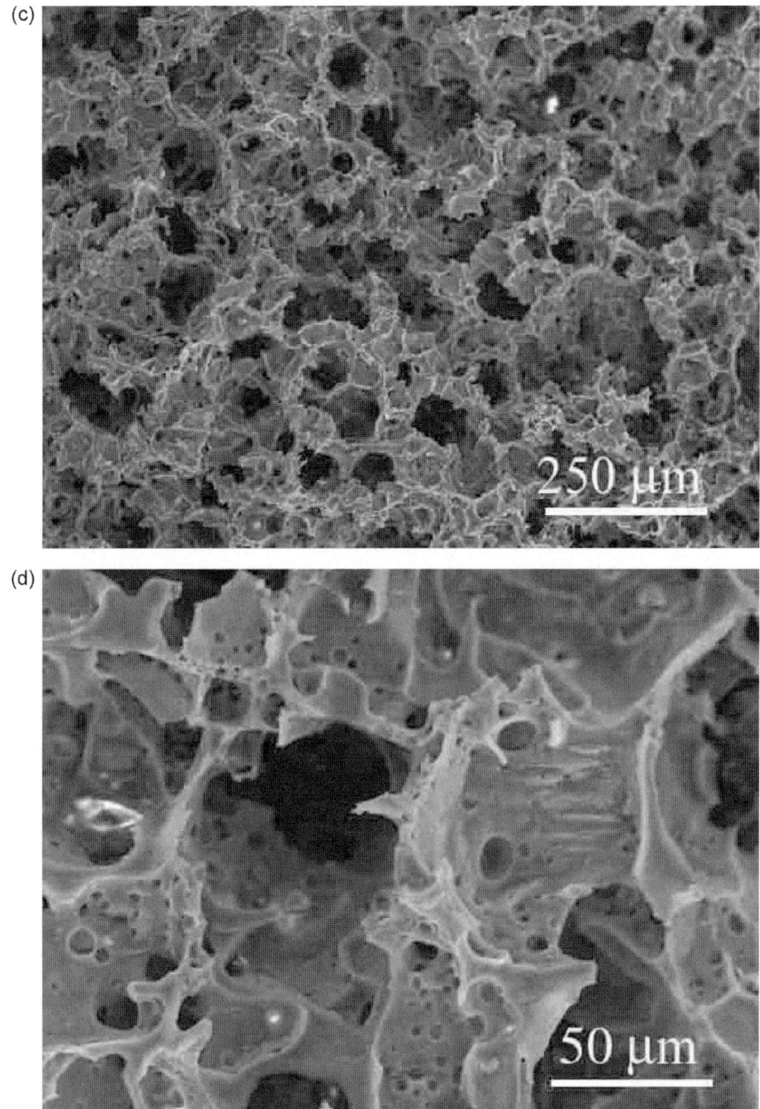

FIGURE 14.4. (*Continued*) (c, d) 5% PDLLA/dioxane/H_2O (dioxane/H_2O = 85/15). (Adapted from [9] with permission. Copyright (1999) John Wiley & Sons.)

without etching the polymeric scaffold materials. The size of the pore is controlled by the dimension of the salt particles and the porosity is determined by the volume ratio of salt to polymer. However, the salt-leaching technique can only fabricate pores with very limited types of shape, unable to match the structural complexity and diversity of tissues and organs.

3D scaffolds with complicated structures were successfully developed recently [11]. For example, one method is to fabricate sugar and salt particles, fibers and disks first, then stack these porogen blocks layer by layer to form complex pore architecture. Polymer solution is poured into the architecture. Porogen blocks are washed away after solidifying polymer. Using this method, scaffolds with various architectures can be prepared. Figures 14.5, 14.6, and 14.7 show SEM pictures of scaffolds prepared by stacking sugar fibers layer by layer in different fashions.

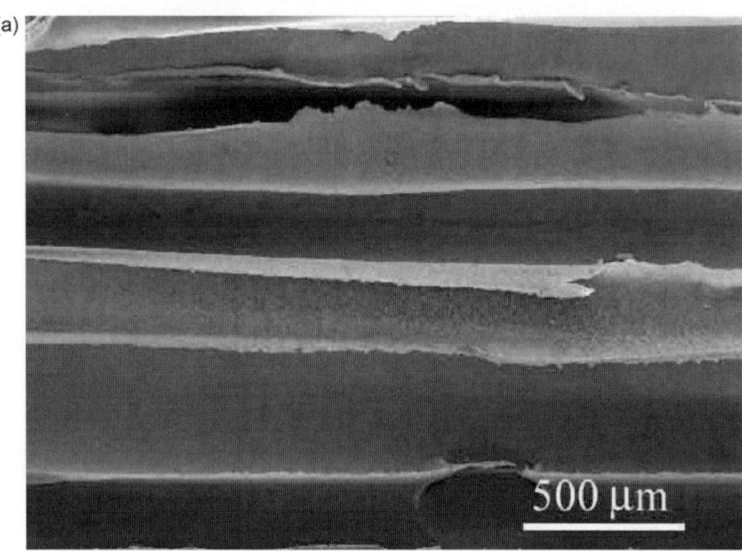

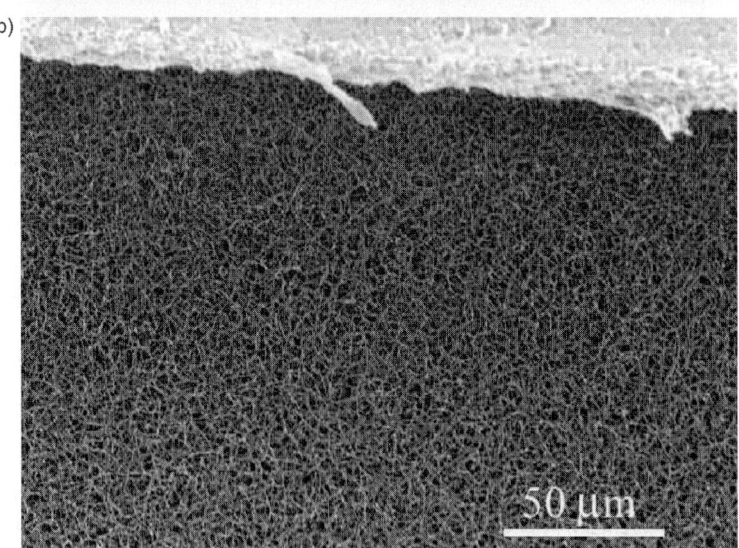

FIGURE 14.5. SEM micrographs of PLLA nanofibrous matrix with uni-axially oriented tubular architecture prepared from PLLA/THF solution and a uni-axially oriented sugar fiber assembly. (Adapted from [11] with permission. Copyright (2000) John Wiley & Sons.)

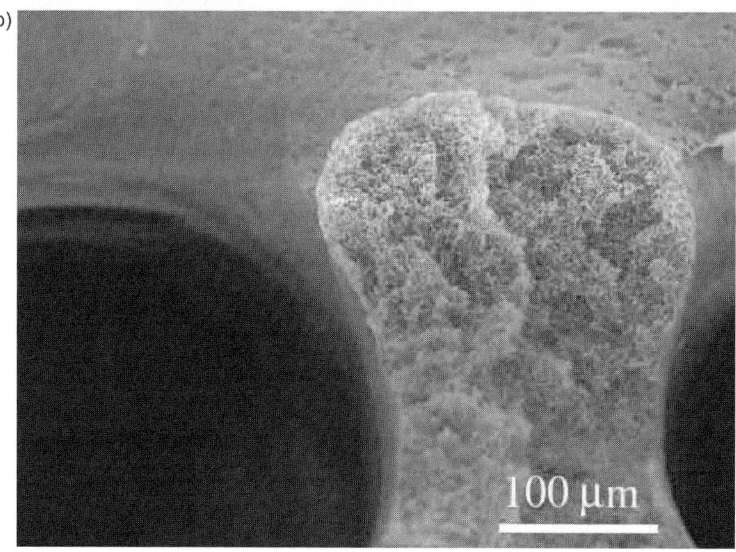

FIGURE 14.6. SEM micrographs of PLLA nanofibrous matrix with orthogonal tubular macropore network prepared from PLLA/THF solution and an orthogonal sugar fiber assembly. (Adapted from [11] with permission. Copyright (2000) John Wiley & Sons.)

2.3. Nanofibers Prepared from Electrospinning

The electrospinning process was first patented about 70 years ago [12]. Electrospinning caught much attention in recent years since the technique can be used to fabricate polymer fibers with diameters in the range of nanometers to micrometers. Numerous articles were published recently on electrospinning process optimization and the applications of thus obtained fibers [13–17].

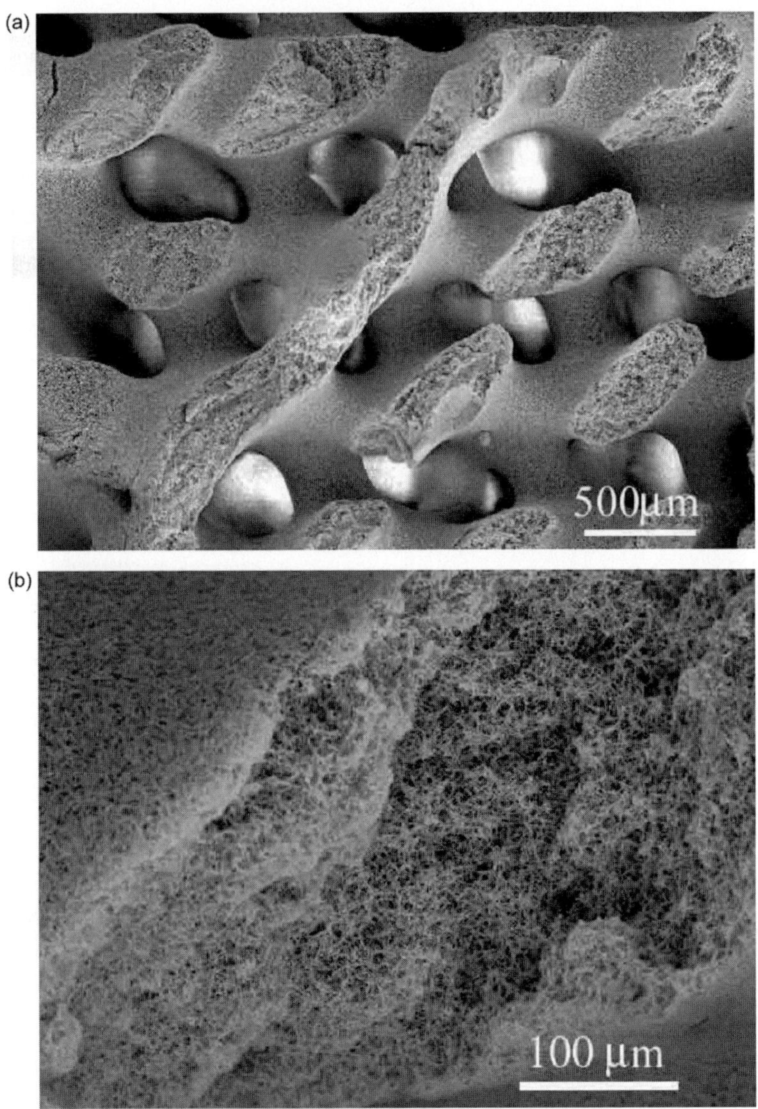

FIGURE 14.7. SEM micrographs of PLLA nanofibrous matrix with helicoidal tubular macropore network prepared from PLLA/THF solution and a helicoidal sugar fiber assembly. (Adapted from [11] with permission. Copyright (2000) John Wiley & Sons.)

In a typical electrospinning setup, one electrode is placed inside spinning solution or melt, and another electrode is connected with a fiber collector. A pipette is placed with the tip facing the collector. Polymer fluid is transferred into the pipette by a pump. Without electric field, the fluid is held inside the pipette by its surface tension. However, under electric field, mutual charge repulsion exerts a force in a direction opposite to the surface tension. The polymer fluid at the tip of the pipette is dragged by the electrostatic force toward the collector. Once the strength of electric field reaches a critical value, the electrostatic force overcomes the surface tension and the fluid is released from the tip of the pipette. The solution then loses the solvent due to fast evaporation. In the case of melt, polymer solidifies upon cooling down. The properties of the polymer fiber is controlled by the parameters like electric potential, distance between the pipette and the collector, molecular weight of the polymer, solution viscosity and conductivity, temperature, humidity, etc. Electrospinning technique has been used to produce scaffolds for tissue engineering [13–16], fibrous polymer membrane for battery applications [18], wound dressing [17], drug delivery devices [19,20] and so forth. A variety of synthetic and natural polymers, including poly(lactic acid) (PLA) [13], gelatin [15,16], silk [21–23], chitosan [14], copolymers [19,20], and composites [16], have been prepared for tissue engineering applications using electrospinning technique.

Among these materials, poly(vinyl alcohol) (PVA) is of special interest. PVA is an inexpensive and biocompatible material. Unlike PLLA, PVA is not biodegradable. It may be used as long-term or permanent scaffolds in tissue engineering. PVA is also one of the most studied electrospinning materials. PVA is usually produced by hydrolysis of poly(vinyl acetate). A major challenge is to electrospin highly hydrolyzed PVA. It has been found that the critical potential needed for electrospinning is dependent on the degree of hydrolysis [24]. Droplets instead of continuous fibers tend to form when aqueous solutions of fully hydrolyzed PVA are used.

It is recognized that the critical electrical voltage increases with the surface tension of the solution. The surface tension of aqueous PVA solutions increases with the degree of hydrolysis of PVA and the increase becomes more significant when the degree of hydrolysis approaches 100% [25]. Hence Yao et al. [24] stated that the difficulty associated with fully hydrolyzed PVA might be caused by the high surface tension. They developed a new process for PVA electrospinning by incorporating Triton X-100 surfactant into electrospinning solutions. Triton X-100 (Sigma-Aldrich) is a nonionic surfactant. The contact angle of 10% (wt) of fully hydrolyzed PVA water solution without Triton X-100 was about 101°. The contact angle decreases with the increase of the concentration of Triton until it reaches about 60° corresponding to the surfactant's concentration of ~0.3 v/w%. It was found that when the concentration of surfactant was <0.06 v/w%, droplets were the dominant form at 2.5 kV/cm. When the surfactant concentration is between 0.1 and 0.2 v/w %, droplets and fibers were both observed. Fibers became dominant if the concentration of Triton reached 0.3 v/w%. The diameters of thus obtained PVA fibers were between 100 and 700 nm (Fig. 14.8).

Electrospinning is another major technique for fiber fabrication besides the phase separation method described in the previous sections. Here, the electrospinning of PVA is presented as an example simply to give readers some idea about how to vary the process parameters to control the quality of fibers. More details about preparing nanofibers by electrospinning can be found from recent review articles [26,27].

FIGURE 14.8. SEM micrograph of PVA mat made by electrospinning. (Adapted from [24] with permission. Copyright (2003) American Chemical Society.)

3. Nanoparticles

3.1. Polymer/Hydroxyapatite Nanocomposite Scaffold for Bone Tissue Engineering

Nature provides a great framework that we can mimic in designing scaffold for tissue regeneration applications. It has been demonstrated that nanofibrous polymeric scaffold greatly improved bone cell attachment and proliferation [7] in vitro partially due to the structural similarity between polymer nanofibers and fibrous collagen, one of the major components of ECM, at the nanometer scale. Natural bone matrix, however, is a composite containing inorganic compounds (mainly partially carbonated HAP in nanometer size) and organic compounds (mainly collagen nanofibers). The nanometer size of both the inorganic component (HAP) and the organic component (collagen) is considered to be important for the unique properties of bone [28]. Mimicking natural bone in both composition and structure has been suggested to facilitate achieving the ultimate goal of functional bone regeneration [29].

Ceramic formulations, such as HAP and calcium phosphates, have been widely used in bone/orthopedic tissue engineering and have shown good osteoconductivity and bone bonding ability. Enhancement of mineralized new bone formation was also demonstrated when ceramic scaffolds were implanted into bone defects. However, the use of these ceramics in orthopedic/dental applications has been largely limited due to the mismatch of mechanical properties (inherent brittleness) to natural bone as well as the difficulty in processing. The combination of polymers and ceramics, as composite scaffolds, would offer the promise to have favorable properties of both polymer and ceramic phases. Compact PLLA/HAP composites (with little porosity) have been developed as implant materials for bone regeneration and have shown some favorable properties over both pure PLLA and pure HAP [30,31]. However, it remains a challenge to develop clinically applicable bone repair materials. The problem has been at least partially due to the inability of these materials to offer sufficient space (high porosity) for the survival and activities of anchorage-dependent bone cells during in vitro culture and in vivo new tissue formation. Biodegradable polymers such as poly(α-hydroxy acids) have great design flexibility to achieve highly porous structures with controllable physical, chemical, and degradation properties, using various scaffolding techniques. Highly porous poly(α-hydroxy acids)/HAP scaffold has been created through a thermally induced phase separation (TIPS) technique (Fig. 14.9) [32]. Phase separation was performed under similar conditions as described previously in Section 2.1 for pure polymer scaffold fabrication, except that HAP particles (micrometer sized, mHAP) were dispersed in the polymer solution before phase separation. Composite scaffolds with porosities such as 95% and open pore structures could be achieved, despite the fact that the addition of mHAP perturbed the phase separation process to some extent [32]. Polymer/mHAP composite scaffolds showed significant improvement in compressive modulus and compressive yield strength over pure polymer scaffolds. Compared to pure polymer scaffolds in which cell ingrowth and tissue matrix formation were limited to the periphery of the scaffold (~240 μm in depth) [33], the composite scaffolds supported uniform cell seeding, cell ingrowth and tissue formation throughout [29], including the very center of the scaffold (Fig. 14.10). Further examination of histology and DNA assay revealed that polymer/HAP scaffolds had a higher osteoblast survival rate, more uniform cell distribution and growth, improved new tissue formation, and enhanced bone specific gene expression in vitro [34,35].

Recent research also suggested that better osteoconductivity would be achieved if synthetic HAP could resemble bone mineral more in composition, size, and morphology [36,37]. The nanosized HAP (nHAP) may also have some special properties due to its small size and huge specific surface area. This leads to the fabrication of polymer/nHAP nanocomposite scaffold for tissue engineering applications. NHAP (around 20 nm) was dispersed in a solution of poly(α-hydroxy acids) and the resulting mixture was phase separated to produce highly porous 3D nanocomposite scaffolds [38]. The scaffolds have high porosity (>90%) and pore size in the range of 50–500 μm. Remarkably, the incorporation of nano-HAP did not alter the scaffold structure significantly. A regular anisotropic pore structure was obtained which was similar to that from plain

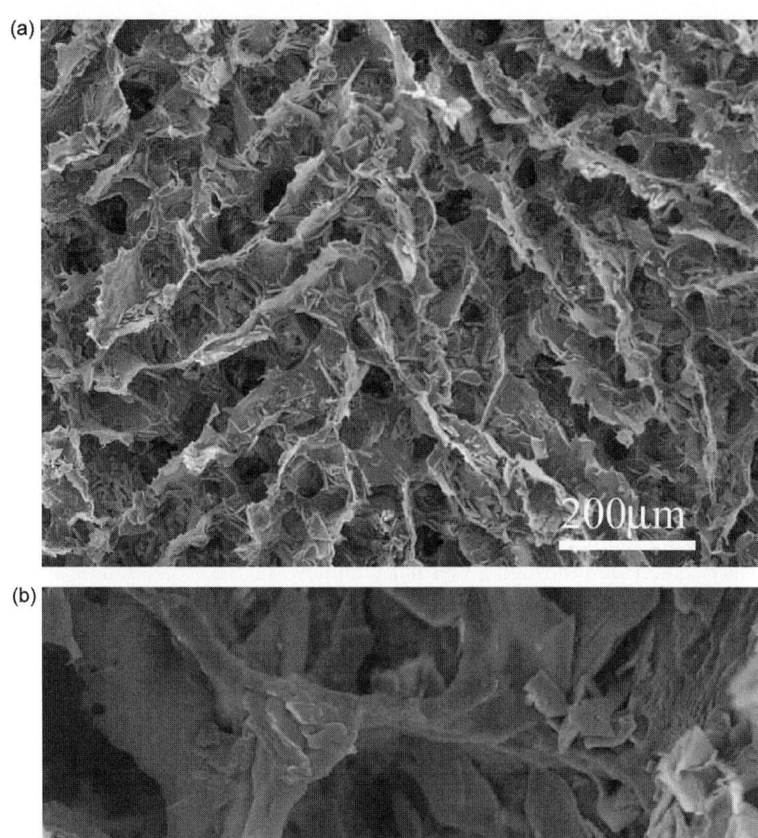

FIGURE 14.9. SEM micrographs of PLLA/mHAP (50:50) scaffold fabricated with TIPS using dioxane as solvent. (Adapted from [32] with permission. Copyright (1999) John Wiley & Sons.)

polymers (Fig. 14.11). NHAP particles were found uniformly dispersed within fibrous and leaf-like porous matrices, using dioxane/methanol and benzene, respectively, as solvent for phase separation (Fig. 14.12). Compared with PLLA/mHAP composite scaffolds prepared under the same conditions (Fig. 14.9), PLLA/nHAP exhibited more regular and uniform morphologies.

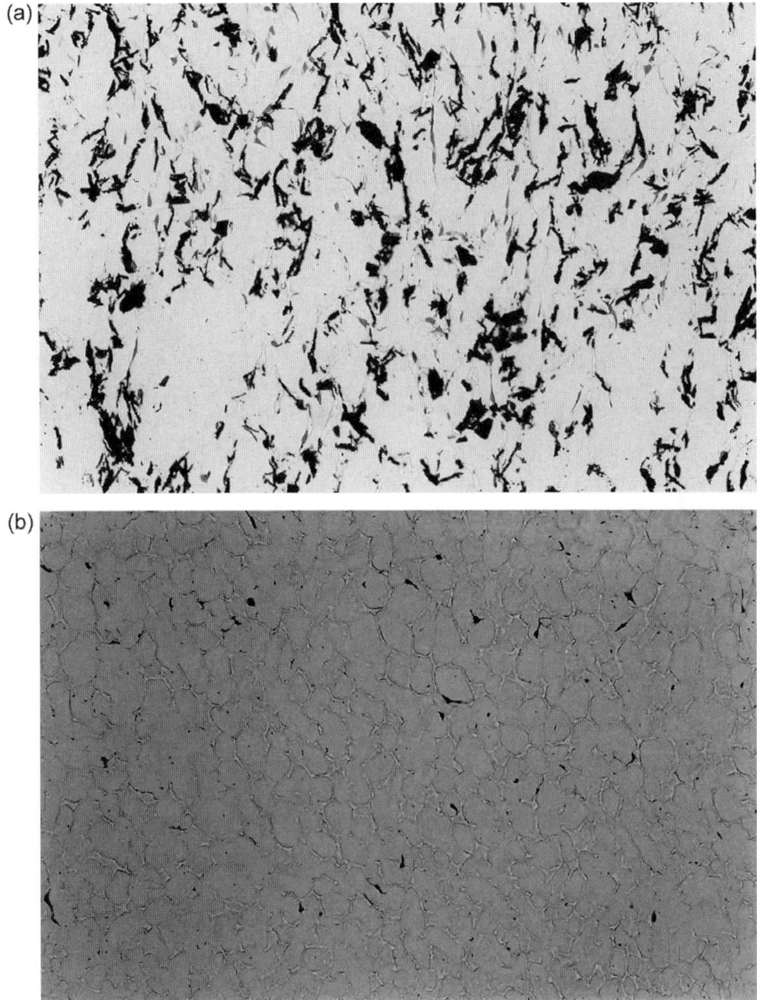

FIGURE 14.10. Osteoblast-PLLA and osteoblast-PLLA/mHAP constructs cultured in vitro for 8 weeks (von Kossa's silver nitrate staining, original magnification × 100). (a) PLLA; and (b) PLLA/mHAP. (Adapted from [29] with permission. Copyright (2001) John Wiley & Sons.)

When phase separation was performed with a uniaxial temperature gradient, oriented microtubular PLLA/nHAP composite scaffolds (Fig. 14.12c) were obtained whereas such pore morphologies could not be obtained when micro-sized HAP (mHAP) was used. Improvement in mechanical properties and serum protein adsorption was observed on the PLLA/nHAP composite scaffolds. Composite scaffolds adsorbed higher amounts of serum proteins than pure polymer scaffolds [35,38], and significantly higher amounts of proteins were

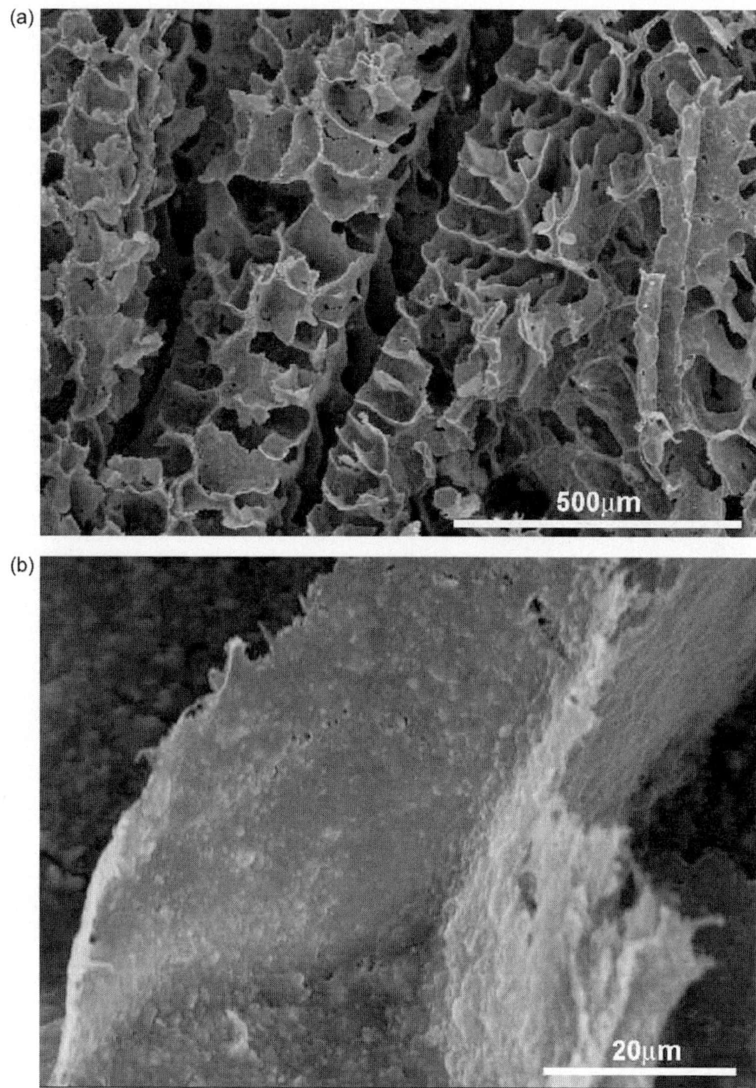

FIGURE 14.11. SEM micrograph of PLLA/nHAP nanocomposite scaffolds using dioxane as solvent. (Adapted from [38] with permission. Copyright (2004) Elsevier.)

adsorbed on PLLA/nHAP scaffolds than on PLLA/mHAP scaffolds. Similar protein adsorption results were also observed on 2D ceramic disks including HAP, titania and alumina. Compared with conventional microsized ceramic formulations, nanophase ceramics adsorbed significantly greater quantities of vitronectin, which subsequently may have contributed to enhanced osteoblast adhesion on nanophase ceramic disks [39,40]. Together with improved mechanical properties and controllable highly porous 3D structure, the bioactive

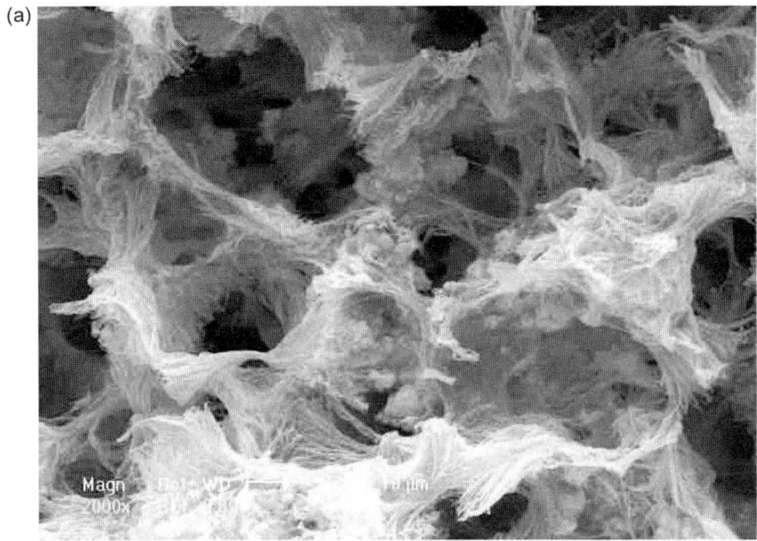

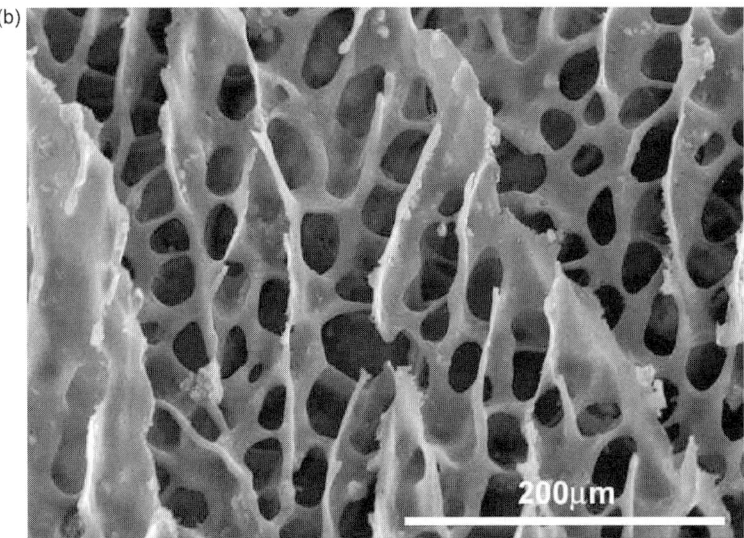

FIGURE 14.12. SEM micrographs of polymer/nHAP nanocomposite scaffolds. (a) PLLA/nHAP (50/50) scaffold using dioxane/water (90:10) mixture solvent; (b) PLGA85/nHAP (70/30) scaffold using dioxane as solvent;

characteristics make the nanocomposite scaffolds favorable for bone tissue engineering applications.

Alternatively, a biomimetic approach has been introduced to prepare bone-like apatite coated composite scaffold [41,42]. In this approach, prefabricated polymeric scaffolds were incubated in a simulated body fluid (SBF) in which bone-like apatite crystals, in sizes of nanometers to micrometers, were nucleated

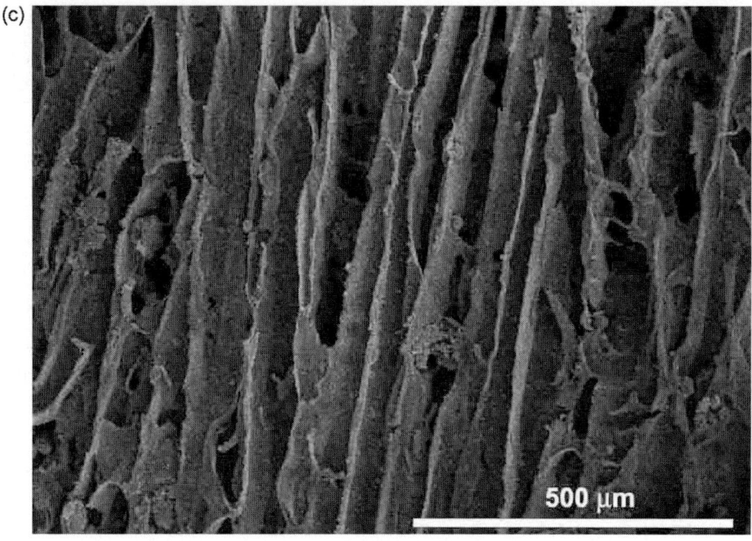

FIGURE 14.12. (*Continued*) (c) PLLA/nHAP (70/30) scaffold using benzene as solvent. (Adapted from [38] with permission. Copyright (2004) Elsevier.)

and grown on the inner pore wall surface of the 3D scaffold (Fig. 14.13). It has been observed that the growth of apatite crystals was affected greatly by the polymer materials, porous structure, ionic concentration of SBF as well as the pH value [42]. Apatite particles grew in size with the increase of incubation time. Besides their similarity in composition and structure to bone minerals as demonstrated by EDX, FTIR, and XRD analyses [41], nanostructured features were also observed on the particle surfaces which may lead to favorable bone cell responses. When macroporous (250–425 µm in diameter resulted from sugar porogen) and nanofibrous scaffolds were investigated for bone-like apatite deposition, a uniform and dense layer of nanoapatite was found to cover the entire inner pore wall surface of the scaffold (Fig. 14.14) [43]. The particle size on macroporous scaffolds was smaller than that on scaffolds without macropores (Fig. 14.13) while the density of apatite particles increased over incubation time. The results indicated that well-interconnected macroporous and nanofibrous structure facilitated the ion transportation within scaffold, which in turn increased the number of apatite crystal nucleation sites. It was also found that nano-HAP had the ability to promote in vitro calcification of composites when nHAP was pre-incorporated into poly(α-hydroxy acids) scaffolds. Significantly greater amounts of apatite particles were grown on nanocomposite scaffolds than on pure polymer scaffolds. Similar results were found for bone-like apatite growth on naturally derived polymer sponges such as gelatin [44] and chitin [45]. Incorporation of HAP can significantly increase the amount and rate of bone-like deposition. Biomimetic deposition of bone-like apatite may be of direct interest for the development of a bone tissue engineering composite scaffold or for assessing the calcification function of existing scaffolds.

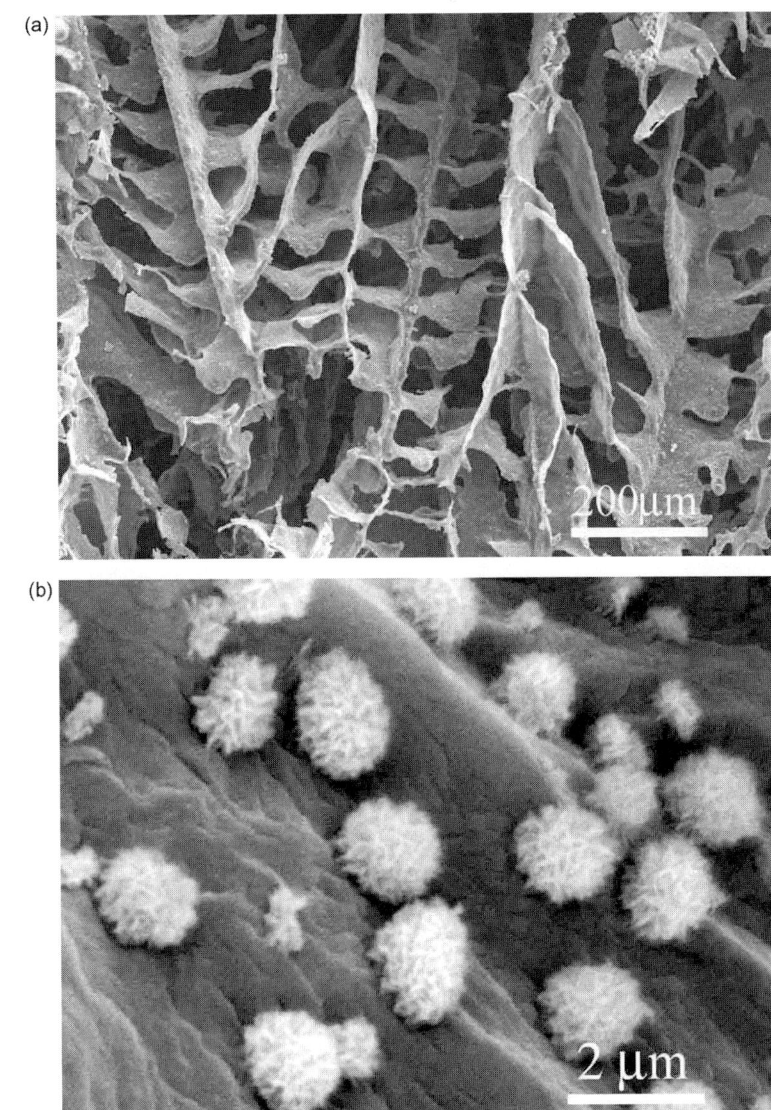

FIGURE 14.13. SEM micrographs of PLLA-apatite composite scaffold prepared by SBF incubation. (a) × 100; and (b) × 10,000. (Adapted from [41] with permission. Copyright (1999) John Wiley & Sons.)

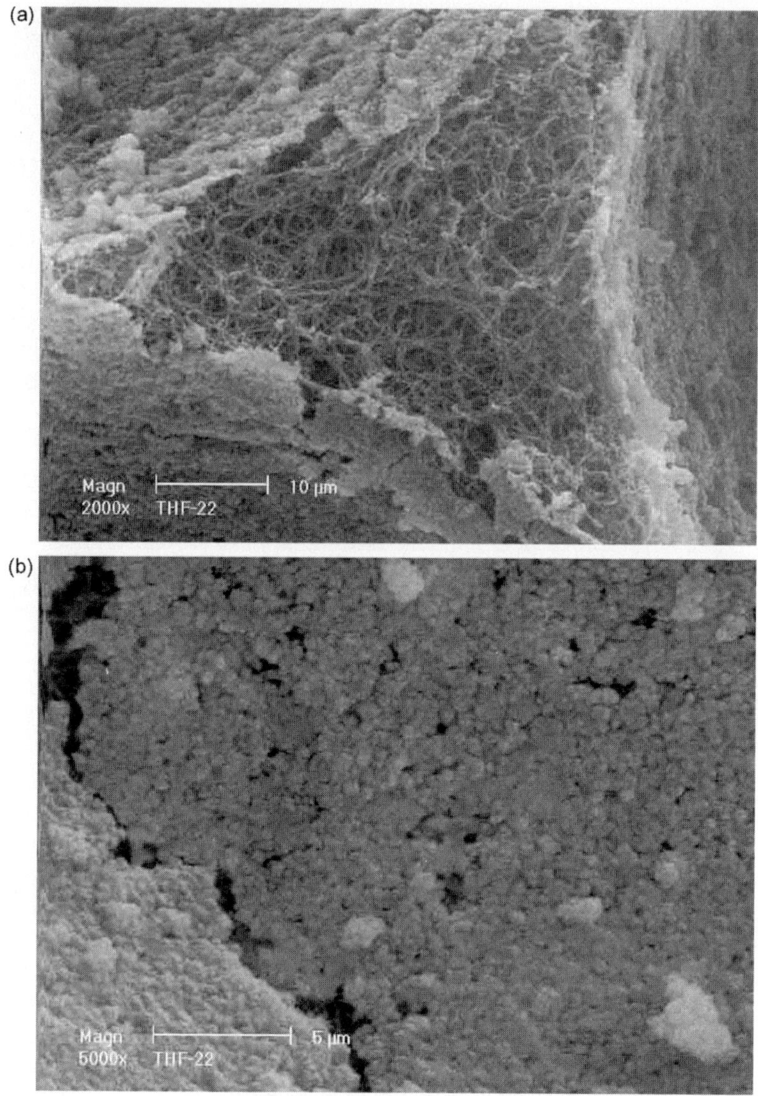

FIGURE 14.14. SEM micrographs of macroporous nanofibrous PLLA-apatite composite scaffolds after 22 days incubation in 1.5× SBF. (a) ×2000; and (b) ×5000.

3.2. Nanoparticle/Nanosphere for Controlled Delivery of Bioactive Factors

Owing to the rapid advance in recombinant technology and the availability of a large scale of purified recombinant proteins, protein drugs such as growth factors have been widely used in recent tissue engineering strategies to stimulate cellular activity and regulate tissue regeneration [5,46,47]. However, protein and peptide drugs in general have short plasma half-lives, are unstable in the gastrointestinal tract and also have low bioavailability due to their relatively large molecular weight and high aqueous solubility. These properties limited their effective clinical applications [48] and a carrier system is required to achieve high therapeutic efficiency of these proteins [49–51]. Polymeric particulate carriers (micro and nanospheres) have been demonstrated to be an effective way to offer controlled release of a contained substance and to protect unstable biologically active molecules from denature and degradation after administration [49]. Among the natural or synthetic polymers used for nanoparticle fabrication, PLA and PLGA were found to be remarkable for their applications in drug delivery due to their excellent biocompatibility and biodegradability through natural pathways [52–54]. Most importantly, the released proteins were able to maintain a high level of biological activity with desired prolonged duration [52,54].

A well-established double emulsion technique has been used to encapsulate various proteins including a model protein (bovine serum albumin: BSA), recombinant human parathyroid hormone (rhPTH) and platelet-derived growth factor (rhPDGF-BB) into PLGA microspheres or nanospheres [54,55]. Microspheres in size of 20–50 μm and nanospheres in size of 200–500 nm were obtained depending on the concentration of surfactant used and the emulsion strength employed during second emulsion process (Fig. 14.15). Increase of surfactant concentration or emulsion strength decreased the size of microspheres effectively. The release of protein from microsphere was controlled in the first stage by diffusion and in the second stage by the degradation of polymer microspheres. The morphology of degraded PLGA50-74K (Mw = 74 kDa) microspheres was monitored by scanning electron microscopy (Fig. 14.16). PLGA50 microspheres with molecular weight of 74 kDa degraded rapidly after a 3-week lag time and were degraded completely within 4 months. In vitro BSA release kinetics from PLGA50-74K microspheres were characterized by a burst release (diffusion controlled) followed by a slow release phase within 1–7 weeks and a second burst release at 8 weeks (degradation controlled), which was consistent with the degradation profiles. By varying the molecular weight and the ratio of LA/GA in PLGA copolymer, protein release over days to months following sustained release kinetics can be achieved [54,55]. The PTH incorporated PLGA50-74K microspheres released detectable PTH in the initial 24 h, and the released PTH was biologically active as evidenced by the stimulated release of cAMP from ROS 17/2.8 osteosarcoma cells in vitro as well as increased serum calcium levels in vivo when injected subcutaneously into mice [54]. Similarly, the released PDGF-BB from PLGA50-6.5K nanospheres was biologically active after 2 weeks and was able to stimulate

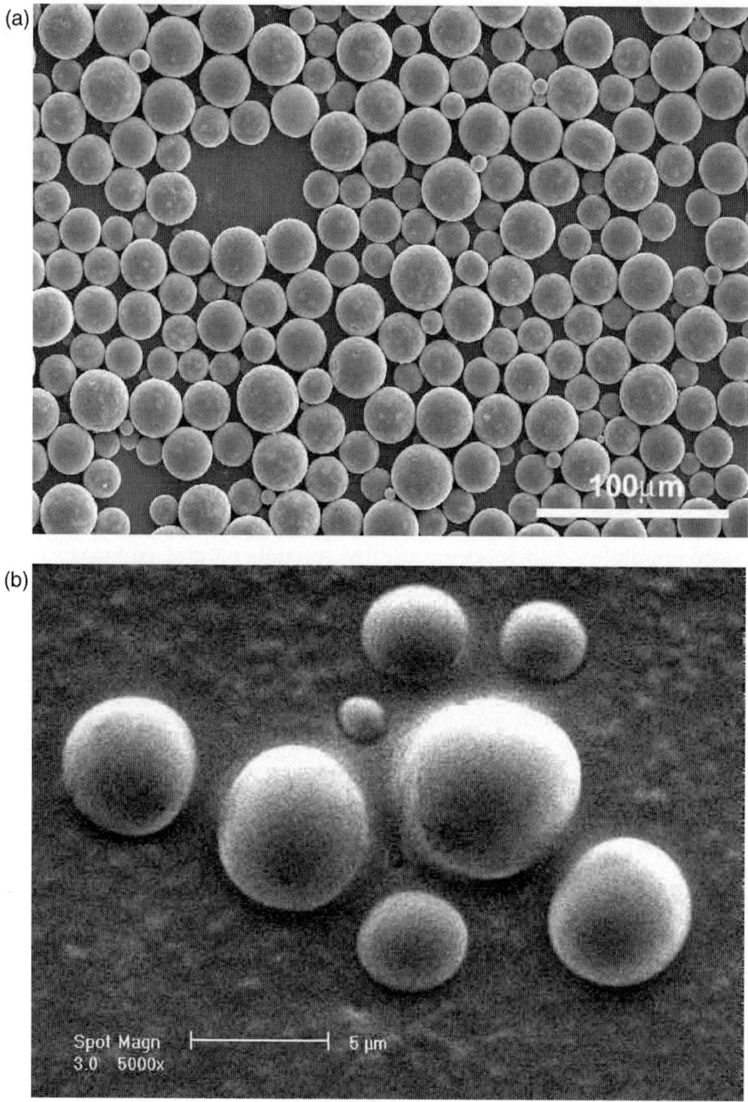

FIGURE 14.15. SEM micrographs of PLGA50-74K microspheres prepared under different conditions. (a) 1% PVA solution, mechanical stirring at 700 rpm; (b) 5% PVA solution, mechanical stirring at 700 rpm;

the proliferation of human gingival fibroblasts (HGF) in vitro [55]. These studies illustrate the feasibility of achieving local delivery of bioactive factors to induce biological cellular responses by a microsphere/nanosphere encapsulation and delivery technique.

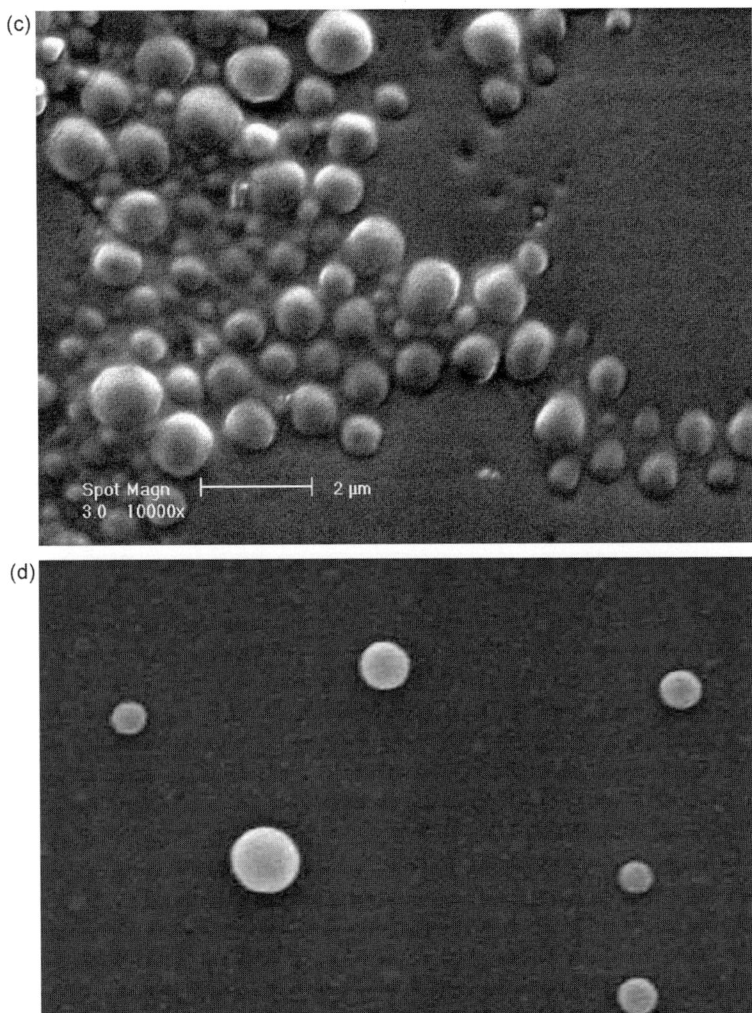

FIGURE 14.15. (*Continued*) (c)1% PVA solution, sonicating at 15 W; (d) 1% PVA solution, sonicating at 30 WC (a–c). (Adapted from [54] with permission. Copyright (2004) Elsevier.)

4. Surface Modification

This section focuses on surface modification of polymeric biomaterials for tissue engineering application, especially on surface engineering of nanofibrous 3D scaffolds. Various surface modification methods are briefly reviewed. A simple and

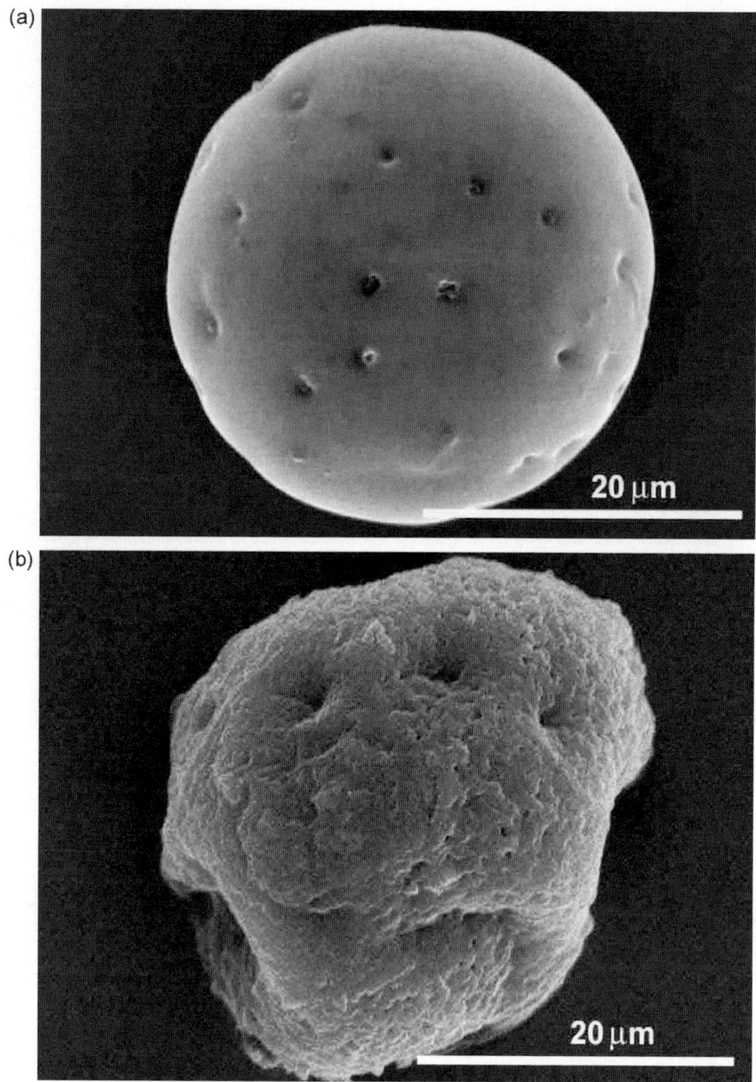

FIGURE 14.16. SEM micrographs of PLGA50-74K microspheres at different degradation stages. (a) 0 week; (b) 7 weeks;

effective surface modification method (surface entrapment) is then introduced. Surface-modified nanofibrous PLLA scaffolds were examined in this study for cell adhesion and proliferation. SEM was utilized to observe both the microstructure of surface-modified nanofibrous PLLA scaffolds and cells morphology in the biomimetic nanofibrous PLLA scaffolds.

FIGURE 14.16. (*Continued*) (c) 11 weeks. (Adapted from [54] with permission. Copyright (2004) Elsevier.)

4.1. Surface Modification Methods for Tissue Engineering

Surface properties as well as scaffolding architecture are important for a desirable scaffold in tissue engineering [56–60]. The interactions of cells with the scaffolding materials takes place on the material's surface; therefore the nature of the surface can directly affect cellular response, ultimately influence the rate and quality of new tissue formation. Although a variety of synthetic biodegradable polymers have been used as tissue engineering scaffolding materials (e.g., PDLLA, PLLA, and PLGA), one disadvantage of these scaffolds is their lack of biological recognition on the material surface. Hydrophobic polymers do not provide the ideal environment for cell–material interactions.

Various surface modification methods have been developed to promote cell–material interactions [61–64]. Surface hydrolysis of poly(glycolic acid) (PGA) scaffold under strong alkaline conditions was used to increase cell seeding density and improve biomaterial–cell interactions [61]. Hydrolysis of ester bonds on the surface of PGA fibers changes the surface properties and results in higher seeding density and more spreading of cells when compared to unmodified PGA scaffolds. The limitation of this method is that it is technique-sensitive and hydrolysis also alters the surface morphology and bulk mechanical properties. Chemical coupling was incorporated to attach RGD peptide to the lysine residue of poly (L-lactic acid-co-lysine) [64]. The peptide content of the copolymer and their resulting chemical and physical characteristics could be varied in a controlled fashion by changing the molar ratio of the peptide to lysine units. These poly(α-hydroxy acid)-based copolymers can be further modified by chemical attachment of a variety of biologically active molecules to meet the specific needs of biomedical and tissue engineering applications. Low pressure ammonia plasma treatment has been utilized for the modification of thin film of poly(3-hydroxybutyrate) (PHB) [63]. The treatment provides a surface functionalized by amide and amino groups, which can further react with proteins needed for the adjustment of interactions between PHB and cells. Plasma exposure is effective and powerful for 2D film surface etching. This technique can be used to introduce the desired groups or chains onto the surface of a material, but this method is not suitable for a 3D scaffold with complex pore morphology and structure.

As discussed above, most of the surface modification work so far has been focused on 2D film surfaces or very thin 3D constructs. True 3D scaffolding (especially nanofibrous 3D scaffolding) surface modification is still a challenge. A simple and effective surface modification method (surface entrapment) is introduced in the following section. Gelatin was incorporated onto the surface of nanofibrous PLLA scaffolds using this method.

4.2. Surface Engineering of Nanofibrous PLLA Scaffolds with Gelatin

A solvent mixture of dioxane and water was introduced to entrap gelatin onto the surface of nanofibrous PLLA scaffolds. The solvent mixture composition was chosen such that gelatin was soluble in the solvent mixture, while the PLLA swelled but did not dissolve in the solvent mixture. In detail, gelatin (0.3 g) was dissolved in dioxane/water (60/40 v/v) solvent mixture (1,000 mL) at 45°C. The PLLA films or scaffolds were immersed in the solution and soaked for a designated time, and then taken out and quickly put into 200 ml ice-water mixture for 10 min. Chemical cross-linking of gelatin with 1-ethyl-3-(3-dimethylamino-propyl) carbodiimide HCl (EDC) and N-hydroxy-succinimide (NHS) was carried out in {2-(N-morpholino)ethanesulfonic acid} hydrate (MES) buffer at 4°C for 24 h. The resulting scaffolds were then washed with distilled water at 4°C for 3×, followed by rinsing at 40°C for 12 h (water was changed every 3 h) to remove nonentrapped gelatins. The surface-modified scaffolds were freeze-dried for 3 days, and then vacuum dried at room temperature for 2 days more.

The nanofibrous pore-wall structure was not affected by the surface modification process (Fig. 14.17). After entrapped with gelatin on the surface, the diameters of the nanofibers were still ranged from 50 to 500 nm, which is the same range of natural collagen fibers.

The density of gelatin entrapped on the nanofibrous PLLA surface was controlled by the solvent mixture composition. The amount of gelatin entrapped on the nanofibrous PLLA surface increased as the solvent ratio dioxane/water in gelatin solution increased. The nitrogen amount increased from 1.1% to 6.7% as the ratio of dioxane/water in solvent mixture increased from 0/100 to 60/40 (v/v). Further increase in dioxane/water ratio led to the difficulty of dissolving gelatin in the solvent mixture. The surface density of gelatin decreased consequently.

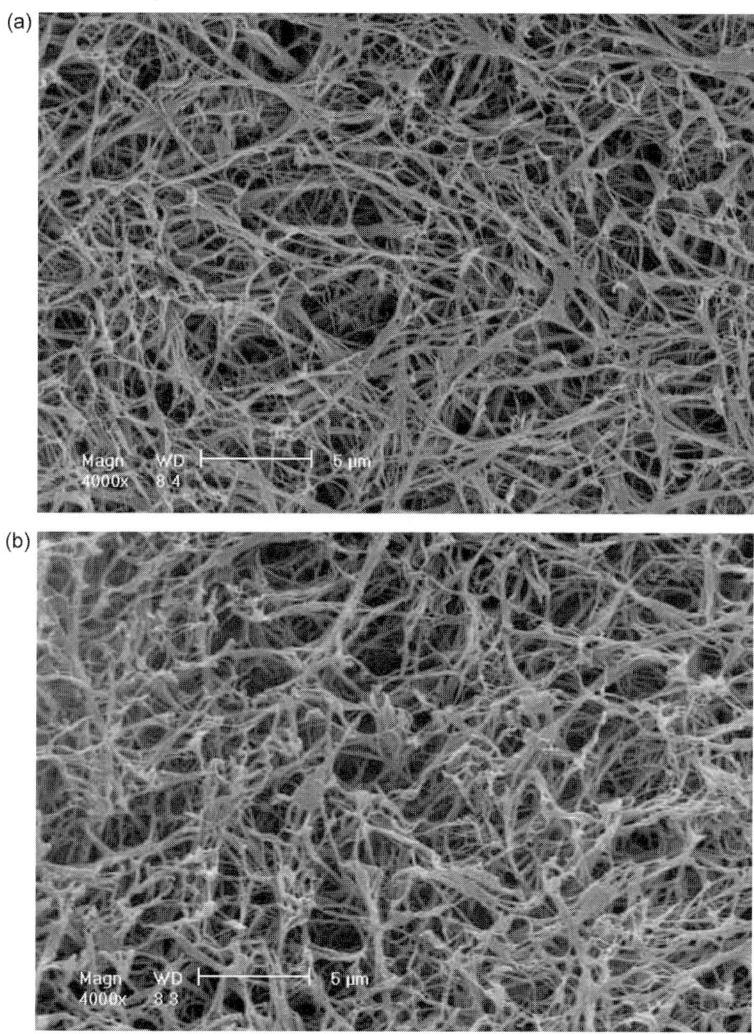

FIGURE 14.17. SEM micrographs of nanofibrous PLLA scaffolds. (a) Before surface entrapped with gelatin; (b) after surface entrapped with gelatin.

Chemical cross-linking after physical entrapment was utilized to further increase the amount of gelatin on the surface. When the polymer sample was quenched in cold water, there were some gelatin molecules, which were not entrapped on the polymer film surface, but were entangled with the entrapped gelatin molecules. These gelatin molecules would be washed away when rinsing in water at 40°C. Chemical cross-linking immobilized these gelatin molecules, increasing the amount of gelatin on the polymer surface.

This surface-entrapped technique is a simple and universally effective method for surface modification. No functional groups in polymer chains are needed for this surface modification method. In contrast to the method of incorporating modified groups by copolymerization, the entrapment method maintains the bulk properties of materials. Since the modified molecules only gather on the surface of materials, the entrapment method also promotes the efficiency of surface modification. In addition, unlike many other surface modification methods the entrapment method theoretically can be used for 3D scaffolds with various geometry, morphology and thickness. Furthermore, the entrapment method allows us to modify the surface in a controlled fashion and various parameters (solvent ratio, gelatin concentration, immersing time, and chemical cross-linking) can be readily adjusted to tailor the polymer surface.

The surface-modified nanofibrous PLLA scaffolds mimic both the chemical composition and the architecture of collagen matrix, and can provide a better environment for cell adhesion and proliferation. First, the highly interconnected pore structure provides channels for mass transport. Second, the surface area-to-volume ratio of nanofibrous PLLA scaffolds is thousands of times higher than that of solid-wall PLLA scaffolds. The nanofibrous architecture of the porous walls provides much larger surface area for cell adhesion and growth than a solid-film architecture. Furthermore, the interfiber spacing (microns) facilitates cell–cell, cell–nutrient, and cell–signal molecules interactions. Third, the incorporation of gelatin onto the surface of nanofibrous PLLA scaffolds promotes cell–material interactions. Statistically, more cells were found on the surface-modified nanofibrous scaffolds than the controls after 2 weeks of cultivation (Fig. 14.18). Nanofibrous scaffolds, regardless of surface modification, have

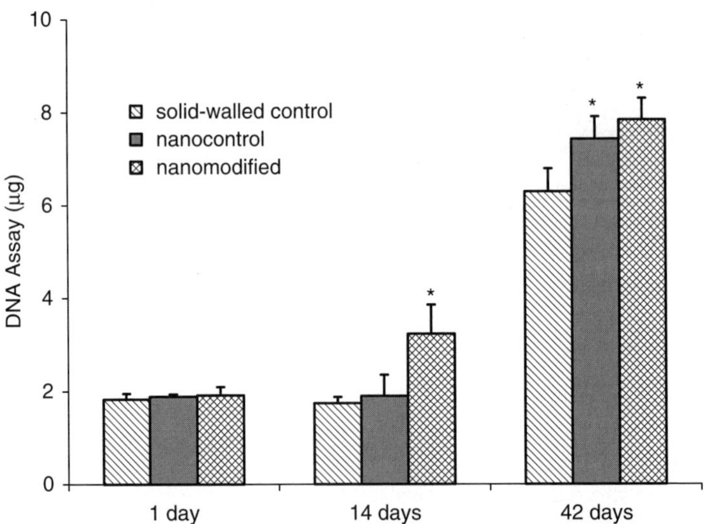

FIGURE 14.18. The DNA amounts in the PLLA scaffolds (control groups) and gelatin-modified nanofibrous PLLA scaffolds according to cultivating time. Cells (5×10^5) were seeded into each scaffold. (Statistical significance: $p < 0.05$ compared with the solid-walled PLLA).

more rapid proliferation rate than the solid-wall PLLA control after 6 weeks of cell cultivation. In addition, the cell number was significantly higher on the surface-modified nanofibrous scaffold than on the nanofibrous scaffold control after 2 weeks of cultivation. SEM images showed much more ECM secretion by cells on the surface-modified nanofibrous scaffolds, which indicated that the metabolism of the cells on surface-modified nanofibrous scaffolds was more vigorous than that on the controls (Figs. 14.19 and 14.20). All the above results

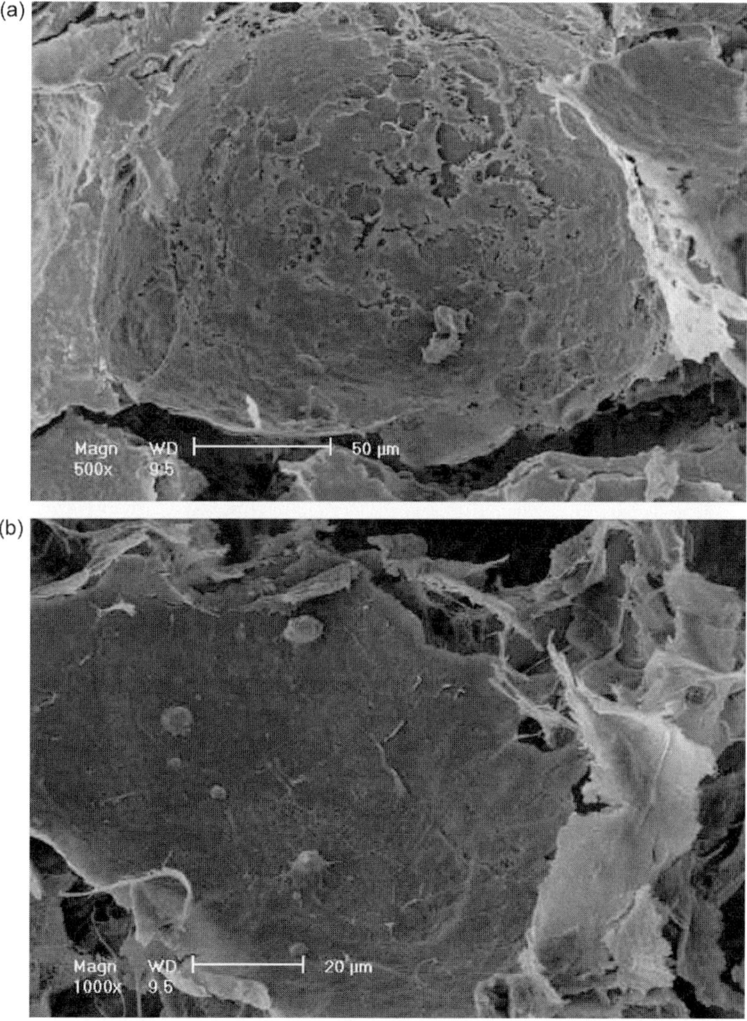

FIGURE 14.19. The SEM images of PLLA scaffolds 2 weeks after cell seeding: (a) solid-walled PLLA; (b) nanofibrous PLLA;

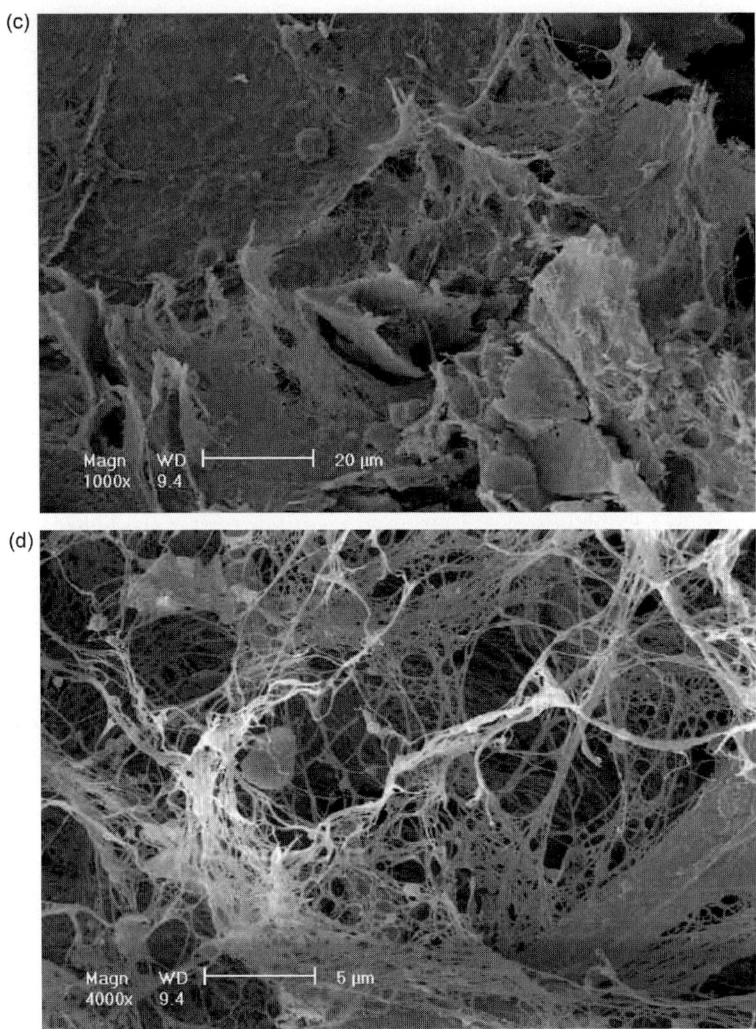

FIGURE 14.19. (*Continued*) (c) surface-modified nanofibrous PLLA; (d) surface-modified nanofibrous PLLA [high magnification of (c)].

demonstrated that the surface gelatin-modified nanofibrous PLLA scaffolds could enhance cells adhesion and proliferation.

5. Summary

This chapter aims to give readers a brief introduction to recent biomaterial research from several aspects: nanofibers, nanoparticles, and surface modifications. It can be seen that SEM is used in each example as a major characterization tool. It provides

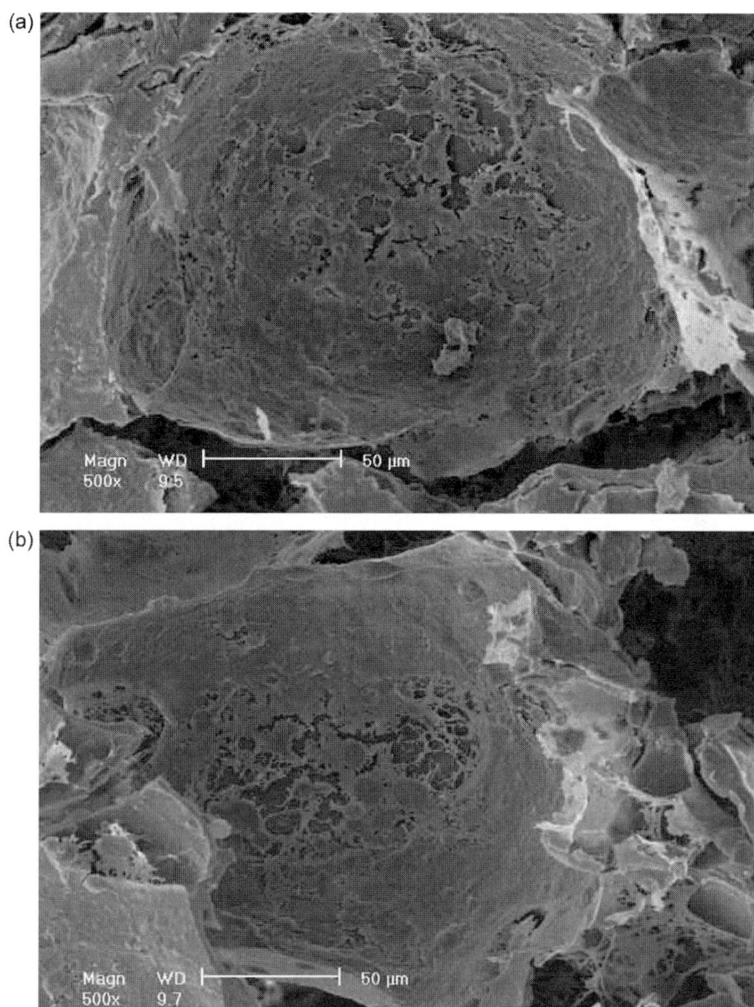

FIGURE 14.20. The SEM images of PLLA scaffolds 6 weeks after cell seeding: (a) solid-walled PLLA; (b) nanofibrous PLLA;

straightforward morphological information about samples. The morphology of biomaterials is one of the key properties that need to be controlled to make the materials fully functional. Since AFM cannot deal with 3D materials, it leaves SEM as the only option. Biomaterial development is a continuously expanding research area. It is doubtless that SEM will continue to play an indispensable role in the biomaterial research.

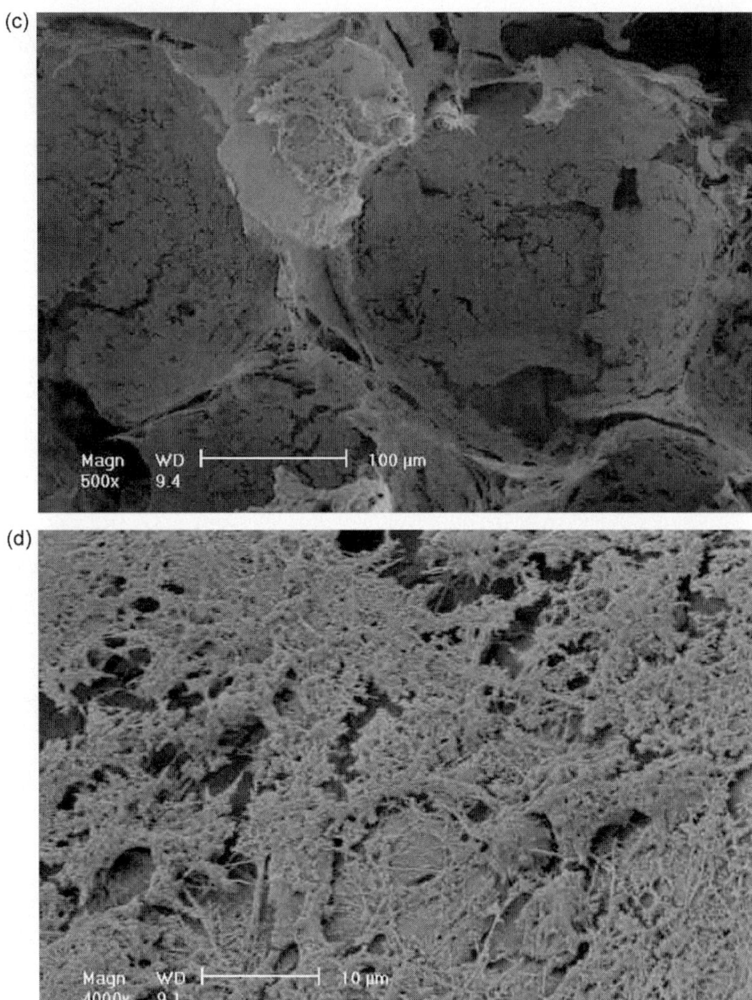

FIGURE 14.20. (*Continued*)(c) surface-modified nanofibrous PLLA; (d) surface-modified nanofibrous PLLA [high magnification of (c)].

References

1. J. W. Boretos and M. Eden (Eds.), *Contemporary Biomaterials: Material and Host Response, Clinical Applications, New Technology, and Legal Aspects*, William Andrew Publishing/Noyes, New York (1984).
2. H. C. Slavkin, *J. Am. Dent. Assoc.*, 127(8) (1996) 1254.
3. J. Black and V. Sholtes, *Orthop. Clin. North. Am.*, 13(4) (1982) 709.
4. A. H. Reddi, *Tissue Eng.*, 6(4) (2000) 351.

5. H. Lee, R. A. Cusick, F. Browne, T. H. Kim, P. X. Ma, H. Utsunomiya, R. Langer, and J. P. Vacanti, *Transplantation*, 73(10) (2002) 1589.

6. G. Wei, G. J. Pettway, L. K. McCauley, and P. X. Ma, *Biomaterials*, 25(2) (2004) 345.

7. K. M. Woo, V. J. Chen, and P. X. Ma, *J. Biomed. Mater. Res.*, 67A(2) (2003) 531.

8. A. Thapa, T. J. Webster, and K. M. Haberstroh, *J. Biomed. Mater. Res.*, 67A(4) (2003) 1374.

9. P. X. Ma and R. Zhang, *J. Biomed. Mater. Res.*, 46(1) (1999) 60.

10. R. W. Baker (Eds.), *Membrane Technology and Applications*, 2nd edition, Wiley, New York (2004).

11. R. Y. Zhang and P. X. Ma, *J. Biomed. Mater. Res.*, 52(2) (2000) 430.

12. A. Formhals, *Process and Apparatus for Preparing Artificial Threads*, US Patent 1975504 (1934).

13. F. Yang, R. Murugan, S. Wang, and S. Ramakrishna, *Biomaterials*, 26(15) (2005) 2603.

14. B. Duan, C. Dong, X. Yuan, and K. Yao, *J. Biomater. Sci. Polym. Ed.*, 15(6) (2004) 797.

15. Z. M. Huang, Y. Z. Zhang, S. Ramakrishna, and C. T. Lim, *Polymer*, 45(15) (2004) 5361.

16. Y. Zhang, H. Ouyang, C. T. Lim, S. Ramakrishna, and Z. M. Huang, *J. Biomed. Mater. Res.*, 72B(1) (2004) 156.

17. X. L. Zong, S. Li, E. Chen, B. Garlick, K. S. Kim, D. F. Fang, J. Chiu, T. Zimmerman, C. Brathwaite, B. S. Hsiao, and B. Chu, *Ann. Surg.*, 240(5) (2004) 910.

18. S. W. Choi, S. M. Jo, W. S. Lee, and Y. R. Kim, *Adv. Mater.*, 15(23) (2003) 2027.

19. K. L. Kim, K. Yen, C. Chang, D. F. Fang, B. S. Hsiao, B. Chu, and M. Hadjiargyrou, *J. Control. Release*, 98(1) (2004) 47.

20. Y. K. Luu, K. Kim, B. S. Hsiao, B. Chu, and M. Hadjiargyrou, *J. Control. Release*, 89(2) (2003) 341.

21. B. M. Min, L. Jeong, Y. S. Nam, J. M. Kim, J. Y. Kim, and W. H. Park, *Int. J. Biol. Macromol.*, 34(5) (2004) 223.

22. B. M. Min, G. Lee, S. H. Kim, Y. S. Nam, T. S. Lee, and W. H. Park, *Biomaterials*, 25(7–8) (2004) 1289.

23. H. J. Jin, S. V. Fridrikh, G. C. Rutledge, and D. L. Kaplan, *Biomacromolecules*, 3(6) (2002) 1233.

24. L. Yao, T. W. Haas, A. Guiseppi-Elie, G. L. Bowlin, D. G. Simpson, and G. E. Wnek, *Chem. Mater.*, 15(9) (2003) 1860.

25. C. A. Finch (Eds.), *Poly(vinyl alcohol): Properties and Applications*, Wiley, New York (1973).

26. D. Li and Y. N. Xia, *Adv. Mater.*, 16(14) (2004) 1151.

27. Z. M. Huang, Y. Z. Zhang, M. Kotaki and S. Ramakrishna, *Composites Sci. Technol.*, 63(15) (2003) 2223.

28. J. Y. Rho, L. Kuhn-Spearing, and P. Zioupos, *Med. Eng. Phys.*, 20(2) (1998) 92.

29. P. X. Ma, R. Y. Zhang, G. Z. Xiao, and R. Franceschi, *J. Biomed. Mater. Res.*, 54(2) (2001) 284.

30. C. M. Flahiff, A. S. Blackwell, J. M. Hollis, and D. S. Feldman, *J. Biomed. Mater. Res.*, 32(3) (1996) 419.

31. C. Verheyen, C. Klein, J. M. A. Deblieckhogervorst, J. G. C. Wolke, C. A. Vanblitterswijn, and K. Degroot, *J. Mater. Sci. Mater. Med.*, 4(1) (1993) 58.

32. R. Y. Zhang and P. X. Ma, *J. Biomed. Mater. Res.*, 44(4) (1999) 446.

33. S. L. Ishaug, G. M. Crane, M. J. Miller, A. W. Yasko, M. J. Yaszemski, and A. G. Mikos, *J. Biomed. Mater. Res.*, 36(1) (1997) 17.

34. K. M. Woo, R. Y. Zhang, H. Y. Deng, and P. X. Ma, *Trans. Soc. Biomater.*, 25 (2002) 605.

35. K. M. Woo, G. Wei, and P. X. Ma, *J. Bone Miner. Res.*, 17(Suppl. 1) (2002) M49.

36. C. Du, F. Z. Cui, Q. L. Feng, X. D. Zhu, and K. D. Groot, *J. Biomed. Mater. Res.*, 42 (1998) 540.
37. C. Du, F. Z. Cui, Q. L. Feng, X. D. Zhu, and K. D. Groot, *J. Biomed. Mater. Res.*, 44 (1999) 407.
38. G. B. Wei and P. X. Ma, *Biomaterials*, 25(19) (2004) 4749.
39. T. J. Webster, L. S. Schadler, R. W. Siegel, and R. Bizios, *Tissue Eng.*, 7(3) (2001) 291.
40. T. J. Webster, C. Ergun, R. H. Doremus, R. W. Siegel, and R. Bizios, *J. Biomed. Mater. Res.*, 51(3) (2000) 475.
41. R. Y. Zhang and P. X. Ma, *J. Biomed. Mater. Res.*, 45(4) (2000) 285.
42. R. Y. Zhang and P. X. Ma, *Macromol. Biosci.*, 4(2) (2004) 100.
43. G. Wei and P. X. Ma, *J. Biomed. Mater. Res.*, 78(2) (2006) 306.
44. A. Bigi, E. Boanini, S. Panzavolta, N. Roveri, and K. Rubini, *J. Biomed. Mater. Res.*, 59 (2002) 709.
45. Y. Zhang and M. Q. Zhang, *J. Biomed. Mater. Res.*, 55 (2001) 304.
46. J. E. Babensee, L. V. McIntire, and A. G. Mikos, *Pharm. Res.*, 17(5) (2000) 497.
47. A. H. Reddi, *Nat. Biotechnol.*, 16(3) (1998) 247.
48. V. H. L. Lee, *CRC Crit. Rev. Ther. Drug Carrier Syst.*, 5(2) (1998) 69.
49. R. Langer, *Science*, 249(4976) (1998) 1527.
50. P. Morley, J. F. Whitfield, and G. E. Willick, *Curr. Pharm. Des.*, 7(8) (2001) 671.
51. N. Ferrara and K. Alitalo, *Nat. Med.*, 5(12) (2001) 1359.
52. J. B. Oldham, L. Lu, X. Zhu, B. D. Porter, T. E. Hefferan, D. R. Larson, B. L. Currier, A. G. Mikos and M. J. Yaszemski, *J. Biomech. Eng. Trans. ASME*, 122(3) (2000) 289.
53. L. Lu, G. N. Stamatas, and A. G. Mikos, *J. Biomed. Mater. Res.*, 50(3) (2000) 440.
54. G. B. Wei, G. J. Pettway, L. K. McCauley, and P. X. Ma, *Biomaterials*, 25(2) (2000) 345.
55. G. Wei, Q. Jin, W. Giannobile, and P. X. Ma, *J. Control, Release,* 112(1) (2006) 103.
56. L. L. Hench and J. M. Polak, *Science*, 295(5557) (2002) 1014.
57. B. D. Boyan, T. W. Hummert, D. D. Dean, and Z. Schwartz, *Biomaterials*, 17(2) (1996) 137.
58. J. A. Hubbell, *Curr. Opin. Biotechnol.*, 10(2) (1999) 123.
59. P. X. Ma and R. Y. Zhang, *J. Biomed. Mater. Res.*, 46(1) (1999) 60.
60. Y. Ito, *Biomaterials*, 20(23–24) (1999) 2333.
61. J. M. Gao, L. Niklason, and R. Langer, *J. Biomed. Mater. Res.*, 42(3) (1998) 417.
62. J. A. Neff, K. D. Caldwell, and P. A. Tresco, *J. Biomed. Mater. Res.*, 40(4) (1998) 511.
63. M. Nitschke, G. Schmack, A. Janke, F. Simon, D. Pleul, and C. Werner, *J. Biomed. Mater. Res.*, 59(4) (1998) 632.
64. Y. H. Hu, S. R. Winn, I. Krajbich, and J. O. Hollinger, *J. Biomed. Mater. Res. A,* 64A(3) (1998) 583.

15
Cryotemperature Stages in Nanostructural Research

Robert P. Apkarian*

1. Introduction

Cryo high-resolution scanning electron microscopy (cryo-HRSEM) also called low temperature (LT)-HRSEM, is the most powerful tool for visualizing 3D topographic ultrastuctural features of solid-state solvated systems. A cryostage provides researchers with the ability to image their cryoimmobilized solvated system, usually aqueous, and image them in the solid-state at cold temperatures with high magnification (5×10^4 to 10^6), such that structural information down to a single nanometer can be obtained. LT-HRSEM is particularly a useful imaging mode for soft materials such as cells, biomolecules, biomaterials, and organic systems. Hydrogels, colloids, suspensions, emulsions, and whole cells all have ultrastructural components that can be imaged on a nanometer scale when the specimens have been LT cooled and transferred onto a cryostage, and then into the HRSEM. Numerous advantages result from low-temperature imaging provided that the sample is prepared as small as possible and kept frost-free. Additionally, cryo-HRSEM is ideally suited for imaging solid structures in nonaqueous or organic solvents such as toluene and TFE, or in organic fluids such as octanol.

Low-temperature "bulk" specimens are prepared for cryo-HRSEM in microliter volume sample sizes. Such samples, in both nonaqueous and aqueous systems, provide vast vistas for acquisition of nanometer size ultrastructural information at high magnification. For instance, a representative aqueous aliquot is prepared that can contain thousands of cells or biomolecular complexes. LT-HRSEM images of nanoparticles can be obtained within the context of complex biological compartments. There is no need to prepare larger samples because the greater mass of a larger specimen will serve only to reduce the cooling rate and result in unwanted crystallized systems.

Dr. Robert P. Apkarian, an outstanding scientist in cryo-SEM and related research, unexpectedly passed away before this book was published. His contribution to this book and the field will be memorialized forever.

Since secondary electron microscopy is a surface imaging electron optical mode, relief of the solid components in a bulk LT aqueous sample is sometimes desirable or necessary. Solid water may be removed by high vacuum (10^{-7} Torr) sublimation. This method is referred to as "etching." A detailed discussion of low temperature preparation follows but the reader should realize that practitioners of LT-HRSEM take great efforts to remove the bulk LT aqueous phase without removal of the hydration shell surrounding molecular structures observed in the nanometer range. Removal of molecular hydration shell is the process of freeze-drying, which is not the aim of LT-HRSEM on bulk specimens. Alternatively, raising the temperature to $-85°C$ under vacuum of 10^{-7} Torr to freeze-dry LT samples has been successful for 1–10 nm structural resolution of thin biomolecular specimens mounted on grids and then imaged by LT-HRSEM [1,2]. It is precisely the HRSEM's ability to image surfaces with topographic contrast on relatively large bulk LT samples, as compared to small cryo-TEM samples, that facilitates its expanded applications in nanostructural studies.

Cryo-HRSEM, just as with the cryo-TEM, can be used to examine single isolated biomolecules and macromolecular complexes (single particle analysis), which are considered as "thin" specimens. Thin specimens are cryoimmobilized by vitrification in a 100 nm thick aqueous layer on carbon-coated grids. The topic of vitrification will be covered in the section dealing with cryoterminology. Conveniently, a thin vitrified specimen can be cryo transferred into a SEM fitted with a scanning transmission electron microscopy (STEM) detector and tandem cryo-HRSEM/cryo-STEM can be performed. It is now appropriate to provide definitions used in cryogenic preparation and imaging.

2. Terminology used in Cryo-HRSEM of Aqueous Systems

Most of all, cryo-HRSEM studies are done in the aqueous phase because all living organisms, cells and molecular components, are in water and for the most part water is the solvent of life. Therefore for the cell biologist-life scientist, cryo-HRSEM is done in a hydrated phase. Practitioners of cryo-TEM and cryo-HRSEM seek to cool their samples without the formation of ice crystals. Logic dictates that the larger the mass, the slower the cooling rate. Specimen mass is the reason why very small samples are prepared. A discussion on specimen cooling is included later in this chapter. When a sample is *slowly frozen* hexagonal ice crystals are formed that usually distort the specimen and make for specimen distortion and erroneous imaging. When a sample is *fast frozen*, minute cuboidal ice crystals (<30 nm), may form however, cubic ice crystals are small and in general do not cause damage to the biological nanostructure [3]. If the term *freezing a sample* is used, it infers that the sample was cooled at a rate slow enough to allow crystal growth and the sample is in *ice*. Ice is crystalline. Conversely if the phrase, *the sample is in ice*, is used, that infers that the sample was frozen at a rate slow enough to allow crystal growth. If one uses the literary term *freezing* a sample, *crystalline ice* is formed.

Three decades ago, it was shown that if thin samples (biomolecules, ribosomes, viruses, and small bacteria) were spread on a TEM grid and the water was wicked away with filter paper, a very thin water layer (~100 nm thick) with the suspended specimen could be *rapidly cooled* by plunging in liquid ethane at its melting temperature (−183.3°C) [3]. The sample is said to be *vitrified*. This means that the aqueous system was cooled at an *ultrafast rate* such that the water in the sample did not have time to nucleate and crystallize. Therefore the sample is said to be in *vitreous water* and not *ice*. Literature has used the term *vitreous ice* for years but efforts are underway to employ the term *vitreous water* to describe the noncrystalline, LT aqueous matrix (J. Dubochet, 2005, personal communication). This infers that the sample was cooled at an appropriately fast rate to avoid *ice* crystal growth. Remember *ice* is crystalline. Vitreous water can be properly maintained in the amorphous or glass-like state by equilibrating the cryostage at maintenance temperature between −120°C and −135°C. Vitrification is the basis for all cryo-TEM preparations and its amorphic nature is verified by electron diffraction analysis, for a complete review see [3]. To say that specimen nanostructure was preserved in *vitreous water* is to infer the image recordings were taken at LT using a cryostage and that electron diffraction had conferred the water was not crystalline.

Unfortunately electron diffraction cannot be performed on *bulk* samples prepared for cryo-HRSEM and therefore absolute proof of the aqueous samples *vitrification* is subject to question. Empirical cryo-HRSEM recordings at high magnification may have no visible evidence of crystallinity, even down to a single nanometer, yet the possibility exists, though unlikely, that the entire specimen is embedded in one large hexagonal crystal. Therefore, special attention to the specific thermodynamics of the *bulk* sample system and the method of cooling must be considered before the claim is made that the sample is *vitrified*. For the purposes of this discussion, I will refer to bulk aqueous and nonaqueous samples prepared for cryo-HRSEM as *LT samples* and most often the cooling method and imaging temperature will be stated. If a method of ultra rapid cooling was used for preparation of a LT-HRSEM specimen and the observed surface appears devoid of large or small ice crystals, I will state that the solid nanometer-sized structures of the sample were embedded in a *LT aqueous phase*. This disclaimer neither states that the LT aqueous sample was *vitrified* nor that crystalline ice was formed. Rather use of the phrase "the sample was observed in the *LT aqueous phase*" infers that the sample appeared to have been appropriately cryoimmobilized such that there was no observable crystallinity or sample distortion. In this fashion quality structural information in the nanometer range may be gathered by cryo-HRSEM recordings.

3. Liquid Water, Ice, and Vitrified Water

No discussion of cooling or freezing of aqueous systems for cryoelectron microscopy can be reasonable without consideration of water and *ice* chemistry. Therefore, a brief discussion of water interaction with biomolecules and LT preparation follows. Hydrogen bonding occurs in a tetrahedral arrangement around a water molecule in ice (Fig. 15.1). Due to the dipole nature of water, strong electrostatic attraction, hydrogen bonding occurs between water molecules. It is because of the nearly tetrahedral arrangement of electrons around the oxygen atom that each water molecule tends to bond to four neighboring water molecules. Cubic and the common hexagonal ice are nearly the same. A water molecule is attached at the summit of the tetrahedron upon cooling, and then the extended structure forms a space filling hexagonal lattice of nearly undistorted hydrogen bonds holding water molecules apart, and explaining why the hexagonal ice crystal is less dense than the liquid. Vitreous water has the same density as cuboidal or hexagonal ice [3]. A vitreous sample on a cold stage in the electron microscope would not *devitrify* and become cuboidal or hexagonal ice at temperatures colder

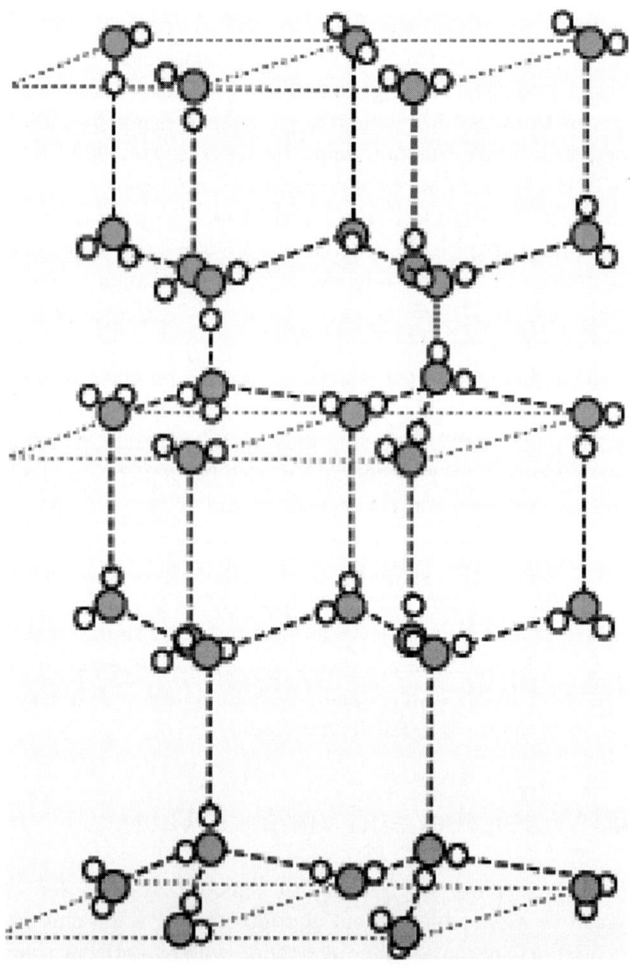

FIGURE 15.1. Hydrogen bonding between adjacent water molecules in ice.

than $-135°C$. However, in the microscope it seems that $-120°C$ is a more useful vitreous to cubic transition temperature.

Let us consider the "cooperative" hydrogen bonding amongst adjacent peptide chains or amongst base pairs in nucleic acids (Fig. 15.2). These biomolecules are amazingly stable in water and their native structure is preserved by their interaction with the water dipole. Since nearly all biological molecules are hydrophilic, even composite proteins and lipid bilayers that have hydrophobic cores and hydrophilic surfaces, then water molecules will orient according to the local surface charge of the macromolecule to form the "hydration shell." Only water molecules contact the macromolecule while media solutes do not penetrate down the surface of the biological matter.

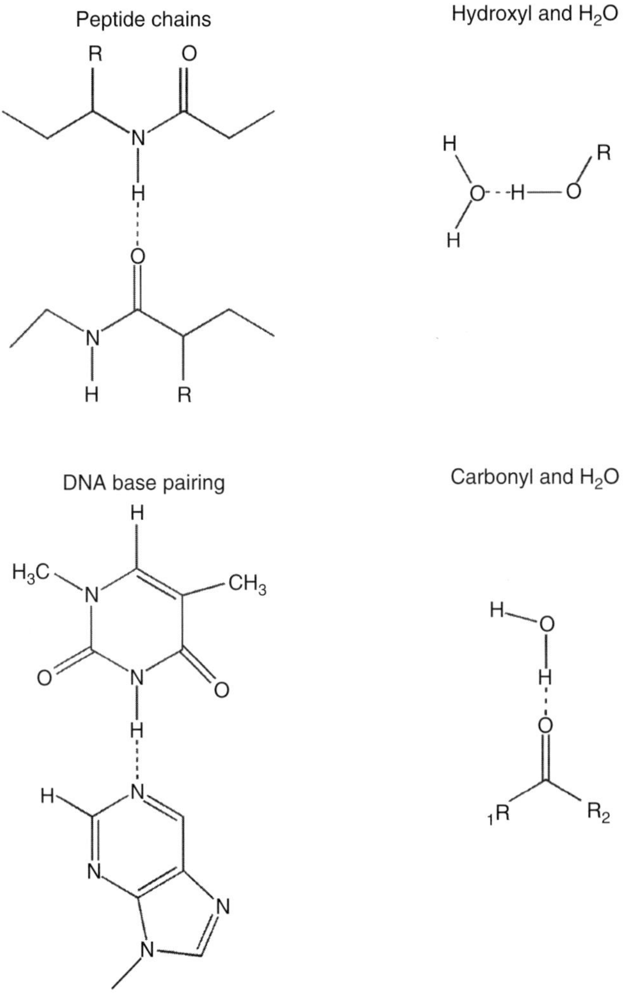

FIGURE 15.2. Hydrogen bonds between some biological molecules.

Ionic solutes such as NaCl change the structure of liquid water because each ion is surrounded by a shell of water dipoles forming more ordered regular structures than hydrogen bonded water molecules. A solute therefore will change the solvent (water) by lowering the freezing point, raising the boiling point, and depressing vapor pressure and also giving the solution an osmotic pressure property.

It is advisable, when planning a cryo-HRSEM study of a bulk specimen such as a microliter size suspension of biomolecules or cells, to consider the hydrogen bonding within bulk water, cooperative hydrogen bonding within biomolecules, and hydrogen bonding surrounding hydrated ions. The concentrations of both ionic salts and biomolecules will greatly influence water structure and therefore its morphology upon low temperature preparation for cryo-HRSEM. Requirements for vitrification of aqueous biological suspensions are not prohibitive because solutes serve as its own natural *cryoprotectant*. Therefore, quite satisfactory results of cells, biomolecules, hydrogels, and high concentration solute suspensions can be routinely obtained on a nanometer scale. Intriguing results of salts and lipid surfactants at relatively lower concentrations have been attained by cryo-HRSEM and efforts to understand their characteristic patterns in the LT aqueous phase are actively being investigated [4].

4. History of Low-Temperature SEM

Many fields of science and engineering have made use of below-lens cryostages on conventional SEMs for nearly four decades [5]. Historically, clays, cements, emulsions, botanical tissue, micro-organisms and insect specimen surfaces have been cryoimmobilized and observed at low temperature, providing microstructural morphologies. In this chapter, we will explore the methods and applications of cryogenic high-resolution SEM for the nanostructural analysis of chemical and biological systems.

As scanning electron microscopes evolved during the 1980s and 1990s, with Schottky and cold cathode field emission (FE) electron sources and analytical immersion lens designs structural biologists advanced the useful range of resolution down to the tens of nanometers. Using a conventional lens system cold cathode FESEM (Hitachi S-800), 10–15 nm features of intramembrane particles (IMP) were observed on the protoplasmic face of frozen fractured yeast cells *Saccharomyces cerevisiae*. Since high atomic number large-grain metal films (Pt and Au) were applied, increased electron scattering took place and significant amounts of BSE/SE-III electrons were generated and collected, thus diffusing secondary electron localization and useful structural feature recognition to >10 nm [6].

By the early 1990s Hermann and Mueller were experimenting with TEM-type in-lens cryostages known for there great stability in cryo-TEM studies [1]. On a quest to provide nanostructural morphologies approaching the limit of instrumental resolution (~1 nm), cryostages immersed in-lens, would support collection of highly localized specimen specific SE-I and SE-II signal components (see Chapter 1) by an above-lens SE detector. Yeast cells had long been the yardstick

(a)   (b)

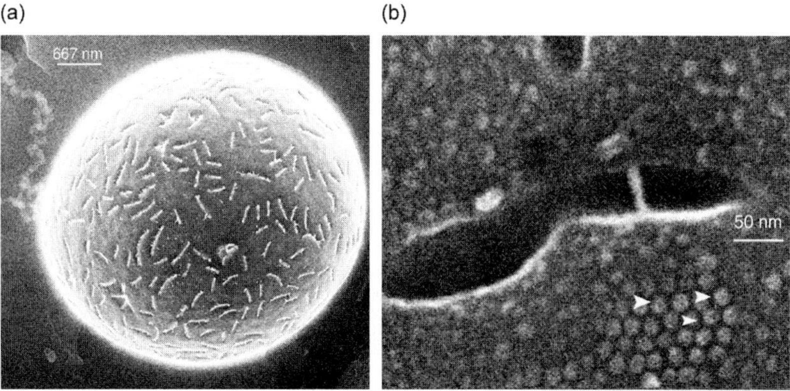

FIGURE 15.3. (a) Cryo-HRSEM images of *Saccharomyces cerevisae* (bakers' yeast) and (b) Note the central dark pixels on the 11 nm intramembrane particles (*arrows*).

by which good freeze fracture platinum replicas were judged for TEM [7]. It was determined that cryopreparation in high vacuum systems $>10^{-6}$ Torr were necessary to reduce the presence of water vapor known to condense and contaminate cryospecimens at lesser vacuums. Quality platinum replicas were produced which showed that the IMPs of yeast membrane had a surface depression resembling a volcano cone with dimensions of a few nanometers. Employing an in-lens cold cathode FESEM (Hitachi S-900) fitted with cryostage, Hermann and Muller observed bulk preparations of frozen fractured yeast membranes coated with a 1 nm W, Pt, or Cr film. IMPs were clearly visible on which some particles appeared volcano-like. By the mid-1990s, TEM-type in-lens cryostages had been developed for a Schottky FESEM/STEM (Topcon DS-130F) [8]. Dark central pixels on Cr-coated yeast membrane IMPs created the appearance of volcano-like structures (Fig. 15.3) [9]. Resolution of the volcano-like IMPs achieved by these two groups demonstrated that cryo-HRSEM provided <5 nm structural resolution and could be directly compared to the resolution achieved by the TEM Pt-replica technique. Yet surpassing the limited areas covered by a Pt-replica for TEM, cryo-HRSEM images provided molecular level resolution from hundreds of cells in a single specimen.

5. Instrumentation and Methods

5.1. In-Lens Cryo-HRSEM

Similar to the TEM-type cryostage used in the S-900 FESEM, the cryostage employed in the Topcon DS-130 FESEM has to be transferred through the atmosphere from the cryoworkstation, where the specimen was fractured, to the coater and then from the coater to the goniometer of the microscope (Fig. 15.4). In each

(a)   (b)

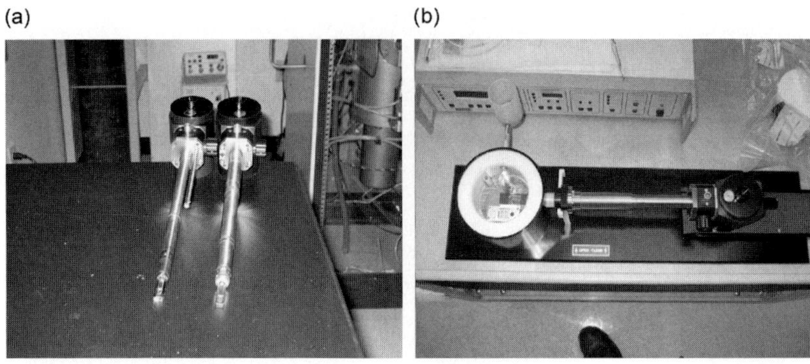

FIGURE 15.4. (a) Stage on right, GatanCT-3500 cryoholder for JEOL 1210 TEM—Identical holder on left for In-lens DS-130 FESEM/STEM. (b) Cryoworkstation with temperature controller and cryostage.

atmospheric transfer, the risk of nanometer-sized ice condensation, from and to each station could result in the artifact of false particulate features. To prevent against this rime frost, modern cryostages like the ones used on the DS-130F are fitted with dual shutters above and below the specimen. Rapid transfer and cryolabs with low relative humidity (20%) are desirable conditions for quality cryospecimen preparation and imaging.

Standard procedure involves loading a LT specimen planchet onto the stage under LN_2 and then fracturing the specimen with a precooled knife to produce a clean surface. Grids supporting vitrified thin specimens are mounted on the cryostage in the same fashion but without fracturing. Although there is vast literature dealing with a multitude of cooling methods, cryogen plunging in liquefied ethane and high pressure freezing are most often employed by this author, see [10] for a complete review of cooling methods. It is advisable to produce the shallowest fracture, usually just a scraping of the cold-knife across the specimen surface. The natural sample surface is in intimate contact with the cryogen during plunging and has the greatest cooling rate and is most likely to be crystalline free.

Stand-alone cryostages and preparation systems such as the one pictured in Figure 15.4b have been fitted to a port on a dedicated high vacuum system for sputter deposition of an ultrathin fine grain metal film such as Cr, Ta, Ti, or W onto the LT specimen [11]. Due to their proximity to one another, the cryostage with shutters closed, is withdrawn from the LN_2 reservoir of the workstation and docked into the coater port simultaneously activating the evacuation of the chamber usually performed in 1–2 s (Fig. 15.5). A background pressure of 1×10^{-7} Torr is achieved in less than 10 min such that the cryostage never drops below −180°C in the time it takes for the stage to be withdrawn from the LN_2 cryoworkstation and high vacuum is achieved in the coating chamber. Even during Ar purging into the chamber, the specimen temperature never varies more than 1°. Since sputter deposition does not generate significant thermal radiation then no temperature variation occurs during metal deposition onto the cryosample. Once the cryosample has

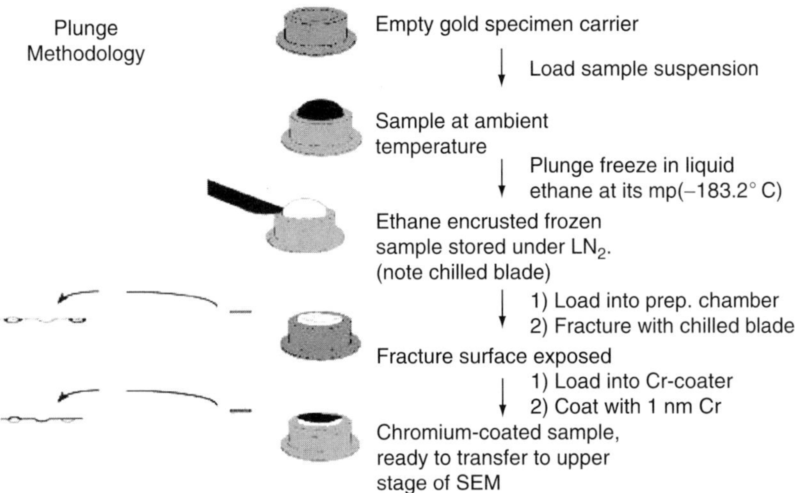

FIGURE 15.5. Rapid transfer of the cryostage from the cryoworkstation (*left*) and into the sputter coating chamber (*right*) is performed in a room with 22% relative humidity.

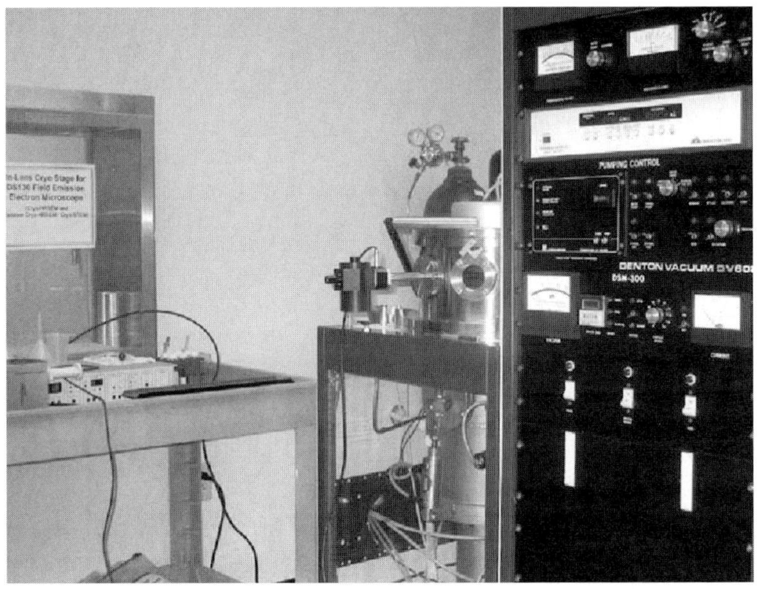

FIGURE 15.6. The cryostage is mounted in-lens of a Topcon DS-130FESEM/STEM.

been sputter coated with a 1–2 nm thick metal film (Cr), the chamber is vented to atmosphere with dry nitrogen gas, the cryostage is removed and shuttled to the goniometer of the in-lens stage on the FESEM. Column pressure at the specimen is 1×10^{-7} Torr and anticontamination cold fingers are essential (Fig.15.6).

The in-lens FESEM pictured in Fig. 15.6 also contains a scanning transmission detector that extends the utility for nanostructure imaging at LT in a FESEM. Biomolecules and macromolecular complexes vitrified on grids may be imaged in both HRSEM and STEM modes.

5.2. Near-Lens Cryo-HRSEM

Just as with the earlier conventional cryo-SEM, cryo-HRSEM has an ever-expanding role in biology, medicine, chemistry, and bio- and inorganic materials analysis. Now by careful specimen preparation and HRSEM imaging methods described in this chapter, cryogenic nanotechnology seeks to describe features of a few nanometers on isolated macromolecules or in the context of complex compartmental organization. Although distinctions exist between "near-lens" and "in-lens" FESEMs fitted with above-lens detectors (Chapter 1) comparable structures can be attained by these two similar FESEM designs. Since near-lens designs are very popular in current model FESEMs (FEI: Quanta FEG ESEM, Hitachi 4700, 4800; JEOL 6700, 7000, and 7401) (Fig.15.7). The following methods and discussion can be applicable to these microscopes as well. These microscopes can be fitted with a cryoworkstation such as the Gatan CT-2500 (Fig. 15.8). This type of cryosystem has its own vacuum system (usually turbopumped) separate from the microscope and allows the investigator to make one atmospheric transfer of the LT specimen from the cooling device into the cryochamber. The cryochamber is permanently installed on the large specimen chamber at the bottom of the electron optical column. A properly cooled sample is loaded into the cryochamber and evacuated into the 10^{-6} Torr range. An external rod attached to a cryoknife is used to fracture the specimen prior to Ar purging and metal film sputter coating. Once the specimen is prepared in the cryochamber a valve is opened through

FIGURE 15.7. A JEOL 7401 near-lens FESEM is designed to accommodate the attachment of a cryochamber.

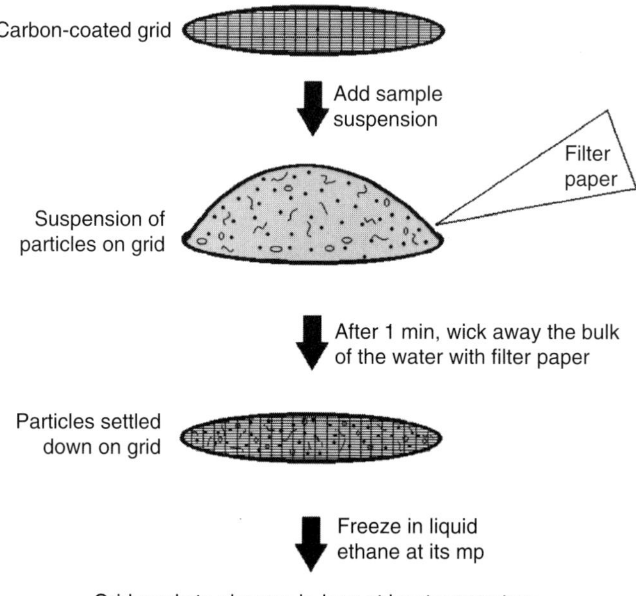

FIGURE 15.8. A Gatan CT-2500 cryochamber and temperature controller provides the necessary workstation for LT specimen mounting, fracturing, and metal film sputter coating. The valve on the right of the chamber gates the specimen into the FESEM chamber. Not pictured are the Ar gas tank and lines, and vacuum pump system.

which the LT specimen is inserted into the microscope chamber. Working distance is set close to the lens (1–5 mm) and the specimen temperature is equilibrated prior to HRSEM imaging at LT.

5.3. Specimen Carriers Used in Low-Temperature Scanned Cryoimaging

Depending on the nature of the specimen a variety of planchets and grids are commercially available for mounting low temperature samples (see Figs. 15.9 and 15.10). Copper mesh grids with carbon support film, lacey carbon film, or quantifoil films with various hole diameters are commercially available and will support a thin aqueous film containing filamentous or particulate macromolecules, viruses, small bacteria or platelets for vitrification (EMS, Washington, PA). A variety of gold and aluminum planchets are available through RMC products (Tucson, AZ).

The following flow chart describes four distinct methods developed for LT-HRSEM employing the same TEM type cryostage (Fig. 15.11).

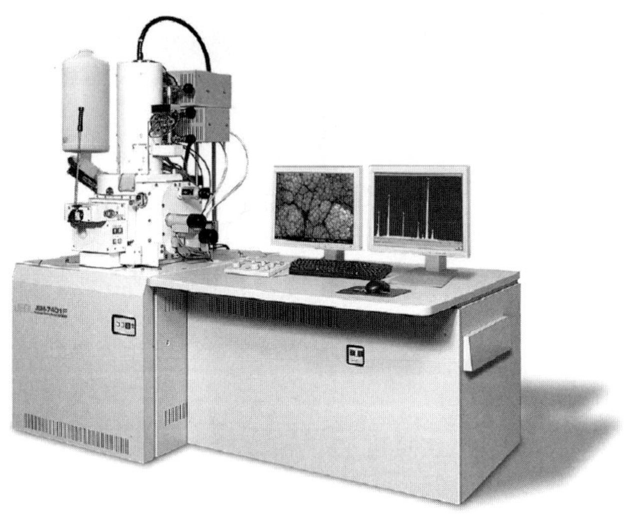

FIGURE 15.9. Plunge cooling methodology for bulk sample LTHRSEM.

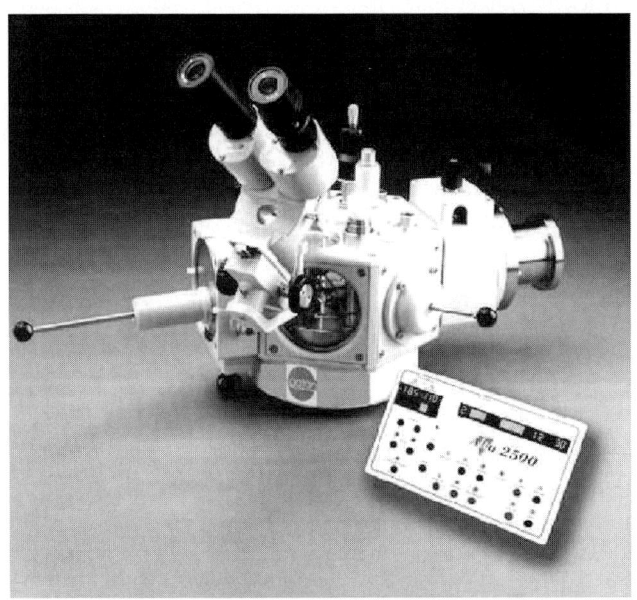

FIGURE 15.10. Thin specimens dispersed from the aqueous phase or organic solvents onto carbon support grids for vitrification in liquid ethane.

5.3.1. Bulk LT Fractured Yeast Cells

Images of bulk LT fractured yeast cells (Fig. 15.3) were obtained by placing a ~5 μL drop onto a planchet and plunging into liquid ethane. Bulk specimens of

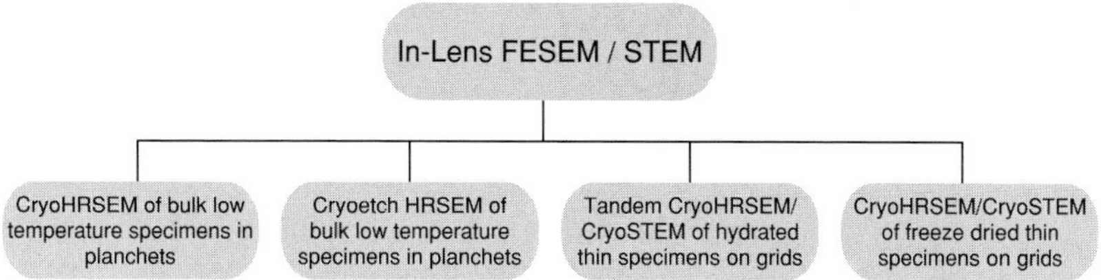

FIGURE 15.11. Flow chart of LT-HRSEM imaging methods.

proteins, nucleic acids, lipid surfactants, extruded phospholipids vesicles, several types of hydrogels and other similarly hydrated specimens can be cryoimmobilized in planchets.

Cryo-HRSEM of bulk low temperature cell suspensions providing molecular resolution in the nanometer range was demonstrated in Fig. 15.3. Low temperature stages have successfully provided nanostructural analysis of macromolecular complexes of lipid, nucleic acid, and protein suspensions in their native hydrated state. Homogeneous suspensions of genetically engineered proteins are routinely characterized from microliter suspensions cryoimmobilized on planchets by plunging into liquid ethane. In this simplified preparation illustrated in Fig. 15.9, a fracture of the surface exposes protein fibrils of elastins, elastin-silks, and potentially any other fibrilular proteins of similar purity. Phospholipids vesicle suspensions, made by 100 nm membrane extrusion, may also be plunge cooled in liquid ethane and cold fractured under LN_2 frequently resulting in the pitted-olive profiles (Fig. 15.12). Transfer of the LN_2 submerged cryorod from the cryoworkstation to the stand-alone high vacuum coater is done in 1–2 s with dual-stage shutters closed to protect the freshly cleaved specimen surface from rime frost condensation.

Resultant transfers provide a clean specimen surface for deposition of an ultrathin 1–2 nm fine grain metal film (see Chapter 1). The specimen temperature during transfer and coating never exceeds −176°C during the coating process. The combination of cold temperature staging and ultrathin film coating optimizes the quality nanostructural details of the sample. The film coating serves several purposes for quality high-resolution imaging. As described in Chapter 1, low mobility fine grain films accurately contour the specimen surface and enrich the SE-I signal. Additionally it serves to seal the surface from entrance of condensed ice crystals that may be produced anytime during transfer or on the microscope stage. These films reduce electron beam irradiation and sublimation of hydrocarbon-based biological specimens.

Cryo-HRSEM analysis was made of a genetically engineered elastin-mimetic hydrogels equilibrated in its collapsed high temperature state (40°C) prior to cryoimmobilization in liquid ethane (−183°C) (Fig. 15.13). Prior to image recording, the cryostage is slowly raised in temperature to −120°C in the microscope and

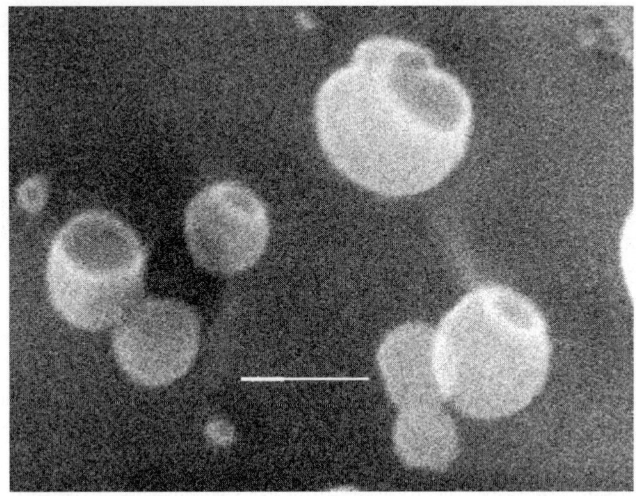

FIGURE 15.12. A pitted-olive appearance characterized an extruded unilamellar vesicle of POPC and cholesterol (3:1) bar = 100 nm [19].

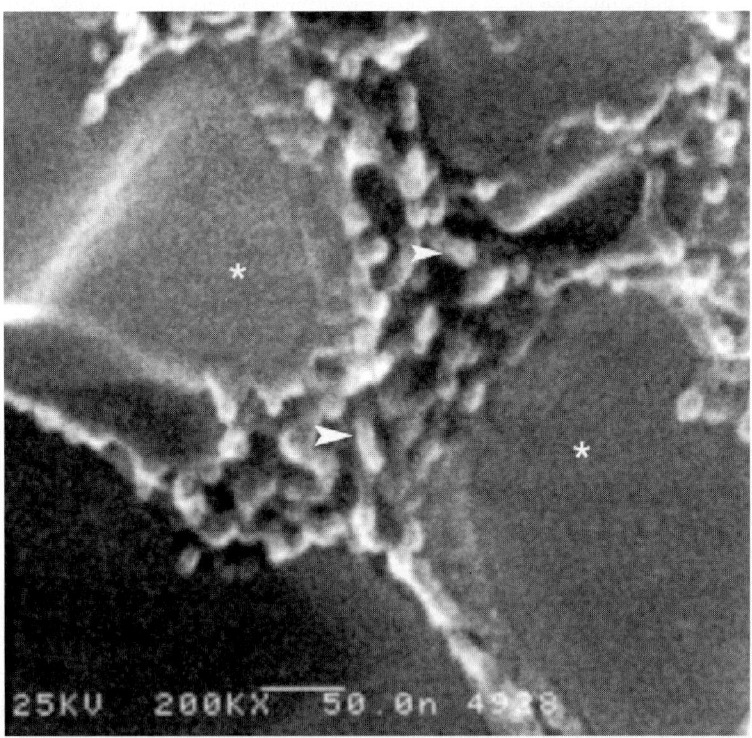

FIGURE 15.13. A high-temperature elastin-mimetic specimen in its collapsed state is characterized by 7 nm diameter filamentous proteins (*arrows*) segregated within bulk low-temperature water (*) [20].

allowed to equilibrate for 30 min. Note the 7 nm fibrils that mimic the diameter of native human elastin. A clear 2 nm wide shoreline is seen demarking the phase separated protein from bulk low temperature water. Several other proteins were similarly characterized with LT-HRSEM. [12,13].

5.3.2. Etching Bulk Samples for Cryo-HRSEM

Etching bulk samples for cryo-HRSEM has proved to be very useful for discerning the nanostructure of lipid, protein, and acrylamide hydrogels as well as particulate and fibrous suspensions. Most biological gels are based on fibers that interact locally and form a sponge-like solid. In the case of biological molecules that do not interact with one another, particles are usually highly mobile globular structures in a "sol" rather than fibrous. Gels from fibrous molecules can form from concentrations as low as 1% (agar) whereas globular molecules require 20–50% concentrations [14]. From these basic concepts and in the context of the previous discussion on hydrogen bonding and formation of hydration shells, protein hydrogels are engineered and extensively exploited for biomaterials and biomedical devices [15]. Illustrations of biomaterial processing of these hydrogels follow.

Hydrogels, and gels formed in organic solvents, can be rapidly cryoimmobilized by loading them onto planchets and plunge cooling them in ethane. Once the low temperature bulk hydrogel specimen in the planchet is loaded onto a cryostage it is transferred into a high vacuum system, either in a stand-alone coating device (Fig. 15.5) or in a dedicated cryochamber (Fig.15.8) attached to a near-lens type FESEM. The etching process is used to remove the bulk low-temperature aqueous phase in order to visualize the solid filamentous components. Etching is temperature and vacuum dependant, such that the rate at which the low-temperature water is removed from the hydrogel or suspension is a function of the vacuum the system can attain. From our earlier discussion of cubic ice formation during the cooling process, nanometer-sized cubic ice crystallization can now be appreciated in the context of the etching process. It will be sublimed away. In a vacuum of 10^{-7} Torr it is estimated that 100Å of low-temperature water is sublimed out of a specimen maintained at $-105°C$. The rate of etching is logarithmic such that at $-110°C$ there is an order of magnitude less sublimation of the low-temperature aqueous phase. Etching due to the nature of each specimen and its hydration shell, as well as its aqueous concentration and ionic environment, the rate of sublimation "etching-time" must be experimentally determined [12].

Elastin-mimetic protein hydrogels from 1 to 25 wt% have been characterized by LT-HRSEM. A 25 wt% elastin-mimetic hydrogel can have complete removal of bulk and loosely bound low-temperature water by high vacuum etching. Even if the low temperature aqueous phase contained cuboidal ice rather than vitreous water the etching would remove all but the bound water without distortion. Figure 15.14 shows a 25 wt% hydrogel. At low magnification, 1,000 μm^3 volume can be observed. At intermediate magnification a very uniform protein scaffold with uniform porosity can easily be observed in 3D. At high magnifications, greater than

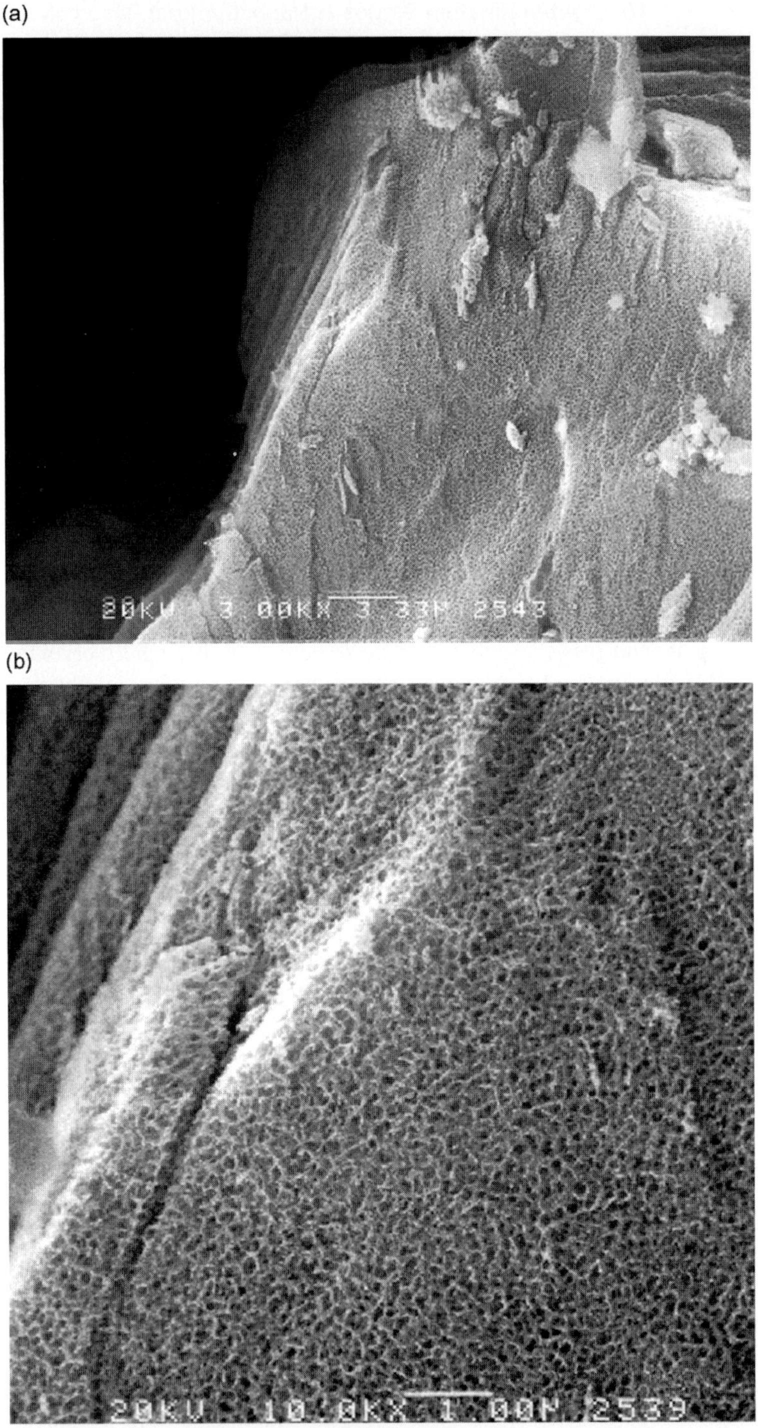

FIGURE 15.14. (a) Reveals the solid components of a cryoetched protein hydrogel surface representing ~1,000 μm^3 volume specimen imaged at low temperature (−120°C). (b) Attests to the quality preservation of a low-temperature protein hydrogel in which homogeneous pore sizes can be observed in 3D.

(c)

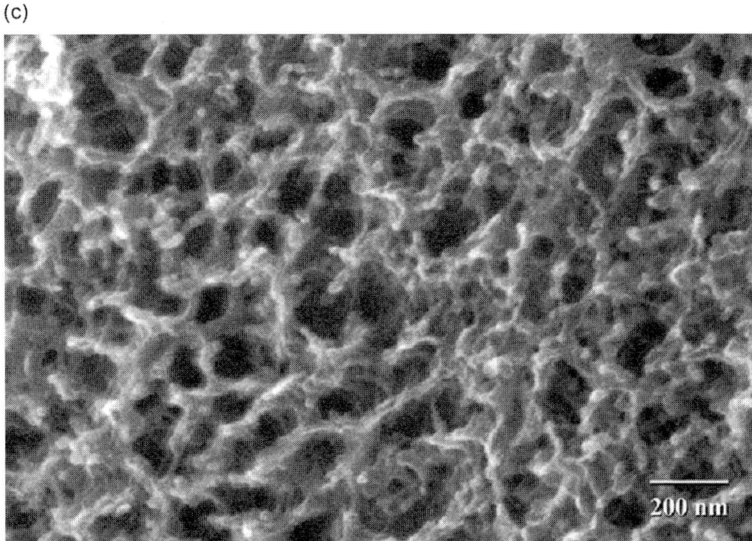

FIGURE 15.14. (*Continued*) (c) Is a high magnification cryoetch HRSEM stereo-image of the same hydrogel, 7 nm filaments and 20 nm nodes compose pore walls in the triblock copolymer. Note there is no trace of the bulk low-temperature aqueous phase.

$10^5\times$, 20 nm diameter protein fibrils were observed to intertwine to form the walls of the hydrogel pores. It appeared that the proteins maintain a hydration shell during etching as opposed to freeze-dried material. We have observed evidence of partial hydrogel collapse when etching times were overextended and contribute this to stripping away of the hydration shell.

A variety of hydrogels have been studied by cryoetch LT-HRSEM, triblock copolymer protein hydrogels have been subsequently processed into biomaterials and then re-characterized by LT-HRSEM. The same copolymer previously studied at 25 wt% has now been recast in water at 10 wt% and rehydrated in phosphate buffered saline (Fig. 15.15). Filamentous proteins can form hydrogels even at 1 wt% and cryoetch LT-HRSEM can be performed to assess their molecular morphologies for biomaterial applications (Fig. 15.16). Since the hydrogel is 99% water then etching the low temperature aqueous phase is done only to relieve the 1% solid component for detail recognition.

Cryoetch-HRSEM has been particularly useful for studies of protein self-assembly. Aβ peptide of Alzheimer's protein (13–21) self-assembles into nanofibrils that mimic and model amyloid disease progression [16]. Fibrils 20 nm in diameter characterize Aβ 13–21 protein while particulate features of the same dimension embedded in the remaining bulk LT-water may represent peptide seeds (Fig. 15.17a). The addition of sulfate ions to an acetylated Aβ 16–22 amyloid sequence has been suggested to effect the aggregation of these amyloid nanotubes (Fig. 15.17b).

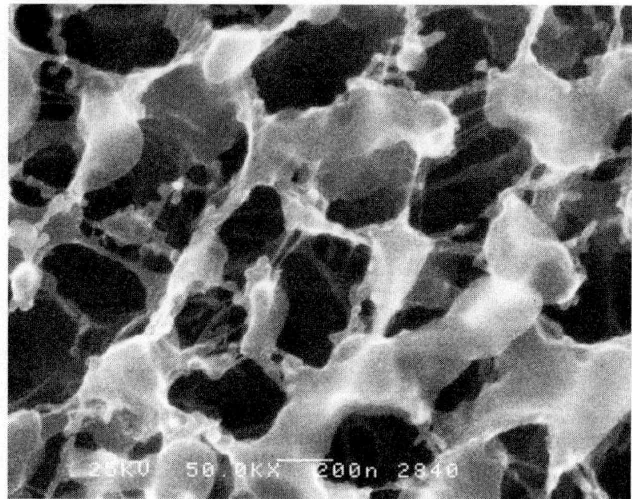

FIGURE 15.15. At the same magnification as Fig. 15.14c the pore dimensions of this 10% water-cast triblock elastin are larger reflecting the lower weight percent of the protein. Note the large cloud-like areas through out the sample. This is due to the incomplete removal of the low-temperature aqueous phase after the etching process. Also note that the ~7 nm protein filaments and 20 nm nodules are still preserved.

(a)   (b)

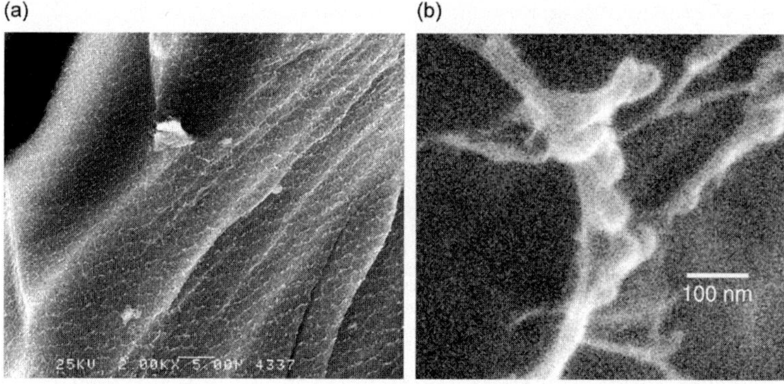

FIGURE 15.16. The 99% aqueous elastin hydrogel contains pore dimensions with greater diameters than those in the 10% hydrogel due to the higher water content, the low-temperature aqueous phase remains within the pore cavities after comparable etching times used at the higher % wt concentrations. Filamentous (~7 nm) and triblock nodules still compose the proteinaceous walls of the hydrogel pores.

5.3.3. Tandem Recordings

Tandem recordings of scanning transmitted signal (STEM) and the high-resolution secondary electron signal (HRSEM) acquired from low-temperature hydrated macromolecules and cells have been developed to correlate morphologies [17,18].

(a) (b)

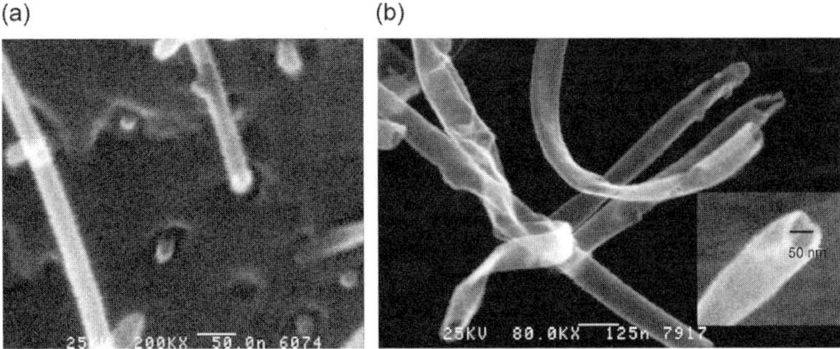

FIGURE 15.17. (a) Aβ13–21 peptide formed 20 nm wide fiber and particulate structures. The background is the bulk LT aqueous phase. (b) Aβ 16–22 peptide form tubelets and tubes. *Inset* shows 50 nm inner diameter and ~6.5 nm wall thickness.

Molecular suspensions dispersed on grids, and blotted with filter paper are plunged into melted ethane in the same fashion as vitrified specimens for cryo-TEM but with the additional preparation step of LT-sample coating with a 1 nm sputtered film of chromium. Although electron diffraction was not available to confirm vitrification, rare crystalline profiles appeared in these samples. Provided that filter paper blotting was sufficient to remove the water from the surface of the molecules prior to plunging, SE-I contrasts produced LT-HRSEM images containing topographic information in the 2–10 nm range. Immediately after the SE-I image was recorded, the STEM image was recorded on the next scan so that beam irradiation was kept at a minimum. The cryo-STEM image revealed accurate molecular dimensions of protein features (Fig. 15.18).

(a) (b)

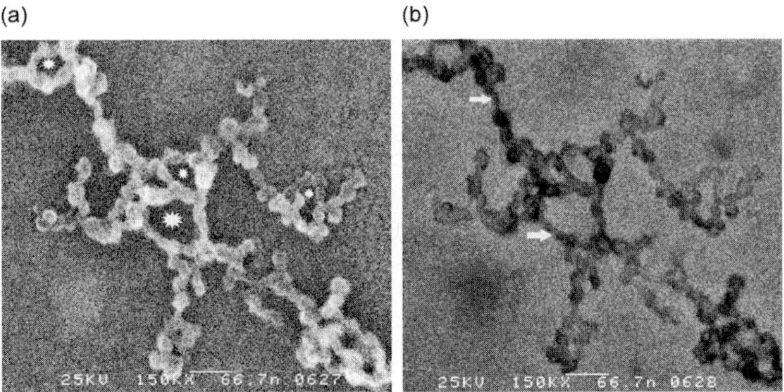

FIGURE 15.18. (a) Cryo-HRSEM image of genetically engineered elastin protein provide topographic contrasts of loop structures (*asterisks*) and (b) Tandem cryo-STEM image of the same protein complex as in (a), provides adequate resolution for filament width measurements (*arrows*).

5.3.4. Low Temperature-HRSEM

Low Temperature-HRSEM can also be of great utility for imaging freeze-dried specimens on grids. As with previous suspension preparations on grids, a drop is placed on a carbon-coated grid, allowed to adhere, blotted nearly dry with filter paper and plunge cooled in liquefied ethane. The grid is transferred onto the cryostage under LN_2 and then the stage is transferred into the high-vacuum chromium coater. Once a vacuum of $\leq 4 \times 10^{-7}$ Torr is achieved the stage shutters are opened and the stage temperature is raised to -85°C so that the sample is freeze-dried. This process is different than cryoetching in that the hydration shell around the organic molecules is effectively stripped off thus drying the sample. The cryostage is returned to -180°C and the specimen is sputter coated with a 1 nm thick continuous Cr film prior to imaging at -180°C in the FESEM. Conversely the cryoetch method described in Section 5.3.2 employs a bulk aqueous suspension with a much greater water content than on a blotted grid. Therefore care is taken during cryoetching preparation at -105°C to sublime away only the bulk and loosely bound LT water so as not to strip away the hydration shell. Synthetic α-helical proteins that are designed to form cable like conduits are made of fine nanometer scale sequences. To provide accurate imaging without chemical cross-linking or alcohol dehydration these proteins have been freeze-dried and observed at low temperature (Fig. 15.19).

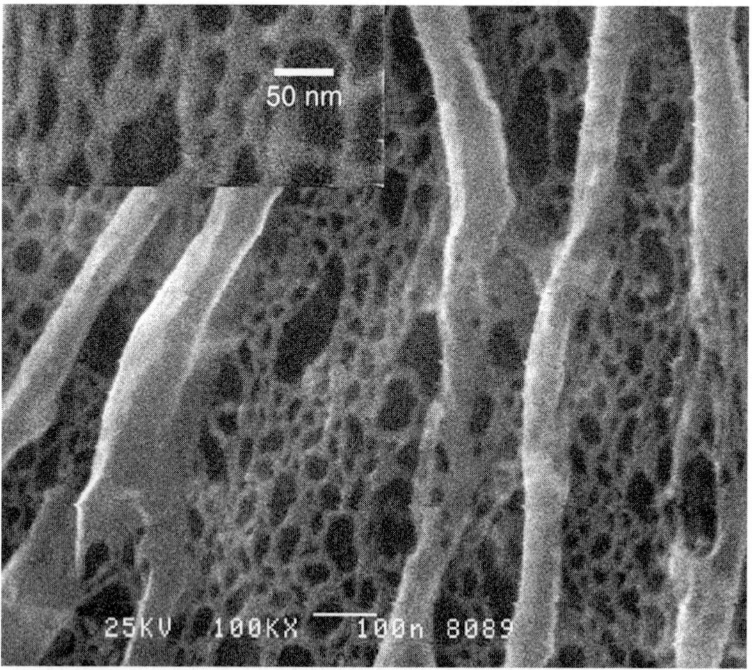

FIGURE 15.19. An α-helical protein was designed to produce fiber structures. Note the fine mesh filamentous network of 7.7 nm average filament widths (*inset*) that comprise the large ~100 nm fibers.

5.3.5. Imaging of Colloids

Although most of the work described in this chapter involves aqueous phase bio-organic specimens, LTHRSEM is not restricted to the low temperature aqueous samples. In fact we have had experiences with imaging colloids in nonaqueous organic solvents. We now observe these systems in bulk at low temperature without etching. Organic solvents all have different vapor pressures and melting points and therefore their sublimation properties under various vacuum conditions will vary greatly. Determinations of etching conditions from nonaqueous solvents are some of the challenges facing microscopists who wish to image nanometer level structures with LT-HRSEM. As methods and experiences evolve, greater employment of cryostages for nanotechnology will undoubtedly occur.

SEM is a complex scientific tool although not as complicated as rocket science. However rocket science too has applications for LT-HRSEM. It is generally unwise to strike an electron beam onto rocket fuel and image its nanostructure. None the less as science seeks to make a safer hydrazine-based rocket fuel its stabilization characteristics need be analyzed and a low temperature holder seemed suitable for the task. A jelled hydrazine sample was produced and a small aliquot was loaded into a planchet that was subsequentially plunge cooled in liquid ethane. As was previously described for aqueous samples the planchet was loaded onto the cryostage, fractured and Cr sputter coated without etching. The sample was then placed in-lens, stabilized at −180°C and imaged at intermediate magnification with low voltage (5 kV) (Fig. 15.20). Cryo-HRSEM images revealed a porous matrix with a solid component containing minute ~10 nm particulate features and voids ranging from 20 to ~100 nm in width.

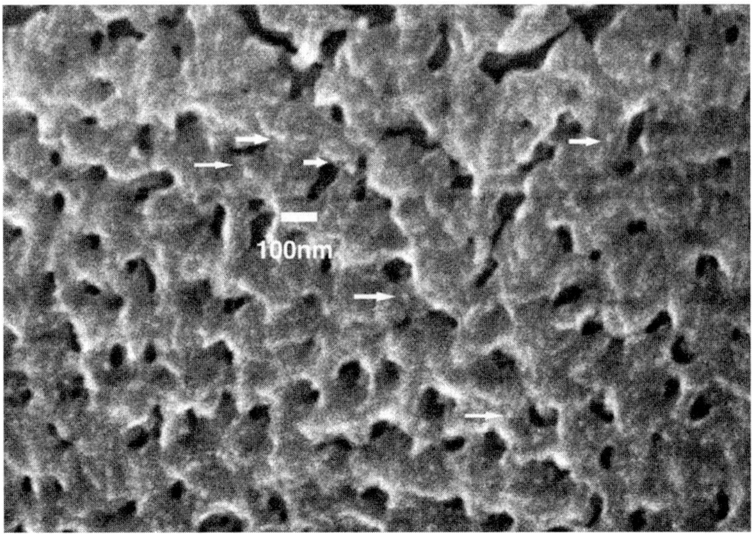

FIGURE 15.20. A hydrazine gel was cryoimmobilized by quenching in liquid ethane followed by sputter coating with a 1 nm Cr film. Note the particulate features comprising the solid matrix (*arrows*).

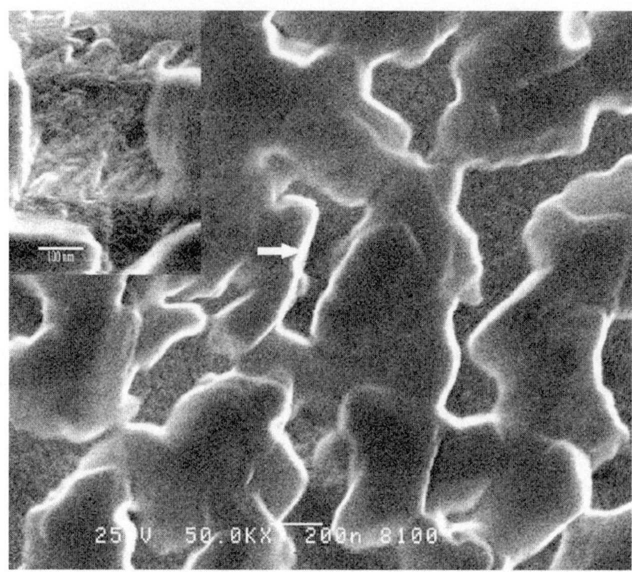

FIGURE 15.21. A low temperature-HRSEM recording of molten (50°C) cyclododecane contained smooth surface masses separated by lakes (*arrow*) that displayed a fine particulate surface roughness (*inset*).

Art conservation is a science with a concern for the chemistry of reagents used to preserve the context of writing on sun or photo bleached papers. The treatment of paper to stabilize or "consolidate" the particulate media (graphite, ink, and pastels) may require the use of a temporary adhesive. Cyclododecane is particularly useful for this application because it sublimes at room temperature, is hydrophobic, and can be introduced as a saturated nonpolar solvent solution or molten solid. The first step for characterizing cyclododecane nanostructure was to cryoimmobilized the compound from its molten state and compare its morphology to room temperature (Fig. 15.21). Since different application methods are employed by art conservationists then cyclododecane prepared in aromatic and hydrocarbon solvents and as well as on paper substrates require further LT-HRSEM studies employing the cryoholder.

Acknowledgment. The author gratefully acknowledges collaborations with Drs. V. P. Conticello, E. Chaikof F. M. Menger, E. R. Wright, K. L. Caran, R. A. McMillan, T. A. T. Lee, J. Dong, D. G. Lynn, W. S. Childers, and R. A. Stein. Images of the hydrazine samples were produced with permission by American Pacific Corporation and Prof. Charles Liotta Group at Georgia Tech. The author greatly appreciates the contributions of S. N. Dublin for graphics used in Chapters 1 and 15. Figures 15.3, 15.6, and 15.9 [9] were used with kind permission of Springer Science and Business Media.

References

1. R. Hermann and M. Muller, *Scanning Microsc.*, 7 (1993) 343–349.
2. D. M. P. Peters, Y. Chen, L. Zardi, and S. Brummel, *Microsc. Microanal.*, 4 (1998) 385–396.
3. J. Dubochet, M. Adrian, J. J. Chang, J. C. Homo, J. Lepault, A. W. McDowall, and P. Q. Schultz, *Rev. Biophys.*, 21 (1988) 129–228.
4. F. M. Menger, A. L. Galloway, M. E. Chlebowski, and R. P. Apkarian, *J. Am. Chem. Soc.*, 126 (2004) 5987–9.
5. P. Echlin, *Philos. Trans. R. Soc. Lond. B Biol. Sci.*, 261 (1971) 51–59.
6. P. Walther, J. Hentschel, P. Herter, T. Muller, and K. Zierold, *Scanning*, 12 (1990) 300–307.
7. O. Kuebler and H. Gross, in: *Electron Microscopy (1978), Vol. II. Biology*, J. M. Sturgess (Ed.), Ninth International Congress, Toronto, Ontario, Canada, 1–9 August 1978, pp. 142–143. (Illustrated Microscopical Society of Canada.)
8. R. P. Apkarian, *54th Ann. Proc. Microsc. Soc. Am.*, (1996), 816–817.
9. R. P. Apkarian, K. L. Caran, and K. A. Robinson, *Microsc. Microanal.*, 5 (1999) 197–207.
10. B. P. M. Menco, *J. Electron Microsc. Tech.*, 4 (1986), 177–240.
11. R. P. Apkarian, *Scanning Microsc.*, 8 (1994) 289–301.
12. R. P. Apkarian, E. R. Wright, V. A. Seredyuk, S. Eustis, L. A. Lyon, V. P. Conticello, and F. M. Menger, *Microsc. Microanal.*, 9 (2003) 286–95.
13. E. R. Wright, V. P. Conticello, and R. P. Apkarian, *Microsc. Microanal.*, 9 (2003) 171–82.
14. E. Kellenberger, in: *Cryotechniques in Biological Electron Microscopy*, R. A. Steinbrecht and K. Zierold (Eds.), Springer-Verlag, New York (1987), p. 39.
15. K. Nagapudi, W. T. Brinkman, J. Leisen, B. S. Thomas, E. R. Wright, C. Haller, X. Wu, R. P. Apkarian, V. P. Conticello, and E. L. Chaikof, *Macromolecules*, 38 (2005) 345–354.
16. J. Dong, R. P. Apkarian, D. G. Lynn, *Bioorg. Med. Chem.*, 13(2005) 5213–5217.
17. R. P. Apkarian, *59th Ann. Proc. Microsc. Soc. Am.*, (2001), 722–723.
18. T. A. T. Lee, A. Cooper, R. P. Apkarian, and V. P. Conticello. *Adv. Mater.*, 12(15) (2000) 1105–1110.
19. F. M. Menger, K. L. Caran, and R. P. Apkarian, *Langmuir*, 16 (2000) 98–101.
20. R. A. McMillan, K. L. Caran, R. P. Apkarian, and V. P. Conticello, *Macromolecules*, 32 (1999) 9067–9070.

Author Index

Wang, X. F., 279
Wang, Y. P., 279
Wang, Y. Q., 279, 424
Wang, Y. W., 279
Wang, Z. L., 1, 150, 279, 304, 355–356, 381, 423–426
Washburn, S., 223–224
Wasserman, E., 426
Watanabe, K., 236
Watanabe, Y., 279
Weber, E., 279, 356, 423
Weber, M., 305
Webster, T. J., 465–466
Wegrowe, J. E., 382
Wehrspohn, R. B., 280, 381
Wei, G. B., 466
Wei, L. W., 278
Wei, Q., 381, 425
Weill, A., 149
Weimann, T., 150
Weitz, D. A., 382
Weller, D., 355
Weller, H., 191
Wells, O. C., 39
Wen, J. G., 279, 356
Wendorff, J. H., 381
Werner, C., 466
West, A., 40
Westra, K. L., 150
Weterings, J. P., 149
Weyland, M., 191
Whang, D., 150, 381
Wharam, D. A., 150
Wheeler, R., 236
White, T., 278
Whitesides, G. M., 355, 383
Whitfield, J. F., 466
Wijnhoven, J. E. G. J., 356
Wiley, J. B., 40, 323, 355, 357, 382
Wilkinson, C. D. W., 150
Willard, M. A., 354
Williams, K. A., 191
Williams, P. A., 224
Willick, G. E., 466
Willmott, P. R., 424
Winn, J. N., 303
Winn, S. R., 466
Wischnitzer, S., 39
Wnek, G. E., 465

Wolf S., 149
Wolf, E., 190
Wolf, S. A., 151
Wolfe, D. E., 424
Wolke, J. G. C., 465
Wong, E. M., 279
Wong, E. W., 423
Wong, S. S., 345, 356
Wong, S., 382
Woo, K. M., 465
Word, M. J., 149
Wright, E. R., 488–489
Wu, C. T., 426
Wu, C., 279
Wu, D., 424
Wu, J. J., 426
Wu, J. S., 426
Wu, P. F., 425
Wu, Q., 278
Wu, X., 489
Wu, Y. Y., 423
Wu, Y., 150, 240, 260, 279, 356, 381, 423
Wu, Z. H., 425

X

Xi, X., 381
Xi, Z. H., 279, 426
Xia, Y. N., 423, 465
Xia, Y., 304, 381, 383
Xiao, G. Z., 465
Xie, C., 279
Xie, S. S., 279
Xie, Y., 279
Xin, X., 381
Xing, Y. J., 279, 426
Xiong, Y., 279
Xu, C., 424
Xu, D., 279, 382
Xu, J. M., 279, 382
Xu, L. B., 355
Xu, L., 40, 355, 382, 425
Xu, N., 278
Xu, Y. J., 254, 279
Xu, Y. Y., 279
Xu, Z., 280, 381
Xue, Z. Q., 426
Xue, Z., 381

Subject Index

Page numbers followed by f and t indicate figures and tables, respectively.